Calculus

Calculus is an area of mathematics in which continuously changing values are studied, and it forms the basis of much of modern mathematics and its applications. It is introduced at the school level and students usually consider it as a set of tricks that they need to memorize. This book provides a thorough reintroduction to calculus, with an emphasis on logical development arising out of geometric intuition. It is written in a conversational style, with motivational discussions preceding every significant result or method. These features aid the student in acquiring the greater mathematical maturity required by a university course in calculus.

Calculus is intended for students pursuing undergraduate studies in mathematics or in disciplines like physics and economics where formal mathematics plays a significant role. For students majoring in mathematics, this book can serve as a bridge to real analysis. For others, it can serve as a base from where they can venture into various applications. After mastering the material in the book, the student would be equipped for higher courses both in the pure (real analysis, complex analysis) and the applied (differential equations, numerical analysis) directions.

Amber Habib is Professor of Mathematics at the Shiv Nadar University, Greater Noida, India. His research interests are in the representation theory of Lie groups and algebras, and mathematical finance. He has contributed to curriculum development at various universities and regularly teaches in the Mathematical Training and Talent Search Programme. He has written one textbook, *The Calculus of Finance* (2011), and co-authored another two: *A Bridge to Mathematics* (2017) and *Exploring Mathematics through Technology* (2022).

Calculus

Amber Habib

CAMBRIDGE
UNIVERSITY PRESS

University Printing House, Cambridge CB2 8BS, United Kingdom

One Liberty Plaza, 20th Floor, New York, NY 10006, USA

477 Williamstown Road, Port Melbourne, VIC 3207, Australia

314 to 321, 3rd Floor, Plot No.3, Splendor Forum, Jasola District Centre, New Delhi 110025, India

103 Penang Road, #05–06/07, Visioncrest Commercial, Singapore 238467

Cambridge University Press is part of the University of Cambridge.

It furthers the University's mission by disseminating knowledge in the pursuit of education, learning and research at the highest international levels of excellence.

www.cambridge.org
Information on this title: www.cambridge.org/9781009159692

First published 2022

Printed in India by Rajkamal Electric Press, Kundli, Haryana.

A catalogue record for this publication is available from the British Library

Library of Congress Cataloging-in-Publication Data

Names: Habib, Amber, author.
Title: Calculus / Amber Habib.
Description: Cambridge, United Kingdom ; New York, NY : Cambridge University Press, 2022. | Includes bibliographical references and index.
Identifiers: LCCN 2021050509 (print) | LCCN 2021050510 (ebook) | ISBN 9781009159692 (paperback) | ISBN 9781009159685 (ebook)
Subjects: LCSH: Calculus.
Classification: LCC QA303.2 .H33 2022 (print) | LCC QA303.2 (ebook) | DDC 515–dc23/eng/20211204
LC record available at https://lccn.loc.gov/2021050509
LC ebook record available at https://lccn.loc.gov/2021050510

ISBN 978-1-009-15969-2 Paperback

To the memory of my aunt and uncle

Shaista and Mohd Mohsin

Contents

Preface

Even now such resort to geometric intuition in a first presentation of the differential calculus, I regard as exceedingly useful, from the didactic standpoint, and indeed indispensable, if one does not wish to lose too much time. But that this form of introduction into the differential calculus can make no claim to being scientific, no one will deny.

— Richard Dedekind, 1872[1]

Calculus is a magical subject. A first encounter in school leads to a radical revision of one's ideas of what is mathematics. We are transported from a rather staid enterprise of counting and measuring to an adventure encompassing change, fluctuation, and a vastly increased ability to understand and predict the workings of the world. At the same time, the student encounters "magic" with both its connotations: awe and wonder on the one hand, mystery and a sense of trickery on the other. Calculus can appear to be a bag of tricks that are immensely useful, provided the apprentice wizard can perfectly remember the spells. As the student pursues mathematics further at university, her instructors may use courses in analysis to persuade her that calculus is a science rather than a mystical art. Alas, all too often the student perceives the new instruction as mere hair-splitting which gives no new powers and may even undermine her previous attainments. The first analysis course is for many an experience that makes them regret taking up higher mathematics.

This book is written to support students in this transition from the expectations of school to those of university. It is intended for students who are pursuing undergraduate studies in mathematics or in disciplines like physics and economics where formal mathematics plays a significant role. Its proper use is in a "calculus with proofs" course taught during the first year of university. The goal is to demonstrate to the student that attention to basic concepts and definitions is an investment that pays off in multiple ways. Old calculations can be done again with a fresh understanding that can not only be stimulating but also protects against error. More importantly, one begins to learn how knowledge can be extended to new domains by first questioning it in familiar terrain. For students majoring in mathematics, this book can serve as a

[1] Translated by W. W. Beman [6].

bridge to real analysis. For others, it can serve as a base from where they can make expeditions to various applications.

The origins of this book lie in our experiments with creating a bouquet of calculus courses at Shiv Nadar University. We found, in our initial years, that quite a few students with excellent marks in school leaving examinations were doing poorly in our introductory calculus course, even though it was not overly formal in its approach. The problem turned out to be that students considered themselves to already be experts and did not pay sufficient attention at the start of the course. Our solution was to rework the course to quickly take students out of their comfort zone and force them to explicitly rethink their assumptions. For example, putting integration at the start helped free them from the belief that it is a mere application of differentiation. Of course, there are other arguments for giving precedence to integration. The fact that it was understood two thousand years before differentiation has to stand for something. It suggests that the process of integration is easier to grasp intuitively, which is good reason to teach it first. As another example, the use of Tarski's version of the Completeness Axiom simplified many proofs and also gave a natural fit with the technique of nested intervals that we have used for the key theorems. In this way, over a few iterations, our shock tactics led us to a presentation with its own internal logic and advantages.

Much of the fascination of calculus comes from its various special topics. We have kept these out of the main presentation so that its structure is uncluttered. However, each chapter closes with one or two sets of thematic exercises, which develop either a particular application or further theory that takes an interested student closer to analysis. They have been given adequate structure and hints to ease the student's progress through them.

A textbook has to go beyond content, to attempt to influence the thinking habits of the reader. For this reason, small exercises called "Tasks" are scattered throughout the main text. These are intended to encourage the student to continually practice and develop her skills. After a while, she may recognize patterns in how each result leads to new questions, and be able to start posing her own questions. Each section is followed by a set of exercises, totalling about 400. Solutions, or hints towards solutions, are provided for all the odd numbered exercises at the end of the book. The solutions have been provided so that the problems do not become roadblocks, and also to provide examples of the expected level of reasoning and communication.

I would be be happy to respond to any corrections, suggestions, or requests for clarification. Please send an email to amberdevhabib@gmail.com.

Acknowledgments

I must thank the two mathematicians who, thirty five years ago, reawakened in me a love for mathematics that had gone dormant under the pressure of crammed syllabi and exams: Shobha Madan and the late O. P. Juneja. I hope their methods of provoking students by well-timed questions have permeated this book, though I have no hope of capturing the accompanying smile or raised eyebrow. Later, I was fortunate to serve as a teaching assistant to Arthur Ogus, Alan Weinstein, Joseph Wolf, and Hung-Hsi Wu. While attending their lectures, and with much assistance from Vinay Kathotia, I realized that my love of analysis had not always improved my understanding of calculus. The bridges between the two realms have to be consciously built.

The driving force behind our experiments with teaching rigorous calculus at Shiv Nadar University was Debashish Bose, my co-instructor for most semesters since 2014. Sushil Singla went well beyond the expected contributions of a teaching assistant, and became a partner in recasting the course and giving form to this book. The supplementary exercises, on special topics, are to a large extent his doing.

The style of presentation is surely influenced by my experience of teaching in the Mathematical Training and Talent Search Programme (MTTS), and by the lectures and writings of S. Kumaresan, in particular. I have learned much about teaching from my fellow MTTS faculty, especially from S. D. Adhikari, A. J. Jayanthan, and S. Somasundaram. MTTS (https://mtts.org.in) is a remarkable programme that has lasted over a quarter of a century and has had significant success in changing students' views on mathematics and giving them the confidence to be mathematicians.

I am grateful to the team at Cambridge University Press for their encouragement, advice and support; especially to Qudsiya Ahmed, Vaishali Thapliyal, Vikash Tiwari and Aniruddha De. The anonymous referees also gave generously of their time and helped improve the exposition in several places.

Introduction

A calculus text written at this point of time can make its plea for existence on the novelty of its exposition and choice of content, rather than any originality in its mathematics. Let me state the case for this book. These explanations are primarily aimed at teachers, but students may also gain some perspective from them if they peruse them *after* going through at least the first two chapters.

The book aims to give a complete logical framework for calculus, with the proofs reaching the same levels of rigour as a text on real analysis. At the same time it eschews those aspects of analysis that are not essential to a presentation of calculus techniques. So it has the completeness axiom for real numbers, but not Cauchy sequences or the theorems of Heine–Borel and Bolzano–Weierstrass. At the other end of the spectrum, it omits the case for the importance of calculus through its applications to the natural sciences or to economics and finance.

The first chapter provides a description of real numbers and their properties, followed by functions and their graphs. For the most part, the material of this chapter would be known to students, but not in such an organized way. I typically use the first class of the course to ask students to share their thoughts on various issues. What are rational numbers? What are numbers? Which properties of numbers are theorems and which are axioms? What is the definition of a point or a line? What do we mean by a tangent line to a curve? These flow from one to another and from students' responses. By the end of the hour, with many students firm in their beliefs but finding them opposed just as firmly by others, I have an opportunity to propose that we must carefully put down our axioms and ways of reasoning so that future discussions may be fruitful. One may still ask whether the abstract approach is overdone; is it necessary to introduce general concepts like field and ordered field? The reason for doing so is that it provides a context within which simple questions can be posed and the student can practice creating and writing small proofs as a warm-up to harder tasks that wait ahead.

The first chapter is also where we introduce the completeness axiom, identified by Dedekind as the key property that sets real numbers apart and makes calculus a rich subject. We have used a version of the completeness axiom that was given by Tarski and is very close to Dedekind's original formulation, instead of the currently popular least upper bound property. Tarski's version is easy to comprehend and simplifies many proofs.

In the second chapter we move directly to integration and its properties, preceding limits and continuity. This is in accord with both history and classic texts such as Apostol [2] and Courant and John [4]. The integration of monotone functions enables the early introduction and study of the logarithmic and exponential functions. The priority given to integration also enables the initial exposure to ϵ arguments to happen in a simplified setting. For example, to prove integrability or to compute an integral, for any $\epsilon > 0$ we have only to find a *single* pair of lower and upper sums that are within ϵ of each other. Whereas, in a limit argument, for each ϵ, one has to find δ such that good things happen for *every* point that is within δ of the base point. The use of Tarski's axiom also assists with the proofs of the properties of integration, giving them the form of direct manipulations of inequalities.

The third chapter takes up limits and continuity, with a fairly conventional presentation. We do make an attempt to confine ϵ-δ arguments to special cases where they are easier to handle, and then reduce the general case to the special case by other means. The major results (intermediate value theorem, boundedness theorem, extreme value theorem, small span theorem) are all proven along the same lines, via nested intervals. This should give the student a clearer sense of the unity of this material and make it easier to assimilate the proofs. With integration already in place, we are able to rigorously develop trigonometric functions. In this way, all the standard functions of calculus become available quite early.

The derivative finally enters the scene in the fourth chapter. We do not adopt the usual approach through tangent lines, as "tangent line" is not a well-defined notion at this stage. Instead, we introduce a "differentiable function" as one which is essentially linear locally. This leads quickly to the usual limit definition. Later in the chapter, we deviate again from the norm by proving results such as a zero derivative implying constancy via Fermat's theorem on local extremes, rather than as consequences of the mean value theorem. This quickens the development of curve sketching as well as techniques of integration.

The fourth chapter contains one half of the fundamental theorem of calculus: differentiation reverses integration. In the fifth, we obtain the other half: antiderivatives give values of definite integrals. This is the key to all techniques of integration. An instructor who is short on time can consider giving less attention to this chapter as students should be quite familiar with its contents and there are no surprises in its exposition. For our part, we have given a complete description of the technique of partial fractions, which is often only exhibited for low degree examples. The section on first-order differential equations raises issues of existence and uniqueness of solutions that students may not have considered earlier.

The place of the mean value theorem in a calculus course has been a matter of debate. The generic textbook gives it a central role. Yet many articles suggest that the theorem is an obstacle for students and propose ways of bypassing it. We have taken a middle path and delayed its appearance until it is absolutely essential. When it does appear, in the sixth chapter, it is immediately put to serious work. It gives us L'Hôpital's rule, enables the use of Riemann sums to carry out surface area and volume calculations, and generates error estimates for Taylor approximations and for numerical integration. The delayed entrance is also a more dramatic one!

In the seventh chapter, we take up sequences and series. Here, we emphasize two techniques which are fundamental both for working out specific examples and proving general theorems. One is comparison with the geometric series. The other is viewing partial sums as lower/upper sums of corresponding integrals. Additionally, the chain of results starting with the Riemann rearrangement theorem and proceeding to the Cauchy product is obtained purely by the use of least upper bounds and simple algebra. These replace somewhat tricky ϵ arguments.

The eighth and final chapter first discusses power series and Taylor expansions. The functions we typically encounter in applications of calculus can be expanded as power series, and once this is done their differentiation and integration becomes easy. Various series can be summed by recognizing them as point values of power series. In the second half of this chapter, we take up two topics that set a path for future studies. Fourier series once caused a revolution in mathematics by forcing us to rethink our fundamental concepts such as sets and functions. They are now a unifying theme across many branches of mathematics and continue to inspire fresh developments. Finally, complex numbers explain many puzzles related to real numbers, and complex series lead to a deeper understanding of the relationship between differentiation and integration.

We have, now, described the core of the book. It can be covered in about 60 lecture hours. However, in many institutions only about 45 lecture hours may be available in a semester. In that case, the first six chapters would form a reasonable goal.

The sets of thematic exercises that follow each chapter can be used for self-study or for student presentations within a course. Instructors of related courses may also find them useful. For example, suppose an introductory course was being taught on ordinary differential equations (ODE). Typical texts for such courses concentrate on implementing techniques and skimp on proofs of existence and uniqueness of solutions. An instructor who wants to show that the standard solutions of a homogeneous second-order ODE with constant coefficients are *all* the solutions may struggle to find an approach that can be easily taught at this level. Such a proof is outlined in the exercises following Chapter 5. It needs only the result that a zero derivative implies constancy.

The exposition has benefited from the literature on alternative approaches to teaching calculus, especially the articles by Bers [48], Swann [58], Taylor [59] and Tucker [61]. A particular inspiration has been the text *Calculus Unlimited* by Marsden and Weinstein [23], which provoked the thought that a simplified approach could be built around Tarski's formulation of the Completeness Axiom. Otto Toeplitz's *Calculus: A Genetic Approach* [35] convinced me that one should gradually increase generality, and at any stage use the generality that is appropriate. This led to the initial emphasis on monotonicity rather than continuity and enabled the rapid development of the standard transcendental functions. Other texts that have influenced this book are Apostol [2] and Spivak [30].

Signage

We have used the following icons to mark certain portions of the text:

✋ A digression or an issue that needs special attention.

💬 A motivational remark preceding a proof.

□ End of an example.

■ End of a proof.

1 | Real Numbers and Functions

Calculus can be described as the study of how one quantity is affected by another, focusing on relationships that are smooth rather than erratic. This chapter sets up the basic language for describing quantities and the relationships between them. Quantities are represented by numbers and you would have seen different kinds of numbers: natural numbers, whole numbers, integers, rational numbers, real numbers, perhaps complex numbers. Of all these, real numbers provide the right setting for the techniques of calculus and so we begin by listing their properties and understanding what distinguishes them from other number systems. The key element here is the completeness axiom, without which calculus would lose its power.

The mathematical object that describes relationships is called "function." We recall the definition of a function and then concentrate on functions that relate real numbers. Such functions are best visualized through their graphs, and this visualization is a key part of calculus. We make a small beginning with simple examples. A more thorough investigation of graphs can only be carried out after calculus has been developed to a certain level. Indeed, the more interesting functions, such as trigonometric functions, logarithms, and exponentials, require calculus for their very definition.

1.1 Field and Order Properties

We begin with a review of the set $\mathbb{R}$ of **real numbers**, which is also called the **Euclidean line**. It is a "review" in that we do not construct the set but just list its key attributes, and use them to derive others. For descriptions of how real numbers can be constructed from scratch, you can consult Hamilton and Landin [11], Mendelson [24], or most books on real analysis. The fundamental ideas underlying these constructions are easy to absorb, but the checking of details can be arduous. You would probably appreciate them more *after* reading this book.

What is the need for this review? Mainly, it is intended as a warm-up session before we begin calculus proper. Many intricate definitions and proofs lie in wait later, and we need to get ready for them by practising on easier material. If you are in a hurry and confident of your basic skills with numbers and proofs, you may skip ahead to the next section, although a patient reading of these few pages would also help in later encounters with linear algebra and abstract algebra.

Field Properties of Real Numbers

Any concept of "numbers" involves rules for combining them to create new ones. We shall use the term **binary operation** to denote a rule for associating a single member of a set to each pair of elements from that set.

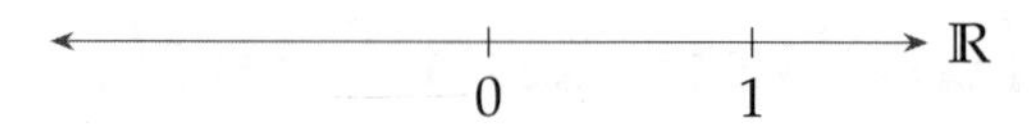

The set $\mathbb{R}$ is equipped with two binary operations, $+$ (addition) and $\cdot$ (multiplication), and has two special elements named zero (0) and one (1), with the following fundamental properties:

R1. *Addition and multiplication are commutative:* $a+b=b+a$ *and* $a\cdot b=b\cdot a$ *for every* $a,b\in\mathbb{R}$.

R2. *Addition and multiplication are associative:* $a+(b+c)=(a+b)+c$ *and* $a\cdot(b\cdot c)=(a\cdot b)\cdot c$ *for every* $a,b,c\in\mathbb{R}$.

R3. *0 serves as identity for addition:* $0+a=a$ *for every* $a\in\mathbb{R}$.

R4. *1 serves as identity for multiplication:* $1\cdot a=a$ *for every* $a\in\mathbb{R}$.

R5. *Each* $a\in\mathbb{R}$ *has an additive inverse* $b\in\mathbb{R}$, *with the property* $a+b=0$.

R6. *Each non-zero* $a\in\mathbb{R}$ *has a multiplicative inverse* $c\in\mathbb{R}$, *with the property* $a\cdot c=1$.

R7. *Multiplication distributes over addition:* $a\cdot(b+c)=(a\cdot b)+(a\cdot c)$ *for every* $a,b,c\in\mathbb{R}$.

The properties R1 to R7 are called the **field axioms** for $\mathbb{R}$. In general, if a set $\mathbb{F}$ has two binary operations $+$ and $\cdot$, such that these seven properties hold (with $\mathbb{R}$ replaced by $\mathbb{F}$ everywhere), then $\mathbb{F}$ is called a **field**. Other familiar examples of fields are the set $\mathbb{Q}$ of rational numbers and the set $\mathbb{C}$ of complex numbers. Each field has its own binary operations and its own special elements called zero and one.

Example 1.1.1

Consider the set $\mathbb{F}_2=\{0,1\}$ of just two elements named 0 and 1. Can we provide it with with binary operations $+$ and $\cdot$ such that it becomes a field in which 0 is the additive identity and 1 is the multiplicative identity? Well, since 0 is to be the additive identity, we must set

$$0+0=0,\quad 0+1=1+0=1.$$

What should be the additive inverse of 1? It obviously cannot be 0, so it must be 1. This gives $1+1=0$. Therefore $+$ is represented by following table.

+	0	1
0	0	1
1	1	0

Similarly, since 1 is to be the multiplicative identity, we must have

$$1 \cdot 1 = 1, \quad 1 \cdot 0 = 0 \cdot 1 = 0.$$

We also compute $0 \cdot 0$ as follows:

$$0 \cdot 0 = 0 \cdot 0 + 0 \cdot 1 = 0 \cdot (0+1) = 0 \cdot 1 = 0.$$

Hence $\cdot$ is represented by the following table.

$\cdot$	0	1
0	0	0
1	0	1

Let us verify that $\mathbb{F}_2$ really is a field. The commutativity of the two operations is clearly built into their definitions. We also see that 0 is the additive identity and 1 is the multiplicative identity. The additive and multiplicative inverses are also present. Only the associative and distributive laws have to be verified. Let $a, b, c \in \mathbb{F}_2$. Consider the following cases:

$$\begin{aligned}
a = 0 &\implies a + (b+c) = 0 + (b+c) = b + c = (0+b) + c = (a+b) + c,\\
b = 0 &\implies a + (b+c) = a + (0+c) = a + c = (a+0) + c = (a+b) + c,\\
c = 0 &\implies a + (b+c) = a + (b+0) = a + b = (a+b) + 0 = (a+b) + c,\\
a = b = c = 1 &\implies a + (b+c) = 1 + (1+1) = (1+1) + 1 = (a+b) + c.
\end{aligned}$$

So $+$ is associative. We ask you to verify in a similar fashion the associativity of $\cdot$ as well as the distributive law. ▢

The set of non-zero real numbers is denoted by $\mathbb{R}^*$. We shall usually abbreviate $a \cdot b$ to ab.

Theorem 1.1.2

The field $\mathbb{R}$ has the following properties.

1. *0 is the only additive identity and 1 is the only multiplicative identity.*
2. *The additive inverse of any real number is unique.*
3. *The multiplicative inverse of any non-zero real number is unique.*

The important thing is to realize that these claims need proof, and then to prove them using only the field axioms.

Proof. Suppose $0'$ and $1'$ are additive and multiplicative identities, respectively. Then we have

$$\begin{aligned}
0' &= 0 + 0' && \text{(because 0 is an additive identity)}\\
&= 0 && \text{(because } 0' \text{ is an additive identity).}
\end{aligned}$$

Similarly, we have $1' = 1' \cdot 1 = 1$. This shows the uniqueness of 0 and 1 as identities for addition and multiplication.

Next, suppose a has additive inverses b and c. Then $a + b = 0$ and $a + c = 0$. Hence,

$$b = b + 0 = b + (a + c) = (b + a) + c = 0 + c = c.$$

This shows the uniqueness of the additive inverse. You can similarly show the uniqueness of the multiplicative inverse. ∎

Having established that the inverses are unique, we can give them special names. We shall denote the additive inverse of a by $-a$ and the multiplicative inverse by $1/a$ or a^{-1}.

Theorem 1.1.3 (Cancellation Laws)

Let $a, b, c \in \mathbb{R}$. Then the following hold:

1. *If $a + b = a + c$ then $b = c$.*
2. *If $ab = ac$ and $a \neq 0$ then $b = c$.*

Proof. The cancellation laws are based on associativity and the existence of inverses.

$$\begin{aligned} a + b = a + c &\implies (-a) + (a + b) = (-a) + (a + c) && \text{(existence of inverse)} \\ &\implies ((-a) + a) + b = ((-a) + a) + c && \text{(associativity)} \\ &\implies 0 + b = 0 + c && \text{(property of inverse)} \\ &\implies b = c && \text{(property of identity).} \end{aligned}$$

If $a \neq 0$ then it has a multiplicative inverse a^{-1} and we have

$$\begin{aligned} ab = ac &\implies a^{-1}(ab) = a^{-1}bc \implies (a^{-1}a)b = (a^{-1}a)c \\ &\implies 1 \cdot b = 1 \cdot c \implies b = c. \end{aligned}$$

You should provide the justification for each step, as we had done for the case of addition. ∎

Theorem 1.1.4

Let $a, b, c \in \mathbb{R}$. Then the following hold:

1. $0 \cdot a = 0$.
2. $-(-a) = a$.
3. *If $a \in \mathbb{R}^*$ then $(a^{-1})^{-1} = a$.*
4. $(-1)a = -a$.
5. $(-1)(-1) = 1$.

6. $(-a)(-b) = ab$.
7. *If* $ab = 0$ *then* $a = 0$ *or* $b = 0$.

Proof.

1. Use $0 = 0 + 0$:

$$a \cdot 0 = a \cdot (0 + 0) = (a \cdot 0) + (a \cdot 0) \implies 0 + (a \cdot 0) = (a \cdot 0) + (a \cdot 0) \implies 0 = a \cdot 0.$$

2. If we let $b = -(-a)$ we have $b + (-a) = 0$. We also have $a + (-a) = 0$. We apply cancellation to get $b = a$.
3. Let $b = (a^{-1})^{-1}$, so that $a^{-1}b = 1$. We also have $a^{-1}a = 1$. Apply cancellation.
4. We have to show that $(-1)a$ is the additive inverse of a. So we add them:

$$(-1)a + a = (-1) \cdot a + 1 \cdot a = ((-1) + 1) \cdot a = 0 \cdot a = 0.$$

5. Substitute $a = -1$ in the previous statement.
6. $(-a)(-b) = \big(a(-1)\big)\big((-1)b\big) = a\Big((-1)\big((-1)b\big)\Big) = a\Big(\big((-1)(-1)\big)b\Big) = a(1 \cdot b) = ab$.
7. We will show that if $a \neq 0$ then we must have $b = 0$. So at least one of $a = 0$ and $b = 0$ must hold.

$$a \neq 0 \implies a^{-1}(ab) = a^{-1}0 \implies (a^{-1}a)b = 0 \implies 1 \cdot b = 0 \implies b = 0.$$

■

Task 1.1.5

Verify that $-(a + b) = (-a) + (-b)$ *and* $(ab)^{-1} = a^{-1}b^{-1}$.

For any $a, b \in \mathbb{R}$, the sum $a + (-b)$ is denoted by $a - b$ and is called the **difference** of a and b. The process of obtaining $a - b$ is called **subtraction**. Similarly, if $b \in \mathbb{R}^*$, the product $a \cdot (1/b)$ is denoted by $\frac{a}{b}$ or a/b and is called the **ratio** of a and b. The process of obtaining a/b is called **division**.

The **square** of a number x is its product with itself and is denoted by x^2.

Task 1.1.6

Show that $(-x)^2 = x^2$.

Task 1.1.7

Use the field axioms of $\mathbb{R}$ *to prove the following:*

(a) $-\frac{a}{b} = \frac{-a}{b} = \frac{a}{-b}$ *if* $b \neq 0$,

(b) $\frac{a}{b} + \frac{c}{d} = \frac{ad + bc}{bd}$ *if* $b, d \neq 0$.

Order Properties of Real Numbers

Since the field axioms of $\mathbb{R}$ are also satisfied by $\mathbb{Q}$ and $\mathbb{C}$, we know that they do not completely determine the real numbers. What else is special about $\mathbb{R}$?

The non-zero real numbers $\mathbb{R}^*$ split into two types: **positive** and **negative**. We shall denote the set of positive real numbers by $\mathbb{R}^+$ and the set of negative real numbers by $\mathbb{R}^-$. The key facts associated with this split are as follows.

R8. *Every non-zero real number is either positive or negative.*

R9. *Zero is neither positive nor negative.*

R10. *No real number is both negative and positive.*

R11. *A real number is negative if and only if its additive inverse is positive.*

R12. *The sum and product of positive numbers are positive.*

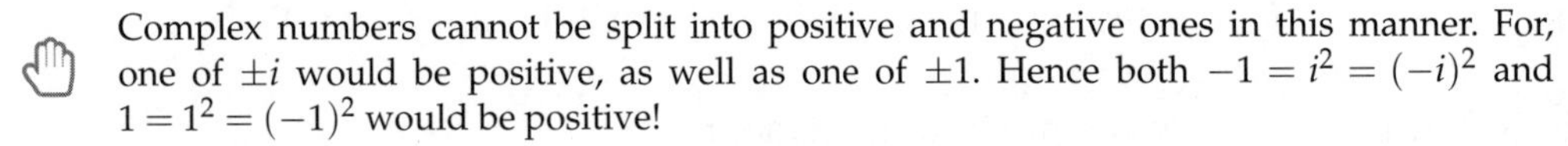

Complex numbers cannot be split into positive and negative ones in this manner. For, one of $\pm i$ would be positive, as well as one of ± 1. Hence both $-1 = i^2 = (-i)^2$ and $1 = 1^2 = (-1)^2$ would be positive!

The properties R8 to R12 are called the **order axioms** of $\mathbb{R}$. Let us see some of their consequences.

Theorem 1.1.8

1. *If* $x, y \in \mathbb{R}^-$ *then* $x + y \in \mathbb{R}^-$.
2. *If* $x, y \in \mathbb{R}^-$ *then* $xy \in \mathbb{R}^+$.
3. *If* $x \in \mathbb{R}^+$ *and* $y \in \mathbb{R}^-$ *then* $xy \in \mathbb{R}^-$.
4. *If* $x \in \mathbb{R}^*$ *then* $x^2 \in \mathbb{R}^+$.
5. $1 \in \mathbb{R}^+$.

Again, these are familiar properties, which you were asked to memorize in school. We wish to convert them to proven facts. We treat the first two to show you the way, and leave the others as exercises.

Proof.
$$\begin{aligned} x, y \in \mathbb{R}^- &\implies -x, -y \in \mathbb{R}^+ \implies (-x) + (-y) \in \mathbb{R}^+ \\ &\implies x + y = -((-x) + (-y)) \in \mathbb{R}^-. \\ x, y \in \mathbb{R}^- &\implies -x, -y \in \mathbb{R}^+ \implies (-x)(-y) \in \mathbb{R}^+ \\ &\implies xy = (-x)(-y) \in \mathbb{R}^+. \end{aligned}$$
■

The split into positive and negative allows us to think of larger and smaller real numbers (an "ordering") as follows. We say that a is **greater** than b, denoted by $a > b$, if $a - b \in \mathbb{R}^+$. In this case, we also say that b is **less** than a and denote that by $b < a$.

Theorem 1.1.9

Let $a,b,c \in \mathbb{R}$. Then the following hold.

1. $\mathbb{R}^+ = \{x \in \mathbb{R} \mid x > 0\}$ *and* $\mathbb{R}^- = \{x \in \mathbb{R} \mid x < 0\}$.
2. *(Trichotomy) Exactly one of the following holds:* $a = b$ *or* $a > b$ *or* $a < b$.
3. *(Transitivity) If* $a > b$ *and* $b > c$ *then* $a > c$.
4. *If* $a > b$ *then* $a + c > b + c$.
5. *Let* $c > 0$. *If* $a > b$ *then* $ac > bc$.
6. *Let* $c < 0$. *If* $a > b$ *then* $ac < bc$.
7. *If* $a < b$ *then* $a < \frac{a+b}{2} < b$.
8. *If* $0 < a < b$ *then* $0 < 1/b < 1/a$.
9. *Suppose* $a,b > 0$. *Then* $a > b \iff a^2 > b^2$.
10. *Suppose* $a,b > 0$. *Then* $a = b \iff a^2 = b^2$.

Proof.

1. We have $x > 0 \iff x - 0 \in \mathbb{R}^+ \iff x \in \mathbb{R}^+$. We similarly obtain the description of $\mathbb{R}^-$.
2. First, we note that $a = b$ implies $a - b = 0$, which rules out $a > b$ as well as $a < b$.

   Now let a,b be distinct real numbers. We have to prove that exactly one of $a > b$ and $a < b$ holds. Since a,b are distinct, $a - b \neq 0$. Therefore $a - b$ belongs to exactly one of $\mathbb{R}^+$ and $\mathbb{R}^-$. Now $a - b \in \mathbb{R}^+$ corresponds to $a > b$ and $a - b \in \mathbb{R}^-$ corresponds to $a < b$.
3. Hint: Consider $a - c = (a - b) + (b - c)$.
4. Hint: Consider $(a + c) - (b + c) = a - b$.
5. Hint: Consider $ac - bc = (a - b)c$.
6. As above.
7. Hint: Add a to both sides of $a < b$ to get one of the inequalities. Add b instead to get the other.
8. Hint: $\frac{1}{a} - \frac{1}{b} = \frac{b-a}{ab}$.
9. Hint: $b^2 - a^2 = (b - a)(b + a)$.
10. As above. ∎

Note that item 7 of this theorem implies that there are infinitely many real numbers between any two distinct ones.

Let A be a subset of $\mathbb{R}$.

- An element $M \in A$ is called the **maximum** of A if $a \leq M$ for every $a \in A$. We write $M = \max(A)$.
- An element $m \in A$ is called the **minimum** of A if $m \leq a$ for every $a \in A$. We write $m = \min(A)$.

A maximum element is also called **greatest** while a minimum element is also called **least**.

Example 1.1.10

Let $A = \{x \in \mathbb{R} \mid x \leq 1\}$. Then 1 is the maximum of A. ▢

Task 1.1.11

Let $A = \{x \in \mathbb{R} \mid x < 1\}$. Show that A has no maximum.

Absolute Value

Let us continue this overview of familiar facts about the real numbers by recalling the definition of the **absolute value** of a real number x,

$$|x| = \begin{cases} x & \text{if } x \geq 0, \\ -x & \text{if } x < 0. \end{cases}$$

We think of a real number as having two aspects: a *direction* determined by whether it is positive or negative, and a *magnitude* given by its absolute value.

$|x|$ $|x|$

$-x$ 0 x $\mathbb{R}$

Theorem 1.1.12

Let $x, y \in \mathbb{R}$. Then we have the following.

1. $|x| \geq 0$.
2. $|x| = 0$ *if and only if* $x = 0$.
3. $|x^2| = |x|^2 = x^2$.
4. $|xy| = |x||y|$.
5. *(Triangle Inequality)* $|x + y| \leq |x| + |y|$.
6. $|x - y| \geq ||x| - |y||$.

Proof. The first two claims are obvious from the definition. Proofs for the others are given below. We make use of the fact that if $a,b \geq 0$ then $a = b \iff a^2 = b^2$.

3. Since $x^2 \geq 0$, we have $|x^2| = x^2$. Further,

$$|x|^2 = \begin{cases} x^2 & \text{if } x \geq 0 \\ (-x)^2 & \text{if } x < 0 \end{cases} = x^2.$$

4. $|xy|^2 = (xy)^2 = x^2y^2 = |x|^2|y|^2 = (|x||y|)^2$.
5. $|x+y|^2 = (x+y)^2 = x^2 + y^2 + 2xy \leq |x|^2 + |y|^2 + 2|x||y| = (|x| + |y|)^2$.
6. $|x-y|^2 = (x-y)^2 = x^2 + y^2 - 2xy \geq |x|^2 + |y|^2 - 2|x||y| = (|x| - |y|)^2 = ||x| - |y||^2$. ■

Task 1.1.13

For any $x,a \in \mathbb{R}$ with $a \geq 0$, prove that $|x| \leq a \iff -a \leq x \leq a$.

Since we think of $|x|$ as the magnitude or size of a real number, $|x-y|$ becomes a measure of the gap between x and y. We call it the **distance** between x and y. The properties of absolute value convert to the following properties of distance.

Theorem 1.1.14

Let $x,y,z \in \mathbb{R}$. Then we have the following:

1. *(Positivity) $|x-y| \geq 0$, and $|x-y| = 0$ if and only if $x = y$.*
2. *(Symmetry) $|x-y| = |y-x|$.*
3. *(Triangle Inequality) $|x-z| \leq |x-y| + |y-z|$.*

Proof. This is left as an exercise for you. ■

Types of Real Numbers

The set of real numbers includes various special types of numbers:

- By repeatedly adding 1 we generate the subset of **natural numbers**,

$$\mathbb{N} = \{1, 2 = 1+1, 3 = 2+1, \dots\}.$$

  By combining (5) of Theorem 1.1.8 and (4) of Theorem 1.1.9 we see that $1 < 2 < 3 < \cdots$.

- By including zero with natural numbers we get **whole numbers**,

$$\mathbb{W} = \mathbb{N} \cup \{0\} = \{0, 1, 2, \dots\}.$$

- By further including the additive inverse of each whole number we get **integers**,

$$\mathbb{Z} = \{\ldots, -2, -1, 0, 1, 2, \ldots\}.$$

- By dividing integers with each other we get **rational numbers**,

$$\mathbb{Q} = \{\, a/b \mid a, b \in \mathbb{Z} \text{ and } b \neq 0 \,\}.$$

  Some examples of rational numbers are $\frac{8}{6}, \frac{12}{9}, \frac{-4}{3}, \frac{4}{-3}$, and $\frac{2}{1}$.

Task 1.1.15

Let $a, b, c, d \in \mathbb{Z}$ with $b, d \neq 0$. Show that $\frac{a}{b} = \frac{c}{d} \iff ad = bc$.

The positive rational numbers will be denoted by $\mathbb{Q}^+$. Any $x \in \mathbb{Q}^+$ can be expressed as $x = p/q$ with $p, q \in \mathbb{N}$.

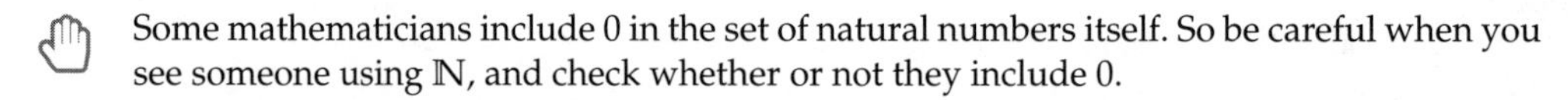
Some mathematicians include 0 in the set of natural numbers itself. So be careful when you see someone using $\mathbb{N}$, and check whether or not they include 0.

Those real numbers that are not rational numbers are called **irrational numbers**. At this point in this text, we still do not know enough about real numbers to be able to say whether there are any irrational numbers! This will be clarified in the next section.

Mathematical Induction

We make a small digression to recall some important facts about natural numbers.

Principle of Mathematical Induction: If A is a subset of $\mathbb{N}$ that contains 1 and is closed under adding 1 then $A = \mathbb{N}$. Alternately: If $P(n)$ is a statement about n (for every natural number n) such that $P(1)$ is true and the truth of $P(n)$ implies the truth of $P(n+1)$, then $P(n)$ is true for every natural number n.

Mathematical induction is used to prove statements that hold for every natural number. As an example, we will use it to better understand integer powers of real numbers.

Example 1.1.16

We define **integer powers** as follows: First, we define $x^0 = 1$ for any $x \in \mathbb{R}$. Then, for any $n \in \mathbb{N}$, we define $x^n = x \cdot x^{n-1}$. If $x \neq 0$, we define $x^{-n} = (x^n)^{-1}$. We will use induction to prove the following:

$$\text{If } x \neq 0 \text{ and } n \in \mathbb{N} \text{ then } (x^{-1})^n = (x^n)^{-1}.$$

Let $P(n)$ be the statement that $(x^{-1})^n = (x^n)^{-1}$. Then $P(1)$ is the statement $x^{-1} = x^{-1}$, which is certainly true. Now assume that some $P(n)$ is true (this is called the "induction hypothesis"). We need to show that this assumption forces $P(n+1)$ to be true. We compute as follows:

$$\begin{aligned}(x^{-1})^{n+1} &= x^{-1}\cdot(x^{-1})^n && \text{(definition of } a^{n+1}\text{)}\\ &= x^{-1}\cdot(x^n)^{-1} && \text{(induction hypothesis)}\\ &= (x\cdot x^n)^{-1} && \text{(because } (a\cdot b)^{-1} = a^{-1}\cdot b^{-1}\text{)}\\ &= (x^{n+1})^{-1} && \text{(definition of } a^{n+1}\text{).}\end{aligned}$$

This shows the truth of $P(n+1)$. Therefore, by the principle of mathematical induction, $(x^{-1})^n = (x^n)^{-1}$ holds for every $n \in \mathbb{N}$. □

Task 1.1.17

Let $x \in \mathbb{R}^*$. Prove that $(x^{-1})^n = (x^n)^{-1}$ for every $n \in \mathbb{Z}$.

Task 1.1.18

Let $x_1, \ldots, x_n \in \mathbb{R}$. Use mathematical induction to show that $\left|\sum_{i=1}^{n} x_i\right| \leq \sum_{i=1}^{n} |x_i|$.

Induction moves from smaller numbers to larger numbers. It may not always be convenient to move in this rigid way from 1 to 2 to 3 and so on. For example, to prove something about 14 we may like to use $14 = 7 \times 2$ rather than $14 = 13 + 1$. It helps to have a more flexible version of mathematical induction, which we give below.

Principle of Strong Mathematical Induction: If A is a subset of $\mathbb{N}$ that contains 1 and contains $n+1$ whenever it contains all of $1, \ldots, n$, then $A = \mathbb{N}$. Alternately: If $P(n)$ is a statement about n (for every natural number n) such that $P(1)$ is true and the truth of $P(1), \ldots, P(n)$ implies the truth of $P(n+1)$, then $P(n)$ is true for every natural number n.

Strong induction has that name because its hypothesis is weaker than ordinary induction. That is, if A satisfies the hypothesis of ordinary induction, it will also satisfy the hypothesis of strong induction. One can see from this that the principle of strong mathematical induction implies the principle of mathematical induction.

Proofs by Contradiction

We shall now discuss a powerful technique for proving theorems. Every theorem claims that if certain assumptions hold then a particular conclusion will automatically follow. To show that the conclusion must follow, we focus instead on what happens if it does not hold. We derive various consequences of the conclusion being false, until we find a statement that is both true and false. This being impossible, the assumption that the conclusion is false must be wrong, and therefore the conclusion must be true.

The next proof may well be the most famous example of a proof by contradiction.

Theorem 1.1.19

There is no rational number whose square is 2.

This may look like a theorem that has no assumptions. However, it is just that they do not need to be stated. The assumptions are the definition and basic properties of rational numbers.

Proof. A proof by contradiction must start by assuming that the conclusion of the theorem is false. So, suppose there *is* a rational number p whose square is 2. We can assume that p is positive (if it is negative, we will replace it with $-p$). We can write $p = a/b$ with $a,b \in \mathbb{N}$. If a,b have any common factors, we cancel them. Now we have $p = m/n$ with $m,n \in \mathbb{N}$ and such that the only common factor of m and n is 1. Let us start exploring the consequences of the square of p being 2:

$$p^2 = 2 \implies \left(\frac{m}{n}\right)^2 = 2 \implies m^2 = 2n^2.$$

It follows that m^2 is even and hence m is even. So we can write $m = 2k$ with $k \in \mathbb{N}$. Again,

$$m^2 = 2n^2 \implies (2k)^2 = 2n^2 \implies 2k^2 = n^2,$$

hence n^2 is even and therefore n is even. This shows that m,n have 2 as a common factor, which contradicts the earlier statement that their only common factor is 1. This impossible situation shows that the assumption $p^2 = 2$ must be false. Hence, there is no rational number whose square is 2. ∎

Exercises for § 1.1

1. Let $\mathbb{F}_3 = \{0,1,2\}$. Complete the following tables for $+$ and $\cdot$ such that $\mathbb{F}_3$ becomes a field.

+	0	1	2
0	0	1	2
1	1		
2	2		

•	0	1	2
0	0		
1	0	1	2
2	2		

2. Put the following numbers in ascending order *without* converting them to decimal form:

$$3, \frac{14}{10}, \frac{17}{12}, 2, -2, -\frac{3}{2}.$$

3. In each case below, find the numbers with the given property and sketch the solutions on the number line.

(a) $x^2 - x > 0$. (b) $3x^2 + 2x - 1 \geq 0$. (c) $x^2 - 5x + 6 < 0$.

4. Find the numbers that meet *all* the conditions of the previous exercise.

5. Let $x, y \in \mathbb{R}$ and $m, n \in \mathbb{Z}$. Prove the following, taking x, y to be non-zero wherever required:

(a) $x^m x^n = x^{m+n}$. (b) $x^m y^m = (xy)^m$. (c) $(x^m)^n = x^{mn}$.

6. The principle of mathematical induction (page 10) is usually taken as an axiom for natural numbers. The principle of strong mathematical induction can be derived from it. Let A be a non-empty subset $\mathbb{N}$ such that $1 \in A$ and $1, \ldots, n \in A \implies n+1 \in A$. Define

$$S = \{ n \in \mathbb{N} \mid k \in \mathbb{N} \text{ and } k \leq n \text{ implies } k \in A \}.$$

Show the following:

(a) $1 \in S$ and $n \in S \implies n+1 \in S$.

(b) $A = \mathbb{N}$.

7. The **Well Ordering Principle** states that every non-empty subset of $\mathbb{N}$ has a least element. The following steps will show that the principle of mathematical induction implies the well ordering principle. First, given a non-empty subset A of $\mathbb{N}$, let

$$S = \{ n \in \mathbb{N} \mid k \in \mathbb{N} \text{ and } k \leq n \text{ implies } k \notin A \}.$$

Now prove the following:

(a) $S \neq \mathbb{N}$.

(b) $S = \varnothing$ if and only if $1 \notin S$.

(c) If $S = \varnothing$ then 1 is the least element of A.

(d) If $S \neq \varnothing$ then there is $N \in \mathbb{N}$ such that $N \in S$ and $N+1 \notin S$. The number $N+1$ is the least element of A.

8. Show that the well ordering principle implies the principle of mathematical induction.

Thus, all three of these principles (induction, strong induction, and well ordering) are equivalent: starting from any one of them, we can prove the others. Their uses are quite different. The induction principles are typically used to prove some relationship involving all natural numbers, while the well ordering principle is used to show the existence of special numbers.

9. Use mathematical induction to prove that for any $x, y \in \mathbb{R}$ and $n \in \mathbb{N}$,

$$x^{n+1} - y^{n+1} = (x - y) \sum_{i=0}^{n} x^i y^{n-i}.$$

Hence, for $x \neq 1$, $\displaystyle\sum_{i=0}^{n} x^i = \frac{1 - x^{n+1}}{1 - x}$.

10. Prove the following for $a, b \in \mathbb{R}$ and $n \in \mathbb{N}$:

(a) $1^n = 1$.

(b) $0 < a < b \implies a^n < b^n$.

11. Prove the following:

(a) $\displaystyle\sum_{k=1}^{n} k = \frac{n(n+1)}{2}$.

(b) $\displaystyle\sum_{k=1}^{n} k^2 = \frac{n(n+1)(2n+1)}{6}$.

12. Recall that the **factorials** of whole numbers are defined by $0! = 1$ and $n! = n \cdot (n-1)!$ for $n \in \mathbb{N}$. Further, the **binomial coefficients** are defined by $\displaystyle\binom{n}{k} = \frac{n!}{k!(n-k)!}$ for $0 \leq k \leq n$. Prove the following for $0 \leq k \leq n$:

(a) $\displaystyle\binom{n+1}{k+1} = \binom{n}{k} + \binom{n}{k+1}$.

(b) $\displaystyle\binom{n}{k} \in \mathbb{N}$.

13. Prove the **binomial theorem**: For any $x, y \in \mathbb{R}$ and $n \in \mathbb{N}$,

$$(x+y)^n = \sum_{k=0}^{n} \binom{n}{k} x^k y^{n-k}.$$

(Note that this requires the definition $0^0 = 1$.)

14. Prove that there is no rational number whose square is 3.

1.2 Completeness Axiom and Archimedean Property

Rational numbers also satisfy the same order axioms as $\mathbb{R}$, and hence the field and order axioms together still do not completely determine the real numbers. For that, we need one more property:

R13. *(Completeness Axiom) Suppose A and B are non-empty subsets of $\mathbb{R}$ such that $a \leq b$ for every $a \in A$ and $b \in B$. Then there is a real number m such that $a \leq m \leq b$ for every $a \in A$ and $b \in B$.*

The completeness axiom completes the list of fundamental properties of $\mathbb{R}$. It has been established that the real numbers form the only system that satisfies the field and order axioms as well as the completeness axiom. (See Hamilton and Landin [11] or Mendelson [24].)

The completeness axiom lends itself to showing the existence of a number with a particular property by locating it between numbers that are too large or too small to have that property. For our first application, we shall prove the existence of square roots of positive real numbers.

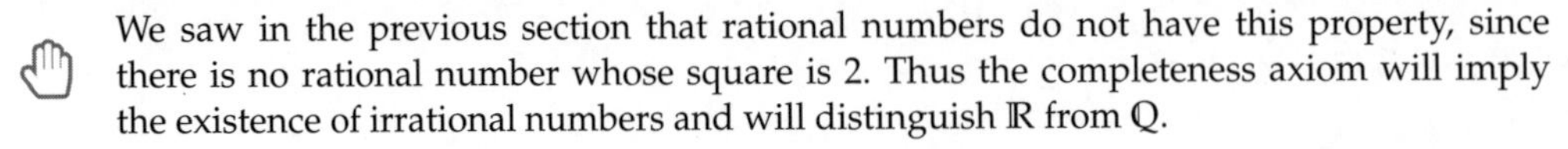

We saw in the previous section that rational numbers do not have this property, since there is no rational number whose square is 2. Thus the completeness axiom will imply the existence of irrational numbers and will distinguish $\mathbb{R}$ from $\mathbb{Q}$.

Theorem 1.2.1

Let $x \in \mathbb{R}^+$. Then there is a unique $y \in \mathbb{R}^+$ such that $y^2 = x$. (We call y the positive square root of x and denote it by $x^{1/2}$ or $\sqrt{x}$.)

This is our first proof which requires considerable effort, so let us discuss the strategy first. To find a number whose square is x, we consider one class of numbers whose squares are greater than x and another class of numbers whose squares are less than x. The completeness axiom gives us a number y that lies between these classes and is a natural candidate for having square equal to x. To show the equality, we can employ trichotomy: If $y^2 > x$ and $y^2 < x$ both lead to contradictions, then we must have $y^2 = x$. For the $y^2 > x$ case we will first use a geometric argument, and then express it algebraically. The $y^2 < x$ case reduces to the $y^2 > x$ one.

Proof. First, we take up existence. We define the following subsets of $\mathbb{R}$:

$$A = \{a \in \mathbb{R}^+ \mid a^2 < x\}, \quad B = \{b \in \mathbb{R}^+ \mid b^2 > x\}.$$

We begin by checking that A and B are non-empty:

- If $x > 1$ then $1 \in A$, while if $x \leq 1$ then $x/2 \in A$.
- In all cases, $x + 1 \in B$.

Now $a \in A$ and $b \in B$ implies that $a^2 < x < b^2$, and hence $a < b$. By the completeness axiom, there is a number $y \in \mathbb{R}^+$ such that $a \leq y \leq b$ for every $a \in A$, $b \in B$.

We note that if $y \in A$ then y is the greatest element of A, while if $y \in B$ then y is the least element of B. Therefore, if we show that A has no greatest member and B has no least member, we will have ruled out both $y^2 < x$ and $y^2 > x$, leaving $y^2 = x$ as the only possibility.

Let us first show that B has no least member. Take any $m \in B$. We need to find an $m' \in B$ such that $m' < m$. Consider the following geometric argument:

Draw a rectangle with sides m and x/m. Then m is the longer side and the area of the rectangle is x. We mark off a square of side x/m in the rectangle as shown below, and bisect the remaining part to form two congruent rectangles.

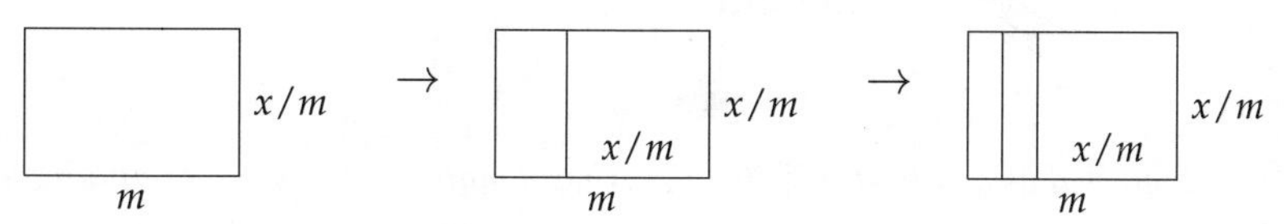

Now we move one of the two newly created strips to an adjacent side of the square, and fill in the missing portion to create a larger square.

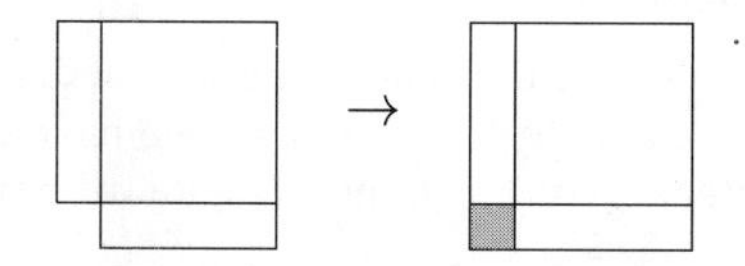

If the side of the final square is m' then it is clear that $m' < m$ while $m'^2 > x$.

The geometric argument given above is easily converted to an algebraic one. We define m' as follows:

$$m' = \frac{1}{2}\left(m + \frac{x}{m}\right) = \frac{x}{m} + \frac{1}{2}\left(m - \frac{x}{m}\right).$$

Then $0 < m' < m$ and $m'^2 > \dfrac{x^2}{m^2} + \dfrac{x}{m}\left(m - \dfrac{x}{m}\right) = x$. Hence B has no least element.

Take any $m \in A$, so that $m > 0$ and $m^2 < x$. Consider the following steps:

1. $\dfrac{x^2}{m^2} > \dfrac{x^2}{x} = x$, hence $\dfrac{x}{m} \in B$.
2. Since B has no least element, we can choose $m' \in B$ such that $m' < \dfrac{x}{m}$.
3. $\dfrac{x^2}{m'^2} < \dfrac{x^2}{x} = x$, hence $\dfrac{x}{m'} \in A$.
4. From $0 < m' < \dfrac{x}{m}$ we see that $\dfrac{x}{m'} > m$.

Hence A has no greatest element. This completes the proof that $y^2 = x$ is the only possibility.

The uniqueness of the positive square root follows from part 10 of Theorem 1.1.9.

■

The completeness axiom can also be used to prove, for any $x \in \mathbb{R}^+$ and $n \in \mathbb{N}$, the existence of the positive n^{th} root $x^{1/n}$. However, we shall wait until we have studied the exponential function, when we will have an extremely simple proof. In the meantime, we will use only square roots and not cube roots or any other n^{th} roots.

Let A be a subset of $\mathbb{R}$.

- An element $M \in \mathbb{R}$ is called an **upper bound** of A if $a \leq M$ for every $a \in A$.
- An element $m \in \mathbb{R}$ is called a **lower bound** of A if $m \leq a$ for every $a \in A$.

If A has an upper bound, we say A is **bounded above**. If it has a lower bound, we say it is **bounded below**. If it is bounded above *and* below, we simply say it is **bounded**. If it is not bounded, we say it is **unbounded**.

Task 1.2.2

Is the empty set bounded as a subset of $\mathbb{R}$? If yes, then what are all its upper and lower bounds?

Theorem 1.2.3 (Archimedean Property of $\mathbb{R}$, Version 1)

The set $\mathbb{N}$ is not bounded above in $\mathbb{R}$.

We may be tempted to say that $\mathbb{N}$ is unbounded because it does not have a largest member. It is true that it does not have a largest member, but we have to take care of the possibility that there may be a non-integer which is larger than all the natural numbers.

Proof. Suppose $\mathbb{N}$ is bounded above. Then the set B of all upper bounds of $\mathbb{N}$ is non-empty. Further, if $a \in \mathbb{N}$ and $b \in B$ then $a \leq b$. Hence, by the completeness axiom, there is a real number α such that $a \leq \alpha \leq b$ for every $a \in \mathbb{N}$, $b \in B$.

Now $\alpha - 1 \notin B$. Hence there is an $N \in \mathbb{N}$ such that $N > \alpha - 1$. But then $N + 1 \in \mathbb{N}$ and $N + 1 > \alpha$, a contradiction. ∎

Task 1.2.4

Show that $\mathbb{Z}$ has neither an upper nor a lower bound in $\mathbb{R}$.

Theorem 1.2.5 (Archimedean Property of $\mathbb{R}$, Version 2)

Let $x, y \in \mathbb{R}^+$. Then there exists $N \in \mathbb{N}$ such that $Nx > y$.

Proof. Consider y/x. By Version 1 of the Archimedean property there is $N \in \mathbb{N}$ such that $N > y/x$, and hence $Nx > y$. ∎

Theorem 1.2.6 (Archimedean Property of $\mathbb{R}$, Version 3)

Let $x, y \in \mathbb{R}^+$. Then there exists $N \in \mathbb{N}$ such that $0 < \dfrac{y}{N} < x$.

Proof. There is $N \in \mathbb{N}$ such that $Nx > y > 0$. Hence $0 < \dfrac{y}{N} < x$. ∎

The next result is the one Archimedes used to prove formulas for lengths, areas and volumes of a variety of shapes, and is the cause of this kind of reasoning being called 'Archimedean'.

Theorem 1.2.7

Suppose $x, y \in \mathbb{R}$ and $M > 0$ such that $y - \dfrac{M}{n} \leq x \leq y + \dfrac{M}{n}$ for every $n \in \mathbb{N}$. Then $y = x$.

Proof. We apply trichotomy. First, suppose $y > x$. Then $0 < y - x$. By Archimedean property Version 3, there is $N \in \mathbb{N}$ such that $0 < M/N < y - x$. Hence $x < y - M/N$. This contradicts the given relationship between x, y, M. So $y > x$ is false. We similarly prove that $y < x$ is false (Try!). Therefore, by trichotomy, $y = x$. ∎

The Archimedean property justifies viewing real numbers as strung out on a line marked off by integers. The first version tells us that the natural numbers can reach out and exceed any real number. The second one says we can change our unit size without losing this property. Any ruler can be used to estimate a distance, no matter how small the ruler and how immense the distance. The third version reassures us that by taking fractional parts of the ruler we can also investigate arbitrarily small distances.

The next step in recovering our mental picture of the Euclidean line is to show that any real number lies between two consecutive integers.

Theorem 1.2.8 (Greatest Integer)

Given a real number x, there is a unique integer m such that $m \leq x < m + 1$.

Proof. Let $A = \{n \in \mathbb{Z} \mid n \leq x\}$ and let B be the set of upper bounds of A. A is non-empty because $\mathbb{Z}$ is not bounded below. B is non-empty because it has x as a member. By the definition of B it is immediate that $a \in A$ and $b \in B$ implies $a \leq b$. Hence, by the completeness axiom, there is an $\alpha \in \mathbb{R}$ such that $a \leq \alpha \leq b$ for every $a \in A, b \in B$.

Now, $\alpha - 1$ is not an upper bound of A, so there is an integer $m \in A$ such that $\alpha - 1 < m \leq x$. From $\alpha - 1 < m$ it follows that $\alpha < m + 1$, hence $m + 1 \notin A$, and therefore $x < m + 1$. We have found the desired m.

For uniqueness, we note that any integer with the given property would be the greatest element of A. ■

The integer m such that $m \leq x < m + 1$ is called the **greatest integer** corresponding to x and is denoted by $[x]$.

Theorem 1.2.9 (Denseness of Rational Numbers)

Let $x, y \in \mathbb{R}$ with $x < y$. Then there is a rational number r such that $x < r < y$.

Proof. Since $y - x > 0$, there is $N \in \mathbb{N}$ such that $1 < N(y - x) = Ny - Nx$ (Archimedean Principle, Version 2).

Take $M = [Nx] + 1$. Then $M - 1 \leq Nx < M$. This also gives $M \leq Nx + 1 < Nx + (Ny - Nx) = Ny$. Hence $Nx < M < Ny$ and $x < M/N < y$. ■

Task 1.2.10

Show that there are infinitely many rational numbers between any two distinct real numbers.

Task 1.2.11

Let t be an irrational number. Show that:

(a) *The numbers $-t$ and $1/t$ are irrational.*

(b) *If r is a rational number then $r + t$ and $r - t$ are irrational.*

(c) *If r is a non-zero rational number then rt and r/t are irrational.*

Theorem 1.2.12 (Denseness of Irrational Numbers)

Let $x, y \in \mathbb{R}$ with $x < y$. Then there is an irrational number t such that $x < t < y$.

Proof. By the denseness of $\mathbb{Q}$, we have $r \in \mathbb{Q}$ such $\sqrt{2}x < r < \sqrt{2}y$. We can arrange for r to be non-zero. Then $t = r/\sqrt{2}$ is an irrational number such that $x < t < y$. ■

When A, B satisfy the hypotheses of the completeness axiom then A is automatically bounded above, and B consists of certain upper bounds of A. Now if we are originally given just a non-empty set A which is bounded above, we can choose B to be the set of all upper bounds of A. The completeness axiom then gives an α which separates A and B. We note the following:

1. Since $a \leq \alpha$ for every $a \in A$, α is an upper bound of A.
2. Since $\alpha \leq b$ for every $b \in B$, α is least among all the upper bounds of A.
3. The number α is the only number with these properties.

Therefore this number α is called the **least upper bound** (or LUB) of A. It is also called the **supremum** (or sup) of A. We have just proved the following:

Theorem 1.2.13 (Least Upper Bound or LUB Property)

Every non-empty subset of $\mathbb{R}$ which is bounded above has a (unique) least upper bound. ■

The supremum of a set can sometimes be built up from the supremums of its subsets with the help of the following result.

Theorem 1.2.14

Let $A, B \subseteq \mathbb{R}$ be non-empty and bounded above. Then

$$\sup(A \cup B) = \max\{\sup(A), \sup(B)\}.$$

Proof. It is clear that $\alpha = \max\{\sup(A), \sup(B)\}$ is an upper bound of $A \cup B$. We will show that for any $\epsilon > 0$, $\alpha - \epsilon$ is not an upper bound of $A \cup B$. We may assume that $\alpha = \sup(A)$. Then there is $a \in A$ such that $a > \sup(A) - \epsilon = \alpha - \epsilon$. But $a \in A \cup B$. ■

Similarly, any non-empty subset A of $\mathbb{R}$ which is bounded *below* has a (unique) **greatest lower bound** (or GLB). This is a number β which is a lower bound of A and which is the greatest among all the lower bounds of A. It is also called the **infimum** (or inf) of A.

The following subsets of real numbers are called **intervals**.

$$(a,b) = \{x \in \mathbb{R} \mid a < x < b\}$$

$$[a,b] = \{x \in \mathbb{R} \mid a \leq x \leq b\}$$

$$[a,b) = \{x \in \mathbb{R} \mid a \leq x < b\}$$

$$(a,b] = \{x \in \mathbb{R} \mid a < x \leq b\}$$

$$(a,\infty) = \{x \in \mathbb{R} \mid a < x\}$$

$$[a,\infty) = \{x \in \mathbb{R} \mid a \leq x\}$$

$$(-\infty,b) = \{x \in \mathbb{R} \mid x < b\}$$

$$(-\infty,b] = \{x \in \mathbb{R} \mid x \leq b\}$$

$$(-\infty,\infty) = \mathbb{R}$$

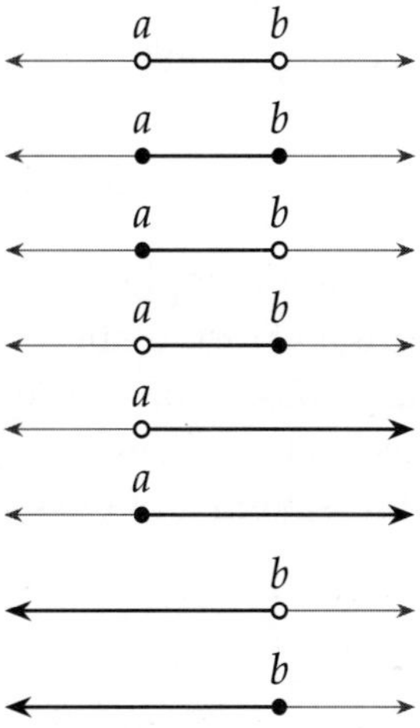

In the above diagrams, a filled black circle is used to indicate that the point (a or b) is included in the interval, while an unfilled circle indicates that it is excluded.

Task 1.2.15

Let $a < b$. Show that the supremum of $[a,b]$ is b, while its infimum is a.

Did you wonder how $(-\infty,\infty)$ could be called both open and closed? We have explained why it is called closed. We must also call it open because it is true that it contains no endpoints. The set of its endpoints is empty, and empty sets have no difficulty in satisfying apparently contradictory requirements!

Task 1.2.16

Let $a < b$. Show that the supremum of (a,b) is b, while its infimum is a.

The supremum and infimum of an interval are called its **endpoints**. All other points of the interval are called its **interior points**.

An interval is called **closed** if it contains its endpoints. Intervals of the form $[a,b]$, $[a,\infty)$, $(-\infty,b]$ and $(-\infty,\infty)$ are closed.

If we consider $[a,\infty)$, we see that a is its only endpoint, and is included. Similarly, $(-\infty,b]$ contains b, which is its only endpoint. Finally, $(-\infty,\infty)$ has no endpoints, so it contains all its endpoints in the sense that no endpoint is excluded! This is the same reasoning by which we declare the empty set to be a subset of every set.

Task 1.2.17

Let I be an interval and $a,b \in I$. Show that $a < x < b$ implies $x \in I$.

Exercises for § 1.2

1. Show that if $x,y \geq 0$ then $\dfrac{x+y}{2} \geq \sqrt{xy}$, with equality if and only if $x = y$.

2. Put the following numbers in ascending order *without* converting them to decimal form:

$$3,\ \sqrt{2},\ \frac{14}{10},\ -2,\ -\frac{3}{2},-\sqrt{2},\ \sqrt{5}.$$

3. Show that the completeness axiom is equivalent to the following (which is Dedekind's original formulation): "If all points of the straight line fall into two classes such that every point of the first class lies to the left of every point of the second class, then there exists one and only one point which produces this division of all points into two classes, this severing of the straight line into two portions."

4. Prove that every non-empty subset of $\mathbb{R}$ which is bounded below has a greatest lower bound.

5. Which intervals are bounded?

6. Show that the following set is unbounded:

$$\left\{1,\ 1+\frac{1}{\sqrt{2}},\ 1+\frac{1}{\sqrt{2}}+\frac{1}{\sqrt{3}},\ 1+\frac{1}{\sqrt{2}}+\frac{1}{\sqrt{3}}+\frac{1}{\sqrt{4}},\ \dots\right\}.$$

7. Show that the following sets are bounded, and find their supremum.

(a) $\{1,\ 1+1/2,\ 1+1/2+1/2^2,\ 1+1/2+1/2^2+1/2^3,\ \dots\}$.

(b) $\{0.1,\ 0.11,\ 0.111,\ 0.1111,\ \dots\}$.

8. Show that the following set is bounded:

$$\left\{\frac{1}{1\cdot 2},\ \frac{1}{1\cdot 2}+\frac{1}{2\cdot 3},\ \frac{1}{1\cdot 2}+\frac{1}{2\cdot 3}+\frac{1}{3\cdot 4},\ \dots\right\}.$$

9. Show that 1 is the supremum of each of the given sets.

(a) $A=[0,1)$.

(b) $B=\left\{1-\frac{1}{n} : n\in\mathbb{N}\right\}$.

10. Let A be a non-empty subset of $\mathbb{R}$. For $c\in\mathbb{R}$, define $cA=\{cx \mid x\in A\}$. Prove the following.

(a) $\sup(cA)=\begin{cases} c\cdot\sup(A) & \text{if } c\geq 0 \text{ and } A \text{ is bounded above,}\\ c\cdot\inf(A) & \text{if } c<0 \text{ and } A \text{ is bounded below.}\end{cases}$

(b) $\inf(cA)=\begin{cases} c\cdot\inf(A) & \text{if } c\geq 0 \text{ and } A \text{ is bounded below,}\\ c\cdot\sup(A) & \text{if } c<0 \text{ and } A \text{ is bounded above.}\end{cases}$

11. Let $A,B\subseteq\mathbb{R}$ be non-empty and bounded above.

(a) Define $A+B=\{a+b \mid a\in A \text{ and } b\in B\}$. Show that $\sup(A+B)=\sup(A)+\sup(B)$.

(b) Define $AB=\{ab \mid a\in A \text{ and } b\in B\}$. Show that if the members of A,B are all non-negative then $\sup(AB)=\sup(A)\sup(B)$.

12. State and prove the results for infimum that correspond to the previous exercise.

13. Produce a rational number and an irrational number that lie between $17/12$ and $\sqrt{2}$.

1.3 Functions

A **function** f from a set X to a set Y is usually described as a rule that associates exactly one element of Y to each element of X. The element of Y associated to $x\in X$ is denoted by $f(x)$. This is not quite a formal definition as one has to wonder what is allowed as a 'rule'. We can do better by stating our requirements purely in terms of membership of sets:

Let X, Y be sets. A **function** f from X to Y is a subset of the Cartesian product $X \times Y$ such that for each $x \in X$ there is exactly one member (x, y) of f, and this member is denoted by $(x, f(x))$.

In the above definitions, X is called the **domain** of f and Y is called the **codomain** of f. The notation $f \colon X \to Y$ is used as shorthand for "f is a function with domain X and codomain Y", and is read as "f is a function from X to Y". The subset of Y consisting of the values actually taken by the function is called its **image** or **range**.

The terms domain, codomain and image are entirely standard in that all authors use them the same way. The term range, however, is sometimes used in the sense of codomain and sometimes in the sense of image. For this reason, we shall strive to use only the term image and not range.

Example 1.3.1

Consider $f \colon \mathbb{R} \to \mathbb{R}$ defined by $f(x) = x^2$. The domain and codomain are both $\mathbb{R}$, but the image of f is $[0, \infty)$. (How do we know that the image of f is all of $[0, \infty)$?) ▢

Consider a function $f \colon X \to Y$. Let $a \in X$ and $b \in Y$ such that $f(a) = b$. Then b is called the **image** of the point a under f, while a is called the **pre-image** of b under f.

Example 1.3.2

Consider the rules depicted by the following pictures. In each pair of ovals, the one on the left represents the domain, while the one on the right represents the codomain. The arrows mark the associations given by the rule. Which diagrams represent functions?

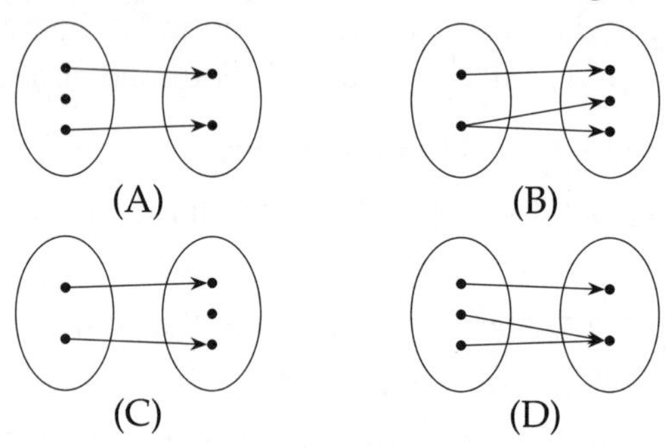

(A) does not represent a function because there is a point in the domain that has no image. (B) also does not represent a function, since there is a point in the domain that has two images. (C) and (D) do represent functions, since it is permitted for points in the *codomain* of a function to have no pre-image as well as to have multiple pre-images. ▢

Task 1.3.3

Consider a binary operation on a set X. Can you describe it as a function with a certain domain and codomain?

We have been using the letter f for a function. This is simply the most commonly used notation for a function (like x is for a variable), but we are free to use any other letter, symbol, or word. Other popular choices are g, h, u, v, F, G, H, η, θ and so on. Functions that are particularly important have their own names such as sin, cos, exp and log.

A function $f\colon X \to Y$ is called **one-one** or **injective** if distinct points in X have distinct images in Y: If $a, b \in X$ and $a \neq b$ then $f(a) \neq f(b)$.

Task 1.3.4

Show that $f\colon X \to Y$ *is one-one if and only if* $f(a) = f(b)$ *implies* $a = b$.

Example 1.3.5

Consider the functions depicted below.

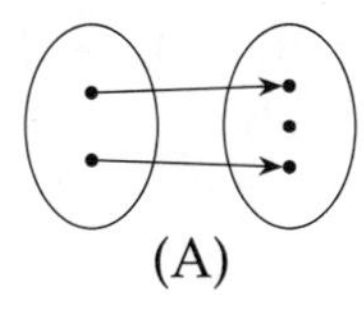

(A)

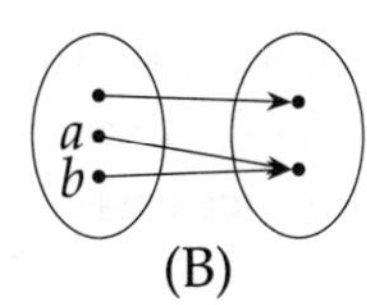

(B)

The function in (A) is one-one because distinct points in the domain are taken to distinct points in the codomain. The function in (B) is not one-one because the points a and b are taken to the same value. □

A function $f\colon X \to Y$ is called **onto** or **surjective** if its image is all of Y, that is, for each $b \in Y$ there exists $a \in X$ such that $f(a) = b$.

Example 1.3.6

Consider the functions depicted below.

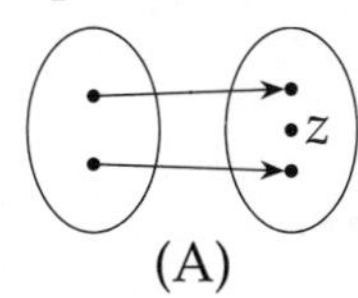

(A)

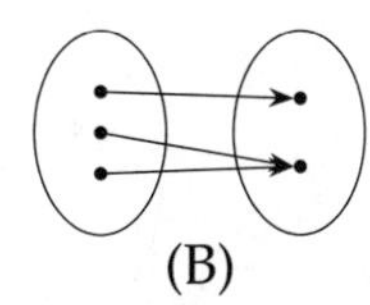

(B)

The function in (A) is not onto because the point z in the codomain has no pre-image. The function in (B) *is* onto. □

Task 1.3.7

Find out whether the following functions are one-one or onto. If a function is not onto, give its image.

(a) $f\colon \mathbb{R} \to \mathbb{R}$, $f(x) = \frac{1}{2}(x + |x|)$.

(b) $g\colon \mathbb{R} \to \mathbb{R}$, $g(x) = x^2$.

(c) $h\colon \mathbb{R} \to \mathbb{R}$, $h(x) = \begin{cases} x^2 + x + 1 & \text{if } x \geq 0, \\ x + 1 & \text{if } x < 0. \end{cases}$

A function $f: X \to Y$ is called a **one-one correspondence** or **bijection** if it is both one-one and onto.

Example 1.3.8

Consider the following functions.

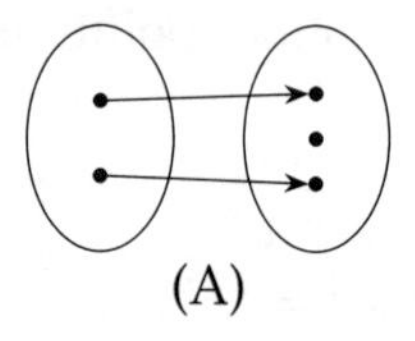
(A)

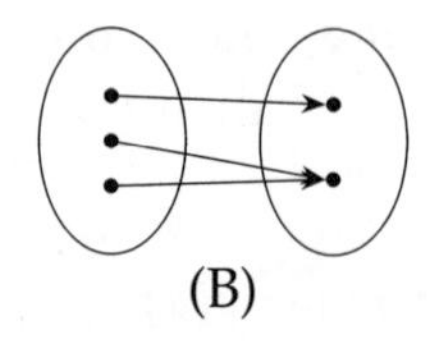
(B)

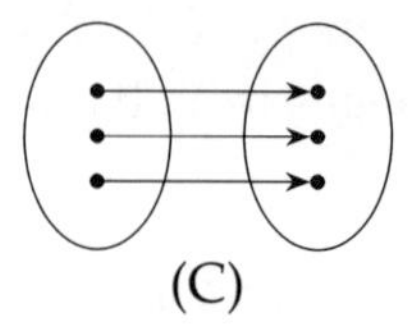
(C)

The function in (A) is one-one but not onto. The function in (B) is onto but not one-one. Finally, the one in (C) is both one-one and onto, hence it is a bijection. □

Let $f: X \to Y$. We ask whether there is a function $g: Y \to X$ which 'reverses' f. That is, if f takes x to y then g takes y to x and conversely.

First, consider f as in the picture below.

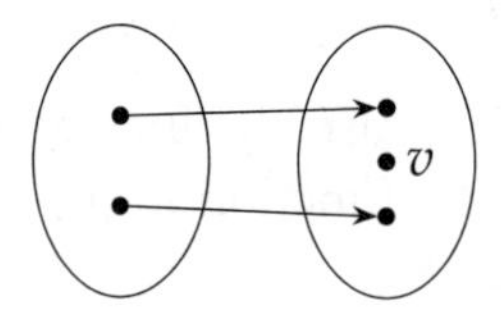

A function which reverses f would be obtained by reversing the arrows in the picture. However, when we reverse the arrows, we find we have not created a function, since no reversed arrow starts at v. Why did we run into this problem? Because f is not onto.

Now consider a different function f.

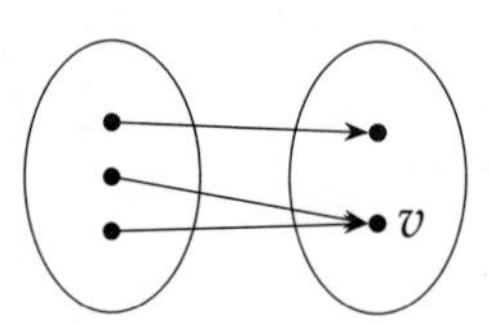

On reversing the arrows, we have difficulty because two of the reversed arrows start at v. Thus we have trouble if f is not one-one. These examples show that our reversing process can only be successful if f is both onto and one-one, that is, it is a bijection.

So, suppose $f: X \to Y$ is a bijection. We define its **inverse function** $f^{-1}: Y \to X$ by $f^{-1}(y) = x \iff f(x) = y$. Starting with any $y \in Y$ we note that since f is a bijection there is exactly one $x \in X$ such that $f(x) = y$. This x is termed $f^{-1}(y)$.

Task 1.3.9

Let $f: X \to Y$ be a bijection. Show that $f^{-1}: Y \to X$ is also a bijection and $(f^{-1})^{-1} = f$.

Let $f\colon X \to Y$ and $g\colon Y \to Z$. Then their **composition** $g \circ f\colon X \to Z$ is defined by

$$g \circ f(x) = g(f(x)), \qquad \text{for every } x \in X.$$

Note that to define $g \circ f$, we have to ensure that the codomain of f equals the domain of g. Only then does $f(x)$ become a valid input for g. (It would actually be enough for the codomain of f to be a *subset* of the domain of g, but that greater generality does not bring any significant improvement in the theory.)

Task 1.3.10

Show that composition of functions is associative: If $f\colon W \to X$, $g\colon X \to Y$ and $h\colon Y \to Z$ then $h \circ (g \circ f) = (h \circ g) \circ f$.

The **identity function** from a set A to itself is denoted 1_A and maps every element to itself: $1_A(a) = a$ for every $a \in A$.

Task 1.3.11

Let $f\colon X \to Y$ and $g\colon Y \to X$. Show that g is the inverse function of f if and only if $g \circ f = 1_X$ and $f \circ g = 1_Y$.

Theorem 1.3.12

Let $f\colon X \to Y$ and $g\colon Y \to Z$ be bijections. Then $(g \circ f)^{-1} = f^{-1} \circ g^{-1}$.

Proof. We start with the following observations.

1. Since f is a bijection it has an inverse $f^{-1}\colon Y \to X$.
2. Since g is a bijection it has an inverse $g^{-1}\colon Z \to Y$.

To verify that $f^{-1} \circ g^{-1}$ is the inverse of $g \circ f$, we carry out the following calculations:

$$\begin{aligned}(f^{-1} \circ g^{-1}) \circ (g \circ f) &= ((f^{-1} \circ g^{-1}) \circ g) \circ f = (f^{-1} \circ (g^{-1} \circ g)) \circ f\\ &= (f^{-1} \circ 1_Y) \circ f = f^{-1} \circ f \ = \ 1_X,\\ (g \circ f) \circ (f^{-1} \circ g^{-1}) &= ((g \circ f) \circ f^{-1}) \circ g^{-1} = (g \circ (f \circ f^{-1})) \circ g^{-1}\\ &= (g \circ 1_X) \circ g^{-1} = g \circ g^{-1} \ = \ 1_Y.\end{aligned}$$

∎

A **constant function** is a function that only takes one value. Suppose $f\colon X \to Y$ only takes one value $c \in Y$, that is, $f(x) = c$ for every $x \in X$. Then we denote f by c and call it 'the constant function c'.

Exercises for § 1.3

The **difference** of two sets A and B is the set consisting of those elements of A which do not belong to B, and is denoted by $A \setminus B$.

1. Find out whether the following functions are one-one or onto. If a function is not onto, give its image.

(a) $f\colon \mathbb{R}\setminus\{0\} \to \mathbb{R}$, $f(x) = \dfrac{1}{x}$.

(b) $g\colon \mathbb{R}\setminus\{1\} \to \mathbb{R}$, $g(x) = \dfrac{x}{1-x}$.

(c) $h\colon \mathbb{R}\setminus\{0,1\} \to \mathbb{R}$, $h(x) = \dfrac{1}{x(1-x)}$.

2. Show that the following functions are bijections, and find their inverses.

(a) $f\colon \mathbb{R}\setminus\{0\} \to \mathbb{R}\setminus\{0\}$, $f(x) = \dfrac{1}{x}$.

(b) $g\colon \mathbb{R}\setminus\{1\} \to \mathbb{R}\setminus\{-1\}$, $g(x) = \dfrac{x}{1-x}$.

(c) $h\colon [1/2,1) \to [4,\infty)$, $h(x) = \dfrac{1}{x(1-x)}$.

3. Give an example to show that composition of functions is not commutative.

4. Consider functions $f\colon X \to Y$ and $g\colon Y \to Z$. Prove the following.

(a) If f and g are injective then $g \circ f$ is injective.

(b) If f and g are surjective then $g \circ f$ is surjective.

(c) If f and g are bijective then $g \circ f$ is bijective.

5. Give a counterexample to the converse of each part of the previous exercise.

6. In each case, give an example of a non-constant function $f\colon \mathbb{R} \to \mathbb{R}$ with the given property.

(a) $f \circ f = f$.

(b) $f \circ f = 1_{\mathbb{R}}$ and $f \neq 1_{\mathbb{R}}$.

(c) $f \circ f = 0$.

7. Give bijections between the given sets.

(a) $\mathbb{N}$ and $\mathbb{W}$.

(b) $\mathbb{N}$ and $\mathbb{Z}$. (Hint: $0,-1,1,-2,2,\ldots$.)

(c) $[0,1)$ and $[0,\infty)$. (Hint:
1
1
)

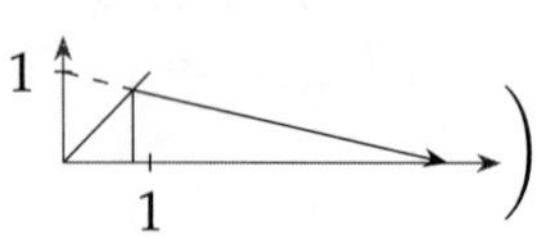

8. Give a bijection between $\mathbb{N}$ and $\mathbb{N} \times \mathbb{N}$. The diagram below gives a hint for one such bijection.

(1,1) (1,2) (1,3) (1,4) ⋯
1↗ 3↗ 6↗
(2,1) (2,2) (2,3) (2,4) ⋯
2↗ 5↗
(3,1) (3,2) (3,3) (3,4) ⋯
4↗
(4,1) (4,2) (4,3) (4,4) ⋯
⋮ ⋮ ⋮ ⋮

(This particular bijection will be useful when we multiply infinite series in Chapters 7 and 8. Your task is to find a formula for it.)

1.4 Real Functions and Graphs

A **real function** is a function whose domain and codomain are subsets of $\mathbb{R}$. In this book, we deal solely with real functions, except in the very last section when we consider complex functions.

Commonly, mathematicians provide only the rule of association that describes a real function and expect the reader or listener to work out the domain. For example, one might write, "Consider the function $f(x) = \sqrt{x}$" without specifying the domain. In this case, the reader is expected to realise that since square roots are defined only for non-negative real numbers, the domain is $[0, \infty)$. In general, if a real function is given only by a rule $f(x)$ the domain is to be taken to consist of all real numbers x such that $f(x)$ is defined. As for the codomain, if it is not explicitly given, it is taken to be $\mathbb{R}$.

Task 1.4.1

Identify the domains of the real functions given by the following rules.

(a) $f(x) = \sqrt{1 - x^2}$.

(b) $g(x) = \dfrac{1}{x}$.

(c) $h(x) = \sqrt{(x-1)(x-2)}$.

(d) $k(x) = \dfrac{1}{\sqrt{1 - x^2}}$.

Suppose f is a real function with domain X. Then the **graph** of f is the subset of the xy-plane which consists of all ordered pairs of the form $(x, f(x))$ with $x \in X$.

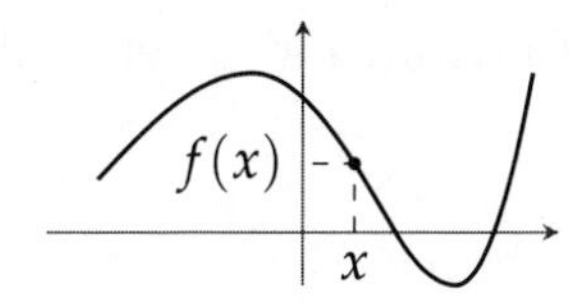

Absolute Value Function

The absolute value $|x|$ of a real number x defines a real function called the **absolute value** or **modulus function**. Its domain is $\mathbb{R}$ and its graph is given below.

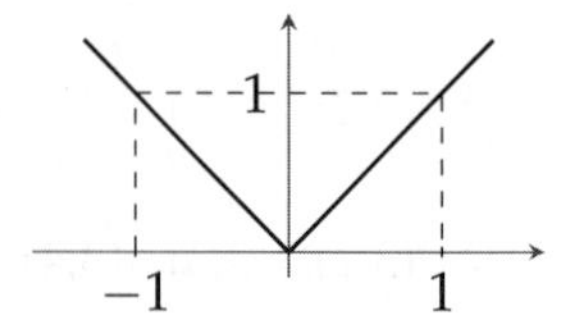

Unit Step Function

The **Heaviside** or **unit step function** is given by $H(x) = \begin{cases} 0 & \text{if } x < 0, \\ 1 & \text{if } x \geq 0. \end{cases}$

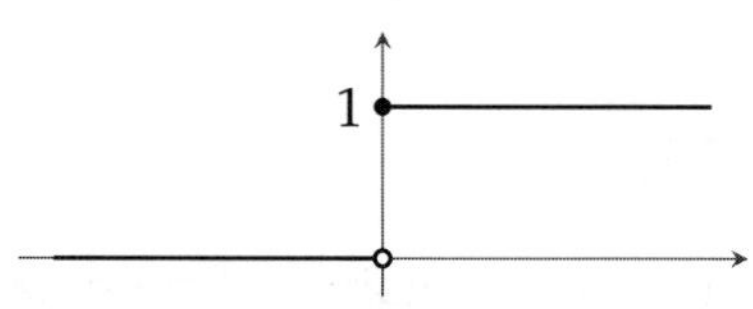

We have utilized a useful convention in depicting this graph. The unfilled circle at the origin indicates that the origin is not part of the graph. The filled circle at $(0,1)$ emphasizes that the function value at $x = 0$ is 1.

Sign Function

The **sign** or **signum function** is defined by $\operatorname{sgn}(x) = \begin{cases} -1 & \text{if } x < 0, \\ 0 & \text{if } x = 0, \\ 1 & \text{if } 0 < x. \end{cases}$

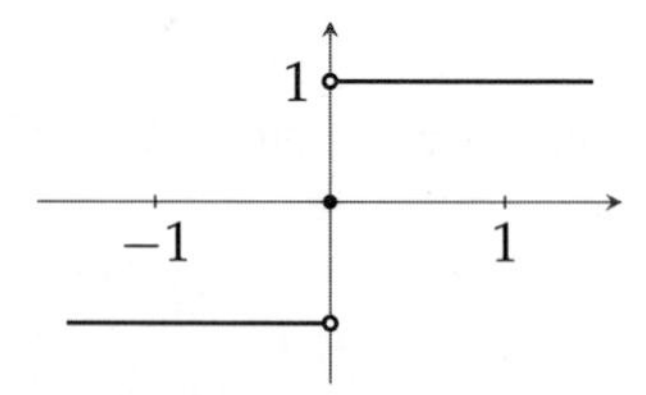

Greatest Integer Function

Recall that for every real number x there is a unique integer $[x]$ such that $[x] \leq x < [x] + 1$. The function which associates $[x]$ to x is called the **greatest integer function**. Sometimes it is called the **floor function** and is denoted by $\lfloor x \rfloor$.

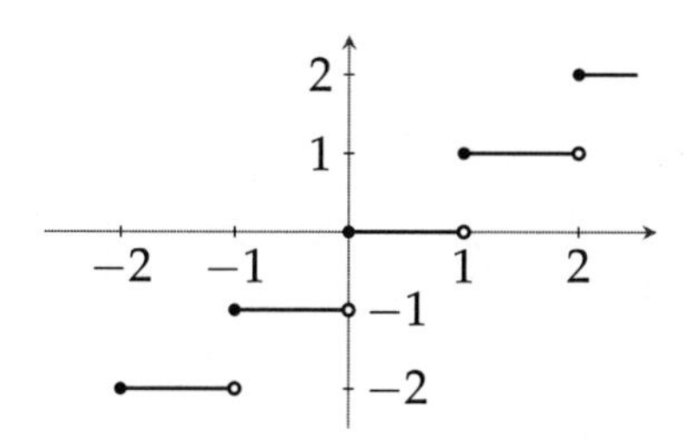

Task 1.4.2

The ***ceiling function****, whose value at x is denoted by $\lceil x \rceil$, is defined by setting $\lceil x \rceil$ to be the least integer which is greater than or equal to x. Draw the graph of the ceiling function.*

Task 1.4.3

Draw the graph of the 'sawtooth' function $r(x) = x - \lfloor x \rfloor$.

Now we will consider some simple ways of creating new functions by modifying existing ones.

Vertical Shift and Scaling

Given a real function f and a real number c, we define a function called $f + c$ by

$$(f + c)(x) = f(x) + c.$$

Adding a constant to a function *shifts* its graph vertically. For example, adding 2 will shift the graph up by 2 units and adding -2 will shift it down by 2 units. The figure below shows the graphs of $f \pm c$ when c is positive.

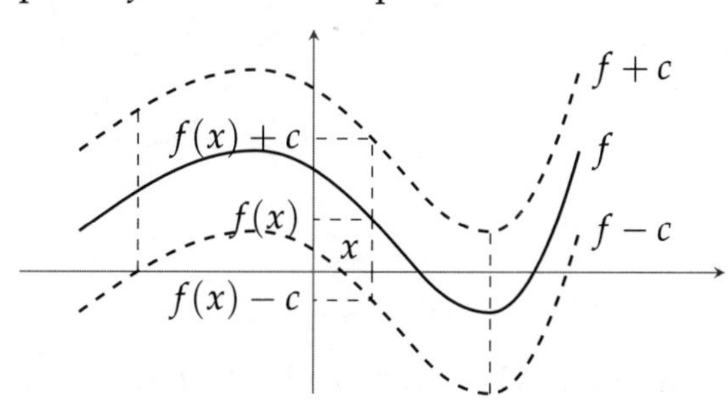

Another modification is to use multiplication to create a function cf defined by

$$cf(x) = c \cdot f(x).$$

Multiplying a function by a constant *scales* its graph vertically. For example, multiplying by 2 will scale the graph vertically by a factor of 2, while multiplying by -2 will further reflect it in the x-axis. The figures below show the graphs of $\pm cf$ when c is positive.

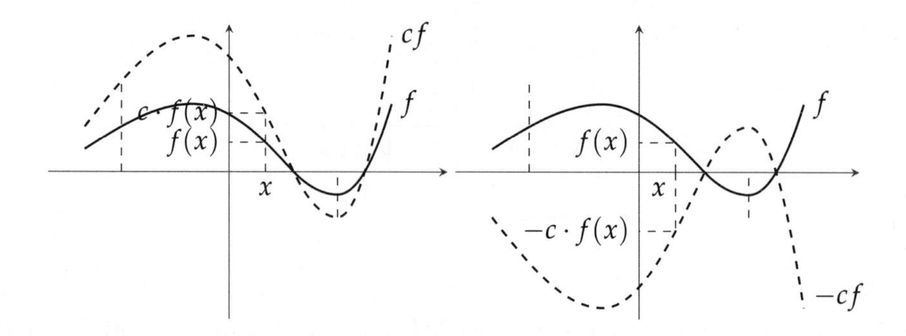

Horizontal Shift and Scaling

We have just seen how the graph of a function changes when a constant is added to or multiplies its *output*. Now we consider what happens when the *input* of the function is changed in this manner. That is, given a function f and a constant c, we consider the functions defined by $g(x) = f(x + c)$ and $h(x) = f(cx)$.

Task 1.4.4

Let f be a real function with domain A and let c be a real number. What are the domains of $g(x) = f(x + c)$ and $h(x) = f(cx)$?

Consider the function $g(x) = f(x + c)$ with $c > 0$. We have $f(x) = f((x - c) + c) = g(x - c)$. That is, the value taken by f at x is taken by g at $x - c$. Thus the graph of g is a horizontal shift to the left of the graph of f.

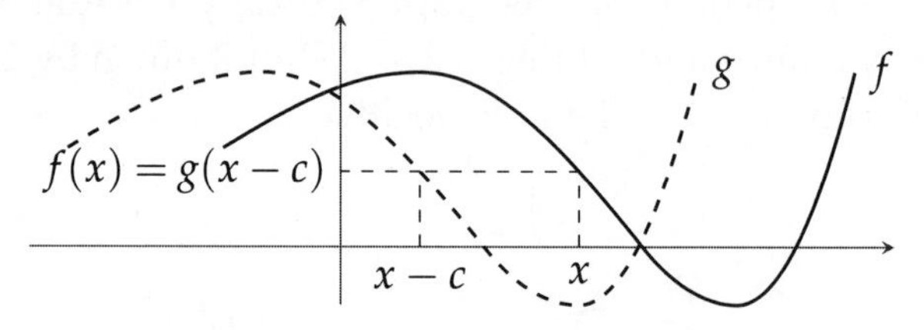

Task 1.4.5

Describe the graph of $g(x) = f(x + c)$ when $c < 0$.

Now consider the function $h(x) = f(cx)$ with $c > 0$. Reasoning as above (try!) we conclude that the value taken by f at x is taken by h at x/c. In this case the graph is scaled horizontally by a factor of $1/c$.

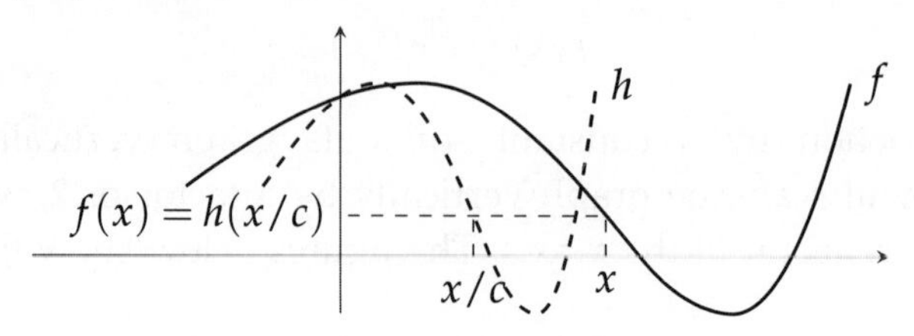

Note that the graph will contract if $c > 1$ and will stretch if $c < 1$.

Task 1.4.6

Describe the graph of $h(x) = f(cx)$ when $c < 0$.

Task 1.4.7

Recall that the graph of $f(x) = x^2$ is an upward opening parabola.

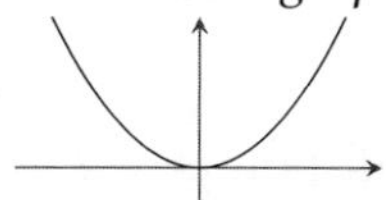

Use your understanding of shifts and scalings to plot the graphs of the following on the same xy-plane.

(a) $g(x) = (x-2)^2 + 1$.

(b) $h(x) = 4x^2 + 12x + 5$.

Even and Odd Functions

A real function $f\colon X \to \mathbb{R}$ is called an **even function** if

1. X is symmetric about 0: $x \in X \iff -x \in X$, and
2. The graph of f is symmetric with respect to the y-axis: $f(-x) = f(x)$ for every $x \in X$.

An example of an even function is given below.

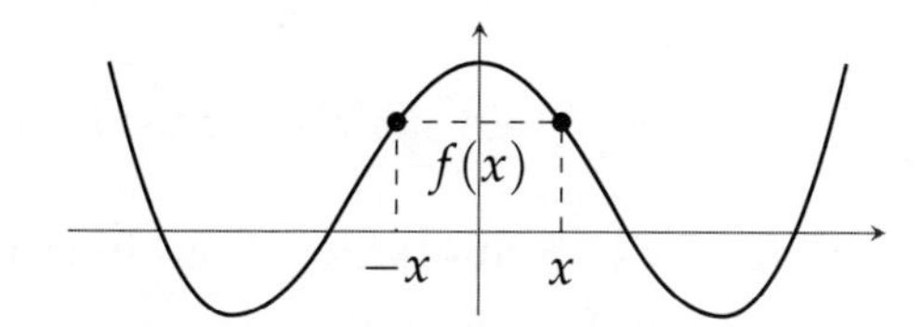

A real function $f\colon X \to \mathbb{R}$ is called an **odd function** if

1. X is symmetric about 0: $x \in X \iff -x \in X$, and
2. The graph of f is symmetric with respect to the origin: $f(-x) = -f(x)$ for every $x \in X$.

Here is an example of an odd function.

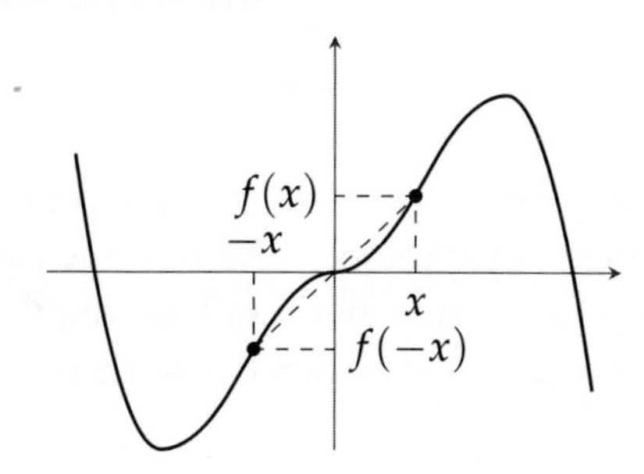

Task 1.4.8

Determine whether the following functions are even, or odd, or neither: $|x|$, $\operatorname{sgn}(x)$ *and* $[x]$.

Task 1.4.9

Can a function be both even and odd?

Monotonic Functions

A real function $f\colon A \to \mathbb{R}$ is called an **increasing function** if, for all points $x, y \in A$, $x \le y$ implies $f(x) \le f(y)$. It is called **strictly increasing** if $x < y$ implies $f(x) < f(y)$.

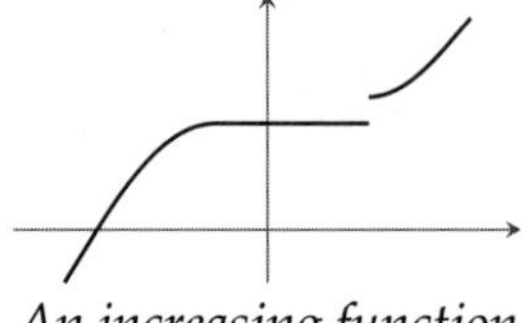

An increasing function

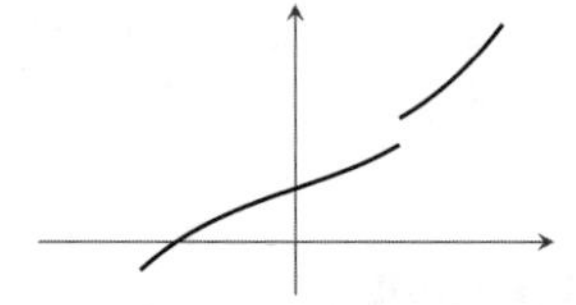

A strictly increasing function

Similarly, a real function f is called a **decreasing function** if $x \le y$ implies $f(x) \ge f(y)$. It is called **strictly decreasing** if $x < y$ implies $f(x) > f(y)$.

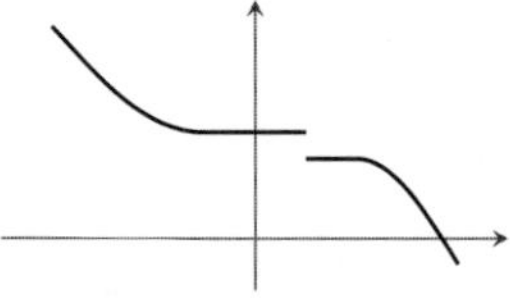

A decreasing function

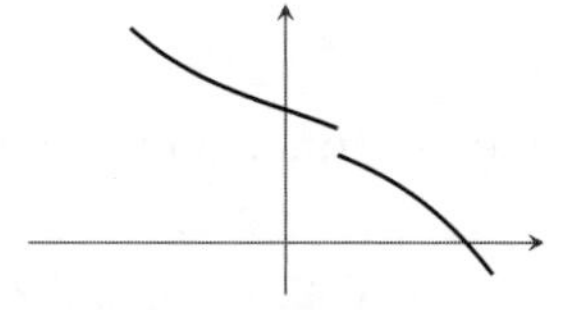

A strictly decreasing function

A real function is called **monotone** or **monotonic** if it is either increasing throughout its domain or decreasing throughout its domain.

Task 1.4.10

Identify which of the given functions are monotonic. If a function is monotonic, state whether it is (strictly) increasing or decreasing.

(a) $f(x) = \operatorname{sgn}(x)$.

(b) $g(x) = [x]$.

(c) $h(x) = x^2$.

(d) $k(x) = -x^3$.

Periodic Functions

We have been looking at functions which are special in having some regularity. Even and odd functions have reflection symmetry. Monotonic functions have a persistent trend of either growth or decay. Another kind of regularity is when a function represents a cyclic phenomenon, one in which the same pattern repeats indefinitely.

A real function is called **periodic** if there is a positive number T such that $f(x + nT) = f(x)$ for every $n \in \mathbb{Z}$ and every x in the domain of f. One can view this as symmetry under horizontal shifts by integer multiples of T. The number T is called a **period** of f.

Task 1.4.11

Show that if T is a period of f then every nT, $n \in \mathbb{N}$, is also a period of f.

Example 1.4.12

Here are some examples of periodic functions.

(a) $f(x) = x - [x]$ has period 1.

(b) $g(x) = \begin{cases} 0 & \text{if } [x] \text{ is even,} \\ 1 & \text{if } [x] \text{ is odd,} \end{cases}$ has period 2.

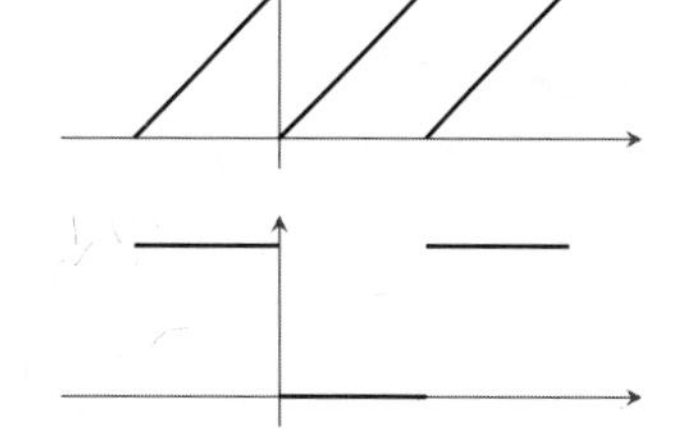

The function f is called the 'sawtooth wave' and g is the 'square wave'. □

Arithmetic of Functions

Let f, g be real functions. We use them to define new functions as follows:

$$(f+g)(x) = f(x) + g(x), \qquad (f-g)(x) = f(x) - g(x),$$

$$(fg)(x) = f(x)g(x), \qquad \frac{f}{g}(x) = \frac{f(x)}{g(x)}.$$

Task 1.4.13

Let f, g be real functions with domains A, B respectively. Describe the domains of the following functions: $f+g$, $f-g$, fg, f/g.

Task 1.4.14

Under the given conditions, are the functions $f+g$, $f-g$, fg, f/g even or odd?

(a) *When f and g are both even.*

(b) *When f and g are both odd.*

(c) *When f is even and g is odd.*

(d) *When f is odd and g is even.*

Inverse Functions

Suppose I, J are subsets of $\mathbb{R}$ and $f: I \to J$ is a bijection. Then we have the inverse function $f^{-1}: J \to I$. We will establish a relationship between the graphs of f and f^{-1}:

$$\begin{aligned}(x,y) \text{ is in the graph of } f &\iff y = f(x)\\ &\iff x = f^{-1}(y)\\ &\iff (y,x) \text{ is in the graph of } f^{-1}.\end{aligned}$$

Now the point (y,x) can be obtained by reflecting (x,y) in the line $y = x$. Therefore the graph of f^{-1} can be obtained by reflecting the graph of f in the line $y = x$.

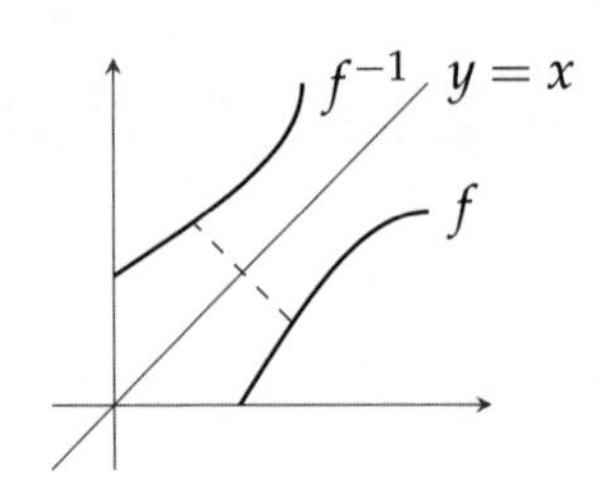

Task 1.4.15

Sketch the graphs of f and f^{-1} on the same coordinate plane.

(a) $f\colon \mathbb{R} \to \mathbb{R}$, $f(x) = 2x + 1$.

(b) $f\colon [0,\infty) \to [0,\infty)$, $f(x) = x^2$.

(c) $f\colon [0,1] \to [0,1]$, $f(x) = \sqrt{1 - x^2}$.

Polynomials

A **monomial** is an expression of the form x^n where $n = 0,1,2,\ldots$. If we let x vary over real numbers, a monomial gives a real function. Here are the graphs of some monomial functions.

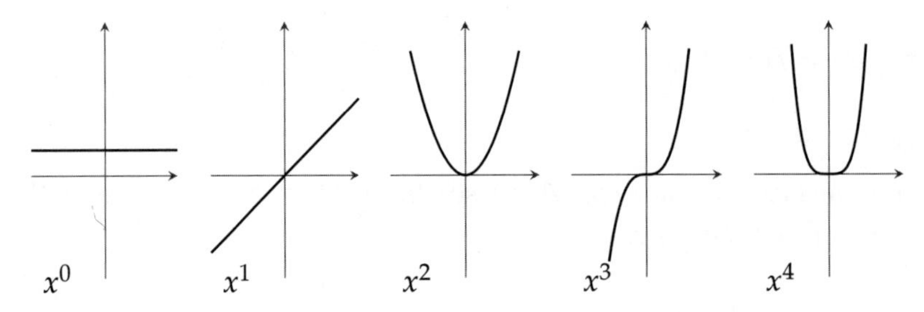

Note the convention that $x^0 = 1$ for every real number, including $0^0 = 1$.

Task 1.4.16

Which monomials are even functions? Which are odd?

A **polynomial** is obtained by scaling and adding monomials. Thus a general polynomial has the form

$$p(x) = a_n x^n + a_{n-1}x^{n-1} + \cdots + a_1 x + a_0 = \sum_{i=0}^{n} a_i x^i,$$

where $n = 0,1,2,\dots$ and each a_i is a real number. If $a_n \neq 0$ we say that $p(x)$ has degree n, and we write $\deg p = n$. If each $a_i = 0$ we call p the **zero polynomial** and write $p = 0$. The degree of the zero polynomial is not defined.

Task 1.4.17

Let p,q be two non-zero polynomials. Show that $\deg(pq) = (\deg p) + (\deg q)$ *and* $\deg(p + q) \leq \max\{\deg p, \deg q\}$.

If we let x vary over real numbers, $p(x)$ gives a real function whose domain is $\mathbb{R}$.

If a polynomial involves only a few monomials, we can combine the graphs of these monomials to understand at least the main features of the graph of the polynomial. For example, consider $p(x) = x^3 - x$.

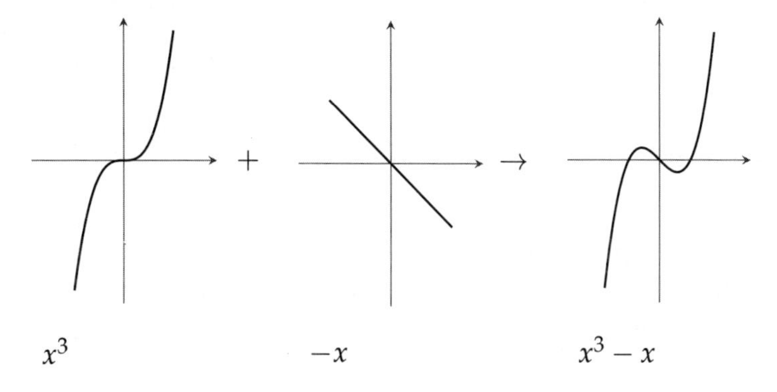

x^3 $-x$ $x^3 - x$

If $p(x)$ is a polynomial and c is a real number such that $p(c) = 0$ then c is called a **zero** or **root** of $p(x)$.

Let us recall the following from our school algebra.

Theorem 1.4.18 (Division Algorithm for Polynomials)

Let $p(x)$ and $q(x)$ be polynomials with $q \neq 0$. Then there are unique polynomials $s(x)$ and $r(x)$ such that $p(x) = s(x)q(x) + r(x)$ and either $\deg r < \deg q$ *or $r = 0$.*

Proof. First we prove existence, by strong induction on the degree of p. If $\deg p < \deg q$ we can take $s = 0$ and $r = p$. Now let $p(x) = \sum_{i=0}^{n} a_i x^i$ and $q(x) = \sum_{i=0}^{m} b_i x^i$ with $n \geq m$ and $a_n, b_m \neq 0$. Define

$$r'(x) = p(x) - \frac{a_n}{b_m} x^{n-m} q(x).$$

We have $\deg r' < \deg p$. By the strong induction hypothesis, $r' = s'q + r$ with $r = 0$ or $\deg r < \deg q$. Then,

$$\begin{aligned} p(x) = r'(x) + \frac{a_n}{b_m} x^{n-m} q(x) &= s'(x)q(x) + r(x) + \frac{a_n}{b_m} x^{n-m} q(x) \\ &= (s'(x) + \frac{a_n}{b_m} x^{n-m}) q(x) + r(x). \end{aligned}$$

As for uniqueness, let $p = sq + r = s'q + r'$ be two such decompositions. Then $(s - s')q = r' - r$. The only way the degrees of the left and right side can match is if $s = s'$ and $r = r'$. ■

Theorem 1.4.19 (Factor Theorem)

Let $p(x)$ be a polynomial and let c be a root of $p(x)$. Then $x - c$ divides $p(x)$. That is, there is a polynomial $q(x)$ such that $p(x) = (x - c)q(x)$.

Proof. Divide $p(x)$ by $x - c$. This will result in a quotient polynomial $q(x)$ and a constant remainder r. Thus $p(x) = (x - c)q(x) + r$. Substituting the value $x = c$ on each side gives $0 = p(c) = (c - c)q(c) + r = r$. Hence $r = 0$ and $p(x) = (x - c)q(x)$. ■

If c is a root of $p(x)$, its **multiplicity** is the largest natural number k such that $(x - c)^k$ divides $p(x)$.

Theorem 1.4.20

Let $p(x)$ be a polynomial of degree $n \geq 1$. Then $p(x)$ has at most n roots, counting multiplicities.

Proof. Let $c_1, \ldots, c_m$ be the distinct roots of $p(x)$ and let $k_1, \ldots, k_m$ be their respective multiplicities. Then $p(x)$ has the form $(x - c_1)^{k_1} \cdots (x - c_m)^{k_m} q(x)$. It follows that $k_1 + \cdots + k_m \leq \deg p(x) = n$. ■

Rational Functions

A **rational function** has the form $\dfrac{p(x)}{q(x)}$ where $p(x)$ and $q(x)$ are polynomials and $q(x)$ is not the zero polynomial.

The simplest interesting example is the function given by $f(x) = 1/x$. Its domain is $\mathbb{R}^*$ and the graph is given on the right.

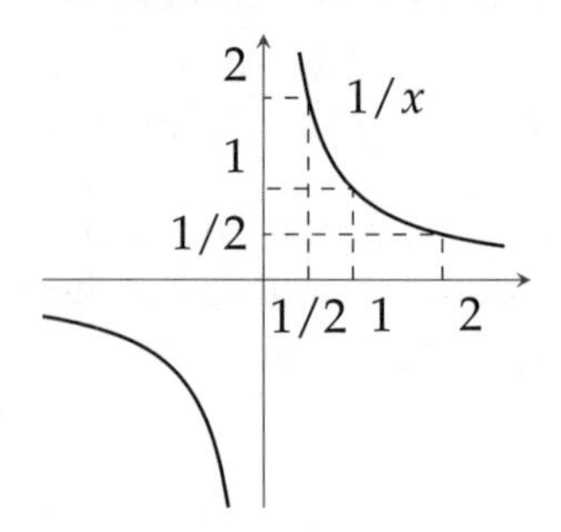

As x increases on the positive side, $1/x$ decreases, and the graph moves towards the x-axis. While, as x decreases towards zero, the graph rises without bound.

Once the graph is obtained for positive x, we obtain it for negative x by symmetry since the function $1/x$ is an odd function.

Example 1.4.21

We shall sketch the graph of $f(x) = \dfrac{1}{(x-1)(x+1)}$. We start by considering the graph of $(x - 1)(x + 1)$.

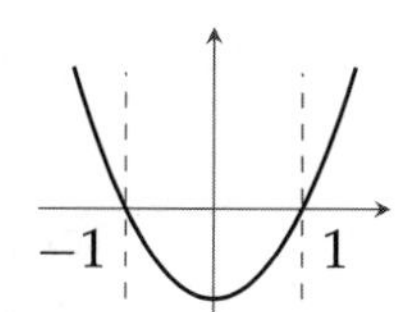

The reciprocal function $f(x)$ will take values of large magnitude where $(x-1)(x+1)$ takes values of small magnitude, and of the same sign. So, as we approach 1 from the right, $f(x)$ will take large positive values, while from the left it will take large negative values. Thus we obtain the following graph.

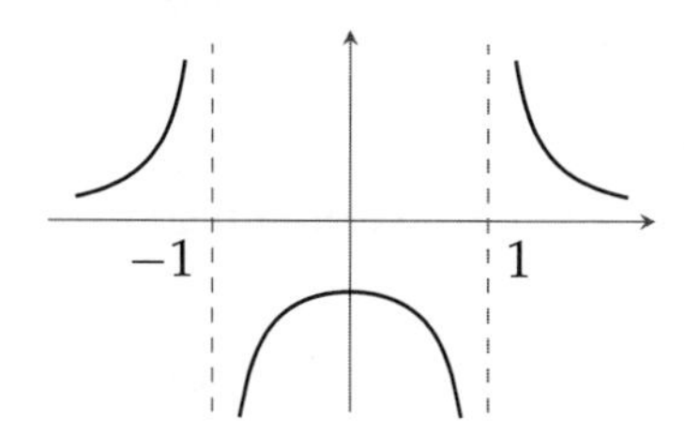

□

Exercises for § 1.4

1. Identify the domains of the real functions given by the following rules:

(a) $f(x) = \sqrt{x^2 - 2}$,

(b) $g(x) = \dfrac{1}{[x]}$,

(c) $h(x) = \sqrt{x(x^2 - 1)}$,

(d) $k(x) = \dfrac{1}{x - [x]}$.

2. Draw the graphs of the following functions, not by plotting points but by transforming the graphs of the standard functions like x, x^2 and $1/x$.

(a) $1 - x^2$.

(b) $x^2 - 4x + 3$.

(c) $\sqrt{x-1}$.

(d) $\dfrac{x}{x-4}$.

3. Graph the following functions.

(a) $f(x) = (x-1)(x-2)(x-3)$.

(b) $g(x) = |x| + |x-1|$.

(c) $h(x) = |x| - |x-1|$.

(d) $k(x) = \dfrac{x}{x^2 - 1}$.

4. Let $f\colon [0,2] \to \mathbb{R}$ be defined as follows: $f(x) = 1$ if $0 \le x \le 1$ and $f(x) = 2$ if $1 < x \le 2$.

(a) Draw the graph of f.

(b) Describe the domain and draw the graph of $g(x) = f(2x)$.

(c) Describe the domain and draw the graph of $h(x) = f(x-2)$.

(d) Describe the domain and draw the graph of $k(x) = f(2x) + f(x-2)$.

5. Graph the following modifications of the greatest integer function over the domain $[-2,2]$.

(a) $f(x) = 2[x]$.

(b) $g(x) = [2x]$.

(c) $h(x) = [x^2]$.

(d) $k(x) = (-1)^{[x]}$.

6. Extend the given graph to a suitable domain so that it represents an even function.

(a) 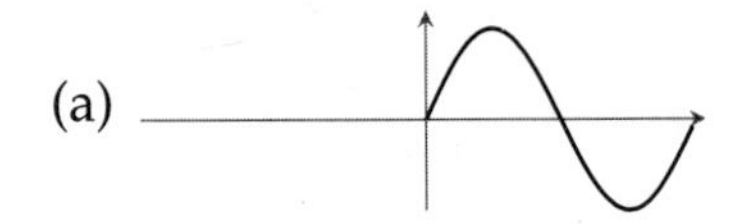

(b) 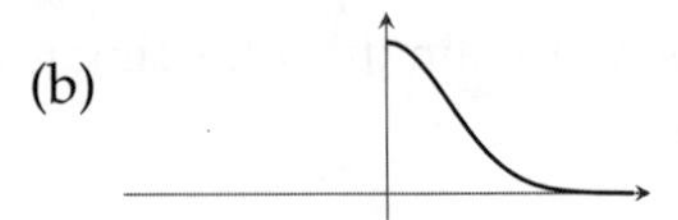

7. Extend the given graph to a suitable domain so that it represents an odd function.

(a) 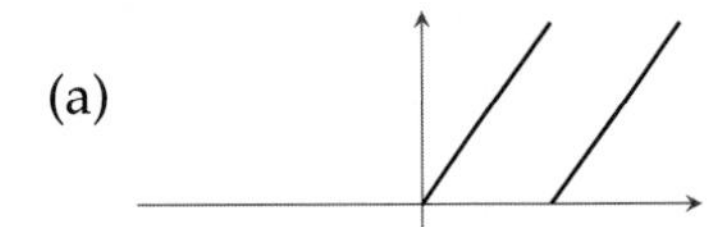

(b) 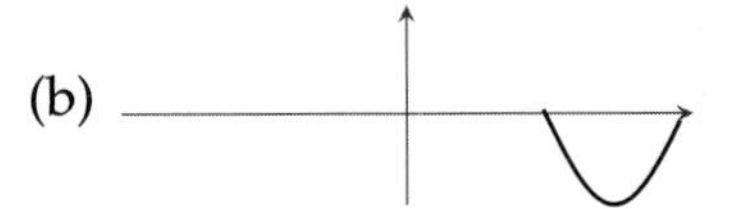

8. Under the given conditions, is the function $f \circ g$ even or odd?

(a) When f and g are both even.

(b) When f and g are both odd.

(c) When f is even and g is odd.

(d) When f is odd and g is even.

9. Prove or give a counter-example.

(a) If f,g are increasing, so is their sum $f+g$.

(b) If f,g are increasing, so is their product fg.

10. Consider a function $f\colon \mathbb{R} \to \mathbb{R}$.

(a) Show that f can be written as the difference of two functions, each of which takes only non-negative values.

(b) Show that f can be written as the sum of an even function and an odd function.

11. Define $x^+ = \max\{0,x\}$. Draw the graphs of the following functions.

(a) $f(x) = x^+$.

(b) $g(x) = (-x)^+$.

(c) $h(x) = x^+ + (-x)^+$.

(d) $k(x) = x^+ - (-x)^+$.

(e) $\ell(x) = (1-x)^+ + (x-2)^+$.

(f) $m(x) = (x-1)^+ - (x-2)^+$.

12. For each $n \in \mathbb{Z}$, define $f_n\colon \mathbb{R} \to \mathbb{R}$ by $f_n(x) = (x-n)^+ - 2(x-n-1/2)^+ + (x-n-1)^+$. (See the previous exercise for the x^+ notation.)

(a) Graph $f_n(x)$.

(b) Graph $F(x) = \sup\{f_n(x) \mid n \in \mathbb{Z}\}$.

13. Let functions f, g have period T. Show that the following functions also have period T: $f+g$, $f-g$, fg, f/g.

14. Find out whether the following functions have any of the properties of being odd/even, increasing/decreasing, or periodic.

(a) $1/x$.

(b) $[x^2]$.

(c) $[x+1/2]-[x]$.

(d) $[x+1/2]-[x]-1/2$.

15. Show that if a function has either of the following pairs of properties, it must be constant.

(a) Monotonic and even.

(b) Monotonic and periodic.

16. The graph of a function is given. Draw the graph of its inverse function.

(a)

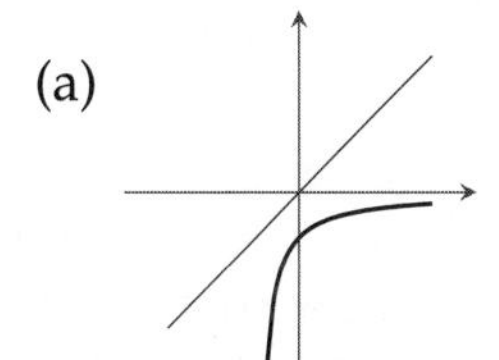

(b)

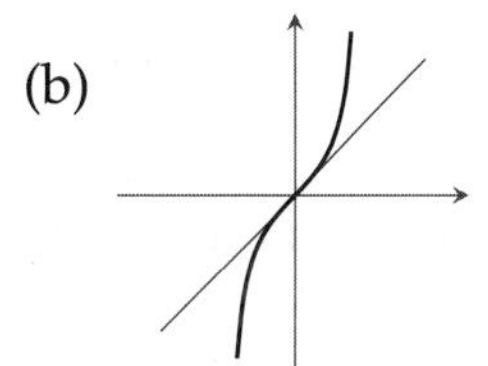

Thematic Exercises

Every chapter of this book is followed by one or two sets of thematic exercises. These either develop applications of the material in that chapter or illustrate theoretical concerns that future courses would take up in detail. It is not essential that you solve them at first sight, but you should at browse through them and keep them in mind as you read on. Chances are you will suddenly recognize a relevant idea and how to apply it here. Or, studying another course, you will see how these exercises support the techniques you are learning there.

Curve Fitting: Interpolation and Least Squares

In the laboratory, we measure a function's values at finitely many points. From these values, we attempt to establish the form of the function. One common practice is to find a polynomial which matches the data and has as low degree as possible. This is called an **interpolating polynomial** for the given data. We shall show below that, given $n+1$ data points, there is a unique interpolating polynomial of degree n or less.

A1. Suppose that $x_0, x_1, \ldots, x_n \in \mathbb{R}$ satisfy $x_i \neq x_j$ when $i \neq j$.

(a) Show that there is a unique polynomial $w_i(x)$ of degree n such that $w_i(x_i) = 1$ and $w_i(x_j) = 0$ when $j \neq i$. (Hint: Use the factor theorem to find the formula of $w_i(x)$.)

(b) Let $y_0, y_1, \ldots, y_n \in \mathbb{R}$. Show that there is a unique polynomial $p(x)$ of degree n or less such that $p(x_i) = y_i$ for each i, and it is given by

$$p(x) = \sum_{i=0}^{n} y_i w_i(x).$$

A2. Find the unique linear or quadratic function that passes through the points $(-h,a)$, $(0,b)$ and (h,c), with $h > 0$.

Actual data has errors and a perfect match to imperfect data has little importance. It is more useful to find a function which is only an approximate match but is easy to work with or allows some special insight. The most common approach is to find a line which passes as close as possible to the data points. This can be done by geometry!

Let $\mathbb{R}^n = \{\vec{x} = (x_1, \ldots, x_n) \mid x_i \in \mathbb{R}\}$. Recall that the dot product on this space is defined by $\vec{x} \cdot \vec{y} = \sum_{i=1}^{n} x_i y_i$. Vectors $\vec{x}$, $\vec{y}$ are perpendicular to each other if and only if $\vec{x} \cdot \vec{y} = 0$. The length of a vector is given by $||\vec{x}|| = (\vec{x} \cdot \vec{x})^{1/2} = (\sum_{i=1}^{n} x_i^2)^{1/2}$. The distance between $\vec{x}$ and $\vec{y}$ is $||\vec{y} - \vec{x}||$.

A3. Prove Pythagoras' theorem: Vectors $\vec{x}$, $\vec{y}$ are perpendicular if and only if $||\vec{x} + \vec{y}||^2 = ||\vec{x}||^2 + ||\vec{y}||^2$.

A4. Let $\vec{u}, \vec{v}$ be distinct non-zero points in $\mathbb{R}^n$, and $\Pi = \{a\vec{u} + b\vec{v} \mid a, b \in \mathbb{R}\}$ be the plane passing through them and origin. Take any fixed vector $\vec{y}$. Show that Π has a unique member $\vec{x}$ which is closest to $\vec{y}$ and that it is given by the equations $(\vec{y} - \vec{x}) \cdot \vec{u} = (\vec{y} - \vec{x}) \cdot \vec{v} = 0$.

Consider $\vec{x} = (x_1, \ldots, x_n)$ and $\vec{y} = (y_1, \ldots, y_n)$. The closeness of a line $y = ax + b$ to the points (x_i, y_i) is measured by the **total squared error**,

$$E(a,b) = \sum_{i=1}^{n} (y_i - ax_i - b)^2.$$

The goal is to find a, b such that $E(a,b)$ is as small as possible. The corresponding line is called the **least squares line** for the data.

A5. Show that the problem of minimizing the total squared error is equivalent to considering the plane $\Pi = \{a\vec{x} + b\vec{v} \mid a, b \in \mathbb{R}\}$ with $\vec{v} = (1, \ldots, 1)$, and finding the member of Π which is closest to $\vec{y}$.

A6. Show that the least squares line $y = ax + b$ is given by

$$a = \frac{n\sum_{i=1}^{n} x_i y_i - (\sum_{i=1}^{n} x_i)(\sum_{i=1}^{n} y_i)}{n\sum_{i=1}^{n} x_i^2 - (\sum_{i=1}^{n} x_i)^2} \quad \text{and} \quad b = \frac{\sum_{i=1}^{n} y_i - a\sum_{i=1}^{n} x_i}{n}.$$

Cardinality

We consider two sets as having the same amount (why did not we say 'number'?) of elements if there is a bijection between them. In this case we say the two sets have the same **cardinality**. A set is called **finite** if it is empty or it has the same cardinality as $\{1,\ldots,n\}$ for some $n \in \mathbb{N}$. Otherwise it is called **infinite**. The most familiar infinite set is $\mathbb{N}$. One can ask if all infinite sets have the same cardinality. The first surprise is that it is not so easy to go beyond $\mathbb{N}$. In Exercises 7 and 8 of §1.3 you were asked to show that $\mathbb{W}, \mathbb{Z}$ and even $\mathbb{N} \times \mathbb{N}$ have the same cardinality as $\mathbb{N}$. A set which is either finite or has the same cardinality as $\mathbb{N}$ is called **countable**.

B1. Prove that the following function $f \colon \mathbb{N} \times \mathbb{N} \to \mathbb{N}$ is a bijection:

$$f(m,n) = 2^{m-1}(2n-1).$$

Since $\mathbb{Q}$ is made of pairs of integers we expect it to be countable. Finding an explicit bijection with $\mathbb{N}$ is a little daunting because a rational number can be written as p/q in many ways and we have to account for this. We present below an elegant bijection which appears to have been first discovered by an undergraduate student in 1960! (McCrimmon [56]) It is based on the **fundamental theorem of arithmetic:** Every natural number greater than one can be written uniquely as a product of prime powers. That is, if $n \in \mathbb{N}$ and $n > 1$, then there is a unique choice of primes $p_1 < \cdots < p_k$ and natural numbers $\alpha_1,\ldots,\alpha_k$ such that $n = p_1^{\alpha_1} \cdots p_k^{\alpha_k}$.

B2. Let $f \colon \mathbb{W} \to \mathbb{Z}$ be a bijection such that $f(0) = 0$. Define $\varphi \colon \mathbb{N} \to \mathbb{Q}^+$ by

$$\varphi(1) = 1 \quad \text{and} \quad \varphi(p_1^{\alpha_1} \cdots p_k^{\alpha_k}) = p_1^{f(\alpha_1)} \cdots p_k^{f(\alpha_k)},$$

for any distinct primes p_i and natural numbers α_i. Show that φ is a bijection.

B3. Give a bijection between $\mathbb{N}$ and $\mathbb{Q}$.

So $\mathbb{Q}$ turns out to be countable. What about $\mathbb{R}$? Cantor gave several proofs that $\mathbb{R}$ is **uncountable** (that is, infinite and not in bijection with $\mathbb{N}$). We present a version of his very first proof, as it is more in line with our general approach. First, we have a simple application of the completeness axiom.

B4. (Nested Interval Property of $\mathbb{R}$) Let $J_n = [a_n, b_n]$ be closed intervals such that

$$[a_1,b_1] \supseteq [a_2,b_2] \supseteq \cdots \supseteq [a_n,b_n] \cdots.$$

Show that $\bigcap_{n=1}^{\infty} [a_n,b_n] \neq \varnothing$.

Now we will show that there cannot be a surjection from $\mathbb{N}$ to $\mathbb{R}$, and hence there certainly cannot be a bijection either. Therefore $\mathbb{R}$ is uncountable.

B5. Let $f \colon \mathbb{N} \to \mathbb{R}$ be a surjection. Construct closed intervals $J_n = [a_n, b_n]$ such that $J_1 \supset J_2 \supset J_3 \cdots$ and $f(n) \notin J_n$. Show that this gives a contradiction. (Hint: J_1 is easily chosen so that $f(1) \notin J_1$. Cut it into three closed subintervals of equal length. At least one of these does not include $f(2)$.)

B6. Is the set of irrational numbers countable or uncountable?

2 | Integration

Calculus has two parts: differential and integral. Integral calculus owes its origins to fundamental problems of measurement in geometry: length, area, and volume. It is by far the older branch. Nevertheless, it depends on differential calculus for its more difficult calculations, and so nowadays we typically teach differentiation before integration.

We shall revert to the historical sequence and begin our journey with integration. Our first reason is that it provides a direct application of the completeness axiom without needing the concept of limits. The second is that important functions such as the trigonometric, exponential, and logarithmic functions are most conveniently constructed through integration. Finally, the student should become aware that integration is not just an application of differentiation or a set of techniques of calculation.

Suppose we wish to find the area of a shape in the Cartesian plane. We can, at least, estimate it by comparing the shape with a standard area, that of a square. We cover the shape with a grid of unit squares and count how many squares touch it, and also how many squares are completely contained in it. This gives an upper and a lower estimate for the area. We can obtain better estimates by taking finer grids with smaller squares. The figures given immediately below illustrate this process of iteratively improving the estimates.

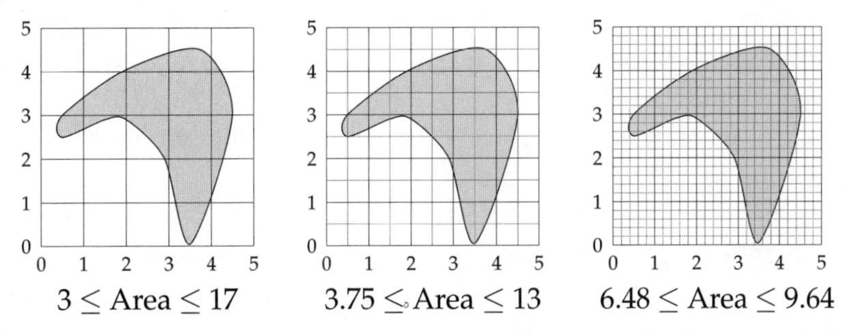

We have said that we are estimating area. But what is our definition of area? In school books you will find descriptions such as "Area is the measure of the part of a plane or region enclosed by the figure." It should be evident that this is not a very useful prescription. It means nothing without a description of the measuring process. In fact, the estimation process described above could become the basis for a meaningful definition of area, by requiring it to be a number that lies *above* all the lower estimates produced by the process, and *below* all the upper estimates. Its existence would be guaranteed by the completeness axiom. This is a promising start, but the sceptic can raise various objections that would have to be answered:

1. Could there be a figure for which multiple numbers satisfy the definition of its area?
2. If we slightly shifted or rotated the grids, could that change our calculation? That is, could moving a figure change its area?

It takes some effort to answer these objections. Indeed, we have to concede the first point, for there are such subsets of the plane. As for the second, we shall content ourselves with working out a method for assigning area only to certain regions, those that are bounded by graphs of functions over an interval $[a,b]$. In this special situation we shall use horizontal rows of rectangles rather than grids of squares to create the bounds. The word "integration" will be used for this process.

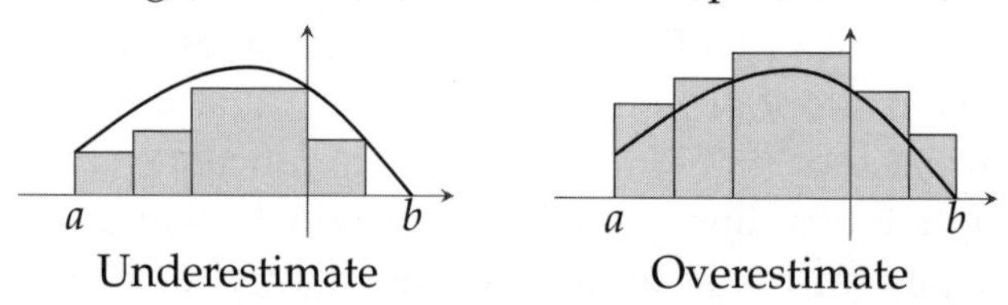

In what follows, it will be important that we use our pre-existing notions about area only to motivate or guess results, and never to justify them. For, area and its properties will formally be defined only by the end of our analysis.

The idea of approximating by rectangles arises very naturally in the context of motion. Imagine a body moving in a straight line with a constant speed s over a time interval $[a,b]$. Then the distance travelled equals $s(b-a)$, which is the area of a rectangle with base $b-a$ and height s. Now, suppose the speed is a function $s(t)$ of time. Let us cut $[a,b]$ into some subintervals and approximate the speed by a constant speed over each subinterval. Then the distance travelled is approximated by the distance travelled under this sequence of constant speeds, and the latter is the sum of the areas of rectangles!

For our final piece of motivation, suppose we know velocity rather than speed, and wish to determine the total displacement. We again cut $[a,b]$ into some subintervals, of widths $\triangle t_i$, and approximate by a constant velocity v_i in each subinterval. Then the total displacement is approximated by $\sum_i v_i \triangle t_i$. Some of the v_i could be negative and then the product $v_i \triangle t_i$ would be negative, and thus not the area of a rectangle. We shall call it the "signed area" instead.

Our development of integration in this chapter will be along the lines of the previous paragraph. We shall use integration to define "signed area" rather than "area," as it is the more general concept. As applications, we will rigorously develop the logarithmic and exponential functions, prove the existence of nth roots, define real powers, and define and estimate π.

2.1 Integration of Step and Bounded Functions

Integration of Step Functions

We shall first develop the notion of 'integral' for especially simple functions. These will take constant values on certain intervals. Let's begin with an example.

Example 2.1.1

Consider the function $f\colon [1,4] \to \mathbb{R}$ defined as follows:

$$f(x) = \begin{cases} 1.1 & \text{if } 1 \le x \le 2, \\ 2.2 & \text{if } 2 < x \le 4. \end{cases}$$

Let us draw the graph of f and shade the region lying between its graph and the x-axis. We see that is made of two rectangles.

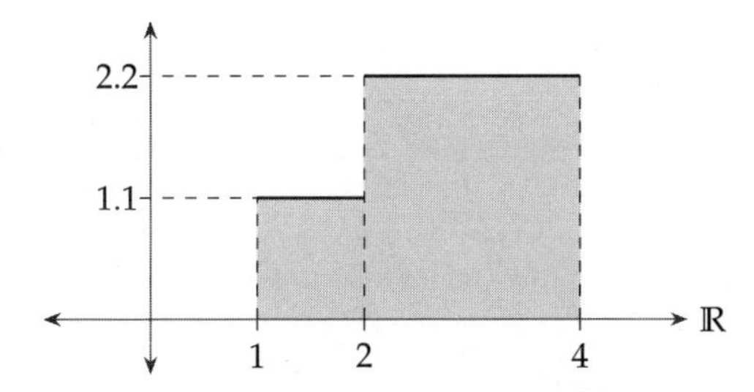

We define the integral of f to be the sum of the areas of these two rectangles:

$$(2-1)\cdot 1.1 + (4-2)\cdot 2.2 = 5.5. \quad \square$$

The calculations in this example required two pieces of information:

1. The intervals over which f has a constant value: $[1,2]$ and $(2,4]$. These can be provided by the list of their endpoints, i.e., by the set $\{1,2,4\}$.
2. The constant values that f has over the intervals $[1,2]$ and $(2,4]$, i.e., 1.1 and 2.2.

Consider a closed and bounded interval $[a,b]$ in $\mathbb{R}$. A **partition** of $[a,b]$ is a set $P = \{x_0, x_1, \dots, x_{n-1}, x_n\}$ such that

$$a = x_0 < x_1 < \cdots < x_{n-1} < x_n = b.$$

Such a partition cuts $[a,b]$ into n subintervals

$$[a,x_1],\ [x_1,x_2],\ \dots\ ,[x_{n-2},x_{n-1}],\ [x_{n-1},b].$$

$a = x_0$ $\quad x_1 \quad x_2 \quad \cdots \quad x_{n-1} \quad x_n = b$ $\quad \mathbb{R}$

As illustrated in the diagram, the subintervals need not have equal lengths.

A function $s\colon [a,b] \to \mathbb{R}$ is called a **step function** if there is a partition $P = \{x_0,\ldots,x_n\}$ of $[a,b]$ such that s is constant on each open subinterval (x_{i-1},x_i):

$$s(x) = s_i \qquad \text{if } x_{i-1} < x < x_i.$$

In this case we say that the partition P is **adapted** to the step function s.

If we consider Example 2.1.1 we see that the set $\{1,2,4\}$ is a partition of $[1,4]$ and is adapted to the function f of that example.

Example 2.1.2

Consider the Heaviside step function, with the domain restricted to $[-1,1]$. It is constant on the intervals $(-1,0)$ and $(0,1)$, hence the partition $P = \{-1,0,1\}$ is adapted to it. On the other hand, the partition $P' = \{-1,0.5,1\}$ is not adapted to it, since the function takes two values on the interval $(-1,0.5)$.

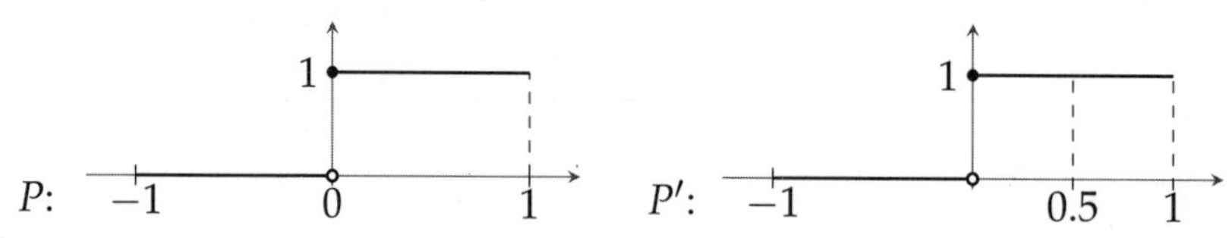

□

Task 2.1.3

Show that each of the following functions is a step function by finding a partition that is adapted to it:

(a) *The signum function, with domain* $[-1,1]$.

(b) *The greatest integer function, with domain* $[0,3]$.

Suppose $s\colon [a,b] \to \mathbb{R}$ is a step function with adapted partition $P = \{x_0,\ldots,x_n\}$, such that $s(x) = s_i$ if $x_{i-1} < x < x_i$. Then the **integral of s from a to b** is defined by

$$\int_a^b s(x)\,dx = \sum_{i=1}^{n} s_i(x_i - x_{i-1}).$$

Example 2.1.4

Consider the Heaviside step function H, with the domain as $[-2,3]$, and the adapted partition $P = \{-2,0,3\}$. It takes values $s_1 = 0$ on $(-2,0)$ and $s_2 = 1$ on $(0,3)$. We calculate its integral as follows:

$$\begin{aligned}\int_{-2}^{3} H(x)\,dx &= s_1(x_1 - x_0) + s_2(x_2 - x_1)\\ &= 0\cdot(0-(-2)) + 1\cdot(3-0) = 3.\end{aligned}$$

□

The integral represents the total **signed area** of the rectangles enclosed by the graph of $s(x)$, the x-axis, and the vertical lines $x = a$ and $x = b$.

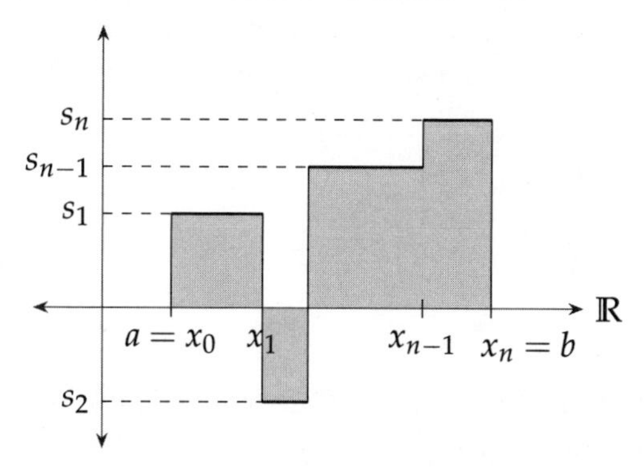

The term "signed area" refers to the area of a rectangle marked by the step function being taken as positive if the rectangle lies above the x-axis and as negative if it lies below the x-axis.

Note that the definition of its integral ignores the step function's values at the partition points. This is acceptable because these values create finitely many vertical line segments, and we view a line segment as having zero area regardless of its length.

Task 2.1.5

Consider the step function $s(x)$ defined by

$$s(x) = \begin{cases} 2 & \text{if } 0 \leq x \leq 1.5, \\ -1 & \text{if } 1.5 < x \leq 2.5, \\ 3 & \text{if } 2.5 < x \leq 3. \end{cases}$$

Calculate $\int_0^3 s(x)\,dx$.

Since the definition of the integral involved a choice of an adapted partition, we need to show that the resulting number is independent of the choice of partition. Let P, P' be partitions of $[a,b]$. We say P' is a **refinement** of P if $P \subseteq P'$.

Task 2.1.6

Suppose a partition P is adapted to a step function s. Show that every refinement of P is also adapted to s.

Note that if P, P' are partitions of $[a,b]$ then $P \cup P'$ is a common refinement for both of them.

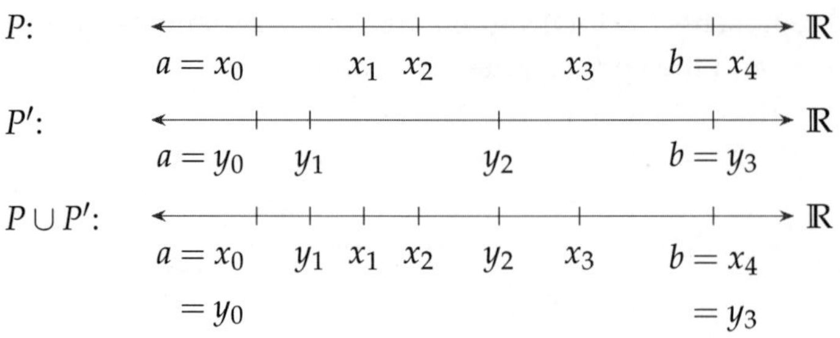

P: $a = x_0$ x_1 x_2 x_3 $b = x_4$ $\mathbb{R}$

P': $a = y_0$ y_1 y_2 $b = y_3$ $\mathbb{R}$

$P \cup P'$: $a = x_0 = y_0$ y_1 x_1 x_2 y_2 x_3 $b = x_4 = y_3$ $\mathbb{R}$

Theorem 2.1.7

Suppose $s\colon [a,b] \to \mathbb{R}$ is a step function and P, Q are partitions of $[a,b]$ that are adapted to s. Then both P and Q lead to the same value of $\int_a^b s(x)\,dx$.

Proof. Let $I(P)$ be the value of the integral of s corresponding to the partition P. We need to prove that $I(P) = I(Q)$. It suffices to prove this when one partition is a refinement of the other. For, if we have proved this, we shall have $I(P) = I(P \cup Q) = I(Q)$.

Next, it suffices to prove $I(P) = I(Q)$ when Q has just one point more than P. Let $P = \{x_0, \ldots, x_n\}$ and $Q = \{x_0, \ldots, x_{k-1}, t, x_k, \ldots, x_n\}$. For each i, let s take the value s_i on (x_{i-1}, x_i). Then

$$\begin{aligned} I(Q) &= \sum_{i=1}^{k-1} s_i(x_i - x_{i-1}) + s_k(t - x_{k-1}) + s_k(x_k - t) + \sum_{i=k+1}^{n} s_i(x_i - x_{i-1}) \\ &= \sum_{i=1}^{k-1} s_i(x_i - x_{i-1}) + s_k(x_k - x_{k-1}) + \sum_{i=k+1}^{n} s_i(x_i - x_{i-1}) \\ &= \sum_{i=1}^{n} s_i(x_i - x_{i-1}) = I(P). \end{aligned}$$

■

Task 2.1.8

Suppose $s, t\colon [a,b] \to \mathbb{R}$ are step functions. Show there is a partition that is adapted to both s and t.

Theorem 2.1.9 (Comparison Theorem)

Let $s, t\colon [a,b] \to \mathbb{R}$ be step functions such that $s(x) \le t(x)$ for every $x \in [a,b]$. Then

$$\int_a^b s(x)\,dx \le \int_a^b t(x)\,dx.$$

Proof. Let $P = \{x_0, \ldots, x_n\}$ be a partition that is adapted to both s and t. Let $s(x) = s_i$ and $t(x) = t_i$ for each $x \in (x_{i-1}, x_i)$. Then $s_i \le t_i$ for each i, and therefore

$$\int_a^b s(x)\,dx = \sum_{i=1}^{n} s_i(x_i - x_{i-1}) \le \sum_{i=1}^{n} t_i(x_i - x_{i-1}) = \int_a^b t(x)\,dx.$$

■

Integration of Bounded Functions

A function $f\colon [a,b] \to \mathbb{R}$ is **bounded** if there is a real number M such that $-M \le f(x) \le M$ for every $x \in [a,b]$. We then say that f is bounded by M.

Let $f\colon [a,b] \to \mathbb{R}$ be bounded by M. Consider a step function $s\colon [a,b] \to \mathbb{R}$ such that $s(x) \le f(x)$ for every $x \in [a,b]$.

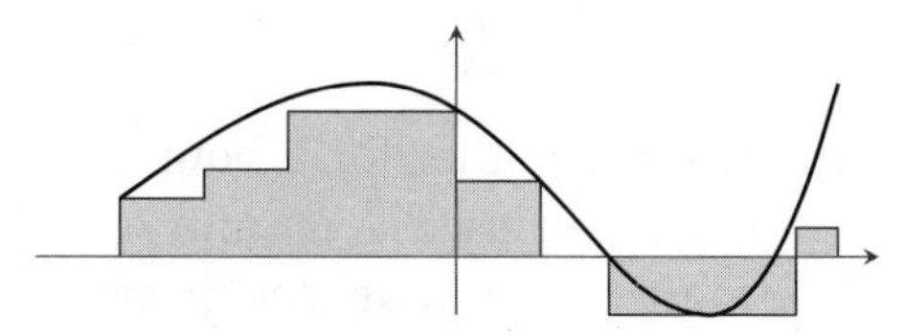

We view the integral of s over $[a,b]$ as being an underestimate of the "signed area" under the graph of f. We can improve the estimate by considering another step function that lies between s and f. Therefore, we consider the collection of all such underestimates:

$$\mathcal{L}_f = \left\{ \int_a^b s(x)\,dx \;\middle|\; s\colon [a,b] \to \mathbb{R} \text{ is a step function and } s(x) \le f(x) \text{ for every } x \in [a,b] \right\}.$$

The set is non-empty because the constant function $-M$ is a step function whose values never exceed those of f.

We can similarly obtain a non-empty collection of overestimates of the signed area under the graph of f:

$$\mathcal{U}_f = \left\{ \int_a^b t(x)\,dx \;\middle|\; t\colon [a,b] \to \mathbb{R} \text{ is a step function and } t(x) \ge f(x) \text{ for every } x \in [a,b] \right\}.$$

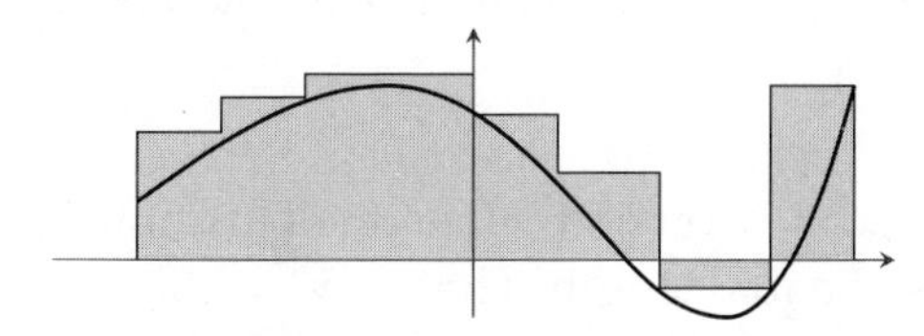

The members of $\mathcal{L}_f$ are called **lower sums** for f, while the members of $\mathcal{U}_f$ are called **upper sums**.

Theorem 2.1.10

Let $f\colon [a,b] \to \mathbb{R}$ be a bounded function. Then $\ell \in \mathcal{L}_f$ and $u \in \mathcal{U}_f$ implies $\ell \le u$.

Proof. Let $s,t\colon [a,b] \to \mathbb{R}$ be step functions such that

(a) $s(x) \leq f(x)$ for every $x \in [a,b]$ and $\int_a^b s(x)\,dx = \ell$.

(b) $t(x) \geq f(x)$ for every $x \in [a,b]$ and $\int_a^b t(x)\,dx = u$.

Then $s(x) \leq f(x) \leq t(x)$ for every $x \in [a,b]$. Hence, by the comparison theorem,

$$\ell = \int_a^b s(x)\,dx \leq \int_a^b t(x)\,dx = u.$$ ■

The completeness axiom informs us that for a bounded function f there will be a number I such that $\ell \leq I \leq u$ for every $\ell \in \mathcal{L}_f$ and $u \in \mathcal{U}_f$. This I is our natural candidate for the value of the signed area under the graph of f. However, we have to face the possibility of there being more than one such I. Here is an example:

Example 2.1.11

Consider the **Dirichlet function** $D\colon [0,1] \to \mathbb{R}$, which is defined by

$$D(x) = \begin{cases} 1 & \text{if } x \in \mathbb{Q} \cap [0,1], \\ 0 & \text{if } x \in \mathbb{Q}^c \cap [0,1]. \end{cases}$$

Let $s\colon [0,1] \to \mathbb{R}$ be a step function such that $s(x) \leq D(x)$ for every $x \in [0,1]$. Each open subinterval of $[0,1]$ contains an irrational number, hence $s(x) \leq 0$ on each open subinterval, and so $\int_a^b s(x)\,dx \leq 0$. Similarly, we see that $\int_a^b t(x)\,dx \geq 1$ if $t\colon [0,1] \to \mathbb{R}$ is a step function such that $t(x) \geq D(x)$ for every $x \in [0,1]$.

Therefore, every number α between 0 and 1 has the property that $\ell \leq \alpha \leq u$ for every $\ell \in \mathcal{L}_D$ and $u \in \mathcal{U}_D$. □

For functions like the Dirichlet function we have to admit that our approach fails to successfully assign a "signed area." For the functions where our approach *does* work we have a special term: a bounded function $f\colon [a,b] \to \mathbb{R}$ is called **integrable** if there is a unique number I such that $\ell \leq I \leq u$ for every $\ell \in \mathcal{L}_f$ and $u \in \mathcal{U}_f$. This unique I is called the **(definite) integral** of f on $[a,b]$ and is denoted by $\int_a^b f(x)\,dx$.

Example 2.1.12

Consider the function $f(x) = x$ on $[0,1]$. Take the partition $P : x_0 < \cdots < x_n$ that cuts $[0,1]$ into n subintervals of equal length. That is, $x_i = i/n$.

P: ⟵ $\mathbb{R}$

$x_1 = 0$ $\quad x_1 = \frac{1}{n}$ $\quad x_{i-1} = \frac{i-1}{n}$ $\quad x_i = \frac{i}{n}$ $\quad x_n = 1$

Corresponding to this partition we define two step functions, s_n and t_n, as follows:

$$s_n(x) = x_{i-1} \quad \text{and} \quad t_n(x) = x_i \qquad \text{if } x_{i-1} \leq x < x_i,$$

and $s_n(1) = t_n(1) = 1$. Then

$$\begin{aligned}\int_0^1 s_n(x)\,dx &= \sum_{i=1}^{n} x_{i-1}\frac{1}{n} = \frac{1}{n}\sum_{i=1}^{n} x_{i-1} = \frac{1}{n}\sum_{i=1}^{n}\frac{i-1}{n} = \frac{1}{n^2}\sum_{i=1}^{n-1} i \\ &= \frac{n-1}{2n} = \frac{1}{2} - \frac{1}{2n}, \\ \int_0^b t_n(x)\,dx &= \sum_{i=1}^{n} x_i\frac{1}{n} = \cdots = \frac{1}{2} + \frac{1}{2n}.\end{aligned}$$

The number $1/2$ is the only number that fits between all these lower and upper sums, and hence it is the integral of f over $[0,1]$. □

Task 2.1.13

Why was it enough to consider only some special partitions and step functions in the above example?

Example 2.1.14

Consider the function $f\colon [-1,1] \to \mathbb{R}$ defined by $f(0) = 1$ and $f(x) = 0$ if $x \neq 0$. This is a step function, and its integral as a step function is zero. But we can also treat it as a bounded function and ask whether it is integrable.

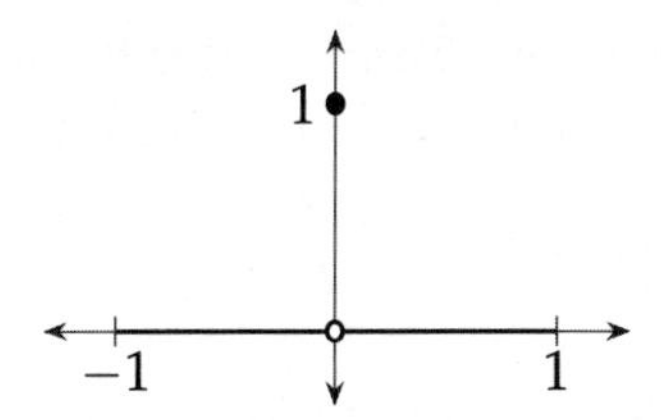

Since f is itself a step function we see that $0 \cdot (0 - (-1)) + 0 \cdot (1 - 0) = 0$ is both a lower sum as well an upper sum for f. So, any number I that lies between $\mathcal{L}_f$ and $\mathcal{U}_f$ must satisfy $0 \leq I \leq 0$. Therefore, $I = 0$ is the only such number. □

Task 2.1.15

Show that every step function is integrable, and the two ways of calculating its integral give the same number.

The following formulation is often useful in establishing the integrability of a function.

Theorem 2.1.16 (Riemann Condition)

Let $f\colon [a,b] \to \mathbb{R}$ be a bounded function. Then f is integrable on $[a,b]$ if and only if for each $\epsilon > 0$ there are step functions $s,t\colon [a,b] \to \mathbb{R}$ such that

1. $s(x) \leq f(x) \leq t(x)$ *for every* $x \in [a,b]$,

2. $\int_a^b t(x)\,dx - \int_a^b s(x)\,dx \leq \epsilon.$

Proof. First, suppose f is not integrable. We have to find an $\epsilon > 0$ for which there are no step functions s,t satisfying the given conditions. Now, since f is not integrable, there are two numbers I_1, I_2 such that

$$\int_a^b s(x)\,dx \leq I_1 < I_2 \leq \int_a^b t(x)\,dx,$$

whenever s,t satisfy the two inequalities in the first condition. Consider $\epsilon = \frac{1}{2}(I_2 - I_1)$. Then any s,t satisfying the first condition will not satisfy the second condition.

Now, suppose f is integrable with integral I. Given any $\epsilon > 0$, consider the interval $(I - \epsilon/2, I + \epsilon/2)$. If this does not contain any lower sum of f, then $I - \epsilon/2$ will also meet the conditions for the integral of f. So, this interval must contain a lower sum $\int_a^b s(x)\,dx$ of f. Similarly, it must contain an upper sum $\int_a^b s(x)\,dx$ of f. And then s,t fulfil all requirements. ■

There is a similar formulation for establishing that a certain number I is the integral of f over $[a,b]$.

Theorem 2.1.17

Let $f\colon [a,b] \to \mathbb{R}$ *be a bounded function. Then* $I \in \mathbb{R}$ *is the integral of* f *over* $[a,b]$ *if and only if for each* $\epsilon > 0$ *there are step functions* $s,t\colon [a,b] \to \mathbb{R}$ *such that*

1. $s(x) \leq f(x)$ *for every* $x \in [a,b]$ *and* $I - \int_a^b s(x)\,dx < \epsilon$,

2. $t(x) \geq f(x)$ *for every* $x \in [a,b]$ *and* $\int_a^b t(x)\,dx - I < \epsilon.$

Proof. First, suppose that f is integrable and the value of its integral is I. Take any $\epsilon > 0$. Then we have $\epsilon/2 > 0$ and the Riemann condition gives us step functions $s,t\colon [a,b] \to \mathbb{R}$ such that $s(x) \leq f(x) \leq t(x)$ for every $x \in [a,b]$ and $\int_a^b t(x)\,dx - \int_a^b s(x)\,dx \leq \epsilon/2$. From the defining properties of I, we have

$$\int_a^b s(x)\,dx \leq I \leq \int_a^b t(x)\,dx.$$

We get the desired conclusions as follows:

$$I - \int_a^b s(x)\,dx \leq \int_a^b t(x)\,dx - \int_a^b s(x)\,dx \leq \epsilon/2 < \epsilon,$$
$$\int_a^b t(x)\,dx - I \leq \int_a^b t(x)\,dx - \int_a^b s(x)\,dx \leq \epsilon/2 < \epsilon.$$

For the converse, suppose a number I has the properties described in our theorem. We shall first show that f is integrable. Let $\epsilon > 0$. Then $\epsilon/2 > 0$. By the assumed properties of I, there are step functions $s,t\colon [a,b] \to \mathbb{R}$ such that $s(x) \leq f(x) \leq t(x)$ for every $x \in [a,b]$, $I - \int_a^b s(x)\,dx < \epsilon/2$ and $\int_a^b t(x)\,dx - I < \epsilon/2$. Then,

$$\int_a^b t(x)\,dx - \int_a^b s(x)\,dx = \left(\int_a^b t(x)\,dx - I\right) + \left(I - \int_a^b s(x)\,dx\right) < \frac{\epsilon}{2} + \frac{\epsilon}{2} = \epsilon.$$

The Riemann condition now gives the integrability of f. To complete the proof, we only have to check that I meets the requirements of being the integral, that it separates the lower and upper sums of f. Consider any $u \in \mathcal{U}_f$. Suppose that $u < I$. Then $\epsilon = I - u > 0$. By the given properties of I, there is an $\ell \in \mathcal{L}_f$ such that $I - \ell < \epsilon$. Then $\ell > I - \epsilon = u$, which is impossible. Hence $u \geq I$. We can similarly prove that $\ell \in \mathcal{L}_f$ implies $\ell \leq I$. Therefore, I meets the requirements for being the integral of f. ∎

We finish this section with two conventions.

- $\displaystyle\int_a^a f(x)\,dx = 0.$
- If $a < b$ then $\displaystyle\int_b^a f(x)\,dx = -\int_a^b f(x)\,dx.$

The first is consistent with line segments having zero area. The second takes into account the direction of travel.

Exercises for § 2.1

1. Let H be the unit step function defined on page 28. Compute the following integrals of step functions:

(a) $\displaystyle\int_{-1}^2 \Big(H(x) + 2H(x-1)\Big)\,dx.$ (b) $\displaystyle\int_0^2 \Big(H(1-x) - H(x-1)\Big)\,dx.$

2. Graph the following step functions on the given domain and compute their integrals:

(a) $[2x-1]$ on $[0,2]$. (b) $[\sqrt{x}]$ on $[0,9]$.

3. Let $s\colon [-a,a] \to \mathbb{R}$ be a step function. Prove the following:

(a) If s is even then $\displaystyle\int_{-a}^0 s(t)\,dt = \int_0^a s(t)\,dt.$

(b) If s is odd then $\displaystyle\int_{-a}^0 s(t)\,dt = -\int_0^a s(t)\,dt.$

4. Let $f(t) = 1$ for $t \in [0,1]$ and $f(t) = 2$ for $t \in (1,2]$. Compute and graph the following function for $x \in [0,2]$:

$$F(x) = \int_0^x f(t)\,dt.$$

5. Calculate the following integrals:

(a) $\int_0^b x\,dx$. (b) $\int_0^b x^2\,dx$.

6. Compute $\int_0^b \sqrt{x}\,dx$ by relating the upper (lower) sums of x^2 to the lower (upper) sums of $\sqrt{x}$, using the following picture:

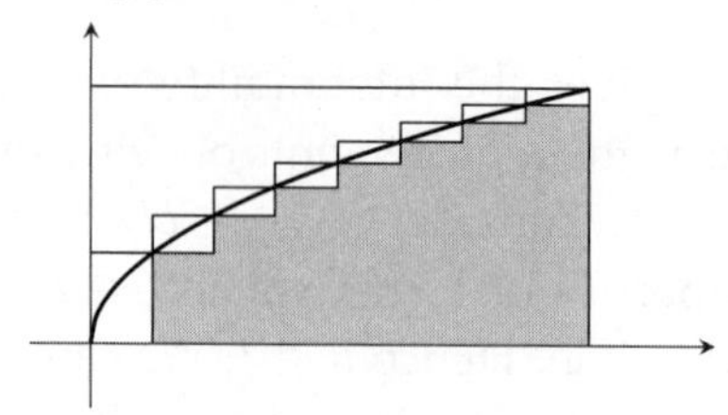

7. Prove the following:

(a) $\sum_{k=1}^{n} k^3 = \left(\frac{n(n+1)}{2}\right)^2$. (b) $\int_0^b x^3\,dx = \frac{b^4}{4}$.

8. Suppose $[c,d] \subseteq [a,b]$. Show that if f is integrable on $[a,b]$ then it is integrable on $[c,d]$.

9. Suppose $f\colon [-a,a] \to \mathbb{R}$ is integrable on $[0,a]$. Show the following:

(a) If f is even then $\int_{-a}^{0} f(x)\,dx = \int_0^a f(x)\,dx$.

(b) If f is odd then $\int_{-a}^{0} f(x)\,dx = -\int_0^a f(x)\,dx$.

10. Let $f\colon [a,b] \to \mathbb{R}$ and let $I \in \mathbb{R}$. Show that $I = \int_a^b f(x)\,dx$ if and only if the following two conditions hold:

(a) Every $u > I$ is an upper sum for f.

(b) Every $\ell < I$ is a lower sum for f.

11. Show that $F(x) = \begin{cases} 1 & \text{if } x = \frac{1}{2}, \frac{1}{3}, \frac{1}{4}, \ldots, \\ 0 & \text{else,} \end{cases}$ is integrable on $[0,1]$.

12. Let $f\colon [a,b] \to \mathbb{R}$ be a bounded function that is integrable on every interval $[c,b]$ with $a < c < b$. Show that f is integrable on $[a,b]$.

2.2 Properties of Integration

We now take up various general properties of integration. We shall first present the corresponding diagrams. The idea is that the general patterns can be intuited from what happens to rectangles.

Comparison: $f \le g$ on $[a,b]$ implies $\int_a^b f(x)\,dx \le \int_a^b g(x)\,dx$.

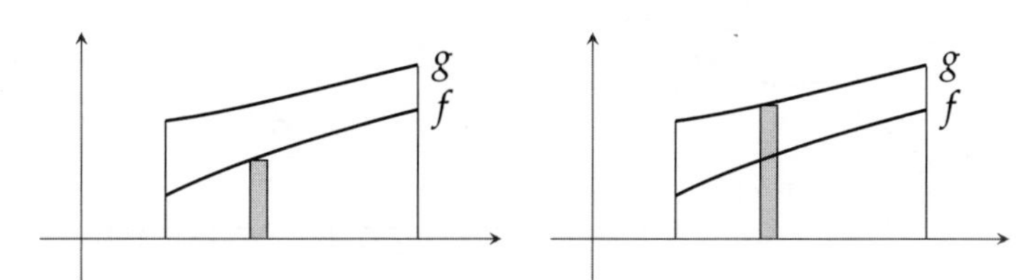

The rectangles for g are higher than those for f.

Homogeneity: $\int_a^b cf(x)\,dx = c\int_a^b f(x)\,dx$.

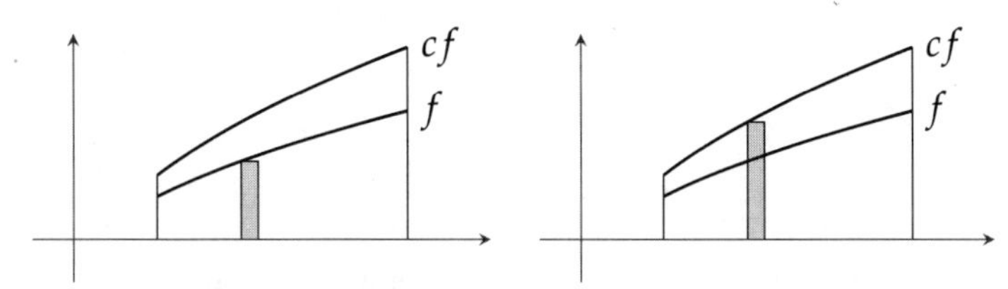

Each rectangle gets scaled vertically by c, hence so does its area.

Additivity: $\int_a^b \Big(f(x)+g(x)\Big)\,dx = \int_a^b f(x)\,dx + \int_a^b g(x)\,dx$.

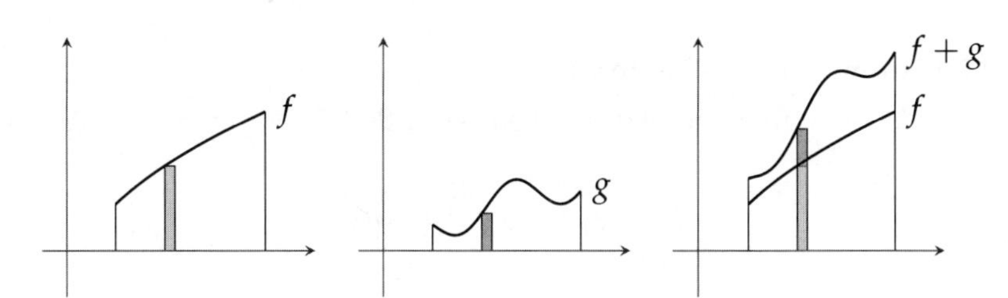

The rectangles for g are placed on top of those for f to get the ones for $f+g$.

Additivity over Intervals: $\int_a^b f(x)\,dx = \int_a^c f(x)\,dx + \int_c^b f(x)\,dx$.

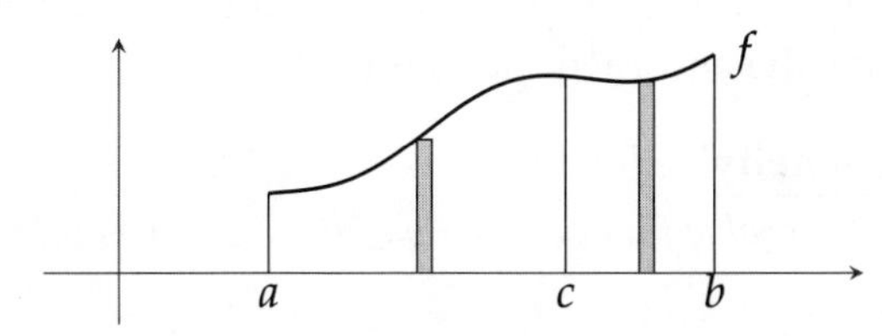

The rectangles for $[a,c]$ and $[c,b]$ are pooled to get all the ones for $[a,b]$.

Shift of Interval of Integration: $\displaystyle\int_{a+k}^{b+k} f(x-k)\,dx = \int_a^b f(x)\,dx.$

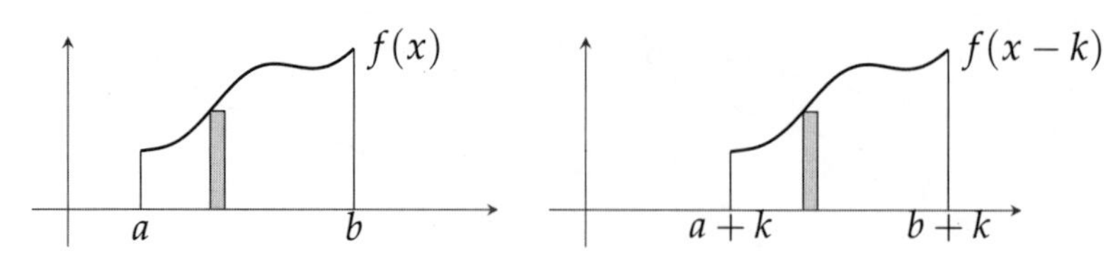

Each rectangle shifts without changing its dimensions.

Scaling of Interval of Integration: $k > 0 \implies \displaystyle\int_{ka}^{kb} f(x/k)\,dx = k\int_a^b f(x)\,dx.$

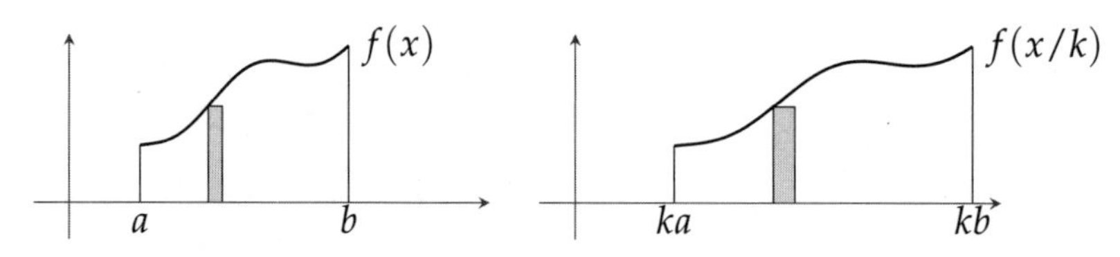

The width of each rectangle is scaled by k, hence so is its area.

The formal proofs continue this theme. The results are first established for step functions, and then generalized to all integrable functions. Our first instance is to generalize the comparison theorem, which we have already proved for step functions.

Theorem 2.2.1 (Comparison Theorem)

Suppose $f,g\colon [a,b] \to \mathbb{R}$ are integrable functions such that $f(x) \leq g(x)$ for every $x \in [a,b]$. Then

$$\int_a^b f(x)\,dx \leq \int_a^b g(x)\,dx.$$

Proof. Suppose that $\displaystyle\int_a^b f(x)\,dx > \int_a^b g(x)\,dx.$

Let $s,t\colon [a,b] \to \mathbb{R}$ be step functions such that $s(x) \leq f(x) \leq t(x)$ for every $x \in [a,b]$. The inequality between f and g gives us $s(x) \leq g(x)$ for every $x \in [a,b]$, and hence $\int_a^b s(x)\,dx \leq \int_a^b g(x)\,dx$. We also have $\int_a^b g(x)\,dx < \int_a^b f(x)\,dx \leq \int_a^b t(x)\,dx$. Thus the integral of g satisfies the defining properties of the integral of f and so must be equal to it. But this contradicts the assumed inequality. ∎

Theorem 2.2.2 (Homogeneity)

Let $f\colon [a,b] \to \mathbb{R}$ be an integrable function and $c \in \mathbb{R}$. Then cf is integrable and

$$\int_a^b cf(x)\,dx = c\int_a^b f(x)\,dx.$$

Proof. First, we prove the result for a step function s. Let $P = \{x_0, \ldots, x_n\}$ be a partition of $[a,b]$, such that $s(x) = s_i$ if $x_{i-1} < x < x_i$. Then $c \cdot s(x) = c \cdot s_i$ if $x_{i-1} < x < x_i$. Hence,

$$\int_a^b cs(x)\,dx = \sum_{i=1}^n (cs_i)(x_i - x_{i-1}) = c\sum_{i=1}^n s_i(x_i - x_{i-1}) = c\int_a^b s(x)\,dx.$$

Second, consider an arbitrary integrable function f and $c > 0$. Let $\epsilon > 0$. Then there are step functions s, t such that $s(x) \leq f(x) \leq t(x)$ for every $x \in [a,b]$ and

$$\int_a^b f(x)\,dx - \int_a^b s(x)\,dx < \frac{\epsilon}{c}, \qquad \int_a^b t(x)\,dx - \int_a^b f(x)\,dx < \frac{\epsilon}{c}.$$

It follows that cs, ct are step functions such that $cs(x) \leq cf(x) \leq ct(x)$ for every $x \in [a,b]$ and

$$c\int_a^b f(x)\,dx - \int_a^b cs(x)\,dx < \epsilon, \qquad \int_a^b ct(x)\,dx - c\int_a^b f(x)\,dx < \epsilon.$$

By Theorem 2.1.17, cf is integrable and $\displaystyle\int_a^b cf(x)\,dx = c\int_a^b f(x)\,dx$.

The $c < 0$ case can be done in a similar fashion. The $c = 0$ case is trivial. ■

Task 2.2.3

Consider the step functions $s(x)$ and $t(x)$ defined by

$$s(x) = \begin{cases} -1 & \text{if } 0 \leq x < 1, \\ 1 & \text{if } 1 \leq x \leq 3, \end{cases} \quad \textit{and} \quad t(x) = \begin{cases} 0 & \text{if } 0 \leq x \leq 1.5, \\ 2 & \text{if } 1.5 < x \leq 2.5, \\ 3 & \text{if } 2.5 < x \leq 3. \end{cases}$$

(a) *Calculate $s(x) + t(x)$.*

(b) *Verify* $\displaystyle\int_0^3 (s(x) + t(x))\,dx = \int_0^3 s(x)\,dx + \int_0^3 t(x)\,dx.$

Theorem 2.2.4 (Additivity)

Let $f, g\colon [a,b] \to \mathbb{R}$ be integrable functions. Then $f + g$ is an integrable function and

$$\int_a^b \Big(f(x) + g(x)\Big)\,dx = \int_a^b f(x)\,dx + \int_a^b g(x)\,dx.$$

Proof. We start by proving this property for step functions. Suppose s, t are step functions on $[a,b]$. If a partition P_s is adapted to s and a partition P_t is adapted to t, then $P = P_s \cup P_t$ is adapted to all three of s, t, and $s + t$. In particular, $s + t$ is a step function. Let $P = \{x_0 \ldots, x_n\}$, and suppose

$$s(x) = s_i \quad \text{if} \quad x_{i-1} < x < x_i,$$
$$t(x) = t_i \quad \text{if} \quad x_{i-1} < x < x_i.$$

Then $s(x)+t(x)=s_i+t_i$ if $x_{i-1}<x<x_i$. Hence,

$$\begin{aligned}\int_a^b \Big(s(x)+t(x)\Big)\,dx &= \sum_{i=1}^{n}(s_i+t_i)(x_i-x_{i-1})\\ &= \sum_{i=1}^{n}s_i(x_i-x_{i-1})+\sum_{i=1}^{n}t_i(x_i-x_{i-1})\\ &= \int_a^b s(x)\,dx+\int_a^b t(x)\,dx.\end{aligned}$$

Now consider any integrable functions $f,g\colon [a,b]\to\mathbb{R}$. Let $\epsilon>0$. We have step functions s_f, s_g, t_f, t_g such that the following hold:

$$\begin{gathered} s_f(x)\le f(x)\le t_f(x) \text{ and } s_g(x)\le g(x)\le t_g(x),\\ \int_a^b f(x)\,dx-\int_a^b s_f(x)\,dx<\frac{\epsilon}{2} \text{ and } \int_a^b g(x)\,dx-\int_a^b s_g(x)\,dx<\frac{\epsilon}{2},\\ \int_a^b t_f(x)\,dx-\int_a^b f(x)\,dx<\frac{\epsilon}{2} \text{ and } \int_a^b t_g(x)\,dx-\int_a^b g(x)\,dx<\frac{\epsilon}{2}.\end{gathered}$$

Then s_f+s_g and t_f+t_g are step functions such that

$$\begin{gathered} s_f(x)+s_g(x)\le f(x)+g(x)\le t_f(x)+t_g(x),\\ \int_a^b f(x)\,dx+\int_a^b g(x)\,dx-\int_a^b\Big(s_f(x)+s_g(x)\Big)\,dx<\epsilon,\\ \int_a^b\Big(t_f(x)+t_g(x)\Big)\,dx-\int_a^b f(x)\,dx-\int_a^b g(x)\,dx<\epsilon.\end{gathered}$$

Now apply Theorem 2.1.17 to see that $I=\int_a^b f(x)\,dx+\int_a^b g(x)\,dx$ satisfies the conditions for being the integral of $f+g$ over the interval $[a,b]$. ∎

Theorem 2.2.5

Suppose $f,g\colon [a,b]\to\mathbb{R}$ such that f is integrable and $g=f$ except at finitely many points. Then g is integrable and $\int_a^b g(x)\,dx=\int_a^b f(x)\,dx$.

Proof. The function $h=g-f$ is zero except at finitely many points. Hence it is a step function, which has integral zero. Hence $g=f+h$ is integrable and

$$\int_a^b g(x)\,dx=\int_a^b f(x)\,dx+\int_a^b h(x)\,dx=\int_a^b f(x)\,dx+0.$$ ∎

Theorem 2.2.6 (Additivity over Intervals)

Let $a<c<b$ and suppose $f\colon [a,b]\to\mathbb{R}$ is a bounded function. Then

1. *f is integrable on $[a,b]$ if and only if f is integrable on both $[a,c]$ and $[c,b]$, and*
2. *if f is integrable on $[a,b]$ then* $\displaystyle\int_a^b f(x)\,dx=\int_a^c f(x)\,dx+\int_c^b f(x)\,dx.$

Proof. Exercise. ■

The equality $\int_a^b f(x)\,dx = \int_a^c f(x)\,dx + \int_c^b f(x)\,dx$ holds for *any* ordering of a,b,c. For example, suppose $a < b < c$. Then

$$\int_a^c f(x)\,dx + \int_c^b f(x)\,dx = \int_a^b f(x)\,dx + \int_b^c f(x)\,dx - \int_b^c f(x)\,dx = \int_a^b f(x)\,dx.$$

Task 2.2.7

Calculate $\int_{-1}^{1} |t|\,dt.$

Theorem 2.2.8 (Shift of Interval of Integration)

Let $f\colon [a,b] \to \mathbb{R}$ *be integrable and let* $k \in \mathbb{R}$. *Then* $f(x-k)$ *is integrable on* $[a+k, b+k]$ *and*

$$\int_{a+k}^{b+k} f(x-k)\,dx = \int_a^b f(x)\,dx.$$

Proof. Exercise. (Hint: If $s(x)$ is a step function with domain $[a,b]$ then $s(x-k)$ is a step function with domain $[a+k,b+k]$.) ■

Theorem 2.2.9 (Scaling of Interval of Integration)

Let $f(x)$ *be integrable on* $[a,b]$ *and let* $k > 0$. *Then* $f(x/k)$ *is integrable on* $[ka,kb]$ *and*

$$\int_{ka}^{kb} f(x/k)\,dx = k\int_a^b f(x)\,dx.$$

Proof. Exercise. (Hint: If $s(x)$ is a step function with domain $[a,b]$ then $s(x/k)$ is a step function with domain $[ka,kb]$.) ■

Task 2.2.10

How will the above theorem be modified if $k < 0$?

Integration of Monotone Functions

Theorem 2.2.11

Let $f\colon [a,b] \to \mathbb{R}$ *be a monotone function. Then* f *is integrable on* $[a,b]$.

Proof. Suppose f is an increasing function. Let $P_n = \{x_0,\dots,x_n\}$ be the partition of $[a,b]$ that cuts it into n equally sized subintervals, each of length $(b-a)/n$. Define step functions s_n and t_n on $[a,b]$ by

$$s_n(x) = f(x_{i-1}) \quad \text{and} \quad t_n(x) = f(x_i) \qquad \text{if } x_{i-1} \le x < x_i,$$

and $s_n(b) = t_n(b) = f(b)$. Then we have $s_n(x) \leq f(x) \leq t_n(x)$ for every $x \in [a,b]$. Further,

$$\begin{aligned}\int_a^b t_n(x)\,dx - \int_a^b s_n(x)\,dx &= \sum_{i=1}^{n} f(x_i)\frac{b-a}{n} - \sum_{i=1}^{n} f(x_{i-1})\frac{b-a}{n} \\ &= \frac{b-a}{n}\sum_{i=1}^{n}\Big(f(x_i) - f(x_{i-1})\Big) \\ &= \frac{b-a}{n}\Big(f(b) - f(a)\Big).\end{aligned}$$

By the Archimedean property, this quantity can be made smaller than any given positive ϵ by taking a large enough n.

The same approach works for decreasing functions. We just switch the definitions of s_n and t_n. ∎

This approach can also succeed in finding the value of the integral of a given function, as we saw earlier for $f(x) = x$.

Task 2.2.12

Suppose that $f\colon [a,b] \to \mathbb{R}$ is an increasing function. Let $x_i = a + (b-a)i/n$. Prove that

$$\frac{b-a}{n}\sum_{i=0}^{n-1} f(x_i) \leq \int_a^b f(x)\,dx \leq \frac{b-a}{n}\sum_{i=1}^{n} f(x_i).$$

A function $f\colon [a,b] \to \mathbb{R}$ is called **piecewise monotone** if there is a partition $x_0 < \cdots < x_n$ of $[a,b]$ such that f is monotone on each open subinterval (x_i, x_{i+1}).

Here are two examples of piecewise monotone functions. Note that the second one is monotone on (x_0, x_1) but not on $[x_0, x_1]$.

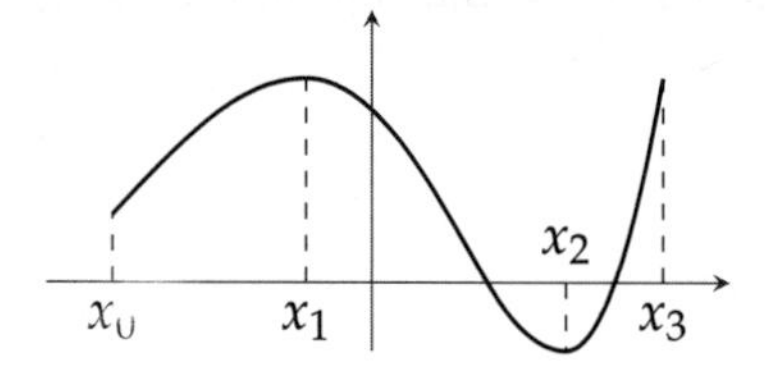

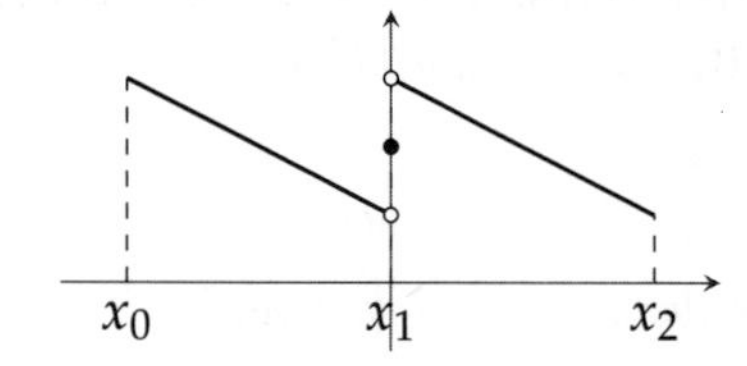

Theorem 2.2.13

If $f\colon [a,b] \to \mathbb{R}$ is bounded and piecewise monotone then it is integrable.

Proof. Let $x_0 < \cdots < x_n$ be a partition of $[a,b]$ such that f is monotone on each (x_i, x_{i+1}). Also, suppose $|f(x)| \leq M$ for every $x \in [a,b]$.

If f is increasing on (x_i, x_{i+1}), define $g_i\colon [x_i, x_{i+1}] \to \mathbb{R}$ by

$$g_i(x) = \begin{cases} -M & \text{if } x = x_i, \\ f(x) & \text{if } x_i < x < x_{i+1}, \\ M & \text{if } x = x_{i+1}. \end{cases}$$

Then g_i is an increasing function. If f is decreasing on (x_i, x_{i+1}), switch the definitions of $g(x_i)$ and $g(x_{i+1})$, and then g_i will be a decreasing function.

In either case, g_i is monotone and hence integrable. Since f equals g_i except perhaps at two points, f is integrable on $[x_i, x_{i+1}]$. Therefore f is integrable on $[a,b]$, by the result on additivity over intervals. ∎

This result suffices to establish the integrability of the functions that are typically encountered in calculus and its applications.

Intermediate Value Property

The Heaviside step function $H(x)$ jumps from 0 to 1 at $x = 0$. We create a new function by integrating it: $F(x) = \int_0^x H(t)\,dt$. We find that this function does not have a jump at $x = 0$.

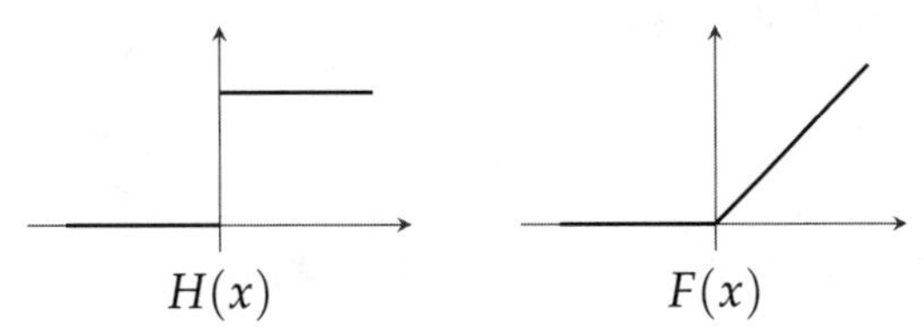

Thus, integration has smoothened out the step function and removed its jump. This is a general phenomenon that we shall explore when we take up continuous functions. For now, we develop an initial result along these lines about integrals of monotone functions.

First, we formalize the intuitive idea of a function not jumping across some value. Let I be an interval and $f\colon I \to \mathbb{R}$. We say f has the **intermediate value property** if, whenever $f(a) < L < f(b)$, there is an α between a and b such that $f(\alpha) = L$.

Theorem 2.2.14

Let I be an interval and $f\colon I \to \mathbb{R}$ a monotone function. Let $a \in I$ and define $F(x) = \int_a^x f(t)\,dt$ for $x \in I$. Then F has the intermediate value property.

Proof. We give the proof for the case when f is decreasing and positive. (See Exercise 14 for the other cases.) We start by noting that F is a strictly increasing function. Let $x < y$. Then

$$F(y) - F(x) = \int_a^y f(t)\,dt - \int_a^x f(t)\,dt = \int_x^y f(t)\,dt$$
$$\geq \int_x^y f(y)\,dt = (y-x)f(y) > 0.$$

Now, suppose $F(b) < L < F(c)$. Define $B = \{x \in I \mid F(x) < L\}$ and $C = \{y \in I \mid F(y) > L\}$. We have $b \in B$ and $c \in C$.

Suppose $x \in B$ and $y \in C$. Then $F(x) < L < F(y)$. Since F is increasing, we must have $x < y$. Hence, by the completeness axiom, there is a real number α such that $x \leq \alpha \leq y$ for every $x \in B, y \in C$. We shall use trichotomy to prove that $F(\alpha) = L$.

First, suppose $F(\alpha) < L$. Then $F(\alpha) = L - \epsilon$ with $\epsilon > 0$. For $\delta > 0$,

$$\begin{aligned} F(\alpha+\delta) &= F(\alpha) + \int_{\alpha}^{\alpha+\delta} f(t)\,dt \leq (L-\epsilon) + \int_{\alpha}^{\alpha+\delta} f(\alpha)\,dt \\ &= (L-\epsilon) + f(\alpha)\delta. \end{aligned}$$

If we choose $\delta = \dfrac{\epsilon}{2f(\alpha)}$, we get $F(\alpha+\delta) \leq L - \epsilon/2 < L$. Hence $\alpha + \delta \in B$, a contradiction.

If $F(\alpha) = L + \epsilon$, we can show in a similar fashion, using $\delta = \dfrac{\epsilon}{2f(a)}$, that $\alpha - \delta \in C$. Therefore, by trichotomy, $F(\alpha) = L$. ■

Exercises for § 2.2

1. For a certain function f, you are given that $\int_0^a f(x)\,dx = a^3$ for every $a \in \mathbb{R}$. Compute the following:

(a) $\int_1^2 f(x)\,dx$.

(b) $\int_{-2}^2 f(x)\,dx$.

(c) $\int_0^2 f(1-x)\,dx$.

(d) $\int_1^2 (f(x) + f(-x))\,dx$.

2. Compute the given integrals:

(a) $\int_1^2 (x-1)(x-2)\,dx$.

(b) $\int_1^2 (x-1)(x-2)(x-3)\,dx$.

3. Find all values of c for which the given equation holds:

(a) $\int_0^c x(1-x)\,dx = 0$.

(b) $\int_0^c |x(1-x)|\,dx = 0$.

4. Find a quadratic polynomial $P(x)$ such that $P(0) = P(1) = 0$ and $\int_0^1 P(x)\,dx = 1$.

5. Calculate $\int_0^2 f(x)\,dx$ where $f(x) = \begin{cases} x^2 & \text{if } 0 \leq x \leq 1, \\ x-2 & \text{if } 1 < x \leq 2. \end{cases}$

6. Suppose $f\colon [-a,a] \to \mathbb{R}$ is integrable on $[0,a]$. Show the following:

(a) If f is even then $\int_{-a}^a f(x)\,dx = 2\int_0^a f(x)\,dx$.

(b) If f is odd then $\int_{-a}^a f(x)\,dx = 0$.

7. Let $f\colon [a,b] \to \mathbb{R}$. Define $f^+(x) = \max\{f(x), 0\}$, $f^-(x) = \max\{-f(x), 0\}$. Prove the following:

(a) $f = f^+ - f^-$ and $|f| = f^+ + f^-$. (Definition: $|f|(x) = |f(x)|$.)

(b) If f is integrable, so are f^+ and f^-. (Hint: If a step function s gives a lower/upper sum for f, consider $s^\pm$.)

(c) If f is integrable, so is $|f|$, and $\left| \int_a^b f(x)\,dx \right| \leq \int_a^b |f(x)|\,dx$.

8. Let $f, g\colon [a,b] \to \mathbb{R}$ be integrable and define $f \vee g$, $f \wedge g$ by

$$(f \vee g)(x) = \max\{f(x), g(x)\} \quad \text{and} \quad (f \wedge g)(x) = \min\{f(x), g(x)\}.$$

Prove that $f \vee g$ and $f \wedge g$ are integrable.

9. Give an example of a function f that is not integrable on $[0,1]$ yet $|f|$ is integrable there.

10. Prove the following:

(a) If step functions s, t satisfy $0 \leq s \leq t \leq M$ on $[a,b]$ then

$$\int_a^b t(x)^2\,dx - \int_a^b s(x)^2\,dx \leq 2M\left(\int_a^b t(x)\,dx - \int_a^b s(x)\,dx \right).$$

(b) If f is integrable on $[a,b]$, so is f^2. (Hint: First assume $f \geq 0$.)

(c) If f and g are integrable on $[a,b]$, so is their product fg.

11. Prove that all polynomials are integrable on any $[a,b]$.

12. Suppose f is integrable on $[-a,a]$ and we define $F(x) = \int_0^x f(t)\,dt$ for each $x \in [-a,a]$. Prove the following:

(a) If f is even then F is odd.

(b) If f is odd then F is even.

13. Estimate $\int_1^2 \frac{1}{x}\,dx$ within an error of no more than 0.1.

14. Complete the proof of Theorem 2.2.14 by following the given hints:

(a) If f is an increasing and positive function, consider $g(x) = f(-x)$.

(b) If f is monotone and is not given to be positive, consider shifts $g(x) = f(x) + K$.

15. Show that $f(x) = x^3$ is a bijection from $\mathbb{R}$ to $\mathbb{R}$. (Hence every real number has a unique cube root.)

16. Consider the function $C\colon \mathbb{R} \to \mathbb{R}$ defined by $C(x) = x^3$. Since it is bijective (see the previous exercise, or the proof of existence of nth roots in the next section), every x has a unique cube root $x^{1/3}$. Now prove that

$$\int_0^b x^{1/3}\,dx = \frac{3}{4}b^{4/3}.$$

17. Let $f\colon \mathbb{R} \to \mathbb{R}$ have period T and be integrable on $[0,T]$.

(a) Show that f is integrable on every interval $[a,b]$.

(b) Show that the function $F(x) = \int_0^x f(t)\,dt$ has period T if $F(T) = 0$.

2.3 Logarithm and Exponential Functions

Natural Logarithm

Consider the function $f(x) = 1/x$ on $(0,\infty)$. It is monotonic and hence integrable on every $[a,b] \subset (0,\infty)$. We use this observation to define the **natural logarithm** function by

$$\log(x) = \int_1^x \frac{1}{t}\,dt \qquad (x > 0).$$

By definition, $\log(1) = 0$.

We shall often just write $\log x$ for $\log(x)$. Using fewer brackets increases readability but we should use brackets whenever ambiguity is a danger. For example, $\log x + y$ should be clearly expressed as either $\log(x+y)$ or $(\log x) + y$, two expressions with very different meanings. The same convention will be used later for any function that has its own name.

Theorem 2.3.1

Let $a,b > 0$. Then

1. $\log(ab) = \log a + \log b$,
2. $\log(a/b) = \log a - \log b$,
3. $\log(a^m) = m\log a$ *for any* $m \in \mathbb{Z}$.

Proof.

1. We use the property of scaling of interval of integration:

$$\begin{aligned}\log(ab) &= \int_1^{ab} \frac{1}{t}\,dt = \int_1^{a} \frac{1}{t}\,dt + \int_a^{ab} \frac{1}{t}\,dt = \log a + \frac{1}{a}\int_a^{ab} \frac{1}{t/a}\,dt \\ &= \log a + \int_1^{b} \frac{1}{t}\,dt \qquad \left(\because \frac{1}{k}\int_{k\alpha}^{k\beta} f(t/k)\,dt = \int_\alpha^\beta f(t)\,dt\right) \\ &= \log a + \log b.\end{aligned}$$

2. $\log(a/b) + \log b = \log\left(\frac{a}{b}\cdot b\right) = \log a.$

3. If $m = 0$, both sides are 0. For any $m \in \mathbb{N}$, we have

$$\begin{aligned}\log(a^m) &= \log(a\cdots a) = \log a + \cdots + \log a = m\log a \\ \log(a^{-m}) &= \log(1/a^m) = -\log(a^m) = -m\log a.\end{aligned}$$

■

Theorem 2.3.2

The function $\log\colon (0,\infty) \to \mathbb{R}$ *is a strictly increasing bijection.*

Proof. Let $b > a > 0$. Then

$$\log b - \log a = \int_1^b \frac{1}{t}\,dt - \int_1^a \frac{1}{t}\,dt = \int_a^b \frac{1}{t}\,dt \geq \int_a^b \frac{1}{b}\,dt = \frac{b-a}{b} > 0.$$

So, log is strictly increasing and hence also one-one.

Consider any $y > 0$. We know that $\log 2 > \log 1 = 0$, hence by the Archimedean property we have $N \in \mathbb{N}$ such that $y < N \log 2 = \log(2^N)$. So,

$$\log 1 < y < \log 2^N.$$

By the intermediate value property of indefinite integrals of monotone functions (Theorem 2.2.14), there is $x \in (1, 2^N)$ such that $\log x = y$.

Now consider $y < 0$. There is $x > 0$ such that $\log x = -y$. Then $\log(1/x) = -\log x = y$. So, log is onto. ∎

Exponential Function

Since $\log\colon (0,\infty) \to \mathbb{R}$ is a bijection, it has an inverse function $\exp\colon \mathbb{R} \to (0,\infty)$, which we call the **exponential function**. We have $y = \exp x$ if and only if $\log y = x$. Since log is strictly increasing, so is exp.

Theorem 2.3.3

Let $a, b \in \mathbb{R}$. *Then*

1. $\exp(a+b) = \exp a \exp b$,
2. $\exp(a-b) = \dfrac{\exp a}{\exp b}$,
3. $\exp(ma) = (\exp a)^m$ *for* $m \in \mathbb{Z}$.

Proof. Since log is one-one, we can prove these identities by applying log to both sides and observing that the resulting values are equal. For example, the first identity is proved as follows:

$$\log(\exp(a+b)) = a + b,$$
$$\log(\exp a \exp b) = \log(\exp a) + \log(\exp b) = a + b.$$ ∎

Euler's Number

The number $e = \exp 1$ is called **Euler's number**. It also satisfies $\log e = 1$. Consider the $y = 1/x$ graph between 1 and e. The area under the graph is $\displaystyle\int_1^e \frac{1}{t}\,dt = \log e = 1$.

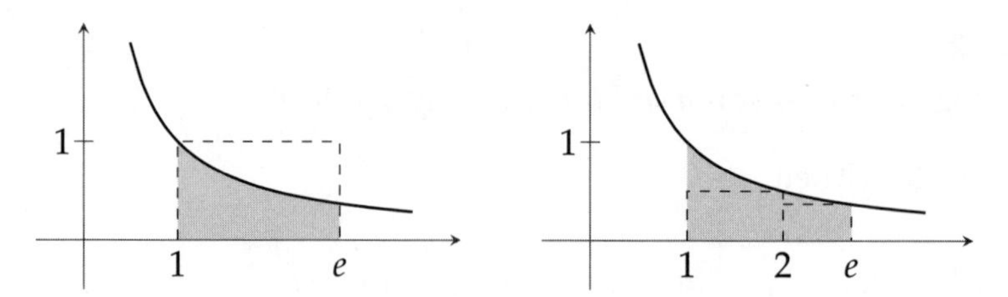

The area of the sketched rectangle in the first diagram is $1(e-1)=e-1$. Since the rectangle has greater area, we see that $e-1>1$ and therefore $e>2$.

Now consider the two inner rectangles shown in the second figure. Adding their areas gives

$$\frac{1}{2}+\frac{e-2}{e}<1 \implies \frac{3e-4}{2e}<1 \implies 3e-4<2e \implies e<4.$$

Later, on page 140, we will see how to get more accurate estimates of e.

Graphs of Log and Exp

The graph of the log function has the following appearance:

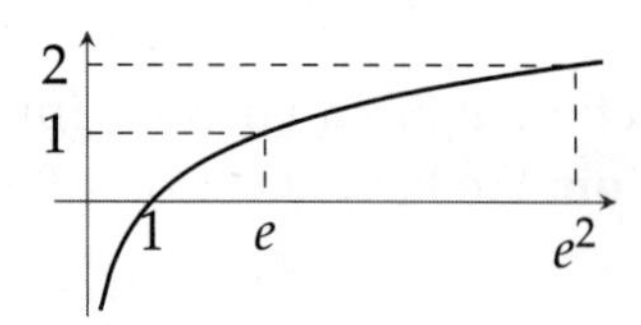

The graph of exp can be obtained by reflecting the log graph in the $y=x$ line:

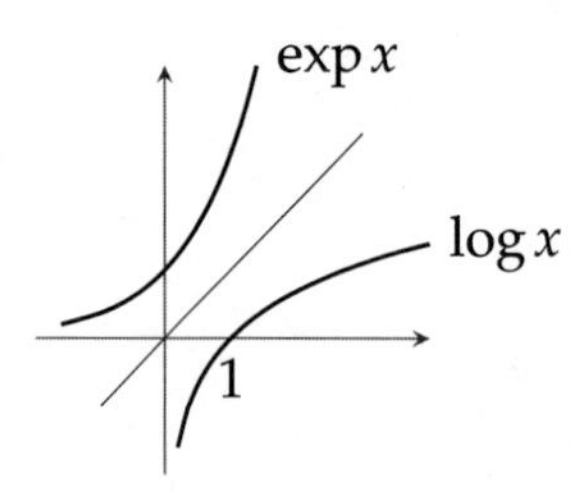

Roots

Theorem 2.3.4

Let $a>0$ and $n\in\mathbb{N}$. Then there is a unique $b>0$ such that $b^n=a$. (This b is called the nth root of a and is denoted by $a^{1/n}$ or $\sqrt[n]{a}$.)

Proof. Since $\exp\colon\mathbb{R}\to(0,\infty)$ is surjective, there is a real number x such that $\exp x=a$. Then $b=\exp(x/n)$ satisfies $b>0$ and $b^n=a$. Uniqueness follows from $b^n-c^n=(b-c)(b^{n-1}+\cdots+c^{n-1})$, as this shows that $b^n-c^n=0 \iff b-c=0$. ∎

Task 2.3.5

Let n be an odd natural number. Show that every real number a has a unique n^{th} root b, that is, $b^n = a$.

Rational Powers

Let $a > 0$ and $r \in \mathbb{Q}$. If $r = m/n$ with $m \in \mathbb{Z}$, $n \in \mathbb{N}$, we define $a^r = a^{m/n} = (a^m)^{1/n}$.

Since a rational number can be expressed as m/n in many different ways, we need to check that our definition of $a^{m/n}$ is independent of these choices.

Task 2.3.6

Let $a > 0$, $m, p \in \mathbb{Z}$, $n, q \in \mathbb{N}$ such that $m/n = p/q$. Show that $(a^m)^{1/n} = (a^p)^{1/q}$. (Hint: Take the nq power of both sides.)

Task 2.3.7

Let $a > 0$, $m \in \mathbb{Z}$, $n \in \mathbb{N}$. Show that $(a^m)^{1/n} = (a^{1/n})^m$.

Theorem 2.3.8

Let $a > 0$, $x \in \mathbb{R}$, and $r \in \mathbb{Q}$. Then

1. $\log(a^r) = r \log a$,
2. $\exp(rx) = (\exp x)^r$.

Proof.

1. Let $r = m/n$ with $m \in \mathbb{Z}$ and $n \in \mathbb{N}$. We have shown that $\log(a^m) = m \log a$ if $m \in \mathbb{Z}$. Replacing m by $n \in \mathbb{N}$ and a by $a^{1/n}$ in this equality, we get $\log a = n \log(a^{1/n})$, and hence $\log(a^{1/n}) = \dfrac{1}{n} \log a$. Combine these calculations:

$$\log(a^r) = \log((a^m)^{1/n}) = \frac{1}{n} \log(a^m) = \frac{m}{n} \log a = r \log a.$$

2. Apply log to both sides. ■

Real Powers

We have seen that rational powers satisfy $\log(a^r) = r \log a$. It is natural to want this to be true for arbitrary real powers, and that makes us *define*:

$$a^x = \exp(x \log a), \qquad \text{if } a > 0 \text{ and } x \in \mathbb{R}.$$

In particular, we have

$$e^x = \exp(x \log e) = \exp x.$$

General powers reduce to powers of e:

$$a^x = e^{x \log a}.$$

With this definition of real powers of a positive real number, we can use the properties of the exponential function to verify the following:

Theorem 2.3.9

Let $a > 0$ and $x, y \in \mathbb{R}$. Then

1. $a^{x+y} = a^x a^y$,
2. $a^{x-y} = a^x / a^y$,
3. $(a^x)^y = a^{xy}$.

Proof. The first two are easily checked using the properties of exp. For the third, we first consider the special case when $a = e$,

$$(e^x)^y = \exp(y \log e^x) = \exp(xy) = e^{xy}.$$

Now we consider the general case,

$$(a^x)^y = (e^{x \log a})^y = e^{xy \log a} = a^{xy}. \qquad \blacksquare$$

For $a > 0$ and $a \neq 1$, the function $\log(a) \cdot x$ is a bijection of $\mathbb{R}$ with $\mathbb{R}$, while e^x is a bijection of $\mathbb{R}$ with $(0, \infty)$. Hence, their composition $a^x = e^{\log(a)x}$ is a bijection of $\mathbb{R}$ with $(0, \infty)$. The inverse function of a^x is called the **logarithm with base** a and is denoted by $\log_a x$: $y = \log_a x \iff a^y = x$. We have $\log x = \log_e x$.

This is different from the notation typically used in schools, where $\log x$ stands for $\log_{10} x$, and the natural logarithm has the special name "ln."

Task 2.3.10

Solve for y in terms of x.

(a) $\log(y+1) - \log(y-1) = 2 \log x$.

(b) $x = \dfrac{2^y + 2^{-y}}{2^y - 2^{-y}}$.

Theorem 2.3.11

Let $a, x > 0$ and $a \neq 1$. Then $\log_a x = \dfrac{\log x}{\log a}$.

Proof. Rearrange the given expression to $\log a \log_a x = \log x$. Now,

$$\begin{aligned}
\log a \log_a x = \log x &\iff e^{\log a \log_a x} = e^{\log x} \\
&\iff (e^{\log a})^{\log_a x} = x \\
&\iff a^{\log_a x} = x \iff x = x.
\end{aligned} \qquad \blacksquare$$

Theorem 2.3.12

Fix $a > 0$, $a \neq 1$. Then the functions a^x and $\log_a x$ are strictly monotone on their domains.

Proof. Apply the previous theorem. ∎

Task 2.3.13

Let $a,b,x > 0$ with $a,b \neq 1$. Show that $\log_a x = \dfrac{\log_b x}{\log_b a}$.

Hyperbolic Functions

Some particular combinations of exponential functions are very convenient in calculations, and also show up directly in some physical problems. These are called **hyperbolic functions** and are defined by

$$\cosh x = \frac{e^x + e^{-x}}{2} \quad \text{and} \quad \sinh x = \frac{e^x - e^{-x}}{2}.$$

These are called the "hyperbolic cosine" and "hyperbolic sine" respectively.

Task 2.3.14

Prove the following:

(a) *The function* cosh *is even and* sinh *is odd.*

(b) *The image of* cosh *is* $[1,\infty)$ *and that of* sinh *is* $\mathbb{R}$.

(c) *For every* $x \in \mathbb{R}$, $(\cosh x)^2 - (\sinh x)^2 = 1$.

This implies that if we take $y = \cosh t$ and $x = \sinh t$, and vary t, we will trace out the upper branch of the hyperbola given by $y^2 - x^2 = 1$. Taking $y = -\cosh t$ and $x = \sinh t$ will trace the lower branch. This explains the choice of names.

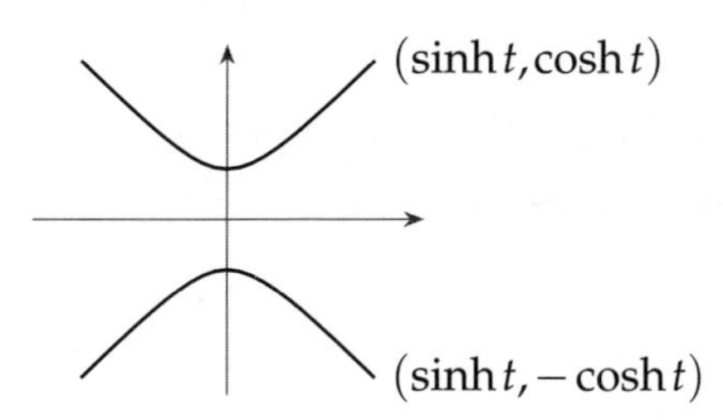

Let us consider the graph of $\cosh x$. Note that $\cosh x > e^x/2$ and $\cosh x$ approaches $e^x/2$ as x becomes larger on the positive side. Whereas, on the negative side, $\cosh x$ approaches $e^{-x}/2$. This gives the following picture, where the graphs of $e^{\pm x}/2$ are shown by dashed curves:

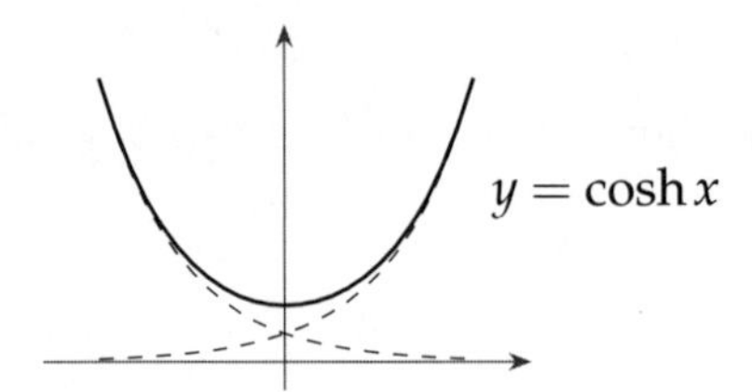

The graph of the hyperbolic cosine function is called the **catenary**. It is important in physics and engineering as it is the shape formed by a freely hanging cable.

Task 2.3.15

Justify the following depiction of the graph of sinh*:*

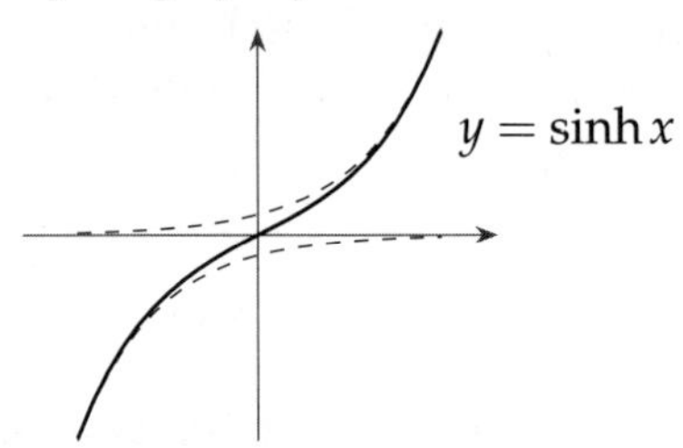

Exercises for § 2.3

1. Estimate $\log 2$ and $\log 3$ within an error of 0.1.

2. You are given some values of the natural logarithm, accurate to two decimal places: $\log 2 = 0.69$, $\log 3 = 1.10$, $\log 5 = 1.61$, and $\log 7 = 1.95$. Use these to find the natural logarithms of the following numbers:

(a) 80, (b) 120, (c) 2.1, (d) $3/35$.

3. Express x in terms of $\log 2$ and $\exp 2$:

(a) $2\exp(5x) = \dfrac{1}{16}$, (b) $12\log(x-3) = 96$.

4. Which is bigger: $\log_2 3$ or $\log_3 5$? (Hint: Compare with $3/2$.)

5. Use the following figures to show that $2.4 < e < 3.5$:

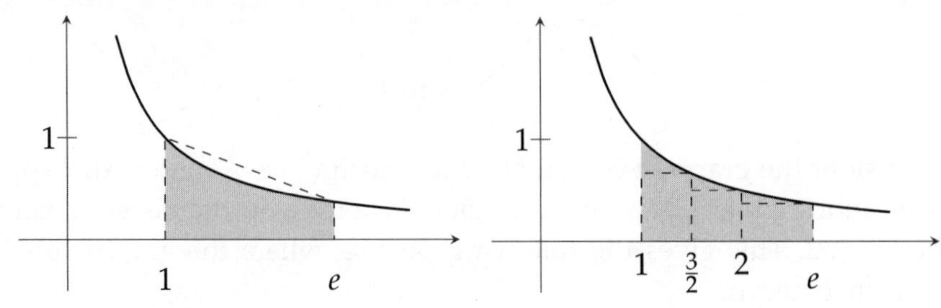

6. Prove the following inequalities for $x > 0$: $1 - \dfrac{1}{x} \leq \log x \leq x - 1$.

7. Plot the graphs of the following functions on the same coordinate plane:

(a) $\log x$, (b) $\log 2x$, (c) $\log x^2$, (d) $\log |x|$.

8. Plot the graphs of the following functions on the same coordinate plane:

(a) $\exp x$, (b) $\exp(1+x)$, (c) $\exp(-x)$, (d) $\exp|x|$.

9. Suppose $a, b > 0$ and $x \in \mathbb{R}$. Show that $a^x b^x = (ab)^x$.

10. Prove that the function a^x is strictly increasing if $a > 1$ and strictly decreasing if $0 < a < 1$.

11. Graph the function a^x for $a = 1/2, 1, 2, e, 3$.

12. Verify the following identities for the hyperbolic functions:

(a) $\cosh(2x) = (\cosh x)^2 + (\sinh x)^2$.

(b) $\sinh(2x) = 2(\cosh x)(\sinh x)$.

(c) $\cosh(x+y) = \cosh x \cosh y + \sinh x \sinh y$.

(d) $\sinh(x+y) = \sinh x \cosh y + \cosh x \sinh y$.

2.4 Integration and Area

We began this chapter with the promise that integration will enable us to calculate areas. Now we shall deliver on that promise. To begin with, consider a bounded function $f\colon [a,b] \to \mathbb{R}$ such that $f(x) \geq 0$ for every $x \in [a,b]$. This creates a region $\mathcal{R}$ in the Cartesian plane, enclosed by the graph of f, the x-axis, and the lines $x = a$, $x = b$.

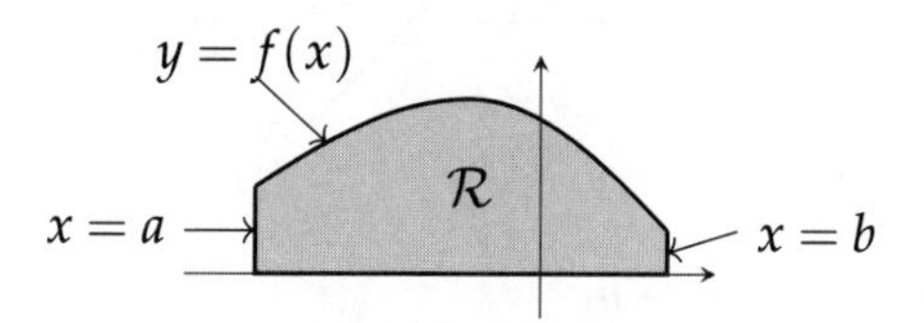

The definite integral $\int_a^b f(x)\,dx$ is our definition of the area of the region $\mathcal{R}$. If f is not integrable, then we fail to define the area of $\mathcal{R}$.

Next, suppose we have two integrable functions $f, g\colon [a,b] \to \mathbb{R}$ such that $f(x) \geq g(x) \geq 0$ for every $x \in [a,b]$. Then we can create a region $\mathcal{S}$ that is enclosed by the graphs of f and g, and by the lines $x = a$, $x = b$.

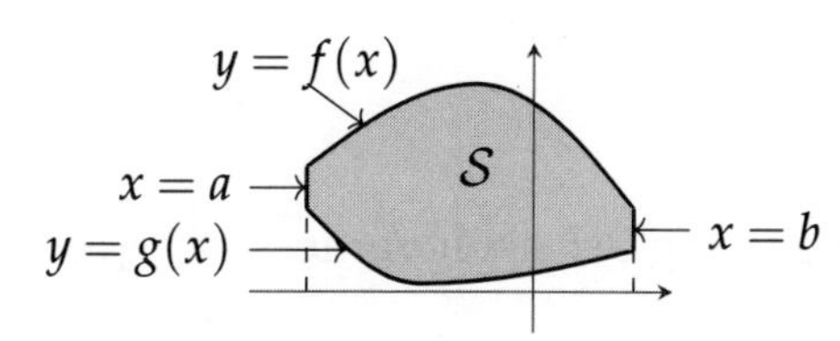

This region can be viewed as the result of removing the region under the graph of g from that which is under the graph of f.

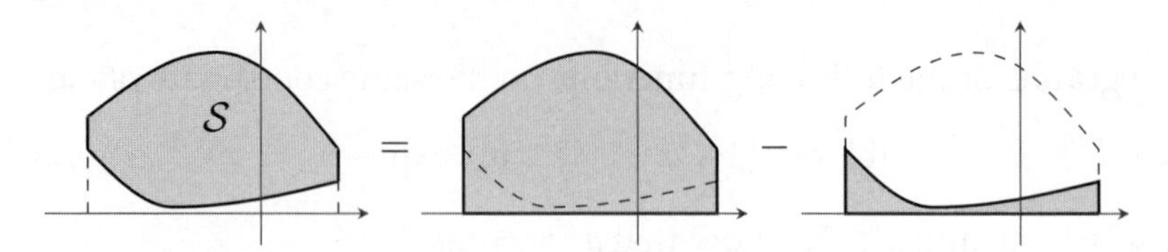

Hence, we define the area of $\mathcal{S}$ to be

$$\int_a^b f(x)\,dx - \int_a^b g(x)\,dx = \int_a^b (f(x) - g(x))\,dx.$$

The assumption $f, g \geq 0$ can be relaxed. Since f, g are bounded, there is a number C such that $f + C, g + C \geq 0$. The region that lies between the graphs of $f + C, g + C \geq 0$ is just a vertical shift of $\mathcal{S}$, so it should have the same area. And that area is

$$\int_a^b (f(x) + C)\,dx - \int_a^b (g(x) + C)\,dx = \int_a^b (f(x) - g(x))\,dx.$$

Now, we shall develop an approach for calculating the area of a polygon whose vertices are given. The building block is the integral corresponding to one side of the polygon. Thus, let f be the function whose graph is the straight line joining (x_1, y_1) to (x_2, y_2), with $x_1 < x_2$.

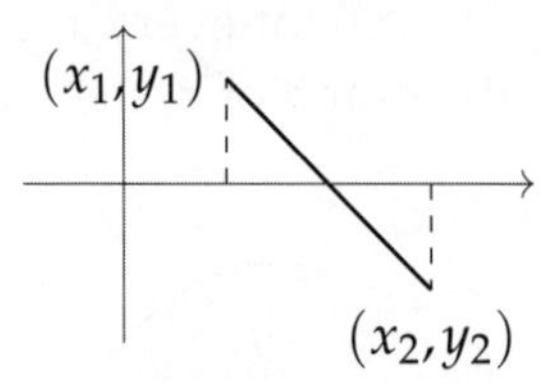

Then
$$\begin{aligned}
\int_{x_1}^{x_2} f(x)\,dx &= \int_{x_1}^{x_2} \left(y_1 + \frac{y_2 - y_1}{x_2 - x_1}(x - x_1)\right) dx \\
&= \int_0^{x_2 - x_1} \left(y_1 + \frac{y_2 - y_1}{x_2 - x_1} x\right) dx \\
&= (x_2 - x_1) y_1 + \frac{y_2 - y_1}{x_2 - x_1} \frac{(x_2 - x_1)^2}{2} \\
&= \frac{1}{2}(x_2 - x_1)(y_2 + y_1).
\end{aligned}$$

Example 2.4.1

Let us find the area of the triangle $\mathcal{T}$ with vertices at (x_1, y_1), (x_2, y_2), and (x_3, y_3) as shown below.

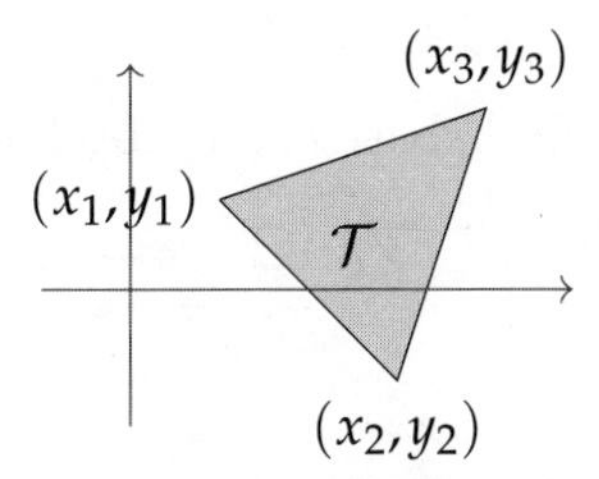

The triangle lies between the graphs of the following functions:

$$f(x) = y_1 + \frac{y_3 - y_1}{x_3 - x_1}(x - x_1) \quad \text{if } x \in [x_1, x_3],$$
$$g(x) = \begin{cases} y_1 + \dfrac{y_2 - y_1}{x_2 - x_1}(x - x_1) & \text{if } x \in [x_1, x_2], \\ y_2 + \dfrac{y_3 - y_2}{x_3 - x_2}(x - x_2) & \text{if } x \in [x_2, x_3]. \end{cases}$$

Now we calculate the area as follows:

$$\begin{aligned} \text{Area}(\mathcal{T}) &= \int_{x_1}^{x_3} f(x)\,dx - \int_{x_1}^{x_3} g(x)\,dx \\ &= \int_{x_1}^{x_3} f(x)\,dx - \int_{x_1}^{x_2} g(x)\,dx - \int_{x_2}^{x_3} g(x)\,dx \\ &= \frac{1}{2}\Big((x_3 - x_1)(y_3 + y_1) - (x_2 - x_1)(y_2 + y_1) - (x_3 - x_2)(y_3 + y_2)\Big) \\ &= \frac{1}{2}\Big((x_1y_2 - x_2y_1) + (x_2y_3 - x_3y_2) + (x_3y_1 - x_1y_3)\Big). \end{aligned}$$ □

This calculation can be used to construct the area formulas for basic polygonal shapes.

Example 2.4.2

Consider a triangle with base b and height h. Place it with base along the x-axis and one vertex at origin.

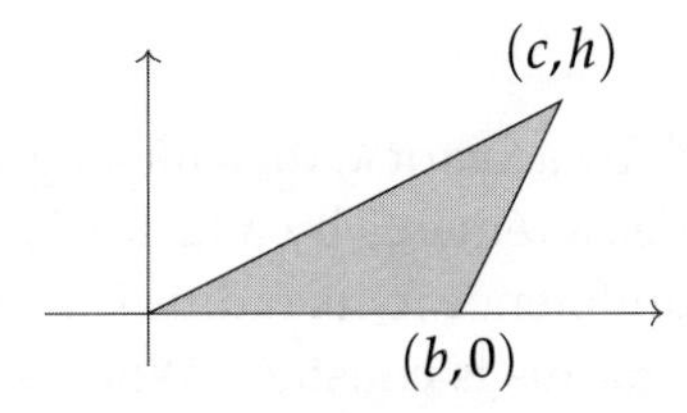

Its area is $\frac{1}{2}\Big((0 \cdot 0 - b \cdot 0) + (b \cdot h - c \cdot 0) + (c \cdot 0 - 0 \cdot h)\Big) = \frac{1}{2}bh.$ □

Example 2.4.3

Consider a trapezium with height h, base b, and top t. Place it as shown below.

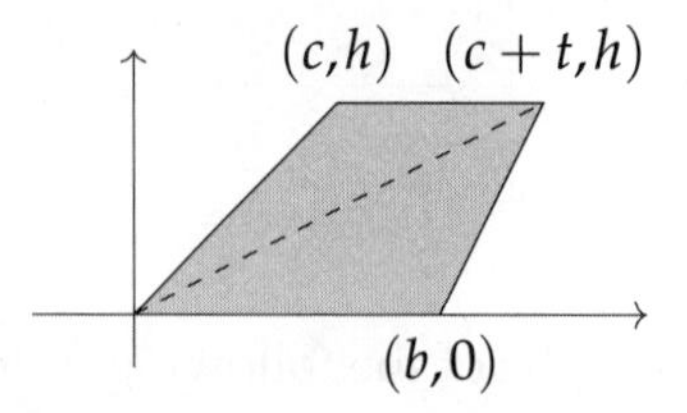

Using the previous example, we find its area to be $\frac{1}{2}bh + \frac{1}{2}th = \frac{b+t}{2}h$. If $t = b$, we get the formula bh for the area of a parallelogram. □

Example 2.4.4

We wish to find the area of the region enclosed by the curves given by $y = 3x$ and $y = 4 - x^2$. The first step is to sketch this region. For this, we first check if the curves meet. Setting $3x = 4 - x^2$ gives $x = 1, -4$. Hence, the curves meet at $(-4,-12)$ and $(1,3)$. The curves plot as follows, and the shaded part is the region that is *enclosed* by the curves.

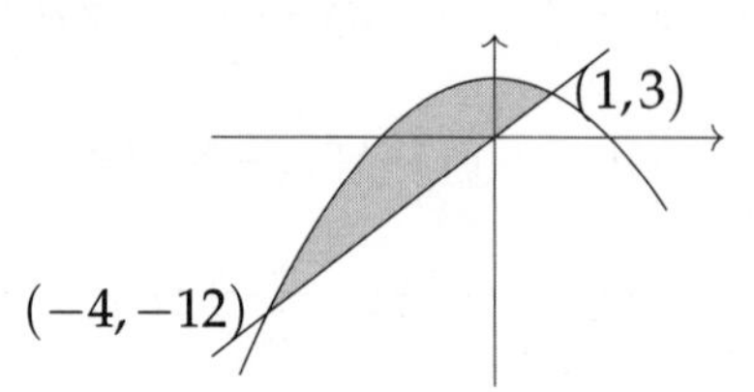

The area of this region is calculated as follows:

$$\begin{aligned}\text{Area} &= \int_{-4}^{1} (4 - x^2 - 3x)\,dx = 4\int_{-4}^{1} 1\,dx - \int_{-4}^{1} x^2\,dx - 3\int_{-4}^{1} x\,dx \\ &= 4 \cdot (1 - (-4)) - \frac{1}{3}(1^3 - (-4)^3) - 3 \cdot \frac{1}{2}(1^2 - (-4)^2) = 20\tfrac{5}{6}.\end{aligned}$$

□

Defining and Estimating π

If we change our unit of length by a factor k, then all length measurements will change by a factor $1/k$ and all area measurements by a factor $1/k^2$. This shows, for example, that the ratio of a circle's circumference to its radius is constant, and also that the ratio of its area to the square of its radius is constant. What is remarkable is that the same constant appears in both cases, in the relationships of circumference equaling π times the diameter and area equaling π times the radius squared.

In this section we study π using area. Eventually, in §6.4, we shall connect π to lengths as well. First, we verify by a direct calculation that the area of a circle is proportional to the square of its radius. Consider a circle with center at origin and radius R. It is given by the equation $x^2 + y^2 = R^2$, and can be seen as the region enclosed by the graphs of the functions $y^+ = \sqrt{R^2 - x^2}$ and $y^- = -\sqrt{R^2 - x^2}$.

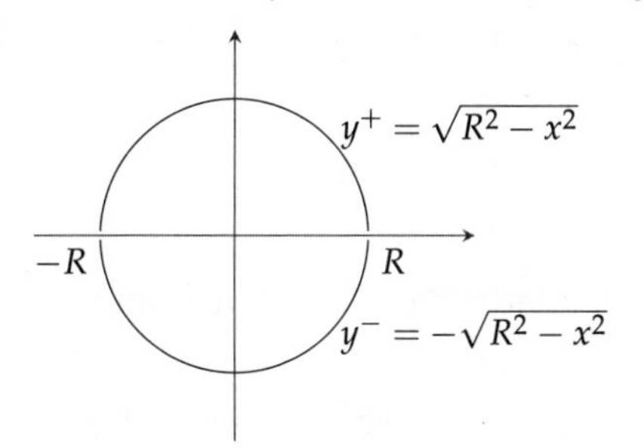

Its area is given by

$$\int_{-R}^{R} \left(\sqrt{R^2 - x^2} - (-\sqrt{R^2 - x^2}) \right) dx = 2 \int_{-R}^{R} \sqrt{R^2 - x^2}\, dx.$$

We apply scaling of the interval of integration (Theorem 2.2.9) to get

$$2 \int_{-R}^{R} \sqrt{R^2 - x^2}\, dx = 2R \int_{-R/R}^{R/R} \sqrt{R^2 - (Rx)^2}\, dx = 2R^2 \int_{-1}^{1} \sqrt{1^2 - x^2}\, dx.$$

Thus, we get the area to be πR^2, where

$$\pi = 2 \int_{-1}^{1} \sqrt{1 - x^2}\, dx = 4 \int_{0}^{1} \sqrt{1 - x^2}\, dx.$$

Since $\sqrt{1 - x^2}$ is a piecewise monotone function on $[-1, 1]$, these integrals exist. We can use lower and upper sums to put bounds on the values of π. For example, suppose we partition $[0, 1]$ into N equal parts. Then we have $x_i = i/N$ and

$$4 \sum_{i=1}^{N} \frac{\sqrt{1 - x_i^2}}{N} < \pi < 4 \sum_{i=1}^{N} \frac{\sqrt{1 - x_{i-1}^2}}{N}.$$

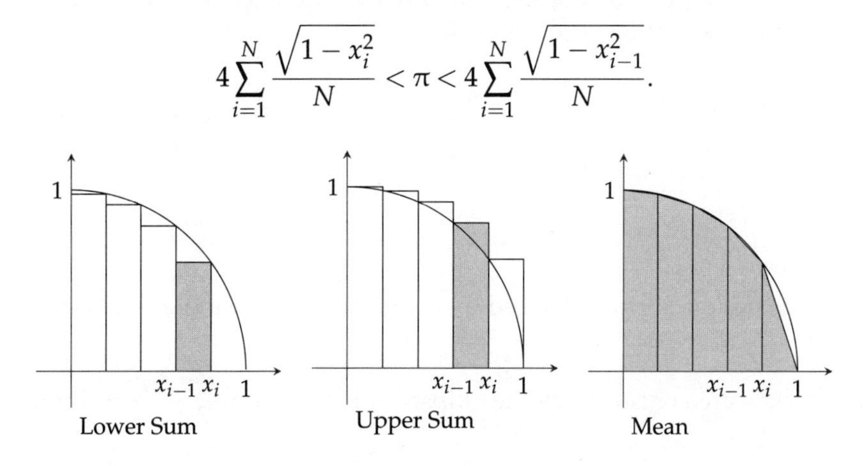

The table shows the bounds and their means for various N:

N	Lower Bound	Upper Bound	Mean
10^2	3.1204	3.1604	3.1404
10^3	3.1395	3.1435	3.1415
10^6	3.1415906	3.1415946	3.1415926

The actual value of π is $3.141592653\ldots$.

Exercises for § 2.4

1. Show that the area of the quadrilateral depicted below is

$$\frac{1}{2}\Big((x_1y_2 - x_2y_1) + (x_2y_3 - x_3y_2) + (x_3y_4 - x_4y_3) + (x_4y_1 - x_1y_4)\Big).$$

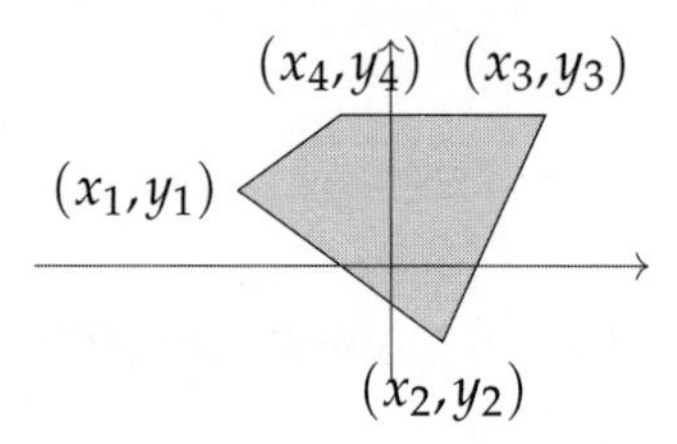

2. Can the result of the previous exercise be generalized to a polygon with n sides?

3. The trapezium drawn below is to be cut into two parts of equal area by a vertical line. How long will it be?

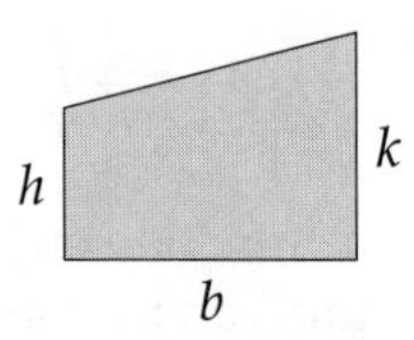

4. Use definite integrals to represent the area of the region enclosed by the given curves (do not try to evaluate the integrals).

(a) 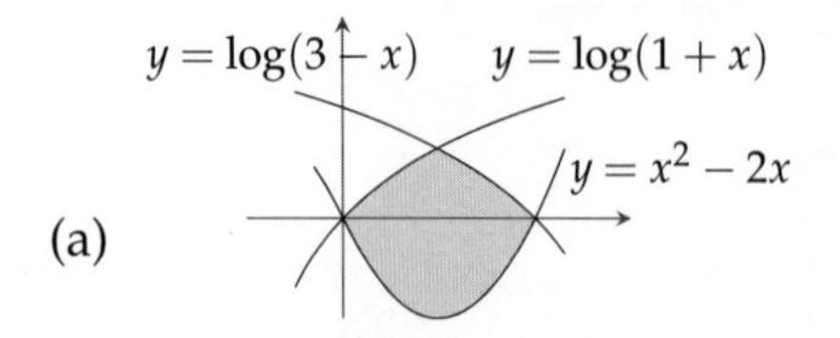

(b) 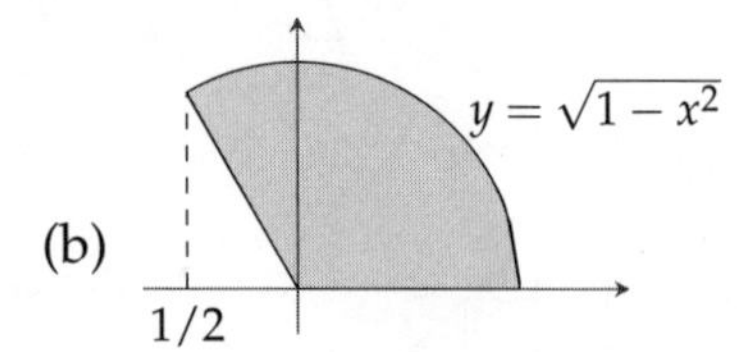

5. Find the area of the region enclosed by the curves $y = x^2$ and $y = x^3$ between $x = 0$ and $x = 1$.

6. Prove that the area enclosed by the ellipse with equation $\dfrac{x^2}{a^2} + \dfrac{y^2}{b^2} = 1$ with $a, b > 0$ is πab.

7. Let f be strictly increasing and have the intermediate value property, and $y = m$, $y = M$ be horizontal lines that cut its graph. Find the point on its graph such that the area depicted below is minimized.

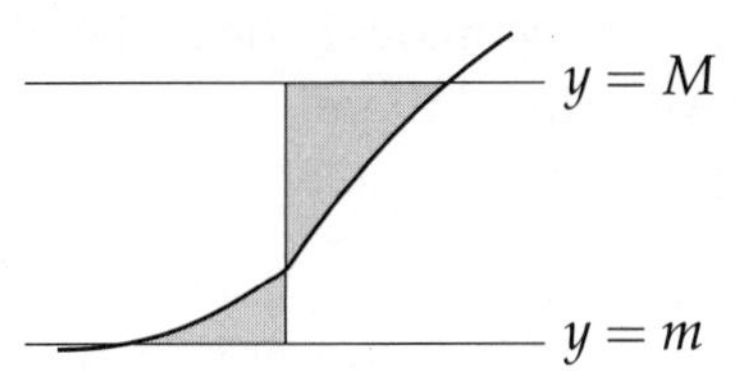

See the article by Khovanova and Radul [54] for the curious history of this problem.

Thematic Exercises

Darboux Integral

Our approach to integration is a minor variation of one due to Gaston Darboux, which he gave in 1875. At this time Joseph Fourier's work had led to a broadening of the idea of a function and so it became important to give integration a formal structure and not rely on intuitions about area. Bernhard Riemann gave the first general approach in 1854, but Darboux's definition is much easier to set up. For completeness, we give the standard version of the Darboux integral below. We also introduce the Riemann integral in the supplementary exercises of Chapter 6.

Consider a bounded function $f\colon [a,b] \to \mathbb{R}$ and a partition $P : x_0 < \cdots < x_n$ of $[a,b]$. Using the LUB property, we define

$$m_i = \inf\{ f(x) \mid x \in [x_{i-1}, x_i] \} \quad \text{and} \quad M_i = \sup\{ f(x) \mid x \in [x_{i-1}, x_i] \}.$$

We then take the lower and upper **Darboux sums** created by these numbers:

$$L(f,P) = \sum_{i=1}^{n} m_i(x_i - x_{i-1}) \quad \text{and} \quad U(f,P) = \sum_{i=1}^{n} M_i(x_i - x_{i-1}).$$

A1. Show that the collection of lower Darboux sums is bounded above and the collection of upper Darboux sums is bounded below.

Hence, we can define the lower and upper **Darboux integrals**:

$$\underline{\int_a^b} f(x)\,dx = \sup\{ L(f,P) \mid P \text{ is a partition of } [a,b] \},$$

$$\overline{\int_a^b} f(x)\,dx = \inf\{ U(f,P) \mid P \text{ is a partition of } [a,b] \}.$$

A2. Find the upper and lower Darboux integrals of the Dirichlet function (Example 2.1.11).

A3. Show that $\underline{\int_a^b} f(x)\,dx \leq \overline{\int_a^b} f(x)\,dx$.

We say f is **Darboux integrable** on $[a,b]$ if its upper and lower Darboux integrals are equal. The common value is called the **Darboux integral** of f.

A4. Show that a bounded function f is Darboux integrable on $[a,b]$ if and only if it is integrable as defined on page 50. Moreover, the value of the Darboux integral is the same as the definite integral we defined there.

3 | Limits and Continuity

Integration can be seen as accumulation, the summing up of local changes to get a global result. In the previous chapter, we set up the general process for achieving this. We also saw how a specific kind of information—monotonicity—could be used to obtain global results. With its help we were able to formally define the natural logarithm and exponential functions, which are usually taken for granted in school mathematics.

Further progress requires a closer look at the local behavior of functions. The more we know of the local behavior, the better our chances of extracting global information. These considerations underlie our development of the notions of limit and continuity in this chapter. As applications, we will rigorously develop angles and their radian measures, followed by the trigonometric functions and their properties.

3.1 Limits

You have seen in school, the notation $\lim_{x \to p} f(x) = L$, which is read as "the limit of $f(x)$ at p is L" and is interpreted as "the values of $f(x)$ approach L as the values of x approach p." We need a clear definition of what we mean by "approaches."

Example 3.1.1

Consider $f(x) = 2x + 5$. What happens if we take values of x that approach 0? Here are some calculations:

x	1	0.1	0.01	0.001	0.0001	0.00001
$f(x)$	7	5.2	5.02	5.002	5.0002	5.00002

We see that as x gets closer to 0, $f(x)$ appears to be getting closer to 5. Can we control this? Can we get the output $f(x)$ close to 5 within any required accuracy level, simply by making the input x appropriately close to 0?

Suppose ϵ is some positive number and we need $f(x) = 2x + 5$ to be within ϵ of 5. We analyze the requirement as follows:

$$|(2x+5) - 5| < \epsilon \iff |2x| < \epsilon \iff |x| < \epsilon/2.$$

Thus, if $|x| < \epsilon/2$, we are guaranteed that $|f(x) - 5| < \epsilon$. □

This example shows what we mean by "limit." We mean that we can control the nearness of the output to a certain number, by controlling the nearness of the input to another number. This is expressed formally as follows:

We say $\lim_{x \to p} f(x) = L$ *if for every* $\epsilon > 0$ *there is a corresponding* $\delta > 0$ *such that* $0 < |x - p| < \delta \implies |f(x) - L| < \epsilon$.

In Example 3.1.1 we have $\delta = \epsilon/2$.

We make three observations about the definition of limit.

(a) It sets up δ as depending on ϵ.

(b) We do not care about the value of $f(p)$, or even whether it is defined.

(c) Since the definition is intended for situations where x can approach p, it should only be applied to such situations. In calculus, this means that we shall only consider the limit of f at p if there is an $\alpha > 0$ such that the open interval $(p - \alpha, p + \alpha)$ is contained in the domain of f, except perhaps for p itself.

We may also write "$f(x) \to L$ as $x \to p$" for $\lim_{x \to p} f(x) = L$.

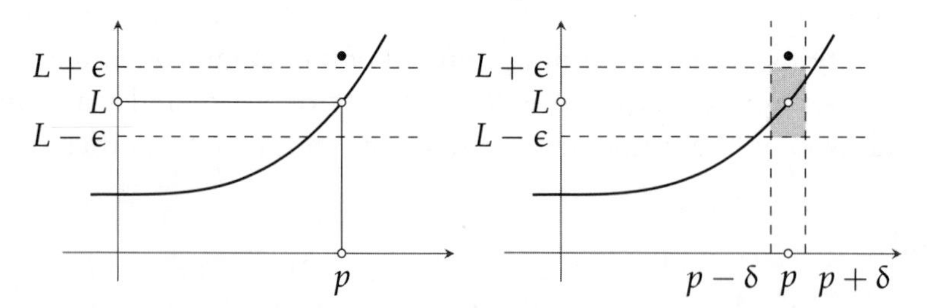

There are two stages in a limit process. In the first stage, we have a requirement to make the output $f(x)$ *lie between* $L - \epsilon$ *and* $L + \epsilon$. *In the second stage, we meet the requirement by finding a* δ *such that input being between* $p - \delta$ *and* $p + \delta$ *guarantees that the output is between* $L - \epsilon$ *and* $L + \epsilon$ *(except perhaps at* p *itself).*

We have defined the limit of a function as a number satisfying certain properties. We face the same issues as we did when we defined the integral of a bounded function: Will there be a number with the given properties, and will it be unique? In the case of integration, existence was guaranteed but not uniqueness. For limits, it is the other way around. Let us begin with the good news.

Theorem 3.1.2

At most one number can satisfy the definition of the limit of a given function at a given point.

Proof. Suppose L, M are two distinct numbers, both of which satisfy the definition of $\lim_{x \to a} f(x)$. Choose $\epsilon = |M - L|/2$. Then there are positive numbers δ_L and δ_M such that

$$0 < |x - a| < \delta_L \implies |f(x) - L| < \epsilon,$$
$$0 < |x - a| < \delta_M \implies |f(x) - M| < \epsilon.$$

Let $\delta = \min\{\delta_L, \delta_M\}$ and x_0 be any point in $(a-\delta, a+\delta)$. Then we have $|f(x_0) - L| < \epsilon$ and $|f(x_0) - M| < \epsilon$. Hence,

$$|M - L| = |M - f(x_0) + f(x_0) - L| \le |M - f(x_0)| + |f(x_0) - L| < \epsilon + \epsilon = |M - L|,$$

which gives the impossible statement $|M - L| < |M - L|$. ■

Example 3.1.3

Consider $\lim_{x \to a} x$. This amounts to asking "What does x approach when x approaches a?" Obviously, our response has to be that it will approach a, that is, $\lim_{x \to a} x = a$. While this is indeed obvious, let us still work it out with the ϵ-δ formulation, for practice.

We start by considering an $\epsilon > 0$. We need to find a $\delta > 0$ such that $|x - a| < \delta \implies |x - a| < \epsilon$. Clearly $\delta = \epsilon$ will work. □

Task 3.1.4

Let $f(x) = c$ be a constant function. Show that $\lim_{x \to p} f(x) = c$.

Example 3.1.5

Consider the limit of the function $y = x^2$ at $x = 2$. A natural guess is that as x approaches 2, x^2 should approach $2^2 = 4$. We can test this guess by trying out some values of $\epsilon > 0$.

For example, suppose $\epsilon = 0.5$. We need a positive δ such that $x \in (2 - \delta,\ 2 + \delta)$ implies $x^2 \in (4 - 0.5,\ 4 + 0.5) = (3.5,\ 4.5)$. We first note that since the function is an increasing one, it maps $(\sqrt{3.5},\ \sqrt{4.5})$ into $(3.5,\ 4.5)$. The interval $(\sqrt{3.5},\ \sqrt{4.5})$ contains 2 but is not centered on it, since $2 - \sqrt{3.5} = 0.129$ while $\sqrt{4.5} - 2 = 0.121$.

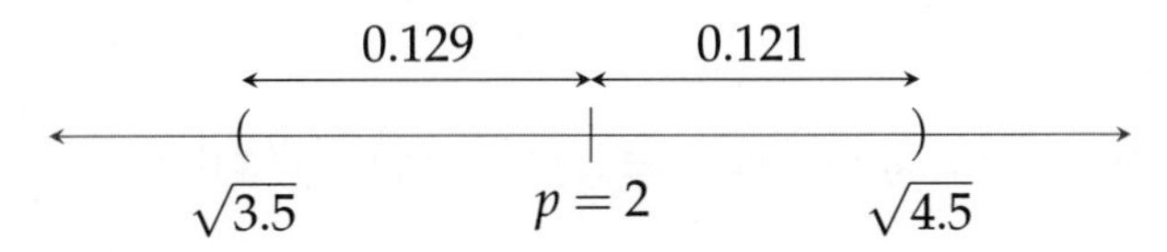

Now, if we take $\delta = \sqrt{4.5} - 2 = 0.121$ (the smaller of the two values), it has the required properties, since then $(2 - \delta,\ 2 + \delta) \subset (\sqrt{3.5},\ \sqrt{4.5})$.

Next, consider $\epsilon = 0.01$. Can you confirm that $\delta = \sqrt{4.01} - 2$ will meet the requirements?

If you have understood the arguments for these two values of ϵ, you are ready to handle *any* choice of positive ϵ. Just take $\delta = \min\{2 - \sqrt{4 - \epsilon},\ \sqrt{4 + \epsilon} - 2\}$. □

This example also demonstrates that δ typically depends on ϵ, with smaller ϵ requiring smaller δ.

Theorem 3.1.6

$\lim_{x \to p} f(x) = L \iff \lim_{x \to p} (f(x) - L) = 0 \iff \lim_{h \to 0} f(p+h) = L.$

Proof. We simply match the definitions of the three limits and see that they are the same:

- $\lim_{x \to p} f(x) = L$: *For every* $\epsilon > 0$ *there is a corresponding* $\delta > 0$ *such that* $0 < |x - p| < \delta \implies |f(x) - L| < \epsilon$.
- $\lim_{x \to p} (f(x) - L) = 0$: *For every* $\epsilon > 0$ *there is a corresponding* $\delta > 0$ *such that* $0 < |x - p| < \delta \implies |(f(x) - L) - 0| < \epsilon$.
- $\lim_{h \to 0} f(p+h) = L$: *For every* $\epsilon > 0$ *there is a corresponding* $\delta > 0$ *such that* $0 < |h| < \delta \implies |f(p+h) - L| < \epsilon$.

The first two are completely identical. The first can be converted to the third, and conversely, by the substitution $x = p + h$. ∎

Task 3.1.7

Let $a, b \in \mathbb{R}$ *with* $a \neq 0$. *Show that* $\lim_{x \to p} f(ax+b) = L \iff \lim_{y \to ap+b} f(y) = L$.

Theorem 3.1.8

$\lim_{x \to p} f(x) = 0 \iff \lim_{x \to p} |f(x)| = 0.$

Proof. Again, just match the definitions:

- $\lim_{x \to p} f(x) = 0$: *For every* $\epsilon > 0$ *there is a corresponding* $\delta > 0$ *such that* $0 < |x - p| < \delta \implies |f(x) - 0| < \epsilon$.
- $\lim_{x \to p} |f(x)| = 0$: *For every* $\epsilon > 0$ *there is a corresponding* $\delta > 0$ *such that* $0 < |x - p| < \delta \implies ||f(x)| - 0| < \epsilon$.

The two definitions are the same because $|f(x) - 0| = |f(x)| = |f(x)| - 0 = ||f(x)| - 0|$. ∎

Theorem 3.1.9

$\lim_{x \to p} f(x) = M \implies \lim_{x \to p} |f(x)| = |M|.$

Proof. We know from the triangle inequality that $||f(x)| - |M|| \leq |f(x) - M|$.

Consider any $\epsilon > 0$. Since $\lim_{x \to p} f(x) = M$, there is a $\delta > 0$ such that $0 < |x - p| < \delta \implies |f(x) - M| < \epsilon$. The same δ works for $|f(x)|$ since $|f(x) - M| < \epsilon$ implies $||f(x)| - |M|| \leq |f(x) - M| < \epsilon$. ∎

Now we consider three examples that illustrate the typical ways in which a limit can *fail* to exist.

Example 3.1.10

Consider the signum function

$$\operatorname{sgn}(x) = \begin{cases} -1 & \text{if } x < 0, \\ 0 & \text{if } x = 0, \\ 1 & \text{if } x > 0. \end{cases}$$

We shall prove by contradiction that $\lim_{x\to 0} \operatorname{sgn}(x)$ does not exist.

Suppose $\lim_{x\to 0} \operatorname{sgn}(x) = L$. Consider $\epsilon = 1$. By the existence of the limit, there is a $\delta > 0$ such that $0 < |x| < \delta \implies |\operatorname{sgn}(x) - L| < 1$.

Both $x = \delta/2$ and $x = -\delta/2$ satisfy the condition $0 < |x| < \delta$.

Hence, $|\operatorname{sgn}(\delta/2) - L| < 1$ and $|\operatorname{sgn}(-\delta/2) - L| < 1$.

Therefore, by triangle inequality,

$$\begin{aligned} |\operatorname{sgn}(\delta/2) - \operatorname{sgn}(-\delta/2)| &= |(\operatorname{sgn}(\delta/2) - L) - (\operatorname{sgn}(-\delta/2) - L)| \\ &\leq |\operatorname{sgn}(\delta/2) - L| + |\operatorname{sgn}(-\delta/2) - L| \\ &< 1 + 1 = 2. \end{aligned}$$

On the other hand, using the definition of $\operatorname{sgn}(x)$, we have

$$|\operatorname{sgn}(\delta/2) - \operatorname{sgn}(-\delta/2)| = |1 - (-1)| = 2.$$

This equality contradicts the previous inequality. So $\lim_{x\to 0} \operatorname{sgn}(x)$ does not exist. □

Example 3.1.11

Define $f\colon \mathbb{R} \to \mathbb{R}$ by $f(0) = 0$ and $f(x) = 1/x$ when $x \neq 0$.

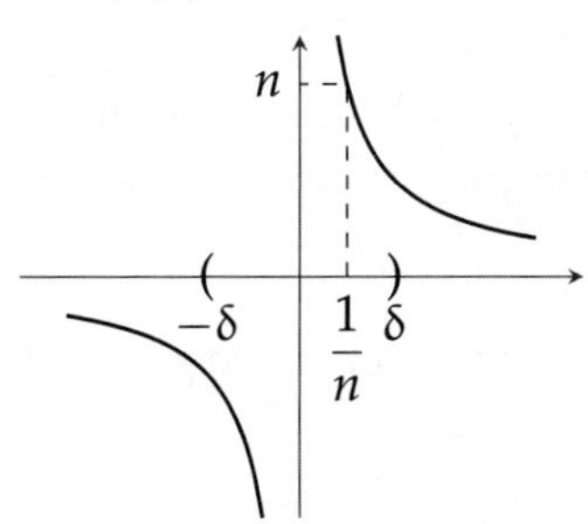

We proceed as in the previous example. Suppose $\lim_{x\to 0} f(x) = L$ and consider $\epsilon = 1/2$. Now consider any $\delta > 0$. By the Archimedean property, the interval $(-\delta, \delta)$ contains points of the form $1/n$ and $1/(n+1)$ with $n \in \mathbb{N}$. Then $f(1/(n+1)) - f(1/n) = 1$ and so it is impossible that both $f(1/(n+1))$ and $f(1/n)$ are within a distance $\epsilon = 1/2$ of L. □

Example 3.1.12

Let $S\colon [-1,1] \to \mathbb{R}$ be defined by $S(1/n) = (-1)^n$ for each $n \in \mathbb{Z} \setminus \{0\}$ and let its graph be a straight line on each interval between these points. Further, let $S(0) = 0$.

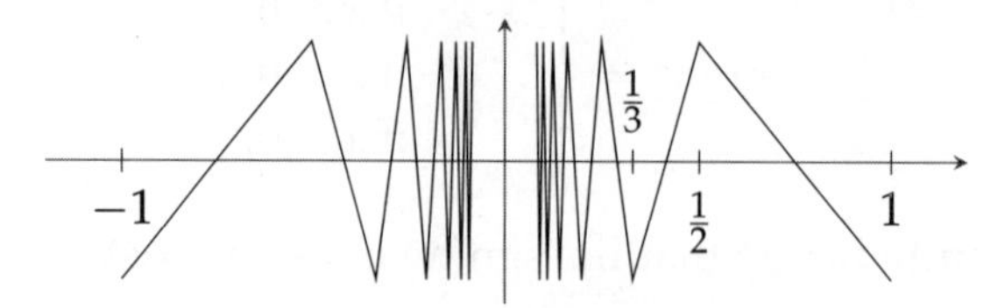

In any $(-\delta,\delta)$ interval, S takes both the values ± 1 and so we can argue as in the previous two examples to show that $\lim_{x\to 0} S(x)$ does not exist. $\square$

Remember our statement that the limit does not have to equal the function's value? Here is an example.

Example 3.1.13

Let $f(x) = 0$ when $x \neq 0$ and $f(0) = 1$. We will show that $\lim_{x\to 0} f(x) = 0$.

Consider any $\epsilon > 0$. Let $\delta = 1$. Then

$$0 < |x-0| < \delta \implies x \neq 0 \implies f(x) = 0 \implies |f(x) - 0| = 0 < \epsilon.$$

Note that the implications are entirely based on $0 < |x|$ and so *any* choice of positive δ would have worked here. $\square$

Limit Theorems

We now take up questions regarding limits of combinations of functions. If we know the limits of two functions at the same point, what can we say about the limits of their sum, product, and other combinations?

We begin by considering the special case when the initial functions have limit zero. The ϵ-δ arguments are much simpler in this situation.

Lemma 3.1.14

Let f,g be real functions such that $\lim_{x\to p} f(x) = \lim_{x\to p} g(x) = 0$. Then

1. $\lim_{x\to p} cf(x) = 0 \quad (c \in \mathbb{R})$,
2. $\lim_{x\to p} (f(x) + g(x)) = 0$,
3. $\lim_{x\to p} f(x)g(x) = 0$,
4. *If* $\lim_{x\to p} h(x) = 1$, *then* $\lim_{x\to p} \dfrac{f(x)}{h(x)} = 0$.

Proof.

1. This is trivial if $c = 0$. So suppose $c \neq 0$. Take any $\epsilon > 0$.

   There is a $\delta > 0$ such that $0 < |x - p| < \delta$ implies $|f(x)| < \epsilon/|c|$.

   Then $0 < |x - p| < \delta$ implies $|cf(x) - 0| = |c||f(x)| < |c|\dfrac{\epsilon}{|c|} = \epsilon$.

2. Take any $\epsilon > 0$.

   First, there is a $\delta_1 > 0$ such that $0 < |x - p| < \delta_1$ implies $|f(x)| < \epsilon/2$.

   Second, there is a $\delta_2 > 0$ such that $0 < |x - p| < \delta_2$ implies $|g(x)| < \epsilon/2$.

   Let $\delta = \min\{\delta_1, \delta_2\}$. Then

$$0 < |x - p| < \delta \implies |f(x) + g(x) - 0| \leq |f(x)| + |g(x)| < \frac{\epsilon}{2} + \frac{\epsilon}{2} = \epsilon.$$

3. Take any $\epsilon > 0$.

   First, there is a $\delta_1 > 0$ such that $0 < |x - p| < \delta_1$ implies $|f(x)| < \sqrt{\epsilon}$.

   Second, there is a $\delta_2 > 0$ such that $0 < |x - p| < \delta_2$ implies $|g(x)| < \sqrt{\epsilon}$.

   Let $\delta = \min\{\delta_1, \delta_2\}$. Then

$$0 < |x - p| < \delta \implies |f(x)g(x)| < \sqrt{\epsilon}\sqrt{\epsilon} = \epsilon.$$

4. Take any $\epsilon > 0$.

   First, there is a $\delta_1 > 0$ such that $0 < |x - p| < \delta_1$ implies $\dfrac{1}{2} < h(x) < \dfrac{3}{2}$.

   Second, there is a $\delta_2 > 0$ such that $0 < |x - p| < \delta_2$ implies $|f(x)| < \dfrac{\epsilon}{2}$.

   Let $\delta = \min\{\delta_1, \delta_2\}$. Then $0 < |x - p| < \delta \implies \left|\dfrac{f(x}{h(x)}\right| < \dfrac{\epsilon/2}{1/2} = \epsilon.$ ∎

Now we take up the general situation. We are able to reduce the calculations to the cases considered in the lemma.

Theorem 3.1.15 (Algebra of Limits)

Let f, g be real functions such that $\lim_{x \to p} f(x) = M$ and $\lim_{x \to p} g(x) = N$. Then

1. $\lim_{x \to p} cf(x) = cM \quad (c \in \mathbb{R})$,
2. $\lim_{x \to p} (f(x) + g(x)) = M + N$,
3. $\lim_{x \to p} (f(x) - g(x)) = M - N$,

4. $\lim_{x\to p} f(x)g(x) = MN$,

5. $\lim_{x\to p} \frac{f(x)}{g(x)} = \frac{M}{N} \quad (N \neq 0)$.

Proof. We use the equivalence $\lim_{x\to p} F(x) = K \iff \lim_{x\to p}(F(x) - K) = 0$.

1. $\lim_{x\to p} \Big(cf(x) - cM \Big) = \lim_{x\to p} c \Big(f(x) - M \Big) = 0$. (By part 1 of Lemma 3.1.14)

2. $\lim_{x\to p} \Big((f(x) + g(x)) - (M + N) \Big) = \lim_{x\to p} \Big((f(x) - M) + (g(x) - N) \Big) = 0.$
(By part 2 of Lemma 3.1.14)

3. Combine parts 1 and 2 of this theorem, using $c = -1$.

4. We use part 3 of Lemma 3.1.14 and parts 1, 2, 3 of this theorem:

$$\begin{aligned}\lim_{x\to p} \Big(f(x)g(x) - MN \Big) &= \lim_{x\to p} \Big((f(x) - M)(g(x) - N) \\ &\quad + Mg(x) + Nf(x) - 2MN \Big) \\ &= \lim_{x\to p} \Big((f(x) - M)(g(x) - N) \Big) \\ &\quad + \lim_{x\to p}(Mg(x)) + \lim_{x\to p}(Nf(x)) - \lim_{x\to p} 2MN \\ &= 0 + MN + NM - 2MN = 0.\end{aligned}$$

5. Due to part 4 of this theorem, it is enough to prove that $\lim_{x\to p} \frac{1}{g(x)} = \frac{1}{N}$.

$$\begin{aligned}\lim_{x\to p} \left(\frac{1}{g(x)} - \frac{1}{N} \right) &= \lim_{x\to p} \frac{N - g(x)}{Ng(x)} = \frac{1}{N^2} \lim_{x\to p} \frac{N - g(x)}{g(x)/N} \\ &= 0. \quad \text{(Part 4 of Lemma 3.1.14)}\end{aligned}$$

■

Let us take up some applications to limit calculations.

Example 3.1.16

Calculate $\lim_{x\to 2}(x^2 + 9)$.

By part 2 of algebra of limits, we have $\lim_{x\to 2}(x^2 + 9) = \lim_{x\to 2} x^2 + \lim_{x\to 2} 9 = \lim_{x\to 2} x^2 + 9$.
By part 4 we have $\lim_{x\to 2} x^2 = (\lim_{x\to 2} x)(\lim_{x\to 2} x) = 2 \cdot 2 = 4$.
Hence, $\lim_{x\to 2}(x^2 + 9) = 4 + 9 = 13$. □

Example 3.1.17

Calculate $\lim_{x\to 2}(7x)^9$.

By part 1 of algebra of limits, we have $\lim_{x\to 2}(7x)^9 = 7^9 \lim_{x\to 2} x^9$.

By part 4 we have $\lim_{x\to 2} x^9 = (\lim_{x\to 2} x)\cdots(\lim_{x\to 2} x) = (\lim_{x\to 2} x)^9 = 2^9$.

Hence, $\lim_{x\to 2}(7x)^9 = 7^9 2^9 = 14^9$. □

Example 3.1.18

Calculate $\lim_{x\to 1} \dfrac{(x-1)^2}{x^2-1}$.

The limit of the denominator is $\lim_{x\to 1}(x^2-1) = \lim_{x\to 1} x^2 - \lim_{x\to 1} 1 = 1^2 - 1 = 0$. So, we cannot apply the rule for ratios. However, we can first simplify the expression and remove this obstacle.

$$\lim_{x\to 1} \frac{(x-1)^2}{x^2-1} = \lim_{x\to 1} \frac{(x-1)^2}{(x-1)(x+1)} = \lim_{x\to 1} \frac{x-1}{x+1}.$$

The cancellation in the last step is allowed because when we calculate $\lim_{x\to 1}$ we work with $x \neq 1$ and hence $x - 1 \neq 0$. This simplified form is easily dealt with:

$$\lim_{x\to 1}(x-1) = 0 \text{ and } \lim_{x\to 1}(x+1) = 2 \implies \lim_{x\to 1} \frac{x-1}{x+1} = \frac{0}{2} = 0.$$ □

Task 3.1.19

Evaluate the following limits:

(a) $\lim_{x\to 2} \dfrac{1}{x^2}$. (b) $\lim_{x\to 3} \dfrac{x^2-6x+9}{x^2-9}$. (c) $\lim_{x\to 0} \dfrac{|x|}{x}$.

Task 3.1.20

Show that if $m \leq f(x) \leq M$ for all x and $\lim_{x\to a} f(x)$ exists, then $m \leq \lim_{x\to a} f(x) \leq M$.

Theorem 3.1.21 (Sandwich or Squeeze Theorem)

Suppose that $f(x) \leq g(x) \leq h(x)$ in an interval $(p-\delta', p+\delta')$, with $\delta' > 0$, except perhaps at p. If $\lim_{x\to p} f(x) = \lim_{x\to p} h(x) = L$, then $\lim_{x\to p} g(x) = L$.

Proof. Let $\epsilon > 0$.

There exists $\delta_f > 0$ such that $0 < |x-p| < \delta_f$ implies $L - \epsilon < f(x) < L + \epsilon$.

There exists $\delta_h > 0$ such that $0 < |x-p| < \delta_h$ implies $L - \epsilon < h(x) < L + \epsilon$.

Let $\delta = \min\{\delta_f, \delta_h, \delta'\}$. Now, if $0 < |x-p| < \delta$, then

- $\delta \le \delta_f \implies L - \epsilon < f(x) < L + \epsilon$,
- $\delta \le \delta_h \implies L - \epsilon < h(x) < L + \epsilon$,
- $\delta \le \delta' \implies f(x) \le g(x) \le h(x)$.

Combining these gives $L - \epsilon < f(x) \le g(x) \le h(x) < L + \epsilon$. Hence, $L - \epsilon < g(x) < L + \epsilon$. Therefore, $\lim_{x \to p} g(x) = L$. ■

Example 3.1.22

The sandwich theorem allows us to calculate a limit without an exact analysis of every part of the expression. For example, suppose we are asked about $\lim_{x \to 0} xS(x)$, where $S(x)$ is the function defined in Example 3.1.12. We have already seen that $\lim_{x \to 0} S(x)$ does not exist. This means that the algebra of limits cannot be applied to the product $xS(x)$. On the other hand, since $S(x)$ takes values between ± 1, it follows that $xS(x)$ takes values between $\pm x$, and this suggests that we take the help of the sandwich theorem.

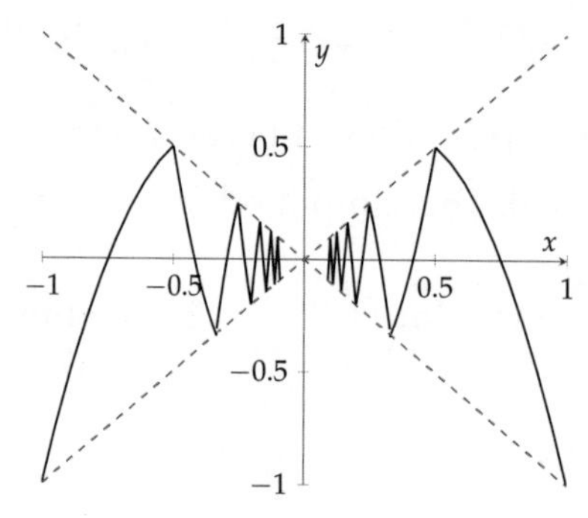

$xS(x)$ lies between $-x$ and x.

In order to avoid the $x > 0$ and $x < 0$ cases we work with $|xS(x)|$.

$$0 \le |S(x)| \le 1 \implies 0 \le |xS(x)| \le |x|.$$

Since $\lim_{x \to 0} |x| = 0$, the sandwich theorem implies that $\lim_{x \to 0} |xS(x)| = 0$. Hence, we have $\lim_{x \to 0} xS(x) = 0$. □

Example 3.1.23

Let $a > 0$ and consider $\lim_{x \to a} \sqrt{x}$. The natural guess for this limit is $\sqrt{a}$. To confirm this, we calculate as follows:

$$0 \le |\sqrt{x} - \sqrt{a}| = \left| \frac{x - a}{\sqrt{x} + \sqrt{a}} \right| \le \frac{|x - a|}{\sqrt{a}}.$$

We have $\lim_{x \to a} \frac{|x - a|}{\sqrt{a}} = 0$. Hence, by the sandwich theorem, $\lim_{x \to a} |\sqrt{x} - \sqrt{a}| = 0$. □

One-sided Limits

We say that $\lim_{x \to p+} f(x) = L$ if for every $\epsilon > 0$ there is a corresponding $\delta > 0$ such that $0 < x - p < \delta \implies |f(x) - L| < \epsilon$. The quantity $\lim_{x \to p+} f(x)$ is called the **right-hand limit** of f at p.

We say that $\lim_{x \to p-} f(x) = L$ if for every $\epsilon > 0$ there is a corresponding $\delta > 0$ such that $0 < p - x < \delta \implies |f(x) - L| < \epsilon$. The quantity $\lim_{x \to p-} f(x)$ is called the **left-hand limit** of f at p.

The right-hand limit at p can be considered if there is an $\alpha > 0$ such that $(p, p + \alpha)$ is in the domain of f. The left-hand limit needs an $\alpha > 0$ such that $(p - \alpha, p)$ is in the domain.

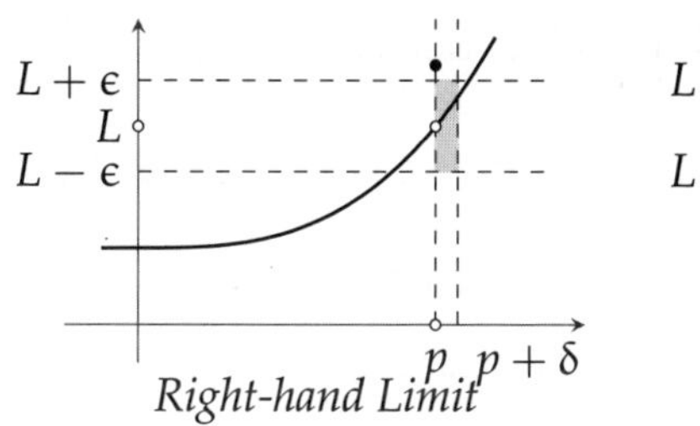

Right-hand Limit

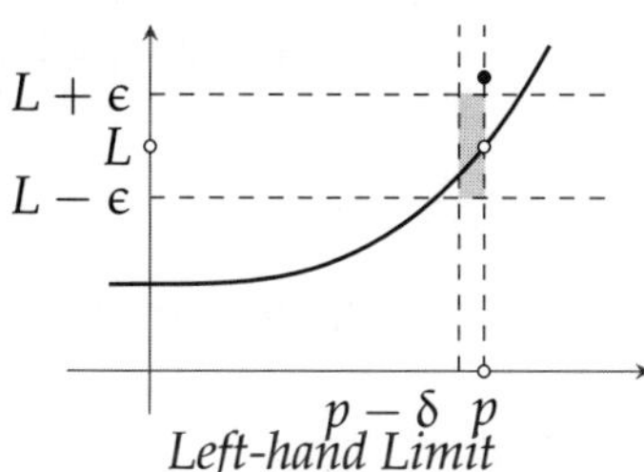

Left-hand Limit

Task 3.1.24

Evaluate the following one-sided limits:

(a) $\lim_{x \to p+} C$.

(b) $\lim_{x \to p-} C$.

(c) $\lim_{x \to 1+} [x]$.

(d) $\lim_{x \to 1-} [x]$.

(e) $\lim_{x \to 0+} \frac{|x|}{x}$.

(f) $\lim_{x \to 0+} \sqrt{x}$.

Theorem 3.1.25

$\lim_{x \to p} f(x) = L$ *if and only if* $\lim_{x \to p+} f(x) = \lim_{x \to p-} f(x) = L$.

Proof. Suppose $\lim_{x \to p} f(x) = L$: Let $\epsilon > 0$. Then there is a $\delta > 0$ such that $0 < |x - p| < \delta \implies |f(x) - L| < \epsilon$. The same δ works for $\lim_{x \to p+} f(x) = L$ and $\lim_{x \to p-} f(x) = L$.

Next, suppose $\lim_{x \to p+} f(x) = \lim_{x \to p-} f(x) = L$. Let $\epsilon > 0$. Then there is a $\delta_1 > 0$ such that $0 < x - p < \delta_1 \implies |f(x) - L| < \epsilon$. There is also a $\delta_2 > 0$ such that $0 < p - x < \delta_2 \implies |f(x) - L| < \epsilon$. Then $\delta = \min\{\delta_1, \delta_2\}$ works for $\lim_{x \to p} f(x) = L$:

$$\begin{aligned} 0 < |x - p| < \delta &\implies 0 < x - p < \delta \text{ or } 0 < p - x < \delta \\ &\implies 0 < x - p < \delta_1 \text{ or } 0 < p - x < \delta_2 \\ &\implies |f(x) - L| < \epsilon. \end{aligned}$$

∎

For many functions, this characterization is a useful way to calculate limits, or to show they do not exist.

Example 3.1.26

Consider the Heaviside step function $H(x) = \begin{cases} 0 & \text{if } x < 0, \\ 1 & \text{if } x \geq 0. \end{cases}$ We calculate the one-sided limits at zero:

$$\lim_{x \to 0+} H(x) = \lim_{x \to 0+} 1 = 1,$$
$$\lim_{x \to 0-} H(x) = \lim_{x \to 0-} 0 = 0.$$

Since the one-sided limits are not equal, $\lim_{x \to 0} H(x)$ does not exist. □

Task 3.1.27

Confirm that the algebra of limits and the sandwich theorem also hold for one-sided limits.

Exercises for § 3.1

1. Suppose $\lim_{x \to p} |f(x)| = L$. Can we conclude that $\lim_{x \to p} f(x) = \pm L$?

2. Consider the function $f(x) = x^3$. For the given a and ϵ, find $\delta > 0$ such that $0 < |x - a| < \delta$ implies $|f(x) - f(a)| < \epsilon$:

(a) $a = 0$, $\epsilon = 1$,

(b) $a = 0$, $\epsilon = 0.1$,

(c) $a = 1$, $\epsilon = 0.1$,

(d) $a = 2$, $\epsilon = 0.1$.

3. Use mathematical induction to prove that if $p(x)$ is a polynomial, then $\lim_{x \to a} p(x) = p(a)$.

4. Compute the limits, explaining which theorem you are using for each step.

(a) $\lim_{x \to 2} \frac{1}{x^2}$.

(b) $\lim_{x \to 2} \frac{x^2 - 4}{x - 2}$.

(c) $\lim_{h \to 0} \frac{(t+h)^2 - t^2}{h}$.

(d) $\lim_{x \to -4} \frac{x^2 + 5x + 4}{x^2 + 3x - 4}$.

(e) $\lim_{x \to 0} \frac{1 - \sqrt{1 - x^2}}{x^2}$.

(f) $\lim_{x \to 1} \frac{x^3 - 1}{x - 1}$.

(g) $\lim_{t \to 0} \left(\frac{1}{t} - \frac{1}{t^2 + t} \right)$.

(h) $\lim_{x \to 7} \frac{\sqrt{x+2} - 3}{x - 7}$.

5. Show the following by means of examples:

(a) $\lim_{x\to a}(f(x)+g(x))$ may exist even when neither $\lim_{x\to a} f(x)$ nor $\lim_{x\to a} g(x)$ exist.

(b) $\lim_{x\to a}(f(x)g(x))$ may exist even when neither $\lim_{x\to a} f(x)$ nor $\lim_{x\to a} g(x)$ exist.

6. Suppose $\lim_{x\to 2}\dfrac{f(x)-5}{x-2}=3$. What can you say about $\lim_{x\to 2} f(x)$?

7. Compute the following limits:

(a) $\lim_{x\to 0} x\exp(x^2)$.

(b) $\lim_{x\to 1}\log x$.

(c) $\lim_{x\to a}\log x$.

(d) $\lim_{x\to 0+} x^2\log x$.

8. Let $n\in\mathbb{N}$. Show that

(a) $\lim_{x\to a} x^{1/n}=a^{1/n}$ if $a>0$,

(b) $\lim_{x\to 0+} x^{1/n}=0$.

9. Compute the following limits:

(a) $\lim_{h\to 0}\dfrac{\log(1+h)}{h}=1$.

(b) $\lim_{h\to 0}\dfrac{\log(x+h)-\log(x)}{h}=\dfrac{1}{x}$.

10. Suppose $\lim_{x\to 0+} f(x)=L$. Show the following:

(a) If f is even, then $\lim_{x\to 0-} f(x)=L$.

(b) If f is odd, then $\lim_{x\to 0-} f(x)=-L$.

11. Suppose $\lim_{x\to a} f(x)=0$ and $g(x)$ is a bounded function defined on an open interval that includes a. Show that $\lim_{x\to a} f(x)g(x)=0$.

3.2 Continuity

A function f is said to be **continuous at** p if

$$\lim_{x\to p} f(x)=f(p).$$

Alternately, f is continuous at p if for every $\epsilon>0$ there is a corresponding $\delta>0$ such that $|x-p|<\delta \implies |f(x)-f(p)|<\epsilon$.

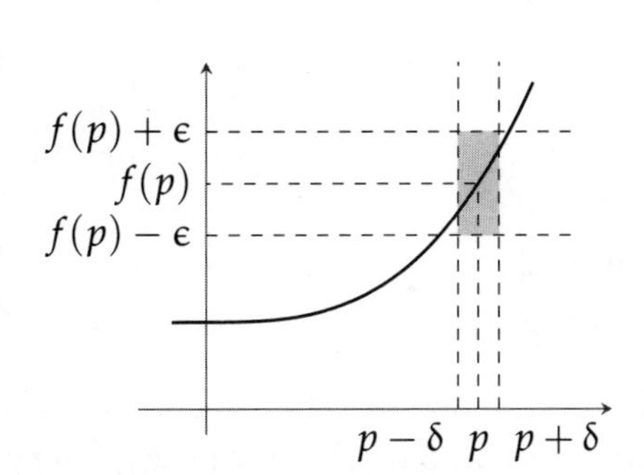

The concept of continuity is only to be applied to points that are in the domain of f. In fact they need to be in an open interval that is completely contained in the domain of f, so that the concept of limit is applicable.

The following functions are not continuous at 0 because their limit does not exist at 0:

$$H(x) = \begin{cases} 0 & \text{if } x < 0, \\ 1 & \text{if } x \geq 0, \end{cases} \quad \text{and} \quad \operatorname{sgn}(x) = \begin{cases} \dfrac{x}{|x|} & \text{if } x \neq 0, \\ 0 & \text{if } x = 0. \end{cases}$$

Next, we have functions that are not continuous at 0 because their limit at 0 exists but does not equal their value at 0:

$$E(x) = \begin{cases} 0 & \text{if } x \neq 0, \\ 1 & \text{if } x = 0, \end{cases} \quad \text{and} \quad F(x) = \begin{cases} x & \text{if } x = 1/n,\ n \in \mathbb{N}, \\ 1 & \text{if } x = 0, \\ 0 & \text{else.} \end{cases}$$

(In both cases the limit is 0 but the function value is 1.)

These functions are continuous at every point of $\mathbb{R}$:

$$f(x) = C, \quad g(x) = x, \quad h(x) = |x|.$$

On the other extreme, the Dirichlet function is not continuous at any point!

Using the algebra of limits we can conclude the following:

Theorem 3.2.1

Let $f(x)$ and $g(x)$ be continuous at p. Then the following functions are also continuous at p:

1. $Cf(x)$.
2. $f(x) \pm g(x)$.
3. $f(x)g(x)$.
4. $\dfrac{f(x)}{g(x)}$, *if* $g(p) \neq 0$.

Proof. We prove the last claim. The others are left as an exercise for the reader.

First note that $\lim_{x \to p} g(x) = g(p) \neq 0$, by continuity of $g(x)$ at $x = p$ and the given condition that $g(p) \neq 0$. So, by the algebra of limits,

$$\lim_{x \to p} \frac{f(x)}{g(x)} = \frac{\lim_{x \to p} f(x)}{\lim_{x \to p} g(x)} = \frac{f(p)}{g(p)}.$$

■

Theorem 3.2.2

Any polynomial is continuous at every point of $\mathbb{R}$.

Proof. Let $a_0, \ldots, a_n \in \mathbb{R}$. The functions $y = a_0$ and $y = x$ are continuous. By repeated application of part 3 of Theorem 3.2.1, every function $y = x^i$ ($i \in \mathbb{N}$) is continuous. By part 1, every function $y = a_i x^i$ is continuous. So by part 2, $\sum_{i=0}^{n} a_i x^i$ is continuous. ■

Recall that a **rational function** has the form $p(x)/q(x)$ where $p(x)$ and $q(x)$ are polynomials and $q(x)$ is not the zero polynomial. The domain of this rational function consists of all real numbers x where $q(x) \neq 0$. Recall that $q(x)$ has only finitely many zeroes. Hence, each point of the domain is the center of an open interval that is contained in the domain, and we can talk about the function's limit at each point in the domain.

Theorem 3.2.3

A rational function is continuous at every point of its domain.

Proof. Combine continuity of polynomials with part 4 of Theorem 3.2.1. ■

One-sided Continuity

A function f is said to be **left-continuous** at p if $\lim_{x \to p-} f(x) = f(p)$. It is called **right-continuous** at p if $\lim_{x \to p+} f(x) = f(p)$.

Example 3.2.4

The greatest integer function is right-continuous at every point. It is left-continuous at all points except the integers.

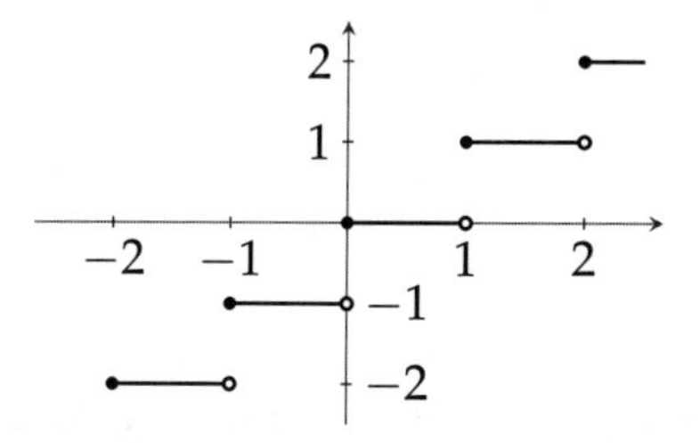

□

Theorem 3.2.5

A function f is continuous at p if and only if it is both left- and right-continuous at p.

Proof. $\lim_{x \to p} f(x) = f(p) \iff \lim_{x \to p+} f(x) = f(p)$ and $\lim_{x \to p-} f(x) = f(p)$. ■

For example, we can argue that the Heaviside step function $H(x)$ is not continuous at $x = 0$ because it is right-continuous but not left-continuous.

Our recognition of different ways in which a discontinuity can happen leads to the following classification:

Removable discontinuity: $\lim_{x \to a} f(x)$ exists but does not equal $f(a)$. We can make f continuous at a by changing its value at a to $\lim_{x \to a} f(x)$.

Jump discontinuity: $\lim_{x \to a+} f(x)$ and $\lim_{x \to a-} f(x)$ exist but are not equal. The quantity $\lim_{x \to a+} f(x) - \lim_{x \to a-} f(x)$ is called the **jump** of f at a.

Essential discontinuity: Either $\lim_{x\to a+} f(x)$ or $\lim_{x\to a-} f(x)$ fail to exist.

A function f is called **continuous on an interval** I if

1. f is continuous at every interior point of I,
2. f is right-continuous at the left endpoint, if the left endpoint is in I, and
3. f is left-continuous at the right endpoint, if the right endpoint is in I.

Continuity of Compositions

Theorem 3.2.6

Let $p \in (a,b)$ and let f and g be real functions such that their composition $g \circ f$ is defined on all of (a,b) except perhaps at p. Let $q = \lim_{x\to p} f(x)$ and suppose g is continuous at q. Then

$$\lim_{x\to p} g(f(x)) = g(q) = g(\lim_{x\to p} f(x)).$$

Proof. Let $\epsilon > 0$. Since g is continuous at q, there is a $\delta' > 0$ such that $|y-q| < \delta'$ implies $|g(y) - g(q)| < \epsilon$. And there is a $\delta > 0$ such that $0 < |x-p| < \delta$ implies $|f(x) - q| < \delta'$. Hence,

$$0 < |x-p| < \delta \implies |f(x) - q| < \delta' \implies |g(f(x)) - g(q)| < \epsilon. \qquad \blacksquare$$

Example 3.2.7

Calculate $\lim_{x\to 1} \sqrt{\dfrac{x^2-1}{x-1}}$.

We first note that $\lim_{x\to 1} \dfrac{x^2-1}{x-1} = 2$. Since the square root function is continuous at 2 (we proved that for $a > 0$, $\lim_{x\to a} \sqrt{x} = \sqrt{a}$), we have

$$\lim_{x\to 1} \sqrt{\frac{x^2-1}{x-1}} = \sqrt{\lim_{x\to 1} \frac{x^2-1}{x-1}} = \sqrt{2}. \qquad \square$$

Theorem 3.2.8

Let f and g be real functions such that their composition $g \circ f$ is defined on an interval (a,b). Let $p \in (a,b)$ such that f is continuous at p and g is continuous at $f(p)$. Then $g \circ f$ is continuous at p.

Proof. $\lim_{x\to p} g(f(x)) = g(\lim_{x\to p} f(x)) = g(f(p)).$ $\blacksquare$

Monotone Functions

Theorem 3.2.9

If I, J are intervals and $f\colon I \to J$ is a surjective monotone function, then f is continuous on I.

Proof. We will focus on the case when J is an open interval. For other intervals we have to do a similar analysis at any end-points that are included in J.

Let $x_0 \in I$ and let $\epsilon > 0$. We may assume that $f(x_0) \pm \epsilon \in J$, by shrinking ϵ if necessary. (A δ that works for a smaller ϵ will also work for the original one.)

The surjectivity of f implies that there are $x_\pm \in I$ such that $f(x_-) = f(x_0) - \epsilon$ and $f(x_+) = f(x_0) + \epsilon$.

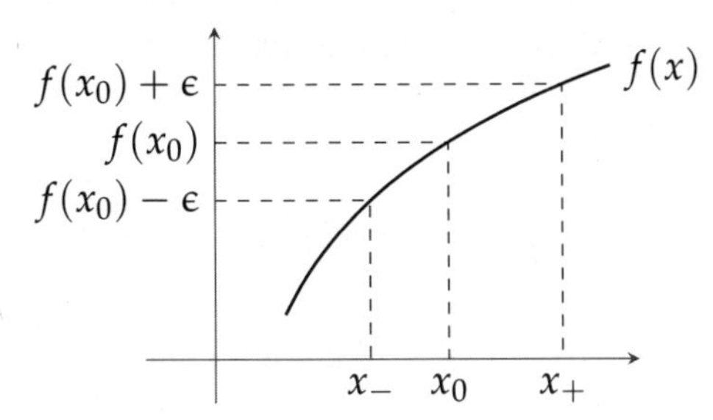

Take $\delta = \min\{x_0 - x_-, x_+ - x_0\}$. ∎

Theorem 3.2.10

All logarithms and exponential functions are continuous.

Proof. They are monotonic bijections between intervals. ∎

Task 3.2.11

Let $r \geq 0$. Show that the function x^r is continuous on $[0, \infty)$.

Indefinite Integrals

Suppose I is an interval and $f\colon I \to \mathbb{R}$ is integrable on every $[\alpha, \beta] \subseteq I$. Fix a point $a \in I$. Then the function

$$F(x) = \int_a^x f(t)\,dt \qquad (x \in I)$$

is called an **indefinite integral** of f.

Example 3.2.12

Calculate the indefinite integral $F(x) = \int_0^x H(t)\,dt$ for the unit step function $H(t)$.

$$x < 0 \implies \int_0^x H(t)\,dt = -\int_x^0 H(t)\,dt = -\int_x^0 0\,dt = 0$$
$$x \geq 0 \implies \int_0^x H(t)\,dt = \int_0^x 1\,dt = x$$

Hence, $F(x) = \begin{cases} 0 & \text{if } x < 0, \\ x & \text{if } x \geq 0. \end{cases}$ □

Task 3.2.13

Show that any two indefinite integrals $F(x) = \int_a^x f(t)\,dt$ and $G(x) = \int_b^x f(t)\,dt$ of the same function, differ only by a constant.

Theorem 3.2.14

Suppose I is an interval, $a \in I$, and $f\colon I \to \mathbb{R}$ is integrable on every $[\alpha, \beta] \subseteq I$. Then the indefinite integral $F(x) = \int_a^x f(t)\,dt$ is continuous on I.

Proof. For any $x, p \in I$ we have

$$F(x) - F(p) = \int_a^x f(t)\,dt - \int_a^p f(t)\,dt = \int_p^x f(t)\,dt.$$

Suppose p is not the right endpoint of I. We shall establish the right-continuity of f at p. First, there is a $\delta > 0$ such that $[p, p+\delta] \subseteq I$. Since f is integrable on $[p, p+\delta]$, it is bounded there. Hence, there is a positive number M such that $-M \leq f(x) \leq M$ for every $x \in [p, p+\delta]$. Now for $p < x < p+\delta$,

$$-M(x-p) = \int_p^x (-M)\,dt \leq \int_p^x f(t)\,dt \leq \int_p^x M\,dt = M(x-p).$$

Therefore,

$$-M(x-p) \leq F(x) - F(p) \leq M(x-p),$$

and so $0 \leq |F(x) - F(p)| \leq M|x-p|$. By the sandwich theorem, we have $\lim\limits_{x \to p+} |F(x) - F(p)| = 0$. Therefore, $\lim\limits_{x \to p+} F(x) = F(p)$.

Similarly, we check that if p is not the left endpoint of I, then f is left-continuous at p. This establishes the continuity of f on I. ■

Exercises for § 3.2

1. Identify the points at which the given functions are *not* continuous. Check one-sided continuity at those points, and also classify the discontinuity as removable, jump, or essential.

(a) $f(x) = x - [x]$.

(b) $g(x) = \operatorname{sgn}(x)\exp(x)$.

(c) $h(x) = S(x)$.

(d) $k(x) = H(x)S(x)$.

(S is the function defined in Example 3.1.12. H is the Heaviside step function.)

2. Compute the following limits:

(a) $\lim_{x \to 1+} \sqrt{\log x}$.

(b) $\lim_{x \to 0} \log \sqrt{x^2+1}$.

(c) $\lim_{h \to 0} (1+h)^{1/h}$.

(d) $\lim_{x \to 0+} x^{x^2}$.

3. Consider $f\colon \mathbb{R} \to \mathbb{R}$ defined by $f(x) = x$ if x is rational and $f(x) = 0$ if x is irrational.

(a) Show that f is continuous only at $x = 0$.

(b) Create a function $g\colon \mathbb{R} \to \mathbb{R}$ that is continuous only at $x = 0$ and $x = 1$.

4. Suppose $f\colon (a,b) \to \mathbb{R}$ is a monotone function. Show that the left and right limits of f exist at every point of (a,b). Thus any discontinuity of f is a jump discontinuity. (Hint: Use the concepts of sup and inf.)

5. Is the following statement correct? "If $f \circ g$ is defined, $\lim_{x \to p} g(x) = q$ and $\lim_{y \to q} f(y) = L$, then $\lim_{x \to p} (f \circ g)(x) = L$."

6. Prove Theorem 3.2.9 for the case $J = (a,b]$.

7. Prove that every monotone and continuous function $f\colon [a,b] \to \mathbb{R}$ has the "intermediate value property": If L lies strictly between $f(a)$ and $f(b)$, then there is a $c \in (a,b)$ such that $f(c) = L$. (Hint: First review the proof of Theorem 2.2.14.)

8. Compute and graph the indefinite integral $\int_0^x f(t)\,dt$ of each $f(t)$ given below:

(a) $\mathrm{sgn}(t)$.

(b) $[t]$.

(c) $H(t)t$.

(d) t^2.

3.3 Intermediate Value Theorem

We have encountered and applied the "intermediate value property." A function is said to have this property if, whenever $f(a) < L < f(b)$, there is a c between a,b with $f(c) = L$. That is, f assumes all intermediate values between any two of its values. In particular, we saw in Theorem 2.2.14 that the indefinite integrals of monotone functions have this property, and this established that the natural logarithm is a surjective function.

Now that we have learned that indefinite integrals are continuous, it is natural to ask if all continuous functions have this property. We start by giving a positive answer when $L = 0$.

Theorem 3.3.1 (Intermediate Value Theorem, ver. 1)

Suppose f is continuous on $[a,b]$ and $f(a)f(b) < 0$. Then there is a number $c \in (a,b)$ such that $f(c) = 0$.

Proof. First, let $a_0 = a$ and $b_0 = b$. Let c_0 be the midpoint of $[a_0, b_0]$. If $f(c_0) = 0$ we have succeeded. If $f(c_0) \neq 0$, define

$$[a_1, b_1] = \begin{cases} [a_0, c_0] & \text{if } f(a_0)f(c_0) < 0, \\ [c_0, b_0] & \text{if } f(b_0)f(c_0) < 0. \end{cases}$$

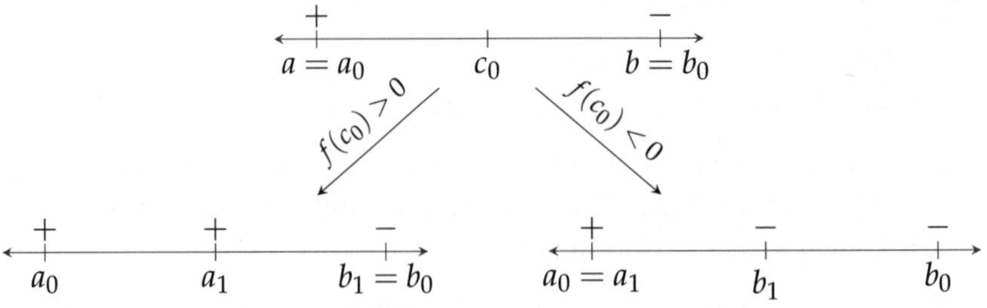

+ −
$a = a_0$ c_0 $b = b_0$
$f(c_0) > 0$ $f(c_0) < 0$
+ + − + − −
a_0 a_1 $b_1 = b_0$ $a_0 = a_1$ b_1 b_0

We have $f(a_1)f(b_1) < 0$, so we repeat this process with $[a_1, b_1]$ replacing $[a_0, b_0]$. Proceeding in this manner, we find a sequence of intervals $[a_n, b_n]$ such that

$$[a_0, b_0] \supset [a_1, b_1] \supset [a_2, b_2] \supset \cdots$$

If we ever come across a midpoint where f is zero, we stop there. If we never encounter such a midpoint, the endpoints will form infinite sequences $a_0, a_1, a_2, \ldots$ and $b_0, b_1, b_2, \ldots$ arranged as follows:

$$a_0 \leq a_1 \leq a_2 \leq \cdots \leq b_2 \leq b_1 \leq b_0.$$

From the completeness axiom we obtain a number c such that $a_n \leq c \leq b_n$ for every n.

Suppose $f(c) > 0$. By continuity, there is a $\delta > 0$ such that $x \in (c - \delta, c + \delta)$ implies $f(x) > 0$.

Now, note that $b_n - a_n = \dfrac{b-a}{2^n} < \dfrac{b-a}{n}$.

Hence, by the Archimedean property, there exists N such that $b_N - a_N < \delta$. Since $c \in [a_N, b_N]$, this implies $[a_N, b_N] \subset (c - \delta, c + \delta)$.

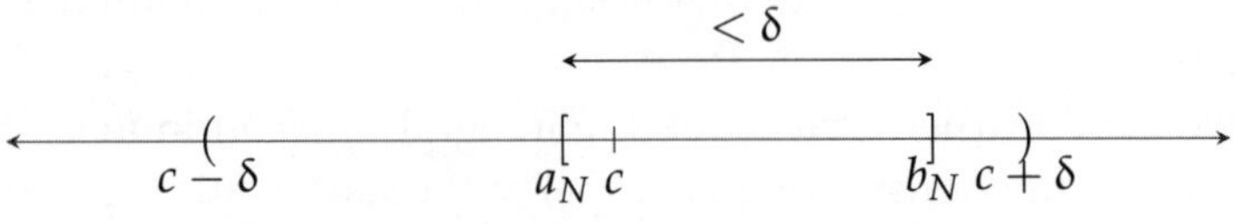

$< \delta$
$c - \delta$ a_N c b_N $c + \delta$

We have a contradiction since f changes sign on $[a_N, b_N]$ but not on $(c - \delta, c + \delta)$.

The $f(c) < 0$ case similarly leads to a contradiction. Hence, $f(c) = 0$. ■

The intermediate value theorem is very useful for showing the existence of special numbers. For example, suppose we wish to show that a certain equation has a solution. By moving all terms to one side of the equality, we put it in the form $f(x) = 0$. If f is continuous, we can try to use the intermediate value theorem.

Example 3.3.2

Consider the equation $x^4 + 4x^3 + x^2 - 6x - 1 = 0$. Since the LHS is a polynomial of degree 4, this equation has atmost 4 distinct real solutions, but it may have fewer, or even none. Let us see how many the intermediate value theorem can help us to locate. We start by calculating the values of $f(x) = x^4 + 4x^3 + x^2 - 6x - 1 = 0$ at various points.

x	-4	-3	-2	-1	0	1	2
$f(x)$	39	-1	-1	3	-1	-1	39

By tracking the sign changes of $f(x)$ we see there are solutions in the intervals $(-4,-3)$, $(-2,-1)$, $(-1,0)$, and $(1,2)$. We can shrink these intervals further by employing the **bisection method**. For example, let us consider the solution that lies in $(1,2)$. We find the value of $f(x)$ at the midpoint of $(1,2)$: $f(1.5) = 10.8$. Therefore, the solution is in $(1,1.5)$. This process can be repeated indefinitely for greater accuracy.

$f(1.25) = 3.3 \implies$ solution is in $(1,1.25)$.

$f(1.125) = 0.81 \implies$ solution is in $(1,1.125)$.

$f(1.0625) = -0.17 \implies$ solution is in $(1.0625,1.125)$.

If we take the next midpoint 1.094 to be an approximate solution, we know it is accurate to within about ± 0.03. □

The intermediate value theorem can be proved for arbitrary L by just vertically shifting the function.

Theorem 3.3.3 (Intermediate Value Theorem, ver. 2)

Suppose f is continuous on $[a,b]$ and L is a value between $f(a)$ and $f(b)$, that is, $f(a) < L < f(b)$ or $f(b) < L < f(a)$. Then there is a number $c \in (a,b)$ such that $f(c) = L$.

Proof. Suppose $f(a) < L < f(b)$. Define $g\colon [a,b] \to \mathbb{R}$ by $g(x) = f(x) - L$. Then $g(a) = f(a) - L < 0$ and $g(b) = f(b) - L > 0$. Hence, there is a number $c \in (a,b)$ such that $g(c) = 0$, and $f(c) = g(c) + L = L$.

The case $f(b) < L < f(a)$ is dealt with in a similar manner. ■

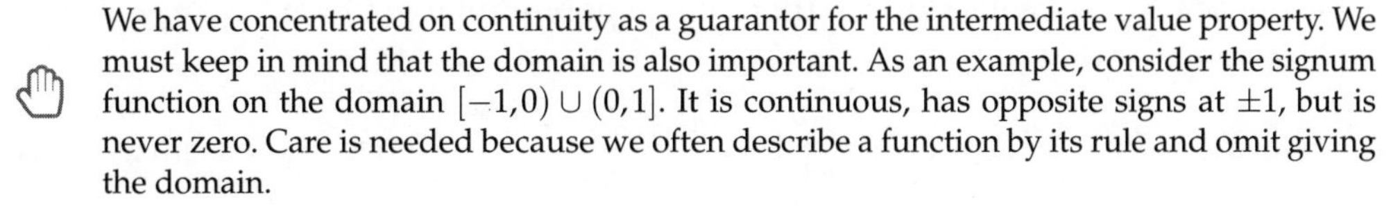

We have concentrated on continuity as a guarantor for the intermediate value property. We must keep in mind that the domain is also important. As an example, consider the signum function on the domain $[-1,0) \cup (0,1]$. It is continuous, has opposite signs at ± 1, but is never zero. Care is needed because we often describe a function by its rule and omit giving the domain.

Exercises for § 3.3

1. Use the intermediate value theorem to find three disjoint intervals, each of which contains a solution of $4x^3 - 6x^2 - 6x + 2 = 0$.

2. Is there a number that is one more than its cube?

3. Use the bisection method to approximate $\sqrt[3]{7}$ to within two decimal places.

4. Fill in the details of this alternate proof of the intermediate value theorem.

(a) Assuming $f(a) < 0 < f(b)$, let $A = \{x \in [a,b] : f(x) < 0\}$. Show that $c = \sup(A)$ exists.

(b) Show that $f(c) > 0$ and $f(c) < 0$ lead to contradictions.

What are the relative merits and demerits of this proof and the original one?

5. Consider the function S of Example 3.1.12, which we know is not continuous. Show that it has the intermediate value property.

6. Can there be a non-constant continuous function $f\colon \mathbb{R} \to \mathbb{Q}$?

7. Let $f\colon [a,b] \to \mathbb{R}$ be continuous and $c_1,\ldots,c_n \in [a,b]$. Then there is a point $c \in [a,b]$ such that $f(c) = \dfrac{f(c_1) + \cdots + f(c_n)}{n}$.

8. Prove the following **fixed point theorem**: If $f\colon [0,1] \to [0,1]$ is continuous then there is a $c \in [0,1]$ such that $f(c) = c$.

9. Suppose $f,g\colon [a,b] \to \mathbb{R}$ are continuous functions such that $f(a) > g(a)$ and $f(b) < g(b)$. Show that there is a $c \in (a,b)$ such that $f(c) = g(c)$.

10. Suppose $f\colon [0,2] \to \mathbb{R}$ is a continuous function with $f(0) = f(2)$. Show that there are $a,b \in [0,2]$ such that $b - a = 1$ and $f(a) = f(b)$.

11. Let $f\colon [a,b] \to \mathbb{R}$ be a continuous and injective function. Assume that $f(a) < f(b)$.

(a) Show that $f(a)$ is the minimum value of f and $f(b)$ is the maximum value of f. Hence, the image of f is $[f(a), f(b)]$.

(b) Show that f is strictly increasing.

(c) Show that $f\colon [a,b] \to [f(a), f(b)]$ has an inverse function that is also strictly increasing and continuous.

12. Show that there cannot be a continuous bijection $f\colon (0,1) \to [0,1]$.

13. The following tasks will establish that a cubic polynomial $p(x) = x^3 + ax^2 + bx + c$ has at least one real root.

(a) Show that for $x \geq 1$, $p(x) \geq x^3 \left(1 - \dfrac{|a| + |b| + |c|}{x}\right)$. Hence, there is an x_1 with $p(x_1) > 0$.

(b) Consider $q(x) = x^3 - ax^2 + bx - c$ and show there is an x_2 with $p(x_2) < 0$.

14. Prove that every polynomial of odd degree has at least one real root. Hence, the image of such a function is all of $\mathbb{R}$.

15. Prove that if a polynomial $p(x) = x^n + a_{n-1}x^{n-1} + \cdots + a_0$ of even degree has a negative value, then it has a real root.

16. Suppose $f\colon [-1,1] \to [-1,1]$ is continuous and satisfies $x^2 + f(x)^2 = 1$ for every x. Prove that either $f(x) = \sqrt{1-x^2}$ for every x or $f(x) = -\sqrt{1-x^2}$ for every x.

3.4 Trigonometric Functions

Let us review the geometric definitions of $\cos t$ and $\sin t$ with $t \in \mathbb{R}$. For this, we first need to describe angles and their measurement. We define an angle to be a region bounded by two rays with a common starting point, as shown below:

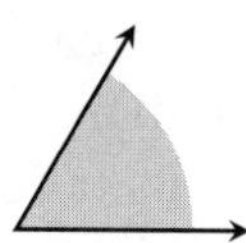

To measure the angle we draw a unit circle whose center is the meeting point of the rays. We take twice the area enclosed by this circle within the angle, and call that the radian measure of the angle. Thus the full circle corresponds to 2π radians while a right angle corresponds to $\pi/2$ radians. (See the discussion of π on page 74.)

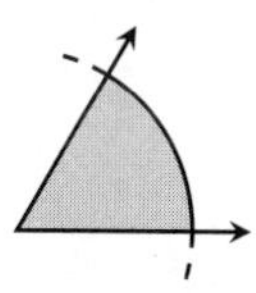

The usual definition of radian is to take the length of the arc of unit radius cut by the angle. However, we have not taken up lengths of curves yet and have to work with areas. We use twice the area to keep our definition compatible with the arc length approach. The association of radians with arc length is achieved later in Example 6.4.8.

At this point, we have associated a real number between 0 and 2π to each angle. We would like to be certain that every such number is the radian measure of some angle. Then we shall have a perfect identification of physical angles with radian measures.

First consider any number x between -1 and 1. We create an angle corresponding to x as shown below:

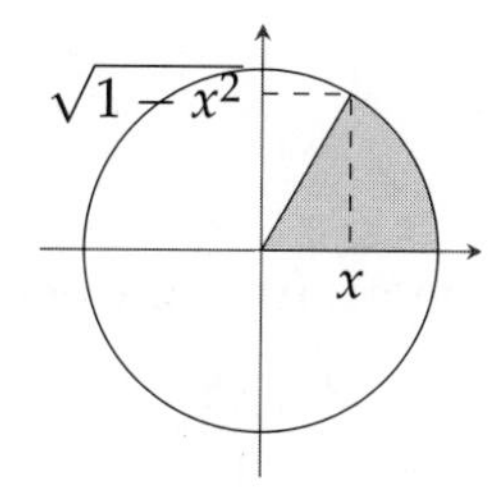

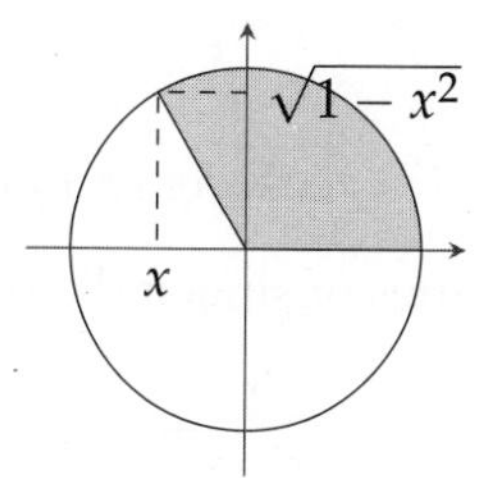

Let $R(x)$ be the radian measure of this angle. We now have a function $R\colon [-1.1] \to [0,\pi]$ defined by

$$R(x) = x\sqrt{1-x^2} + 2\int_x^1 \sqrt{1-t^2}\,dt.$$

Now R is continuous, $R(-1) = \pi$ and $R(1) = 0$. By the intermediate value theorem, R takes every value between 0 and π. Thus every number between 0 and π is the radian measure of an angle.

Task 3.4.1

Show that every number between π and 2π is the radian measure of an angle.

Now that we know how to identify angles with real numbers, we are in a position to define the trigonometric functions.

Consider the ray in the xy-plane created by rotating the positive x-axis counterclockwise through an angle of t radians. This ray cuts the unit circle with center at origin at exactly one point (x,y). We then define $\cos t = x$ and $\sin t = y$ (cos is an abbreviation of "cosine" while sin is an abbreviation of "sine").

The figures below illustrate the definitions for an acute and an obtuse angle respectively:

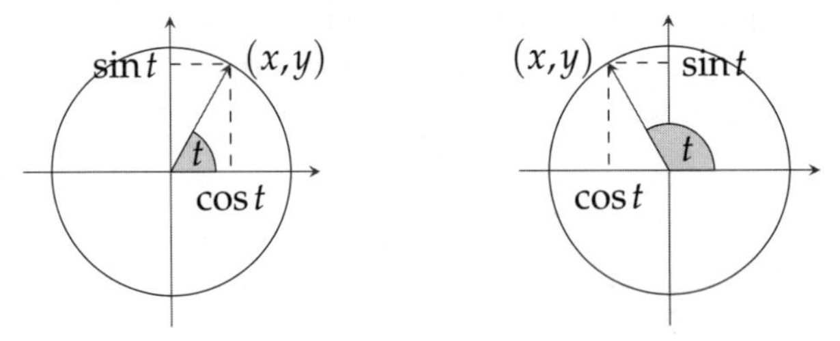

It follows from this definition that $\sin, \cos : [0,2\pi] \to [-1,1]$ are onto. For, let $-1 \leq y \leq 1$. Then $1 - y^2 \geq 0$ and $x = \sqrt{1-y^2}$ is defined. Note that $x^2 + y^2 = 1$ and hence (x,y) is on the unit circle with center at origin. Let t be the angle between the positive x-axis and the ray emanating from origin and passing through (x,y). Then, by the definition of the sine function, $\sin t = y$. Therefore, sine is onto. Similarly, cosine is also onto.

Task 3.4.2

Show that $\sin^2 t + \cos^2 t = 1$ *for every* $t \in [0,2\pi]$.

Task 3.4.3

Show that $\sin(\pi/2 - t) = \cos t$ *for every* $t \in [0,\pi/2]$.

The following values of sine and cosine are obvious from the definitions:

x	0	$\pi/2$	π	$3\pi/2$	2π
$\sin x$	0	1	0	-1	0
$\cos x$	1	0	-1	0	1

Task 3.4.4

Show that $\sin(\pi/4) = \cos(\pi/4) = 1/\sqrt{2}$.

These observations indicate that the graph of cosine over the interval $[0,\pi/2]$ is likely to be as follows:

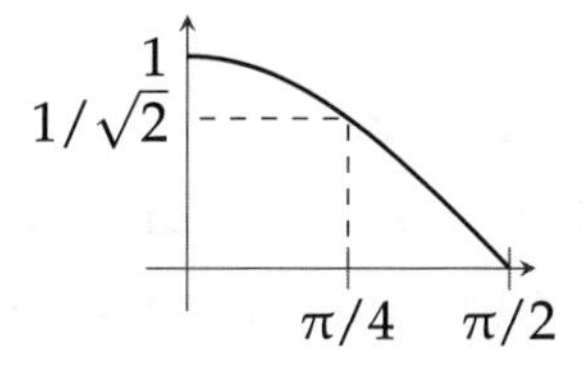

As we learn more calculus, we will be able to confirm that the graph indeed looks like this. The following identities also follow directly from the definitions:

$$\cos(\pi - t) = \cos(\pi + t) = -\cos t, \text{ for every } t \in [0,\pi].$$

With their help we can visualize the graph over the entire interval $[0.2\pi]$, using the piece for $[0,\pi/2]$ as the building block.

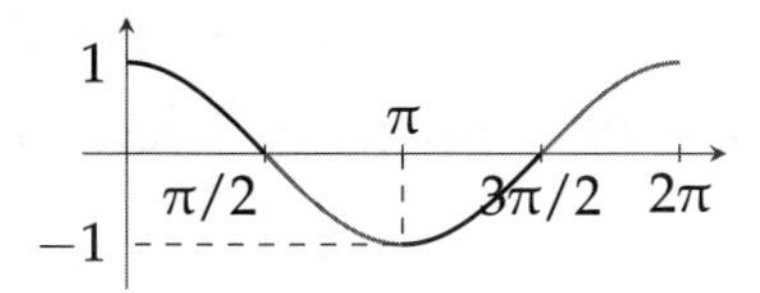

We have similar identities for the sine function:

$$\sin(\pi - t) = -\sin(\pi + t) = \sin t.$$

These generate the graph of the sine function:

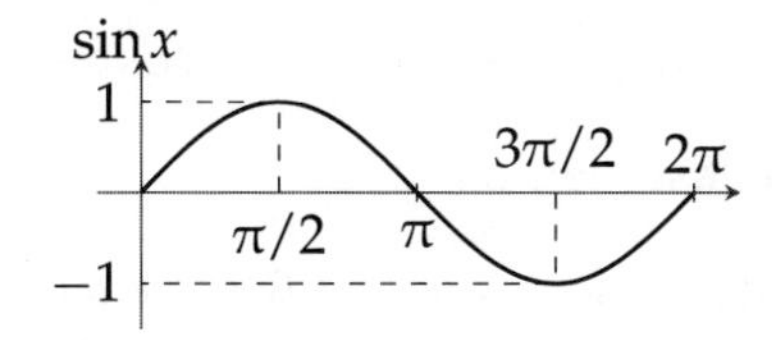

We notice that as the input changes from 0 to 2π, the sine and cosine functions return to their initial values. Thus the function domain can be extended on each side by just repeating the function values, using $\sin(x + 2\pi) = \sin x$ and $\cos(x + 2\pi) = \cos x$:

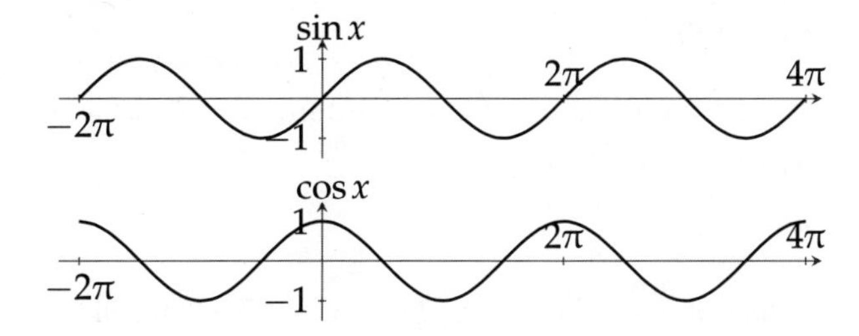

The following properties of $\sin, \cos \colon \mathbb{R} \to [-1,1]$ are obvious from the definitions:

1. The sine and cosine functions are **periodic**, with a period of 2π: $\sin(t+2\pi) = \sin t$ and $\cos(t+2\pi) = \cos t$ for every t.
2. We have $\sin^2 t + \cos^2 t = 1$ for every $t \in \mathbb{R}$.
3. The sine function is odd while the cosine function is even.

We need one last property of trigonometric functions, which will be essential in later work. This is the set of identities involving sums and differences of angles.

In this figure, CP and QR are perpendicular to OA, while CQ is perpendicular to OB. We have the following calculations:

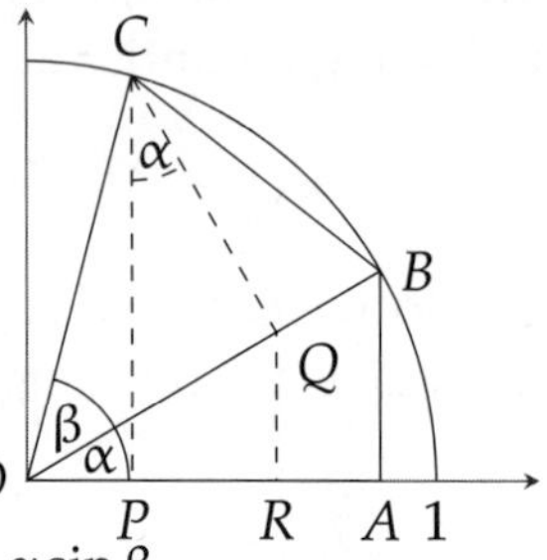

$$OQ = \cos\beta \implies OR = \cos\alpha\cos\beta,$$
$$CQ = \sin\beta \implies PR = \sin\alpha\sin\beta.$$

Therefore,

$$\cos(\alpha+\beta) = OP = OR - PR = \cos\alpha\cos\beta - \sin\alpha\sin\beta.$$

Our figure is only valid for $0 \le \alpha, \beta$ and with $\alpha + \beta \le \pi/2$. The identity can be extended to arbitrary α, β by other appropriate figures.

The other sum of angle identities can be obtained from this one. First, replacing β by $-\beta$ gives

$$\cos(\alpha-\beta) = \cos\alpha\cos\beta + \sin\alpha\sin\beta.$$

Alternately, replacing α by $\pi/2 - \alpha$ and β by $-\beta$ gives

$$\sin(\alpha+\beta) = \sin\alpha\cos\beta + \cos\alpha\sin\beta.$$

Substituting β by $-\beta$ in the last identity gives

$$\sin(\alpha-\beta) = \sin\alpha\cos\beta - \cos\alpha\sin\beta.$$

Task 3.4.5

Show that $\sin\pi/6 = \cos\pi/3 = 1/2$ *and* $\cos\pi/6 = \sin\pi/3 = \sqrt{3}/2$.

Task 3.4.6

Prove the ***half-angle formulas****:*

$$\cos 2x = \cos^2 x - \sin^2 x = 1 - 2\sin^2 x = 2\cos^2 x - 1,$$
$$\sin 2x = 2\cos x \sin x.$$

Task 3.4.7

Compute numerical values of $\dfrac{\sin x}{x}$ *for* $x = \pi/2^n$, $n = 1,2,3$. *(You can use a calculator for the arithmetic operations and square roots, but do not use the inbuilt sine and cosine functions.)*

Theorem 3.4.8 (Law of Sines)

Consider a triangle whose sides have lengths a, b, c, and the corresponding opposite angles are α, β, γ. Then

$$\frac{\sin\alpha}{a} = \frac{\sin\beta}{b} = \frac{\sin\gamma}{c}.$$

Proof. Take the side a as the base of the triangle and let h be the height.

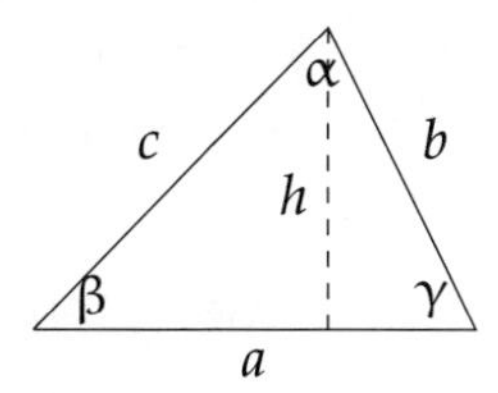

We have $h = c\sin\beta = b\sin\gamma$, hence $\dfrac{\sin\beta}{b} = \dfrac{\sin\gamma}{c}$.

Similarly, taking b as the base, we get $\dfrac{\sin\gamma}{c} = \dfrac{\sin\alpha}{a}$. ■

Theorem 3.4.9 (Law of Cosines)

Consider a triangle whose sides have lengths a, b, c, and the corresponding opposite angles are α, β, γ. Then

$$c^2 = a^2 + b^2 - 2ab\cos\gamma.$$

Proof. Take the side a as the base of the triangle and drop a perpendicular from the opposite vertex. We show two cases below, depending on whether one of the base angles is obtuse:

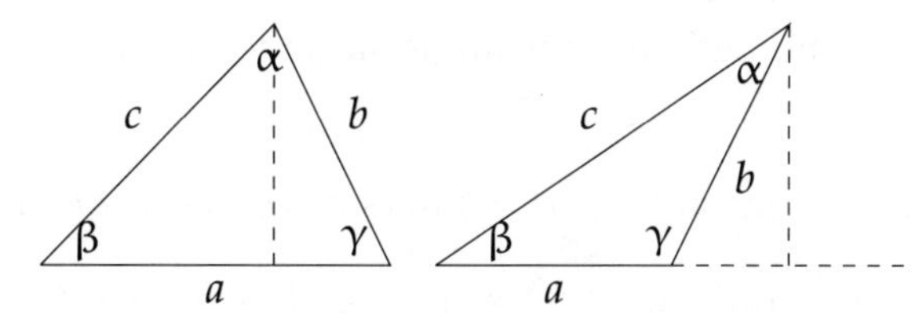

We have $a = c\cos\beta + b\cos\gamma$ in both cases. Hence, $a^2 = ac\cos\beta + ab\cos\gamma$. Similarly,

$$b^2 = ab\cos\gamma + bc\cos\alpha,$$
$$c^2 = ac\cos\beta + bc\cos\alpha.$$

Hence,

$$a^2 + b^2 = 2ab\cos\gamma + ac\cos\beta + bc\cos\alpha = c^2 + 2ab\cos\gamma.$$ ■

Limits and Continuity

Now we will verify that the trigonometric functions are continuous. We compute their limits at zero, and then use the identities for $\sin(x+y)$ and $\cos(x+y)$ to transfer these calculations to other points.

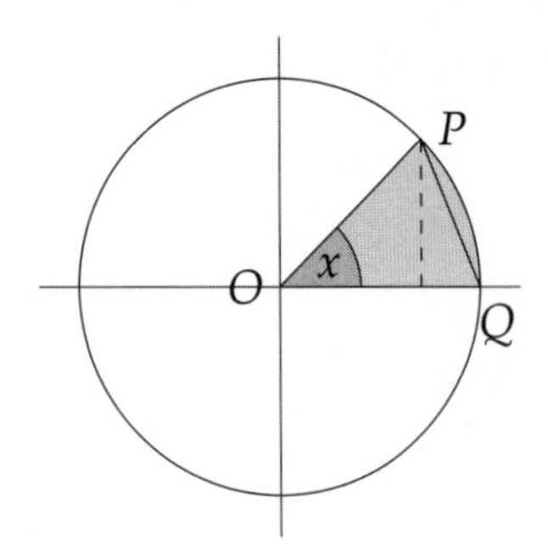

In the figure on the left, the circle has radius 1. Consider the areas of the triangle $\triangle OPQ$ and the sector $\triangledown OPQ$. The triangle is fully contained in the sector, hence has smaller area. So for $0 < x < \pi/2$ we have

$$\begin{aligned}
& 0 < \text{Area}(\triangle OPQ) < \text{Area}(\triangledown OPQ) \\
\Longrightarrow\ & 0 < \frac{1}{2}\sin x < \frac{x}{2} \\
\Longrightarrow\ & 0 < \sin x < x.
\end{aligned}$$

Applying the sandwich theorem gives $\lim\limits_{x \to 0+} \sin x = 0$. Since $\sin x$ is an odd function, we get

$$\lim_{x \to 0-} \sin x = -\lim_{x \to 0+} \sin x = 0.$$

Both the one-sided limits being 0, we have $\lim\limits_{x \to 0} \sin x = 0$.

Now use the half-angle formula:

$$\lim_{x \to 0} \cos x = \lim_{x \to 0} \left(1 - 2\sin^2 \frac{x}{2}\right) = 1.$$

The limits at 0 can be combined with the angle sum identities to compute the limits at other points:

$$\begin{aligned}
\lim_{x \to a} \sin x &= \lim_{h \to 0} \sin(a+h) = \lim_{h \to 0} (\sin a \cos h + \cos a \sin h) = \sin a, \\
\lim_{x \to a} \cos x &= \lim_{h \to 0} \cos(a+h) = \lim_{h \to 0} (\cos a \cos h - \sin a \sin h) = \cos a.
\end{aligned}$$

Thus the sine and cosine functions are continuous on $\mathbb{R}$.

Task 3.4.10

Calculate $\lim\limits_{x \to 1} \sin\left(\dfrac{x^2 - 2x + 1}{x^2 - 1}\right)$.

Let us now recall the other four trigonometric functions:

$$\tan x = \frac{\sin x}{\cos x}, \qquad \cot x = \frac{\cos x}{\sin x}, \qquad \sec x = \frac{1}{\cos x}, \qquad \csc x = \frac{1}{\sin x}.$$

By the properties of continuity, these functions are continuous at every point of their domains.

Two Fundamental Limits

We shall derive two limits that are especially important in calculus. One of their contributions will be to enable the differentiation of the sine and cosine functions in the next chapter.

We have seen that $0 < x < \pi/2 \implies \sin x < x$. Hence,

$$0 < x < \pi/2 \implies \frac{\sin x}{x} < 1.$$

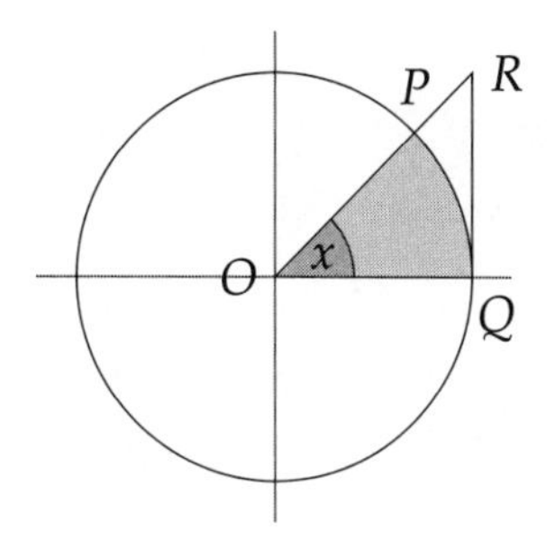

Now consider the unit circle drawn on the left, and compare the areas of the sector $\triangledown OPQ$ and the triangle $\triangle ORQ$. The sector has less area than the triangle, and so

$$\frac{x}{2} < \frac{1}{2}\tan x, \text{ or } \cos x < \frac{\sin x}{x}.$$

Hence, for $0 < x < \pi/2$, we have

$$\cos x < \frac{\sin x}{x} < 1.$$

Therefore, by the sandwich theorem,

$$\lim_{x\to 0^+} \frac{\sin x}{x} = 1.$$

Since $(\sin x)/x$ is an even function, we also get $\lim_{x\to 0^-} \frac{\sin x}{x} = 1$. We have reached

$$\lim_{x\to 0} \frac{\sin x}{x} = 1 \tag{3.1}$$

Another important limit is of $(1 - \cos x)/x$ at 0. We again use the half-angle formula:

$$\lim_{x\to 0} \frac{1-\cos x}{x} = \lim_{x\to 0} \frac{2\sin^2(x/2)}{x} = \lim_{x\to 0} \sin(x/2) \lim_{x\to 0} \frac{\sin(x/2)}{x/2} = 0 \cdot 1 = 0. \tag{3.2}$$

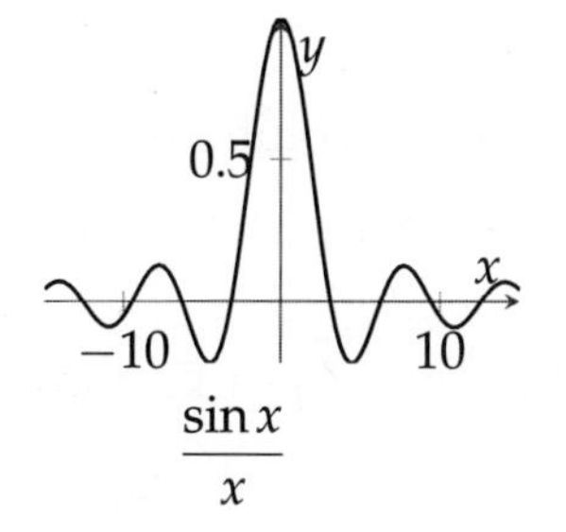

$\dfrac{\sin x}{x}$

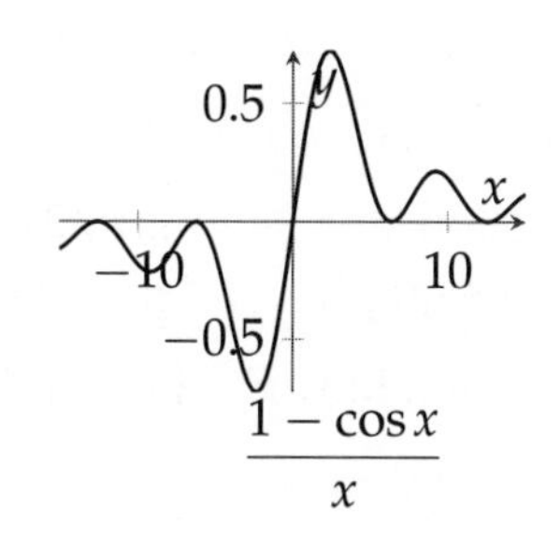

$\dfrac{1-\cos x}{x}$

Task 3.4.11

Calculate $\lim_{x\to 1} \frac{\sin(x^2-1)}{x-1}$.

At this point, all the standard continuous functions of calculus are available to us: polynomials, rational functions, roots, real powers, exponential, logarithm, sine, cosine.

Exercises for § 3.4

1. Graph the following functions:

(a) $f(x) = \sin(x-1)$.

(b) $g(x) = \sin 2x$.

(c) $h(x) = \sin|x|$.

(d) $k(x) = \sin(x^2)$.

2. Graph the following functions:

(a) $p(x) = |\sin x|$.

(b) $q(x) = \sin^2 x$.

(c) $r(x) = \sin x + \cos x$.

(d) $s(x) = [\sin x]$.

3. Compute the following limits:

(a) $\lim_{x\to 0} \frac{\sin 5x - \sin 3x}{x}$.

(b) $\lim_{x\to a} \frac{\sin x - \sin a}{x-a}$.

(c) $\lim_{x\to 0} \frac{\tan 2x}{\sin x}$.

(d) $\lim_{x\to \pi/2} (\sec x - \tan x)$.

4. Compute the following limits:

(a) $\lim_{x\to 0} \frac{1-\cos x}{x^2}$.

(b) $\lim_{x\to 0} \frac{1-\cos(1-\cos x)}{x^4}$.

(c) $\lim_{x\to 0} x \sin\frac{1}{x}$.

(d) $\lim_{x\to 1} \sin\sqrt{\frac{x^2-1}{x-1}}$.

5. Graph the given functions and identify where they are continuous:

(a) $f(x) = \sin(1/x)$ if $x \neq 0$ and $f(0) = 0$.

(b) $g(x) = x\sin(1/x)$ if $x \neq 0$ and $g(0) = 0$.

6. Show that the equation $\cos x = x^3$ has at least one solution.

7. Consider the function $\tan\colon (-\pi/2, \pi/2) \to \mathbb{R}$. Show that it is odd, strictly increasing, and surjective. Plot its graph.

8. Consider the function $\tan\colon \mathbb{R} \setminus \{\, (2n+1)\frac{\pi}{2} \mid n \in \mathbb{Z} \,\} \to \mathbb{R}$. Show that it has period π. Plot its graph.

9. Find the domains and plot the graphs of the functions $\cot x$, $\sec x$, $\csc x$.

10. Prove that a linear combination $A\sin x + B\cos x$ can be expressed in the form $R\sin(x+\phi)$ for some R,ϕ. (Hint: First do the case when $A^2+B^2=1$.)

11. Prove the following identities:

(a) $\cos x + \cos y = 2\cos\left(\frac{x+y}{2}\right)\cos\left(\frac{x-y}{2}\right)$.

(b) $\sin x + \sin y = 2\sin\left(\frac{x+y}{2}\right)\cos\left(\frac{x-y}{2}\right)$.

12. Consider the following identities, which we have already proved:

$$\sin^2(\pi/2-\theta) = 1-\sin^2\theta \quad \text{and} \quad \sin^2\theta = \frac{1-\sin(\pi/2-2\theta)}{2}.$$

(a) Starting with the known values of $\sin\theta$ for $\theta = \pi/6, \pi/4$, and $\pi/3$, use these identities to find the sine values for $\theta = \pi/12$ and $5\pi/12$.

(b) Continue the above process to find the sine values for θ starting at zero and increasing in steps of $\pi/24$ to $\pi/2$.

(This process is described in the work *Pancha-Siddhantika* by Varahamihira, written in the sixth century CE. Varahamihira went one step further and calculated in steps of $\pi/48$ or $3°45'$.)

13. We will develop a rational function that is a close approximation to $\sin x$ over the interval $[0,\pi]$.

(a) Find a quadratic polynomial p such that $p(x) = \sin x$ for $x = 0, \frac{\pi}{2}, \pi$.

(b) Find a quadratic polynomial q that is symmetric about $\pi/2$ and satisfies $q(x) = \frac{p(x)}{\sin x}$ for $x = 0, \pi/6$.

(c) Plot $\sin x$ and $r(x) = p(x)/q(x)$ over $[0,\pi]$ using a graphing software.

(The approximation $r(x)$ was first developed by Bhaskara in the seventh century CE, of course using degrees rather than radians.)

3.5 Continuity and Variation

We have defined continuity point-by-point, thus it is a "local" property of a function. The question arises as to whether functions that are continuous at every point of their domain will have some special characteristics. That is, whether the local property will have global consequences. The intermediate value theorem is an example of such a global consequence.

In this section, we shall establish three other fundamental properties of continuous functions, which occur when the domain is also special – a closed and bounded interval $[a,b]$. The first two relate to the extreme values of the function. First, we shall observe that the range of the function must be bounded. Second, that it will achieve maximum and minimum values. The third relates to how quickly a continuous function changes values. It says that, given any ϵ, we can partition the domain $[a,b]$ so that over each interval of the partition the extreme values of the function are within ϵ of each other. This will be useful to us when we study the integrability of continuous functions in the next section.

We will prove all three results using the same technique that worked for the intermediate value theorem. We successively halve the intervals until they are small enough for continuity to pay off. We hope that the use of similar proofs will make their assimilation easier for the student.

Theorem 3.5.1 (Boundedness Theorem)

Let $f\colon [a,b] \to \mathbb{R}$ be continuous. Then the image of f is bounded.

Proof. We shall carry out a proof by contradiction. Let us assume that f is unbounded on $[a,b]$.

Let $a_1 = a$, $b_1 = b$ and define $c_1 = (a_1 + b_1)/2$. Then f is unbounded on at least one of the intervals $[a_1, c_1]$ and $[c_1, b_1]$. Let that one be called $[a_2, b_2]$. Applying the same process to $[a_2, b_2]$, and so on, we get a sequence of intervals

$$[a,b] = [a_1,b_1] \supset [a_2,b_2] \supset \cdots \supset [a_n,b_n] \supset \cdots,$$

so that f is unbounded on each of them, and with $b_n - a_n = (b-a)/2^{n-1}$. Note that

$$a = a_1 \le a_2 \le \cdots \le b_2 \le b_1 = b.$$

By the completeness axiom, there is an $\alpha \in \mathbb{R}$ such that $a_i \le \alpha \le b_i$ for every i. Hence, α is in each $[a_n, b_n]$. Since f is continuous at α, there is a $\delta > 0$ such that $|x - \alpha| < \delta$ implies $|f(x) - f(\alpha)| < 1$. In particular, f is bounded on $(\alpha - \delta, \alpha + \delta)$.

Since the length of the intervals $[a_n, b_n]$ is halved at each stage, for large enough n we will have $[a_n, b_n] \subset (\alpha - \delta, \alpha + \delta)$, implying that f is bounded on $[a_n, b_n]$. This is a contradiction. ∎

Theorem 3.5.2 (Extreme Value Theorem)

Let $f\colon [a,b] \to \mathbb{R}$ be continuous. Then there are points $c, d \in [a,b]$ such that

$$f(c) = \max\{f(x) \mid x \in [a,b]\} \quad \text{and} \quad f(d) = \min\{f(x) \mid x \in [a,b]\}.$$

Proof. We will prove the existence of c. We shall use $\sup_I f$ to denote $\sup\{f(x) \mid x \in I\}$. By the boundedness theorem, $M = \sup_{[a,b]} f$ exists. We have to show that f actually takes the value M at some point c.

Let $a_1 = a$, $b_1 = b$ and define $c_1 = (a_1 + b_1)/2$. By Theorem 1.2.14, $M = \sup_{[a_1,c_1]} f$ or $\sup_{[c_1,b_1]} f$. Let the subinterval that gives equality be called $[a_2, b_2]$. Applying the same process to $[a_2, b_2]$, and so on, we get a sequence of intervals

$$[a,b] = [a_1,b_1] \supset [a_2,b_2] \supset \cdots \supset [a_n,b_n] \supset \cdots,$$

with $b_n - a_n = (b-a)/2^{n-1}$ and $M = \sup_{[a_n,b_n]} f$ for each n. By the completeness axiom, there is a point c that is in each $[a_n,b_n]$. We claim that $f(c) = M$. If $f(c) < M$, then, by the continuity of f, there is a $\delta > 0$ such that $x \in (c-\delta, c+\delta)$ implies $f(x) < \dfrac{M+f(c)}{2}$. Since the length of the intervals $[a_n,b_n]$ is halved at each stage, for large enough n we will have $[a_n,b_n] \subset (\alpha-\delta,\alpha+\delta)$, implying that the values of f on $[a_n,b_n]$ are bounded above by $\dfrac{M+f(c)}{2} < M$. This is a contradiction to the choice of $[a_n,b_n]$, hence $f(c) = M$. ∎

Let $f\colon [a,b] \to \mathbb{R}$ be a bounded function. The **span** of f on $[a,b]$ is $\sup\{|f(x)-f(y)| : x,y \in [a,b]\}$.

Task 3.5.3

Find the spans of the following functions on $[0,1]$: $\operatorname{sgn}(x)$, $\sin \pi x$.

Task 3.5.4

Let $f\colon [a,b] \to \mathbb{R}$ *be a continuous function. Show that the span of* f *on* $[a,b]$ *equals* $\max\{f(x) \mid x \in [a,b]\} - \min\{f(x) \mid x \in [a,b]\}$.

Theorem 3.5.5 (Small Span Theorem)

Let $f\colon [a,b] \to \mathbb{R}$ *be continuous. For every* $\epsilon > 0$ *there is a partition* P *of* $[a,b]$ *such that the span of* f *is less than* ϵ *on every subinterval of* P.

Proof. Suppose there is an $\epsilon > 0$ such that no such partition of $[a,b]$ exists. Let $a_1 = a$, $b_1 = b$ and define $c_1 = (a_1+b_1)/2$. Then at least one of the intervals $[a_1,c_1]$ and $[c_1,b_1]$ fails to have such a partition. Let that one be called $[a_2,b_2]$.

Applying the same process to $[a_2,b_2]$, and so on, we get a sequence of intervals

$$[a,b] = [a_1,b_1] \supset [a_2,b_2] \supset \cdots \supset [a_n,b_n] \supset \cdots$$

none of which has such a partition, and with $b_n - a_n = (b-a)/2^{n-1}$. In particular, the span of f exceeds ϵ on every $[a_i,b_i]$.

Note that

$$a = a_1 \le a_2 \le \cdots \le b_2 \le b_1 = b.$$

By the completeness axiom, there is an $\alpha \in \mathbb{R}$ such that $a_i \le \alpha \le b_i$ for every i. Hence, α is in each $[a_n,b_n]$. Since f is continuous at α, there is a $\delta > 0$ such that $|x-\alpha| < \delta$ implies $|f(x)-f(\alpha)| < \epsilon/2$. And then $x,y \in (\alpha-\delta,\alpha+\delta) \implies |f(x)-f(y)| < \epsilon$.

By taking large enough n we can ensure that $[a_n,b_n] \subset (\alpha-\delta,\alpha+\delta)$. Then we would have span of f being less than ϵ on $[a_n,b_n]$, which contradicts our earlier observation about these intervals. ∎

Exercises for § 3.5

1. Prove the following functions are not continuous, by showing they lack some property of continuous functions (and not by showing discontinuity at any particular point):

(a) $f\colon [0,1] \to \mathbb{R}$, $f(x) = 1/x$ if $x \neq 0$ and $f(0) = 0$.

(b) $g\colon [0,1] \to \mathbb{R}$, $g(x) = \sqrt{x}$ if $x \neq 1$ and $g(1) = 0$.

(c) $h\colon [0,1] \to \mathbb{R}$, $h(x) = [e^x]$.

2. Give an example of each of the following:

(a) An unbounded continuous function with a bounded domain.

(b) A bounded continuous function that does not have a maximum value.

3. Consider a non-constant polynomial $p(x) = x^n + a_{n-1}x^{n-1} + \cdots + a_0$ of even degree. Prove the following:

(a) For any real number y there is a real number $R(y) \geq 0$ such that $|x| > R(y)$ implies $p(x) > y$.

(b) Let m be the minimum value of $p(x)$ over the interval $[-R(a_0), R(a_0)]$. Show that m is the minimum value of $p(x)$ over the entire real line.

(c) Show that the image of p is $[m, \infty)$.

4. Complete the following sketch of an alternate proof of the boundedness theorem:

(a) If a function is bounded on sets A and B, then it is bounded on $A \cup B$.

(b) Given a continuous function $f\colon [a,b] \to \mathbb{R}$, define

$$A = \{\, x \in [a,b] \mid f \text{ is bounded on } [a,x] \,\}.$$

Show that $\alpha = \sup A$ exists.

(c) If $\alpha < b$, use the continuity of f at α to show that f is bounded on $[a, \alpha + \delta]$ for some $\delta > 0$. Hence, $\alpha = b$.

(d) Use the continuity of f at b to show that f is bounded on $[a,b]$.

5. Suppose f is continuous and positive on $[a,b]$. Show there is a $\delta > 0$ such that $f(x) \geq \delta$ for every $x \in [a,b]$.

6. Complete the following sketch of an alternate proof of the extreme value theorem:

(a) Show $A = \{\, f(x) \mid x \in [a,b] \,\}$ has a least upper bound M.

(b) Assuming $f(x)$ never equals M, show that the image of $g(x) = \dfrac{1}{M - f(x)}$ has an upper bound, say R.

(c) Show that $M - 1/R$ is an upper bound of A, a contradiction.

3.6 Continuity, Integration and Means

Theorem 3.6.1 (Integrability of Continuous Functions)

Let $f\colon [a,b] \to \mathbb{R}$ be continuous. Then f is integrable on $[a,b]$.

The plan is to apply the Riemann condition. For it to work, we need to find upper and lower sums that are close to each other. The small span theorem helps out by giving a partition where the function fluctuates little on each subinterval.

Proof. Let $\epsilon > 0$. By the small span theorem, there is a partition $P = \{x_0, \ldots, x_n\}$ of $[a,b]$ such that

$$x_{i-1} \le x, y \le x_i \implies |f(x) - f(y)| < \frac{\epsilon}{b-a}.$$

Let $m_i = \min\{\, f(x) \mid x \in [x_{i-1}, x_i] \,\}$ and $M_i = \max\{\, f(x) \mid x \in [x_{i-1}, x_i] \,\}$. Then $M_i - m_i < \epsilon/(b-a)$ for every i. Define step functions $s \in \mathcal{L}_f$ and $t \in \mathcal{U}_f$ by $s(x_i) = t(x_i) = f(x_i)$ for each i and

$$\begin{aligned} s(x) &= m_i \qquad \text{if } x_{i-1} < x < x_i, \\ t(x) &= M_i \qquad \text{if } x_{i-1} < x < x_i. \end{aligned}$$

Then $s \le f \le t$ on $[a,b]$ and

$$\int_a^b t(x)\,dx - \int_a^b s(x)\,dx = \sum_{i=1}^n (M_i - m_i)(x_i - x_{i-1}) < \frac{\epsilon}{b-a} \sum_{i=1}^n (x_i - x_{i-1}) = \epsilon.$$

By the Riemann condition, f is integrable on $[a,b]$. ∎

A function $f\colon [a,b] \to \mathbb{R}$ is called **piecewise continuous** if there is a partition $P = \{x_0, \ldots, x_n\}$ of $[a,b]$ such that f is continuous on each open interval (x_{i-1}, x_i) and the one-sided limits of f exist at the x_i's.

Theorem 3.6.2

Let $f\colon [a,b] \to \mathbb{R}$ be a piecewise continuous function. Then f is integrable.

Proof. Exercise. See the proof of Theorem 2.2.13. ∎

Let us recall that the **average** or **mean** of numbers $x_1, \ldots, x_n$ is defined by $\bar{x} = \frac{1}{n}\sum_{i=1}^n x_i$. This notion of average can be generalized from finitely many numbers to the values of integrable functions. First consider an interval $[a,b]$ with a partition $P = \{x_0, \ldots, x_n\}$ of equally spaced points. Consider a step function $s\colon [a,b] \to \mathbb{R}$ such that $s(x) = s_i$ for $x \in (x_{i-1}, x_i)$. Then

$$\int_a^b s(x)\,dx = \sum_{i=1}^n s_i(x_i - x_{i-1}) = \frac{b-a}{n} \sum_{i=1}^n s_i = (b-a)\bar{s}$$

$$\implies \bar{s} = \frac{1}{b-a} \int_a^b s(x)\,dx.$$

This motivates the following. If $f\colon [a,b] \to \mathbb{R}$ is integrable, we define its **average** by

$$\bar{f} = \bar{f}_{[a,b]} = \frac{1}{b-a}\int_a^b f(x)\,dx.$$

The average of a function has the same basic properties as the average of a collection of numbers. For example, the average lies between the extremes.

Task 3.6.3

Show that if f has upper and lower bounds M and m respectively, then $m \le \bar{f} \le M$.

Further, recall that if $x_1,\dots,x_n$ have average $\bar{x}$ and $y_1,\dots,y_m$ have average $\bar{y}$, then the pooled collection $x_1,\dots,x_n,y_1,\dots,y_m$ has average $\dfrac{n}{m+n}\bar{x} + \dfrac{m}{m+n}\bar{y}$.

Task 3.6.4

Suppose $a < b < c$ and $f\colon [a,c] \to \mathbb{R}$ is integrable. Show that

$$\bar{f}_{[a,c]} = \frac{b-a}{c-a}\bar{f}_{[a,b]} + \frac{c-b}{c-a}\bar{f}_{[b,c]}.$$

Finally, if we acquire new data whose values are lower than previous values, then the average decreases.

Task 3.6.5

Suppose that f is a decreasing function. Show that $\bar{f}_{[a,x]}$ is also a decreasing function of x.

Now let us see a phenomenon that is special to averages of functions. The average of a collection of numerical data is usually not a member of that data set. However, the average of a continuous function is a value of that function.

Theorem 3.6.6 (Mean Value Theorem for Integration)

Consider a continuous function $f\colon [a,b] \to \mathbb{R}$. There is a number $c \in [a,b]$ such that

$$f(c) = \frac{1}{b-a}\int_a^b f(x)\,dx.$$

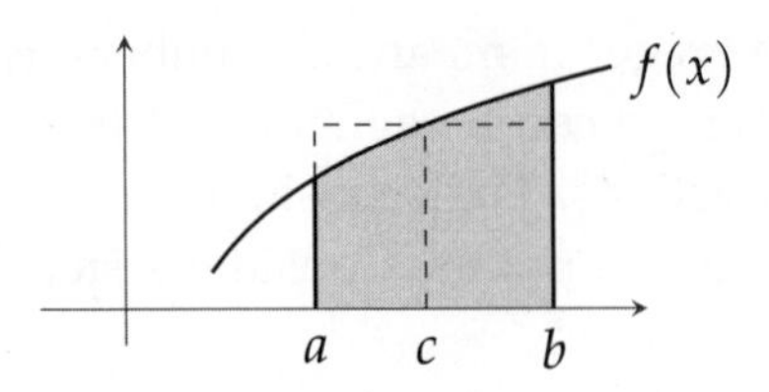

Proof. Let m, M be the minimum and maximum values, respectively, of $f(x)$ on $[a,b]$. Then there exist $a', b' \in [a,b]$ such that $f(a') = m$ and $f(b') = M$. Further,

$$f(a') = m \le \frac{1}{b-a}\int_a^b f(x)\,dx \le M = f(b').$$

By the intermediate value theorem, there is a number c between a' and b' with $f(c) = \dfrac{1}{b-a}\displaystyle\int_a^b f(x)\,dx$. ■

A **weighted mean** or **weighted average** of numbers $x_1, \ldots, x_n$ is a combination $\sum_{i=1}^n w_i x_i$ where each $w_i \geq 0$ and $\sum_{i=1}^n w_i = 1$. The concept of weighted average generalizes that of ordinary average by allotting different levels of importance (or weights) to each number. If we set each $w_i = 1/n$, we get the original $\bar{x}$.

The analogue for integration is to define the weighted average of an integrable function f to be $\int_a^b f(x)g(x)\,dx$ where g is non-negative on $[a,b]$ and $\int_a^b g(x)\,dx = 1$. This definition requires fg to be integrable. For that, see Exercise 10 of §2.2.

When we calculate a weighted average of a *continuous* function, we again find that it equals one of the values of the function.

Theorem 3.6.7 (Mean Value Theorem for Weighted Integration)

Consider functions $f, g\colon [a,b] \to \mathbb{R}$ where f is continuous, while g is integrable and $g \geq 0$ on $[a,b]$. Then there is a number $c \in (a,b)$ such that

$$f(c)\int_a^b g(x)\,dx = \int_a^b f(x)g(x)\,dx.$$

Proof. Let m, M and a', b' be as in the proof of the previous theorem. Then

$$\begin{aligned} m \leq f(x) \leq M &\Longrightarrow mg(x) \leq f(x)g(x) \leq Mg(x) \\ &\Longrightarrow m\int_a^b g(x)\,dx \leq \int_a^b f(x)g(x)\,dx \leq M\int_a^b g(x)\,dx. \end{aligned}$$

If $\int_a^b g(x)\,dx = 0$, these inequalities give $\int_a^b f(x)g(x)\,dx = 0$, and then any c will work. If $\int_a^b g(x)\,dx \neq 0$, we have $m \leq \dfrac{\int_a^b f(x)g(x)\,dx}{\int_a^b g(x)\,dx} \leq M$. Then intermediate value theorem gives the desired c. ■

Exercises for § 3.6

1. Let $f\colon [a,b] \to \mathbb{R}$ be strictly increasing and continuous with $a, f(a) \geq 0$. Prove that

$$\int_{f(a)}^{f(b)} f^{-1}(y)\,dy + \int_a^b f(x)\,dx = bf(b) - af(a).$$

2. Let $f\colon [0,1] \to \mathbb{R}$ be continuous and $\int_0^1 f(x)\,dx = 1$. Show there is a $c \in (0,1)$ such that $f(c) = 1$.

3. Suppose f is bounded on $[a,b]$ and continuous on $(a,b]$. Show that f is integrable on $[a,b]$.

4. Suppose f is continuous and non-negative on $[a,b]$. If $\int_a^b f(x)\,dx = 0$ then $f = 0$ on $[a,b]$. (Hint: If $f(c) > 0$, then $f(x) > f(c)/2$ for x near c.)

5. Suppose f is continuous on $[a,b]$ and $\int_a^b f(x)^2\,dx = 0$. Show that $f = 0$ on $[a,b]$.

6. Suppose f is continuous on $[a,b]$ and $\int_a^b f(x)\,dx = 0$. Show that $f(c) = 0$ for some $c \in (a,b)$.

7. Use the mean value theorem for weighted integration to prove:

$$\frac{1}{3\sqrt{2}} \leq \int_0^1 \frac{x^2}{\sqrt{1+x}}\,dx \leq \frac{1}{3}.$$

3.7 Limits Involving Infinity

In calculus, the word "infinity" does not indicate an actual amount but a process of getting larger and larger, without any bound. We can have functions whose inputs can be arbitrarily large and we would want to know what happens to the output as the input increases without bound. Or it could be the output that shoots up near a particular input. Here are some examples of such situations. Keep these pictures in mind to better understand the formal definitions that follow.

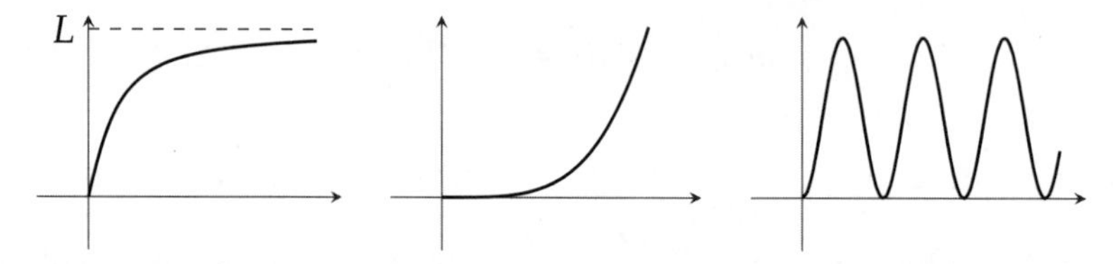

In the set of three examples above, we see various situations as the input x increases without bound, which we denote by $x \to \infty$. In the first graph, the values of f are approaching a particular number L and we would say $\lim_{x\to\infty} f(x) = L$. In the second, the values of f are also increasing without bound, so we would say $\lim_{x\to\infty} f(x) = \infty$. In the third, the values of f oscillate and we would just say that $\lim_{x\to\infty} f(x)$ does not exist.

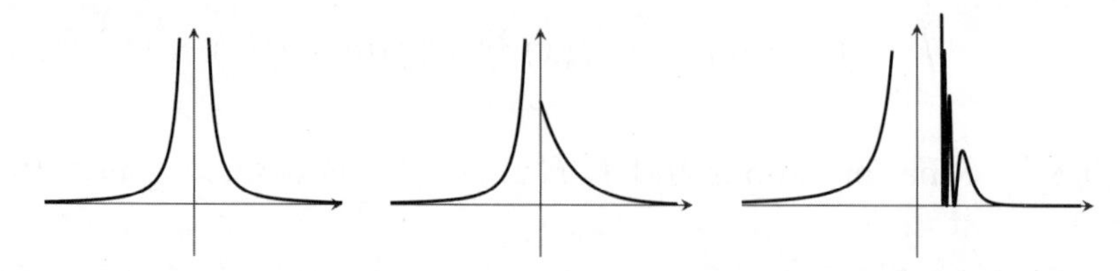

The next three examples show different ways of a function "blowing up" around a point, which in these cases is the origin. In the first graph, the function values increase without bound as we approach origin and we say that $\lim_{x\to 0} f(x) = \infty$. In the second case, this happens only from the left and we say $\lim_{x\to 0-} f(x) = \infty$ but not $\lim_{x\to 0} f(x) = \infty$. In the third graph, while the function does take larger values as it approaches origin, it also keeps taking small values and so we would *not* say that $\lim_{x\to 0} f(x) = \infty$.

It should be noted that $\lim_{x\to a} f(x) = \infty$ and $\lim_{x\to\infty} f(x) = \infty$ are actually forms of the limit *not* existing.

Limits at Infinity

We say $\lim_{x\to\infty} f(x) = L$ to indicate that as x gets larger, the values $f(x)$ approach L. Formally, $\lim_{x\to\infty} f(x) = L$ means that for every $\epsilon > 0$ there is an $M \in \mathbb{R}$ such that $x > M$ implies $|f(x) - L| < \epsilon$. Similarly, $\lim_{x\to-\infty} f(x) = L$ means that for every $\epsilon > 0$ there is an $M \in \mathbb{R}$ such that $x < M$ implies $|f(x) - L| < \epsilon$.

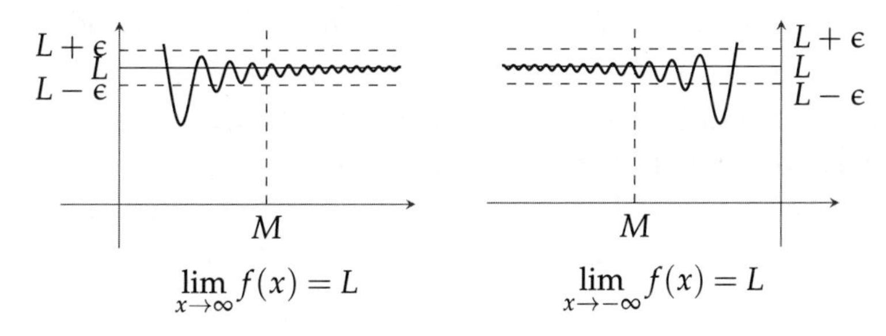

$\lim_{x\to\infty} f(x) = L$ $\qquad$ $\lim_{x\to-\infty} f(x) = L$

Example 3.7.1

The basic limit for calculations at infinity is: $\lim_{x\to\infty} \frac{1}{x} = 0$. This is easily proved by taking $M = 1/\epsilon$. □

Task 3.7.2

Show that $\lim_{x\to\infty} e^{-x} = \lim_{x\to-\infty} e^{x} = 0$.

Theorem 3.7.3 (Monotone Convergence Theorem)

Suppose f is increasing on $[a,\infty)$. Then $\lim_{x\to\infty} f(x)$ exists if and only if f is bounded above, and then $\lim_{x\to\infty} f(x) = \sup\{f(x) \mid x \geq a\}$.

Similarly, the limit at infinity of a decreasing function f exists if and only if f is bounded below, and it equals the infimum of its values.

Proof. Let f be increasing on $[a,\infty)$. Define $A = \{f(x) \mid x \geq a\}$.

First, suppose that f is bounded above. We shall show that $L = \sup(A)$ is the desired limit. Consider any $\epsilon > 0$. $L - \epsilon$ is not an upper bound of A, hence there is an x_0 such that $f(x_0) > L - \epsilon$. Since f is increasing, we also have $f(x) > L - \epsilon$ for each $x \geq x_0$. Hence,

$$x \geq x_0 \implies L - \epsilon < f(x) \leq L < L + \epsilon.$$

Conversely, suppose that $\lim_{x \to \infty} f(x)$ exists. Let its value be L. If any $f(x)$ is greater than L, then for every $y \geq x$ we have $f(y) - L \geq f(x) - L > 0$, which contradicts L being the limit of f. Hence, L is an upper bound of A and f is bounded above. Further, given any $\epsilon > 0$ there is an M such that $x > M$ implies $L - \epsilon < f(x) < L$. Hence, $L = \sup(A)$. ■

Theorem 3.7.4

The algebra of limits and the sandwich theorem hold for limits at infinity.

Proof. The proofs are minor modifications of the earlier ones. We illustrate the required adjustments with the proof for the sandwich theorem in this setting. Suppose $f(x) \leq g(x) \leq h(x)$ on an interval (a, ∞) and $\lim_{x \to \infty} f(x) = \lim_{x \to \infty} h(x) = L$. Consider an $\epsilon > 0$.

There exists $M_f > a$ such that $x > M_f$ implies $L - \epsilon < f(x) < L + \epsilon$.

There exists $M_g > a$ such that $x > M_g$ implies $L - \epsilon < h(x) < L + \epsilon$.

Let $M = \max\{M_f, M_g\}$. Then for $x > M$, we have $L - \epsilon < f(x) \leq g(x) \leq h(x) < L + \epsilon$. This proves $\lim_{x \to \infty} g(x) = L$. ■

Example 3.7.5

A simple application of the algebra of limits:

$$\lim_{x \to \infty} \frac{3x^2 + x + 5}{x^2 - 5} = \lim_{x \to \infty} \frac{3 + 1/x + 5/x^2}{1 - 5/x^2} = \frac{3 + 0 + 5 \cdot 0^2}{1 - 5 \cdot 0^2} = 3.$$ □

Example 3.7.6

Consider $\lim_{x \to \infty} \frac{\sin x}{x}$. We use the sandwich theorem, after observing that we can take $x > 0$ in this limit calculation.

$$-1 \leq \sin x \leq 1 \implies -\frac{1}{x} \leq \frac{\sin x}{x} \leq \frac{1}{x}.$$

We know that $\lim_{x \to \infty} \pm \frac{1}{x} = 0$. Hence, $\lim_{x \to \infty} \frac{\sin x}{x} = 0$. □

Theorem 3.7.7

Let f and g be real functions such that their composition $g \circ f$ is defined on an interval (a, ∞). Let $q = \lim_{x \to \infty} f(x)$ and suppose g is continuous at q. Then

$$\lim_{x \to \infty} g(f(x)) = g(q) = g(\lim_{x \to \infty} f(x)).$$

Proof. Let $\epsilon > 0$. Since g is continuous at q, there is a $\delta > 0$ such that $|y - q| < \delta$ implies $|g(y) - g(q)| < \epsilon$. And there is $M > a$ such that $x > M$ implies $|f(x) - q| < \delta$. Hence,

$$x > M \implies |f(x) - q| < \delta \implies |g(f(x)) - g(q)| < \epsilon.$$

■

Example 3.7.8

An application of the algebra of limits combined with continuity:

$$\begin{aligned}\lim_{x\to\infty}(\sqrt{x+1} - \sqrt{x}) &= \lim_{x\to\infty}\frac{(\sqrt{x+1} - \sqrt{x})(\sqrt{x+1} + \sqrt{x})}{\sqrt{x+1} + \sqrt{x}} \\ &= \lim_{x\to\infty}\frac{1}{\sqrt{x+1} + \sqrt{x}} \\ &= \lim_{x\to\infty}\frac{\sqrt{1/x}}{\sqrt{1+1/x}+1} = \frac{\sqrt{0}}{\sqrt{1+0}+1} = 0.\end{aligned}$$

□

We shall explore how the logarithmic function grows with x. We begin by showing it is much slower than linear growth.

Example 3.7.9

For $x > 1$, we have

$$0 < \log x = \int_1^x \frac{dt}{t} \le \int_1^x 1\,dt = x - 1.$$

Therefore,

$$x > 1 \implies 0 < \frac{\log x}{x^2} \le \frac{1}{x} - \frac{1}{x^2}.$$

Replacing x by $\sqrt{x}$, we have $x > 1$ implies $0 < \frac{1}{2}\frac{\log x}{x} \le \frac{1}{\sqrt{x}} - \frac{1}{x}$. By the sandwich theorem, we get $\lim_{x\to\infty}\frac{\log x}{x} = 0$.

□

Task 3.7.10

Show that for any $n \in \mathbb{N}$, $\lim_{x\to\infty}\frac{\log x}{x^{1/n}} = 0$.

When we interpret this limit being zero as meaning that $\log x$ grows much more slowly than $x^{1/n}$, we should keep in mind that this is true only eventually and not all along. For example, consider the graphs of $\log x$ and $x^{1/4}$. Initially, log shoots ahead and it is only at about $x = 5500$ that $x^{1/4}$ catches up and overtakes it.

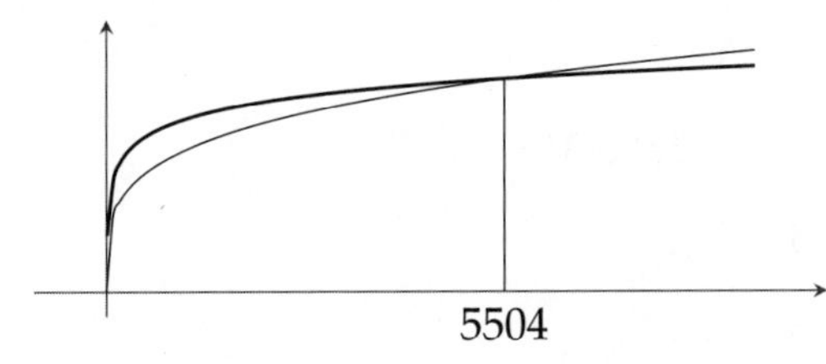

Limits at infinity can be converted to limits at zero through a substitution. This connection provides another way of extending results about limits at real numbers to limits at infinity.

Theorem 3.7.11

Let $a > 0$ and $f\colon (a,\infty) \to \mathbb{R}$. Then $\lim_{x\to\infty} f(x) = L$ if and only if $\lim_{x\to 0+} f(1/x) = L$.

Proof. Suppose $\lim_{x\to\infty} f(x) = L$. Let $\epsilon > 0$. Then there is $M > a$ such that $y > M$ implies $|f(y) - L| < \epsilon$. Take $\delta = 1/M$. Then $0 < x < \delta$ implies $0 < M < 1/x$ and hence $|f(1/x) - L| < \epsilon$. Therefore $\lim_{x\to 0+} f(1/x) = L$.

The steps are easily reversed to prove the converse as well. ∎

Task 3.7.12

Find $\lim_{x\to 0+} x \log x$ and $\lim_{x\to 0+} x^x$.

Limits Equal to Infinity

- We say $\lim_{x\to a} f(x) = \infty$ if for every $N \in \mathbb{R}$ there is a $\delta > 0$ such that $0 < |x - a| < \delta$ implies $f(x) > N$.
- We say $\lim_{x\to\infty} f(x) = \infty$ if for every $N \in \mathbb{R}$ there is an $M \in \mathbb{R}$ such that $x > M$ implies $f(x) > N$.

There are similar definitions for limits involving $-\infty$ and for one-sided limits.

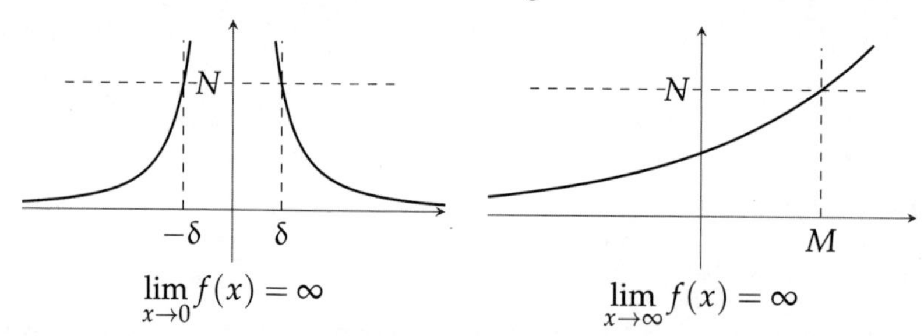

The fundamental examples of infinite limits are as follows:

1. $\lim_{x\to 0+} \dfrac{1}{x} = \infty$: Take $\delta = \dfrac{1}{N}$.
2. $\lim_{x\to\infty} x = \infty$: Take $M = N$.
3. $\lim_{x\to\infty} x^p = \infty$ when $p > 0$: Take $M = N^{1/p}$.
4. $\lim_{x\to\infty} e^x = \infty$: Take $M = \log N$.
5. $\lim_{x\to\infty} \log x = \infty$: Take $M = e^N$.

The **algebra of limits** partially applies to limits that are infinite:

Theorem 3.7.13

Let a stand for a real number or for the symbol ∞. Let f,g be functions defined in an open interval around a (when $a \in \mathbb{R}$) or on an interval (b,∞) (when $a = \infty$). Let $c \in \mathbb{R}$. The limits of f,g as $x \to a$ obey the following rules:

1. $f \to \infty$ *and* $c > 0$ *implies* $cf \to \infty$.
2. $f \to \infty$ *and* $c < 0$ *implies* $cf \to -\infty$.
3. $f \to \infty$ *and* $g \to c$ *implies* $f + g \to \infty$.
4. $f,g \to \infty$ *implies* $f + g \to \infty$.
5. $f,g \to \infty$ *implies* $fg \to \infty$.
6. $f \to \infty$ *and* $g \to c > 0$ *implies* $fg \to \infty$.
7. $f \to \infty$ *implies* $1/f \to 0$.

Proof. We shall prove the third implication and leave the others to you. Let $N \in \mathbb{R}$. Also suppose $a \in \mathbb{R}$.

Since $g \to c$, we have $|g| \to |c|$. Hence, there is $\delta' > 0$ such that $0 < |x - a| < \delta'$ implies $|g(x)| < |c| + 1$.

Since $f \to \infty$, there is $\delta' > 0$ such that $0 < |x - a| < \delta''$ implies $f(x) > N + |c| + 1$.

Let $\delta = \min\{\delta', \delta''\}$. Then $0 < |x - a| < \delta$ implies

$$f(x) + g(x) \geq f(x) - |g(x)| > N + |c| + 1 - (|c| + 1) = N.$$

If $a = \infty$, there are M', M'' such that $x > M'$ implies $|g(x)| < |c| + 1$ and $x > M''$ implies $f(x) > N + |c| + 1$. Now take $M = \max\{M', M''\}$. ■

Example 3.7.14

Consider a non-constant polynomial of the form $p(x) = x^n + a_{n-1}x^{n-1} + \cdots + a_0$. Let us compute its limit at infinity. We begin with

$$p(x) = x^n \left(1 + \frac{a_{n-1}}{x} + \cdots \frac{a_0}{x^n}\right).$$

We repeatedly apply the algebra of limits to x^k with $k > 0$:

$$\begin{aligned} \lim_{x\to\infty} x^k = \infty \implies \lim_{x\to\infty} x^{-k} = 0 \implies \lim_{x\to\infty} \frac{a_{n-k}}{x^k} = 0 \\ \implies \lim_{x\to\infty} \left(1 + \frac{a_{n-1}}{x} + \cdots \frac{a_0}{x^n}\right) = 1 \\ \implies \lim_{x\to\infty} p(x) = \infty. \end{aligned}$$

□

There is no general rule for $f-g$ or f/g when $f,g \to \infty$. These are therefore called **indeterminate forms** of type $\infty - \infty$ and ∞/∞ respectively. Each such limit has to be worked out individually without recourse to a general formula.

Example 3.7.15

Consider $\lim_{x\to\infty}(x-\log x)$. This is an $\infty-\infty$ form. We can use the properties of the log function:

$$x - \log x = x\Big(1 - \frac{\log x}{x}\Big) \to \infty. \qquad \square$$

Similarly, there is no general formula for the limit of fg when $f \to \infty$ and $g \to 0$. This is called an indeterminate form of type $\infty \cdot 0$.

The sandwich theorem is replaced by a one-sided result.

Theorem 3.7.16 (Comparison Theorem)

If $f(x) \le g(x)$ for every x, and $f(x) \to \infty$ as $x \to a$, then $g(x) \to \infty$ as $x \to a$. Here a can be a real number or $\pm\infty$.

Proof. Exercise. ∎

Theorem 3.7.17

Let $\lim_{x\to\infty} g(x) = L$ where L is a real number or ∞, and $\lim_{x\to a} f(x) = \infty$ where a is a real number or ∞. Then $\lim_{x\to a} g(f(x)) = L$.

Proof. Similar to Theorem 3.7.7. ∎

Example 3.7.18

Let us explore how the exponential function grows with x. We start with the known limits $\lim_{x\to\infty} e^x = \infty$ and $\lim_{x\to\infty} \frac{\log x}{x} = 0$. Substituting the first in the second gives us

$$\lim_{x\to\infty} \frac{x}{e^x} = \lim_{x\to\infty} \frac{\log(e^x)}{e^x} = 0.$$

Let us see how the exponential function compares with higher powers of x. Our first observation is that for $n \in \mathbb{N}$,

$$\lim_{x\to\infty} \frac{x}{e^{x/n}} = n \lim_{x\to\infty} \frac{x/n}{e^{x/n}} = n \lim_{y\to\infty} \frac{y}{e^y} = 0.$$

Hence, $\lim_{x\to\infty} \frac{x^n}{e^x} = \left(\lim_{x\to\infty} \frac{x}{e^{x/n}}\right)^n = 0$. Thus, the exponential function grows faster than any power of x, and therefore faster than any polynomial. Here is a comparison with x^4. Note how quickly *both* grow!

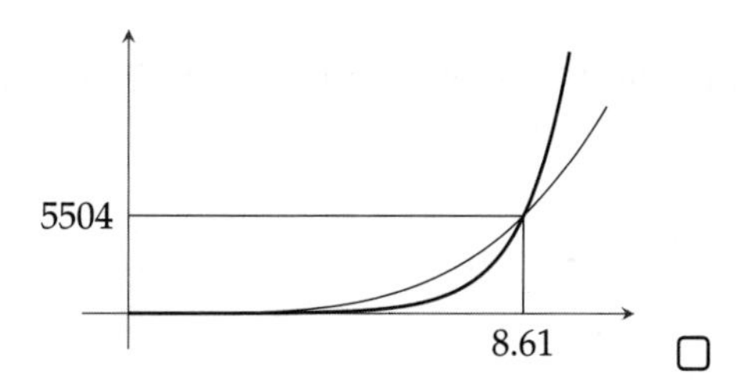

□

Task 3.7.19

Suppose $f(x) \to \infty$ and $f(x)g(x) \to 0$ as $x \to a$, where a could be real or $\pm\infty$. Show that $g(x) \to 0$ as $x \to a$.

Asymptotes

One can visualize $\lim_{x\to\infty} f(x) = L$ as saying that the graph of f merges with the horizontal line $y = L$ as x becomes large. This line is called a **horizontal asymptote** of f. In the same way, $\lim_{x\to-\infty} f(x) = L$ also gives a horizontal asymptote, in this case the merging happens in the negative direction.

Similarly, if either one-sided limit is infinite at $x = a$, we call the line $x = a$ a **vertical asymptote** of the function.

The graph may also approach a slanted line as $x \to \pm\infty$. We say that $y = ax + b$ is a **slant asymptote** of $y = f(x)$ as $x \to \infty$ if

$$\lim_{x\to\infty} (f(x) - ax - b) = 0.$$

We have the analogous definition of a slant asymptote as $x \to -\infty$. The concept of slant asymptote includes that of horizontal asymptote as a special case ($a = 0$).

Theorem 3.7.20

The line $y = ax + b$ is a slant asymptote of the function $y = f(x)$ if and only if $a = \lim_{x\to\infty} \frac{f(x)}{x}$ and $b = \lim_{x\to\infty} (f(x) - ax)$.

Proof. First, suppose $\lim_{x\to\infty} (f(x) - ax - b) = 0$. Then $\lim_{x\to\infty} x\left(\frac{f(x)}{x} - a - \frac{b}{x}\right) = 0$. Since $x \to \infty$, this is only possible if $\frac{f(x)}{x} - a - \frac{b}{x} \to 0$, which gives $\frac{f(x)}{x} \to a$. The formula for b is a simple rearrangement of $\lim_{x\to\infty} (f(x) - ax - b) = 0$.

The converse is trivial. ■

Example 3.7.21

Consider the upper branch of the hyperbola $y^2 - x^2 = 1$. It is the graph of the function $f(x) = \sqrt{x^2+1}$. We have

$$a = \lim_{x\to\infty} \frac{f(x)}{x} = \lim_{x\to\infty} \sqrt{1+1/x^2} = 1,$$

$$b = \lim_{x\to\infty} (\sqrt{x^2+1} - 1\cdot x) = \lim_{x\to\infty} \frac{1}{\sqrt{x^2+1}+\cdot x} = 0.$$

Hence, the line $y = x$ is a slant asymptote as $x \to \infty$. Similarly, $y = -x$ is a slant asymptote as $x \to -\infty$. The graph is

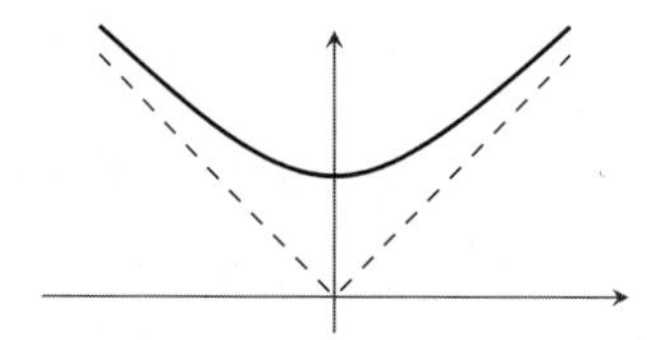

□

Example 3.7.22

The function $y = \tan x$ is periodic and non-constant so it cannot have slant asymptotes. It is continuous on its domain, which is all of $\mathbb{R}$ except the points $n\pi + \pi/2$, with $n \in \mathbb{Z}$. Therefore, these are the points where it may have vertical asymptotes.

$$\lim_{x\to\pi/2-} \tan x = \lim_{x\to\pi/2-} \frac{\sin x}{\cos x} = \lim_{t\to 0+} \sqrt{\frac{1}{t^2} - 1} = \infty,$$

$$\lim_{x\to\pi/2+} \tan x = \lim_{x\to\pi/2+} \frac{\sin x}{\cos x} = -\lim_{t\to 0+} \sqrt{\frac{1}{t^2} - 1} = -\infty.$$

So, there is a vertical asymptote at $x = \pi/2$. Since tan has a period of π, it also has vertical asymptotes at every $x = n\pi + \pi/2$.

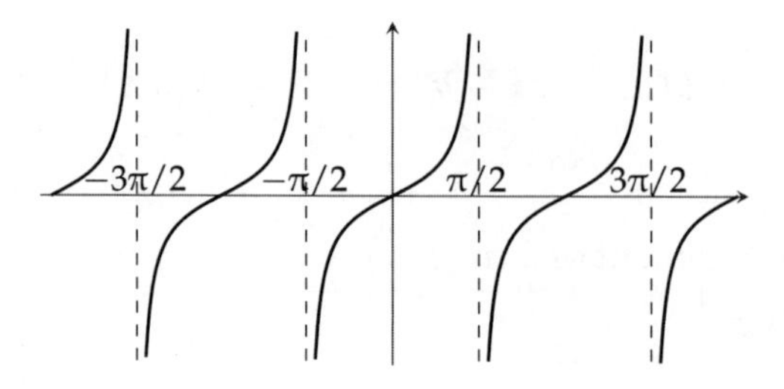

□

Exercises for § 3.7

1. Verify the following limits:

(a) $\lim_{x\to\infty} \arctan x = \pi/2$.

(b) $\lim_{x\to\infty} \dfrac{\sinh x}{\cosh x} = 1$.

(c) $\lim_{x\to 0+} \cot x = \infty$.

(d) $\lim_{x\to\infty} \dfrac{(\log x)^n}{x} = 0 \quad (n \in \mathbb{N})$.

2. Find the following limits:

(a) $\lim_{x\to\pi/2-} \sec x$.

(b) $\lim_{x\to\infty} \cos x$.

(c) $\lim_{x\to\infty} (x - \sqrt{x})$.

(d) $\lim_{x\to\infty} \dfrac{x+1}{\sqrt{4x^2+1}}$.

(e) $\lim_{x\to\infty} e^{-x^2}$.

(f) $\lim_{x\to 0} e^{-1/x^2}$.

(g) $\lim_{x\to\infty} x \sin \dfrac{1}{x}$.

(h) $\lim_{x\to\infty} \dfrac{x \sin x}{x^2+1}$.

(i) $\lim_{x\to\infty} (\sqrt{x^2+x} - x)$.

(j) $\lim_{x\to\infty} \left(\sqrt{x+\sqrt{x}} - \sqrt{x-\sqrt{x}}\right)$.

3. True or False?

(a) If $\lim_{x\to a} f(x) = 0$ then $\lim_{x\to a} \dfrac{1}{f(x)} = \infty$.

(b) If $\lim_{x\to a+} f(x) = 0$ then $\lim_{x\to a+} \dfrac{1}{f(x)} = \infty$.

(c) If $f(x) > 0$ for every x and $\lim_{x\to a} f(x) = 0$ then $\lim_{x\to a} \dfrac{1}{f(x)} = \infty$.

4. Let p, q be polynomials. Describe how $\lim_{x\to\infty} \dfrac{p(x)}{q(x)}$ and $\lim_{x\to-\infty} \dfrac{p(x)}{q(x)}$ depend on the degrees of these polynomials.

5. Consider $p(x) = ax^2 + bx + c$, $b > 0$. What happens to the roots of p if $a \to 0+$ with b, c kept fixed?

6. Let $f\colon [a,\infty)$ be a continuous function such that $\lim_{x\to\infty} f(x) = L \in \mathbb{R}$. Show that f is bounded.

7. Let $f, g\colon [a,\infty) \to \mathbb{R}$ and g is bounded. Prove the following:

(a) If $\lim_{x\to\infty} f(x) = \infty$ then $\lim_{x\to\infty} (f(x) + g(x)) = \infty$.

(b) If $\lim_{x\to\infty} f(x) = 0$ then $\lim_{x\to\infty} (f(x)g(x)) = 0$.

8. Let α be any real number. Show that $\lim_{x\to\infty} e^{-x} x^{\alpha} = 0$.

9. Find all asymptotes (horizontal, vertical, slant) of the given functions:

(a) $f(x) = \dfrac{1}{x-1}$.

(b) $g(x) = e^{1/x} - 1$.

(c) $h(x) = \dfrac{x^2+1}{x+1}$.

(d) $k(x) = (2x^2 - x^3)^{1/3}$.

Thematic Exercises

Continuity and Intervals

We take a closer look at intervals and their relationship with continuity. The first exercise gives a simple characterization of intervals. It is the converse of Task 1.2.17.

A1. Let $A \subseteq \mathbb{R}$ such that for any $a, b \in A$ if $a < x < b$, then $x \in A$. Show that A is an interval. (Hint: Consider cases of whether A is bounded above or below, and whether its supremum and infimum belong to it.)

The interval $[a,a]$ just has the single element $\{a\}$. We could even write intervals such as (a,a) that would be empty. Such intervals are called "degenerate." Intervals with at least two points are called "proper." When we refer to a continuous function on an interval, we will naturally consider only proper intervals as domains.

A2. Let f be a continuous function on an interval I.

(a) Show that $f(I)$ is an interval.

(b) Show that if I is closed and bounded, then so is $f(I)$.

(c) Are other types preserved? For example, if I is bounded, will $f(I)$ be bounded? If I is closed, will $f(I)$ be closed? Try different combinations as well.

A3. Let f be an injective continuous function on an interval I.

(a) Show that f is strictly monotonic. (Hint: Start by proving that f maps any $[a,b]$ to $[f(a), f(b)]$ or $[f(b), f(a)]$.)

(b) Show that $f^{-1} \colon f(I) \to \mathbb{R}$ is continuous.

(c) If I is open, so is $f(I)$.

(d) Can there be a continuous $f \colon \mathbb{R} \to \mathbb{R}$ such that $f \circ f(x) = -x$ for every x?

In the ϵ-δ definition of continuity of a function at a point a, the δ depends on both ϵ and a. If there is a choice of δ that depends only on ϵ and works throughout the function's domain, we say the function is **uniformly continuous**. That is, $f \colon A \to \mathbb{R}$ is uniformly continuous if for every $\epsilon > 0$, there is a $\delta > 0$ such that $x, y \in A$ and $|x - y| < \delta$ implies that $|f(x) - f(y)| < \epsilon$.

A4. Show that the following continuous functions are not uniformly continuous.

(a) $f \colon \mathbb{R} \to \mathbb{R}$, $f(x) = x^2$.

(b) $g \colon (0,1) \to \mathbb{R}$, $g(x) = 1/x$.

A5. Show that if $f\colon [a,b] \to \mathbb{R}$ is continuous, then it is uniformly continuous. (Hint: Use the small span theorem.)

A6. Show that $f\colon (0,1] \to \mathbb{R}$ extends to a continuous function on $[0,1]$ if and only if it is uniformly continuous. (Hint: Use the nested interval property to identify the suitable value of $f(0)$.)

A7. Show that if $f\colon [a,\infty) \to \mathbb{R}$ is continuous and $\lim_{x\to\infty} f(x)$ exists then f is uniformly continuous.

4 | Differentiation

In this chapter, we take a closer look at the idea that local information about functions should help us resolve integration problems. We already saw that continuity guarantees integrability. Another application of continuity combined with integration was in enabling the definition of the trigonometric functions. On the other hand, continuity did not give us new tools to calculate integrals. In this chapter we shall study the stronger property of differentiability, which will eventually give us techniques for calculating integrals. It also has its own significance, independent of integration. We shall use it for a better understanding of the shapes of graphs of functions.

Among the continuous functions, the ones that are easiest to integrate are the "piecewise linear" ones. Their graphs consist of line segments, such as in the example below:

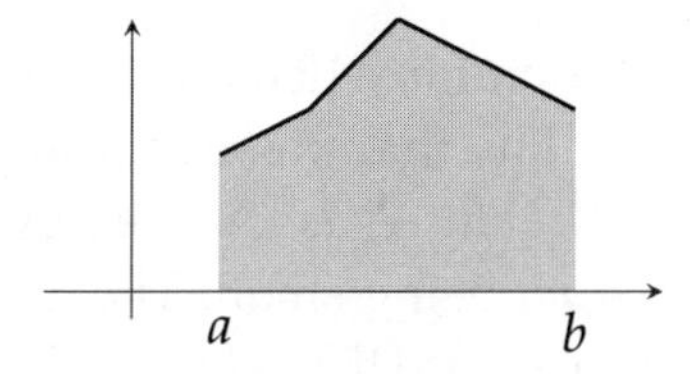

This suggests that we try to locally approximate functions by straight line segments. If we can get good approximations of this type, we can use them to assess the integral. We shall give the name "differentiable" to functions that can be locally approximated by straight lines. Most of this chapter is devoted to identifying these functions and to calculating the corresponding straight lines. Then we make the first connection between the processes of differentiation and integration, the so-called first fundamental theorem of calculus. Finally, we see that differentiation has a life of its own, and we use it to explore the problems of finding the extreme values and sketching the graph of a function.

4.1 Derivative of a Function

Let us consider what happens if we zoom in for a closer look at the graph of a function such as $y = x^2$, near a point such as $(1,1)$.

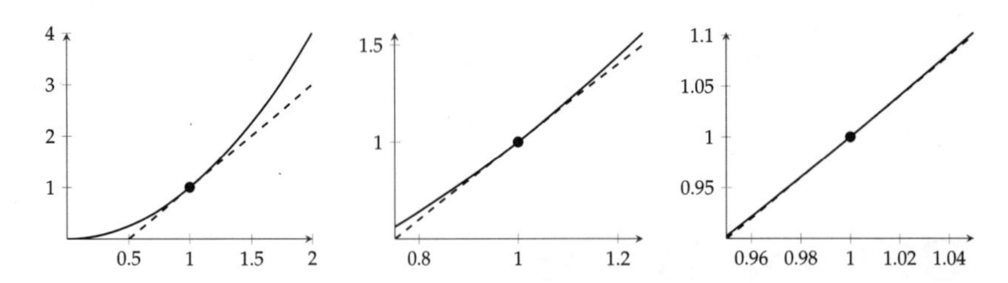

We see that the graph of the function $y = x^2$ looks more like the line $y = 2x - 1$ as we zoom in towards $(1,1)$, and at some stage becomes indistinguishable from it.

This can happen even for functions with rapid oscillations. Let us look at the function defined by $y = x^2 \cos(1/x)$ if $x \neq 0$, and $y = 0$ if $x = 0$, near the origin.

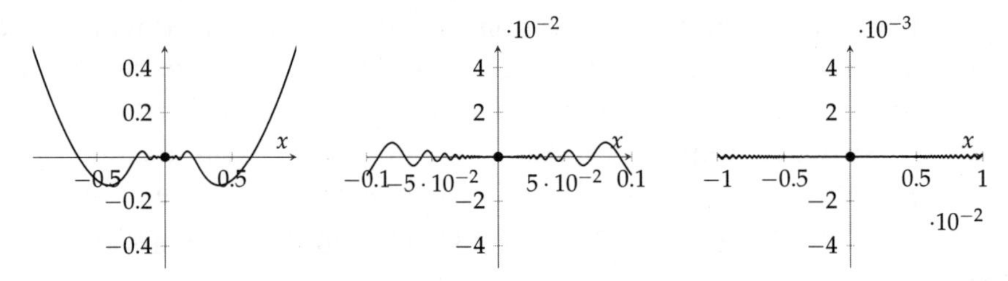

No matter how much we zoom in, the function has infinitely many oscillations. Nevertheless, their amplitudes decrease and in that sense the function becomes closer to the line $y = 0$ as we zoom in. (We have kept a constant ratio between the unit lengths in the x and y directions.)

We wish to set up a clear criterion for when a function $y = f(x)$ can be considered to merge, on zooming in, with a line that passes through $(a, f(a))$ and has slope m. This line has equation $y = f(a) + m(x - a)$. A "nearby" line would have equation $y = f(a) + m'(x - a)$ with $|m' - m|$ being small. The graph of f will merge with the given line if for any $\epsilon > 0$, we can ensure that $f(x)$ lies between $f(a) + (m \pm \epsilon)(x - a)$ for x close enough to a. This leads to the following definition:

We say that a function $f \colon I \to \mathbb{R}$, where I is an open interval, has **derivative** m **at a point** $a \in I$ if for each $\epsilon > 0$ there is a $\delta > 0$ such that $|x - a| < \delta$ implies $|f(x) - f(a) - m(x - a)| \leq \epsilon |x - a|$.

- If f has a derivative at a, we say that f is **differentiable** at a. The act of finding the derivative is called **differentiation**.
- If f has derivative m at a, the line $y = f(a) + m(x - a)$ is called the **tangent line** to the graph of f at $(a, f(a))$.
- If f has derivative m at a, we use the notation $f'(a)$ or $\dfrac{df}{dx}(a)$ or $\left.\dfrac{df}{dx}\right|_{x=a}$ for m.

Task 4.1.1

Consider a linear function $y = mx + c$. Show that its derivative at any point is m. (Hence, the derivative of a constant function is zero.)

The next sequence of graphs illustrates this definition for $y = x^2$, $a = 1$, $m = 2$ and $\epsilon = 0.1$. The curve is $y = x^2 - 1 - 2(x-1) = x^2 - 2x + 1$ and the shaded zone is bounded by the lines $y = \pm 0.1(x-1)$. We see that $\delta = 0.05$ works for these values, and brings the curve inside the shaded zone.

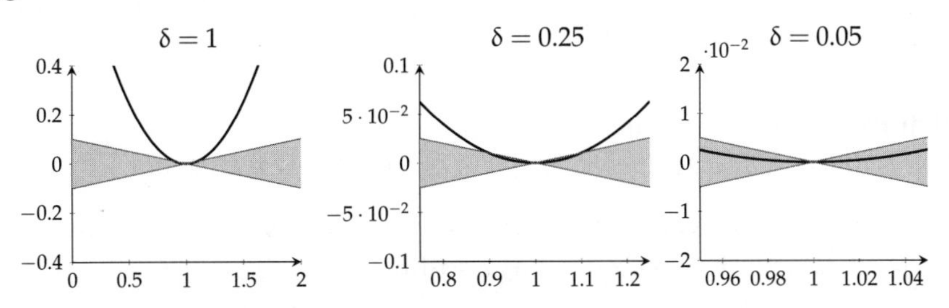

Functions can fail to be differentiable at a given point. The next set of diagrams shows functions whose graph passes through the origin but never resembles a line no matter how much we zoom in.

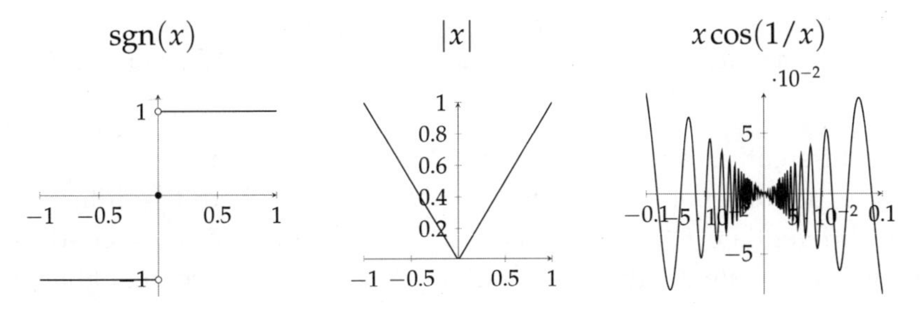

Our definition of derivative can be rephrased as follows:

Theorem 4.1.2

A function f has derivative $f'(a)$ at a if and only if there is a function φ such that $f(x) - f(a) - f'(a)(x-a) = \varphi(x)(x-a)$ and $\lim_{x \to a} \varphi(x) = \varphi(a) = 0$.

Proof. First, suppose the derivative $f'(a)$ exists. Clearly, the only option for defining φ is

$$\varphi(x) = \begin{cases} \dfrac{f(x) - f(a)}{x - a} - f'(a) & \text{if } x \neq a, \\ 0 & \text{if } x = a. \end{cases}$$

Consider any $\epsilon > 0$. The definition of derivative gives $\delta > 0$ such that $|x - a| < \delta$ implies $|f(x) - f(a) - f'(a)(x-a)| \leq \epsilon |x - a|$. Then $0 < |x - a| < \delta$ implies $|\varphi(x)| \leq \epsilon$, which corresponds to the desired $\lim_{x \to a} \varphi(x) = 0$.

The steps can be reversed to obtain the converse. ∎

Example 4.1.3

Let us check that the derivative of $f(x) = x^2$ at $a = 1$ is 2.

$$x^2 - 1^2 - 2(x-1) = \varphi(x)(x-1) \implies \varphi(x) = x - 1 \implies \lim_{x\to 1} \varphi(x) = 0 = \varphi(1).$$

□

Theorem 4.1.4

If a function is differentiable at a point, then it is continuous at that point.

Proof. Suppose f has derivative $f'(a)$ at $x = a$. Then there is a function φ such that $f(x) - f(a) - f'(a)(x-a) = \varphi(x)(x-a)$ and $\lim_{x\to a} \varphi(x) = 0$. Hence,

$$\lim_{x\to a} f(x) = \lim_{x\to a} \Big(f(a) + f'(a)(x-a) + \varphi(x)(x-a) \Big) = f(a).$$ ■

Task 4.1.5

Give an example of a function that is continuous at every point but fails to be differentiable at some point.

When we differentiate a real function $f\colon D \to \mathbb{R}$ we create a new function $f'\colon D' \to \mathbb{R}$ where D' is the subset of D consisting of all the points where f is differentiable. We can further differentiate f' to get a function $f'' = (f')'$, called the **second derivative** of f. Then we can create the **third derivative** $f''' = (f'')'$, and so on. Other choices of notation are:

$$\begin{aligned}
f^{(0)}(x) &= f(x), \\
f^{(1)}(x) &= f'(x) = \frac{df}{dx}(x), \\
f^{(2)}(x) &= f''(x) = \frac{d^2 f}{dx^2}(x), \\
f^{(3)}(x) &= f'''(x) = \frac{d^3 f}{dx^3}(x), \\
&\vdots \\
f^{(n)}(x) &= \frac{d^n f}{dx^n}(x).
\end{aligned}$$

The function $f^{(n)}$, obtained by differentiating f successively n times, is called the n**th derivative** of f.

Derivative via Limits

We have established a connection between derivatives and limits. Let us make it more explicit.

Theorem 4.1.6

Let $f\colon I \to \mathbb{R}$ where I is an open interval. Then f has derivative $f'(a)$ at $a \in I$ if and only if

$$f'(a) = \lim_{x \to a} \frac{f(x) - f(a)}{x - a} = \lim_{h \to 0} \frac{f(a+h) - f(a)}{h}.$$

Proof. We begin with the first equality.

$$\begin{aligned} f'(a) = m &\iff \text{there is } \varphi \text{ such that } f(x) - f(a) - m(x-a) = \varphi(x)(x-a) \\ &\qquad \text{and } \lim_{x \to a} \varphi(x) = \varphi(a) = 0 \\ &\iff \lim_{x \to a} \frac{f(x) - f(a) - m(x-a)}{x-a} = 0 \\ &\iff \lim_{x \to a} \frac{f(x) - f(a)}{x - a} = m. \end{aligned}$$

The second equality follows from Theorem 3.1.6. ■

This result gives us another perspective on the nature of the derivative. Imagine that x represents time and $f(x)$ is a position on the number line. Then the ratio $\dfrac{f(x) - f(a)}{x - a}$ represents the average velocity over the time interval $[a, x]$. Letting x approach a gives us a better idea of the velocity in the immediate vicinity of a, and the limit is seen as defining the *instantaneous velocity* at a. In general, the derivative of any function f is called the **(instantaneous) rate of change** of f.

Task 4.1.7

Show (again) that a constant function will have zero derivative.

Our original definition of derivative is useful for conceptualizing and proving abstract results. For example, it gives the right starting point for discussing differentiation in higher dimensions. On the other hand, the limit expression is convenient for calculations. Let us see an example.

Example 4.1.8 (Power Rule)

Consider the function x^n, for a fixed $n \in \mathbb{N}$. Its derivative can be calculated as follows:

$$(x^n)' = \lim_{y \to x} \frac{y^n - x^n}{y - x} = \lim_{y \to x} \sum_{i=0}^{n-1} y^i x^{n-1-i} = \sum_{i=0}^{n-1} x^i x^{n-1-i} = \sum_{i=0}^{n-1} x^{n-1} = nx^{n-1}.$$

The second equality uses the identity

$$y^n - x^n = (y - x)(y^{n-1} + y^{n-2}x + \cdots + yx^{n-2} + x^{n-1}) = (y - x) \sum_{i=0}^{n-1} y^i x^{n-1-i}.$$

In particular, $x' = 1$, $(x^2)' = 2x$, and so on. □

Here is an example of using limits to show that a certain function is not differentiable:

Example 4.1.9

Let $f(x) = |x|$. Let us try to calculate $f'(0)$:

$$f'(0) = \lim_{x \to 0} \frac{f(x) - f(0)}{x - 0} = \lim_{x \to 0} \frac{|x| - |0|}{x - 0} = \lim_{x \to 0} \frac{|x|}{x}.$$

The last limit does not exist, since the right-hand limit is 1 while the left-hand limit is -1. □

One-sided Derivatives

The concept of one-sided limits can be applied to derivatives to obtain one-sided derivatives.

- $f'_+(a) = \lim_{x \to a+} \frac{f(x) - f(a)}{x - a}$ is the **right derivative** of f at a.
- $f'_-(a) = \lim_{x \to a-} \frac{f(x) - f(a)}{x - a}$ is the **left derivative** of f at a.

Task 4.1.10

Show that a function f is differentiable at $x = a$ if and only if the left and right derivatives of f at a exist and are equal.

We say f is **differentiable on an interval** I if it is differentiable at every interior point of I, and has the appropriate one-sided derivative at any end-point that is included in I. Further, we shall denote the one-sided derivative at an end-point c by $f'(c)$ for simplicity.

Theorem 4.1.11

Suppose I is an interval and $f\colon I \to \mathbb{R}$ is a differentiable function. Then the following hold:

1. *If f is an increasing function, then $f'(a) \geq 0$ for every $a \in I$.*
2. *If f is a decreasing function, then $f'(a) \leq 0$ for every $a \in I$.*

Proof. We shall prove that if f is an increasing function and a is not the right end-point of I, then $f'_+(a) \geq 0$:

$$\begin{aligned} x > a &\implies \frac{f(x) - f(a)}{x - a} \geq 0 \\ &\implies f'_+(a) = \lim_{x \to a+} \frac{f(x) - f(a)}{x - a} \geq 0. \end{aligned}$$

The other cases are proved in a similar fashion. ■

Task 4.1.12

Suppose I is an interval and $f: I \to \mathbb{R}$ is a differentiable function. If f is strictly increasing, can we conclude that $f'(x) > 0$ for every $x \in I$?

Graph of Derivative

A useful skill is to be able to sketch the graph of f' from that of f, without actually calculating f'. We can do this by observing where the tangent slopes appear to be 0, positive, or negative. As an example, let $y = f(x)$ have the "bell-shaped" graph shown below:

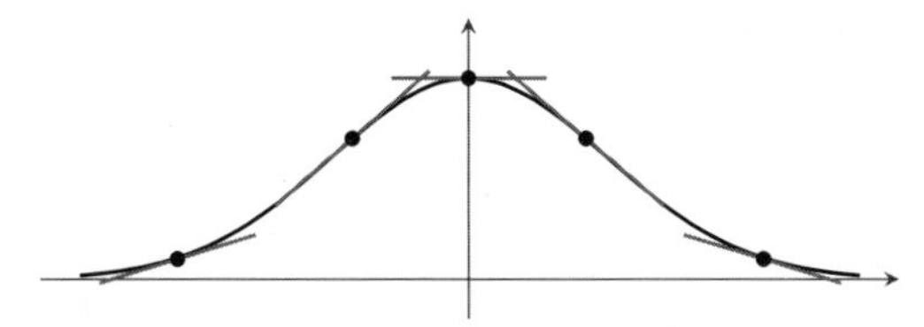

We see that the tangent line at $x = 0$ is horizontal, by symmetry, and hence has slope 0. Thus $f'(0) = 0$.

As we move to the right from $x = 0$, the tangents have negative slope. We also see that for a while their steepness increases but then they start flattening out. Thus, as x increases, $f'(x)$ at first takes more and more negative values but then starts moving up towards zero. Similarly, as we move to the left from $x = 0$, $f'(x)$ at first takes more and more positive values but then starts moving down towards zero. The plot below shows the graph of $f'(x)$ according to these observations:

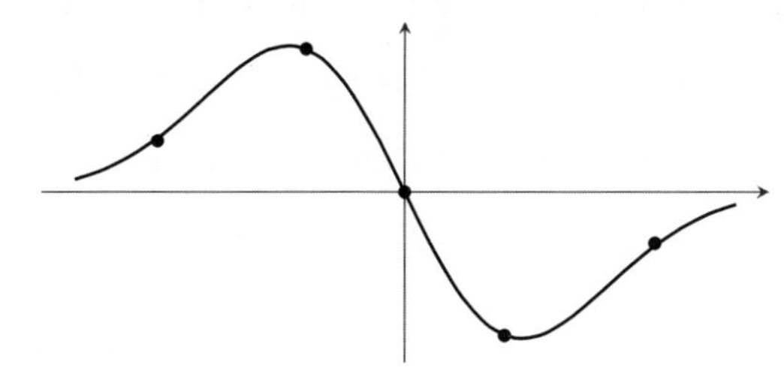

Exercises for § 4.1

1. Differentiate each function at $a = 0$.

(a) $f(x) = 2x + 1$.

(b) $g(x) = 2x^2 + 1$.

(c) $h(x) = x^2 + x + 1$.

(d) $k(x) = x|x|$.

2. Show that the following functions are not differentiable at $a = 0$:

(a) $f(x) = \operatorname{sgn}(x)$.

(b) $g(x) = \sqrt{|x|}$.

3. Consider the function $f\colon \mathbb{R} \to \mathbb{R}$ defined by $f(x) = x^2$ when $x \in \mathbb{Q}$ and $f(x) = 0$ when $x \notin \mathbb{Q}$. Show that f is differentiable only at $x = 0$.

4. Prove the following:

(a) The function $f(x) = \begin{cases} x\sin(1/x) & \text{if } x \neq 0, \\ 0 & \text{if } x = 0, \end{cases}$ is continuous but not differentiable at $x = 0$.

(b) The function $g(x) = \begin{cases} x^2\sin(1/x) & \text{if } x \neq 0, \\ 0 & \text{if } x = 0, \end{cases}$ is differentiable at $x = 0$.

5. Let $n \in \mathbb{N}$ and consider the function $f(x) = x^{1/n}$ with domain $x > 0$. Show that $f'(x) = \dfrac{1}{n}x^{(1/n)-1}$.

6. Prove the following:

(a) $\sin'(0) = 1$.

(b) $\cos'(0) = 0$.

7. Suppose a function f is differentiable at $x = a$. Prove that the function $g(x) = xf(x)$ is also differentiable at $x = a$, with $g'(a) = af'(a) + f(a)$.

8. Suppose $f\colon (-a,a) \to \mathbb{R}$ is differentiable at every point, so that we have $f'\colon (-a,a) \to \mathbb{R}$. Prove the following:

(a) If f is even then f' is odd.

(b) If f is odd then f' is even.

9. Let $f(x) = x^n$ for $n \in \mathbb{N}$. Prove that $f^{(n)}(x) = n!$ and $f^{(n+1)}(x) = 0$.

10. Suppose f is an even function that is differentiable at 0. Show that $f'(0) = 0$.

11. Let $f\colon \mathbb{R} \to \mathbb{R}$ have period T and be differentiable. Show that f' has period T.

12. Match the graphs of f in (a), (b), (c) with the graphs of f' in (i), (ii), (iii).

(a) 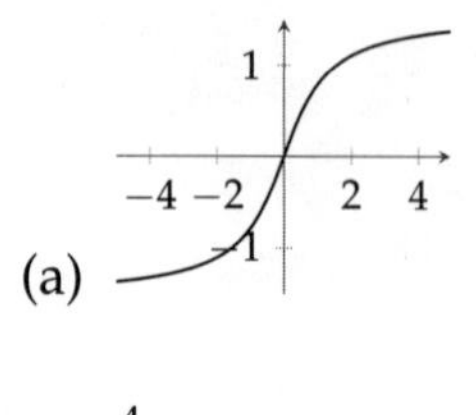

(b) 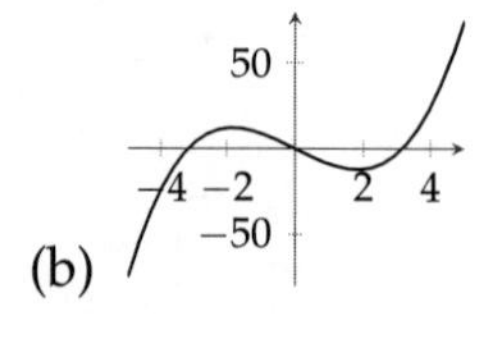

(c) 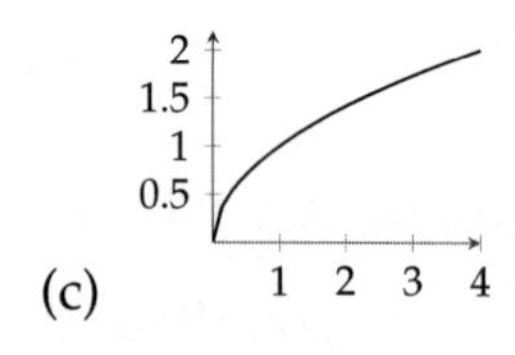

(i) 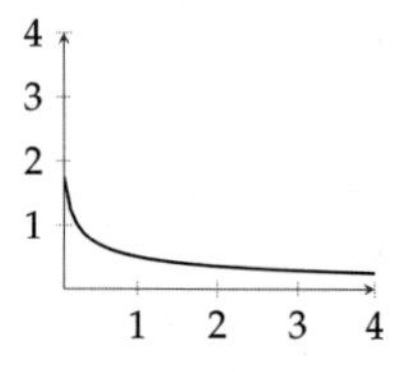

(ii) 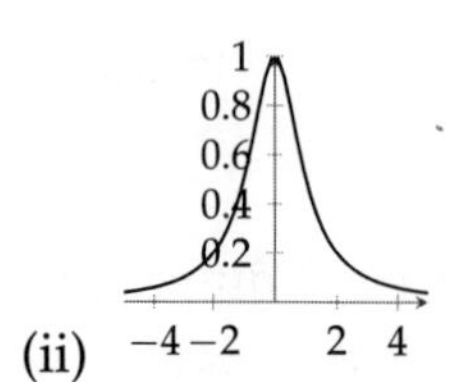

(iii) 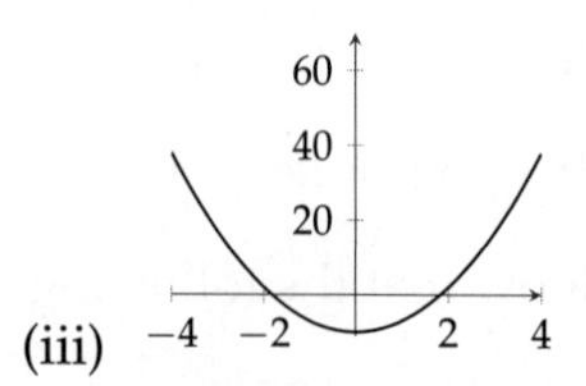

4.2 Algebra of Derivatives

In this section, we shall develop the rules for differentiating arithmetic combinations of functions. These are straightforward applications of the corresponding rules for limits, although their final form may differ.

Theorem 4.2.1 (Algebra of Derivatives)

Let f and g be differentiable at p, and let $C \in \mathbb{R}$. Then their combinations satisfy the following rules:

1. *(Scaling)* $(Cf)'(p) = Cf'(p)$.
2. *(Sum Rule)* $(f+g)'(p) = f'(p) + g'(p)$.
3. *(Difference Rule)* $(f-g)'(p) = f'(p) - g'(p)$.
4. *(Product Rule)* $(fg)'(p) = f'(p)g(p) + f(p)g'(p)$.
5. *(Reciprocal Rule)* $\left(\frac{1}{f}\right)'(p) = -\frac{f'(p)}{f(p)^2}$, *if* $f(p) \neq 0$.
6. *(Quotient Rule)* $\left(\frac{g}{f}\right)'(p) = \frac{g'(p)f(p) - g(p)f'(p)}{f(p)^2}$, *if* $f(p) \neq 0$.

Proof. We apply the algebra of limits.

1. Scaling:
$$(Cf)'(p) = \lim_{x \to p} \frac{Cf(x) - Cf(p)}{x - p} = C \lim_{x \to p} \frac{f(x) - f(p)}{x - p} = Cf'(p).$$
2. Sum Rule:
$$\begin{aligned}(f+g)'(p) &= \lim_{x \to p} \frac{f(x) + g(x) - f(p) - g(p)}{x - p} \\ &= \lim_{x \to p} \frac{f(x) - f(p)}{x - p} + \lim_{x \to p} \frac{g(x) - g(p)}{x - p} = f'(p) + g'(p).\end{aligned}$$
3. Difference Rule: Combine the sum rule with scaling by $C = -1$.
4. Product Rule:
$$\begin{aligned}(fg)'(p) &= \lim_{x \to p} \frac{f(x)g(x) - f(p)g(p)}{x - p} \\ &= \lim_{x \to p} \frac{f(x)g(x) - f(x)g(p) + f(x)g(p) - f(p)g(p)}{x - p} \\ &= \lim_{x \to p} f(x)\frac{g(x) - g(p)}{x - p} + \lim_{x \to p} \frac{f(x) - f(p)}{x - p} g(p) \\ &= f(p)g'(p) + f'(p)g(p).\end{aligned}$$

(Since $f'(p)$ exists, f is continuous at p and we have $\lim_{x\to p} f(x) = f(p)$.)

5. Reciprocal Rule: By continuity, $f(x) \neq 0$ for x near p. Hence,

$$\left(\frac{1}{f}\right)'(p) = \lim_{x\to p} \frac{1/f(x) - 1/f(p)}{x-p} = \lim_{x\to p} \frac{f(p) - f(x)}{f(x)f(p)(x-p)} = -\frac{f'(p)}{f(p)^2}.$$

(Since $f'(p)$ exists, we have $\lim_{x\to p} f(x) = f(p)$.)

6. Quotient Rule: Combine the product rule and reciprocal rule. ■

With these rules we can differentiate polynomials and rational functions. For example,

$$\begin{aligned}
(x^{45} + 7x^4 + 99)' &= (x^{45})' + (7x^4)' + (99)' && \text{(sum rule)} \\
&= (x^{45})' + (7x^4)' && (C' = 0) \\
&= (x^{45})' + 7(x^4)' && \text{(scaling)} \\
&= 45x^{44} + 28x^3. && \text{(power rule)}
\end{aligned}$$

Task 4.2.2

Differentiate the given functions and identify the points where the derivative exists:

(a) $\dfrac{x}{x-1}$. (b) $[x]$. (c) $x^{-n}, n \in \mathbb{N}$.

The algebra of derivatives gives us ways to deal with combinations of functions. Its effectiveness requires a collection of "elementary" functions whose derivatives are already known. We now begin building that collection by working out the derivatives of the trigonometric and logarithmic functions.

Trigonometric Functions

To differentiate the trigonometric functions, we use the two "fundamental limits" calculated earlier, namely

$$\lim_{x\to 0} \frac{\sin x}{x} = 1 \quad \text{and} \quad \lim_{x\to 0} \frac{1-\cos x}{x} = 0.$$

Theorem 4.2.3

For every $x \in \mathbb{R}$, $\sin' x = \cos x$ and $\cos' x = -\sin x$.

Proof. We differentiate the sine function, and leave the cosine for the reader.

$$\begin{aligned}
\sin' x &= \lim_{h\to 0} \frac{\sin(x+h)-\sin x}{h} \\
&= \lim_{h\to 0} \frac{\sin x\cos h + \cos x \sin h - \sin x}{h} \\
&= \lim_{h\to 0} \left(\frac{\cos h - 1}{h}\sin x + \frac{\sin h}{h}\cos x \right) \\
&= 0\cdot \sin x + 1\cdot \cos x \\
&= \cos x.
\end{aligned}$$

■

Task 4.2.4

Use the reciprocal and quotient rules to show that

$$\sec' x = \sec x \tan x, \qquad \csc' x = -\csc x \cot x,$$
$$\tan' x = \sec^2 x, \qquad \cot' x = -\csc^2 x.$$

Logarithms

In order to obtain the derivative of $\log x$ we begin with the following simple inequalities:

Theorem 4.2.5

For $x > 0$, $1 - \dfrac{1}{x} \le \log x \le x - 1$.

Proof. For $x \ge 1$ these inequalities are obtained from

$$\int_1^x \frac{1}{x}\,dt \le \int_1^x \frac{1}{t}\,dt \le \int_1^x 1\,dt.$$

Substituting $1/x$ for x gives the inequalities for $0 < x \le 1$. ■

Theorem 4.2.6

For every $x > 0$, $\log' x = \dfrac{1}{x}$.

Proof. We apply the limit definition of the derivative.

$$\log' x = \lim_{y\to x} \frac{\log y - \log x}{y - x} = \lim_{y\to x} \frac{\log(y/x)}{y-x} = \lim_{h\to 1} \frac{\log(hx/x)}{hx - x} = \frac{1}{x}\lim_{h\to 1} \frac{\log h}{h-1}.$$

For $h > 1$, we have $\dfrac{1}{h} \le \dfrac{\log h}{h-1} \le 1$ from Theorem 4.2.5. The sandwich theorem gives $\lim\limits_{x\to 1+} \dfrac{\log x}{x-1} = 1$. If $h < 1$, the inequalities reverse and again give $\lim\limits_{h\to 1-} \dfrac{\log h}{h-1} = 1$. ■

Task 4.2.7

Let $a > 0$ and $a \neq 1$. Show that $\log_a' x = \dfrac{1}{x\log a}$.

The limit calculation that we carried out in the last proof can also be expressed as

$$\lim_{h\to 0}\frac{\log(1+h)}{h} = 1 \quad \text{or} \quad \lim_{h\to 0}\log((1+h)^{1/h}) = 1.$$

Applying the exponential function, and recalling that it is continuous, we get

$$\lim_{h\to 0}(1+h)^{1/h} = e.$$

We can use this limit to get better estimates of e. Let us take a closer look at the behavior of $(1+h)^{1/h}$ for $h > 0$. First, since $\log(1+h)^{1/h} = \dfrac{1}{h}\displaystyle\int_1^{1+h}\frac{dx}{x}$ is the average of $1/x$ over the interval $[1, 1+h]$ and $1/x$ is a decreasing function, so is $\log(1+h)^{1/h}$ (Task 3.6.5). Hence, $(1+h)^{1/h}$ is also a decreasing function. Therefore, $(1+h)^{1/h}$ is an underestimate of e when $h > 0$. Similarly, it is an overestimate when $h < 0$. So, we can get bounds for e by taking small h of both signs.

h	$(1+h)^{1/h}$	$(1-h)^{-1/h}$
$1/2$	2.25	4
$1/10$	2.59	2.87
$1/10^3$	2.717	2.719

Thus, by taking $h = 0.001$ we already know that $e \approx 2.718$. The actual value of e when rounded to 6 decimal places is 2.718282.

Exercises for § 4.2

1. Differentiate the following functions:

(a) $\dfrac{x^2-1}{x^2+1}$.

(b) $\sin 2x$.

(c) $x\log x - x$.

(d) $\log|x|$.

2. Given the formula $1 + x + x^2 + \cdots + x^n = \dfrac{x^{n+1}-1}{x-1}$, determine formulas for the following by differentiation

(a) $1x + 2x^2 + 3x^3 + \cdots + nx^n$.

(b) $1^2x + 2^2x^2 + 3^2x^3 + \cdots + n^2x^n$.

3. (Leibniz Rule) Let u, v be real functions with a common domain, each being differentiable n times. Show that

$$(uv)^{(n)} = \sum_{k=0}^{n}\binom{n}{k}u^{(k)}v^{(n-k)}.$$

4. Let p be a polynomial of degree n. Show that $p^{(n+1)} = 0$.

5. Can there be a polynomial $p(x)$ such that $p(x) = \sin x$ on some interval (a,b)?

6. Let functions $f_1,\dots,f_n$ have derivatives $f_1',\dots,f_n'$.

(a) Find a rule for differentiating the product $g = f_1 \cdots f_n$ and prove it by mathematical induction.

(b) Show that if $f_i(x) \neq 0$ for every i then

$$\frac{g'(x)}{g(x)} = \frac{f_1'(x)}{f_1(x)} + \cdots + \frac{f_n'(x)}{f_n(x)}.$$

7. Let f be a differentiable function. Show that $(f(x)^n)' = nf(x)^{n-1}f'(x)$ for each $n \in \mathbb{Z}$.

8. Let $r \in \mathbb{Q}$ and consider the function $f(x) = x^r$ with domain $x > 0$. Show that $f'(x) = rx^{r-1}$.

9. We say that a is a zero with multiplicity k ($k \in \mathbb{N}$) of a polynomial p if $p(x) = (x-a)^k q(x)$, where q is a polynomial such that $q(a) \neq 0$. Show that a is a zero of p, with multiplicity k, if and only if $p(a) = p'(a) = \cdots = p^{(k-1)}(a) = 0$ and $p^{(k)}(a) \neq 0$.

10. Prove that $e^x \geq 1 + x$ for each $x \in \mathbb{R}$.

11. Prove that if $a, b \in \mathbb{N}$ satisfy $a < b$ and $b^a = a^b$ then $a = 2$ and $b = 4$. (Hint: Take $b = a(1+t)$ and apply the previous exercise.)

4.3 Chain Rule and Applications

The next task is to be able to differentiate compositions of functions. Let us first see what happens in the simplest case, when the functions are linear. Consider $f(x) = mx + c$ and $g(x) = nx + d$. Then $f \circ g(x) = f(nx+d) = m(nx+d) + c = mnx + md + c$. Therefore, $(f \circ g)'(x) = mn$, the product of the individual derivatives of f and g. This motivates the following result:

Theorem 4.3.1 (Chain Rule)

Let g be differentiable at a and let f be differentiable at $b = g(a)$. Then the composition $f \circ g$ is differentiable at a and the derivative is given by

$$(f \circ g)'(a) = f'(g(a))g'(a).$$

Proof. Let $f'(g(a)) = m$ and $g'(a) = n$. By Theorem 4.1.2, the differentiability of g and f gives functions φ and ψ such that the following hold:

1. $g(x) - g(a) = (n + \varphi(x))(x - a)$ and $\lim_{x \to a} \varphi(x) = \varphi(a) = 0$.
2. $f(y) - f(b) = (m + \psi(y))(y - b)$ and $\lim_{y \to b} \psi(y) = \psi(b) = 0$.

$$\begin{aligned}\text{Hence, } f(g(x)) - f(g(a)) &= \Big(m + \psi(g(x))\Big)(g(x) - b) \\ &= \Big(m + \psi(g(x))\Big)\Big(n + \varphi(x)\Big)(x - a) \\ &= mn(x-a) + E(x)(x-a),\end{aligned}$$

where $E(x) = m\varphi(x) + n\psi(g(x)) + \psi(g(x))\varphi(x)$. Now, $E(a) = 0$ and

$$\begin{aligned}\lim_{x\to a} E(x) &= m\lim_{x\to a}\varphi(x) + n\psi(\lim_{x\to a} g(x)) + \psi(\lim_{x\to a} g(x))\lim_{x\to a}\varphi(x) \\ &= m\varphi(a) + n\psi(b) + \psi(b)\varphi(a) = 0.\end{aligned}$$

This establishes that $(f \circ g)'(a) = mn = f'(g(a))g'(a)$. ■

Task 4.3.2

Differentiate the given functions and identify the points where the derivative exists:

(a) $f(x) = (x^2+1)^{10}$.

(b) $g(x) = |\cos x|$.

(c) $h(x) = \cos|x|$.

(d) $k(x) = \dfrac{\sin^2 x}{\sin x^2}$.

Implicit Differentiation

We have been studying relationships of the form $y = f(x)$ between two variables x, y. Sometimes the relationship is not of such a simple form. For example, it may be $x^2 + y^2 = 1$. Clearly this shows a dependence: for any $x \in [-1,1]$ we can solve for corresponding values $y = \pm\sqrt{1-x^2}$. We say that $x^2 + y^2 = 1$ defines y **implicitly** in terms of x. In fact this implicit relation can be separated into two explicit functions $y^+ = \sqrt{1-x^2}$ and $y^- = -\sqrt{1-x^2}$.

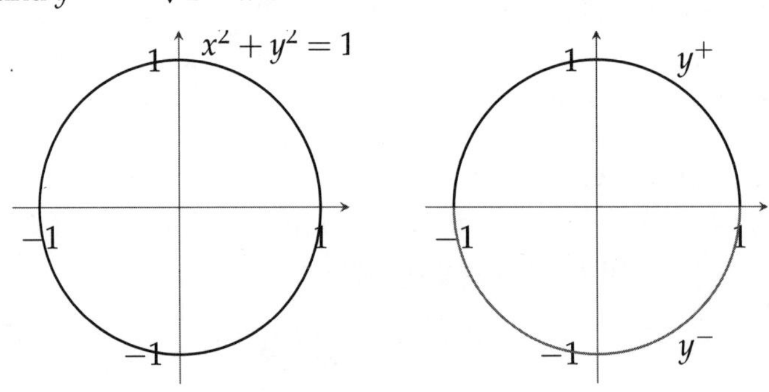

The chain rule allows us to calculate dy/dx without solving explicitly for y:

$$x^2 + y^2 = 1 \implies 2x + 2yy' = 0 \implies y' = -x/y \quad (\text{if } y \neq 0).$$

This works simultaneously for both cases of $y^{\pm} = \pm\sqrt{1-x^2}$!

The real advantage of this process of **implicit differentiation** is that often one may be unable to solve for y as an explicit function of x, and then this is the only approach available.

Example 4.3.3 (Folium of Descartes)

Consider the relation $x^3 + y^3 = 6xy$. Its solutions plot as follows:

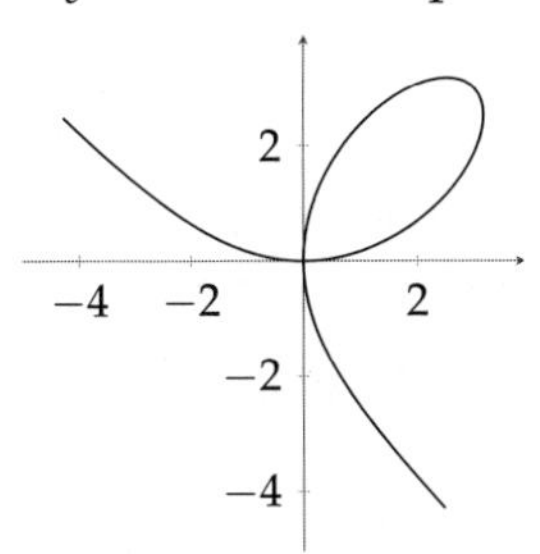

It is hard to separate this into explicit functions, but easy to differentiate implicitly:

$$\begin{aligned} x^3 + y^3 = 6xy &\implies 3x^2 + 3y^2 y' = 6y + 6xy' \\ &\implies (y^2 - 2x)y' = 2y - x^2 \implies y' = \frac{2y - x^2}{y^2 - 2x}. \end{aligned}$$

Suppose we wish to find a point on the curve where the tangent line is horizontal. We have

$$\begin{aligned} y' = 0 &\implies 2y - x^2 = 0 \implies y = x^2/2 \\ &\implies x^3 + x^6/8 = 3x^3 \implies x^3 = 16 \implies \begin{matrix} x = 2^{4/3} \\ y = 2^{5/3} \end{matrix}. \end{aligned}$$ □

Example 4.3.4 (Tangent to Ellipse)

The equation of an ellipse in standard form is

$$\frac{x^2}{a^2} + \frac{y^2}{b^2} = 1.$$

Implicit differentiation gives

$$\frac{2x}{a^2} + \frac{2y}{b^2} y' = 0.$$

If (x_0, y_0) is a point on the ellipse, the slope m of the tangent line there is given by

$$\frac{2x_0}{a^2} + \frac{2y_0}{b^2} m = 0 \quad \text{or} \quad m = -\frac{x_0\, b^2}{y_0\, a^2}.$$

Hence, the equation of the tangent line at (x_0, y_0) is

$$y = y_0 - \frac{x_0\, b^2}{y_0\, a^2}(x - x_0) \quad \text{or} \quad \frac{yy_0 - y_0^2}{b^2} + \frac{xx_0 - x_0^2}{a^2} = 0 \quad \text{or} \quad \frac{yy_0}{b^2} + \frac{xx_0}{a^2} = 1.$$ □

This technique of implicit differentiation takes the following for granted: that the given relation between x and y breaks into parts, each of which implicitly defines y as a function of x, and each of these functions is differentiable. Ideally, we should have criteria for deciding whether these assumptions hold. Such criteria exist but involve the study of functions $f(x,y)$ of two variables. They can be found in texts on analysis such as Apostol [1], under the name "Implicit Function Theorem."

Derivatives of Inverse Functions

Suppose f is a differentiable function such that its inverse exists and is also differentiable. Then, using the chain rule and assuming $f(a) = b$, we have

$$(f \circ f^{-1})(y) = y \implies (f \circ f^{-1})'(b) = 1 \implies f'(a)(f^{-1})'(b) = 1$$
$$\implies (f^{-1})'(b) = \frac{1}{f'(a)}.$$

That was easy. But it leaves something unanswered. Is there a *guarantee* that f^{-1} will indeed be differentiable?

Theorem 4.3.5

Let f be a continuous and monotonic bijection between two intervals. Let $f'(a)$ exist and be non-zero. Then f^{-1} is differentiable at $b = f(a)$ and the derivative is given by

$$(f^{-1})'(b) = \frac{1}{f'(a)}.$$

Proof. We begin by noting that if a line with slope $m \neq 0$ is reflected in the $y = x$ line, the resulting line has slope $1/m$. The following picture now represents a proof:

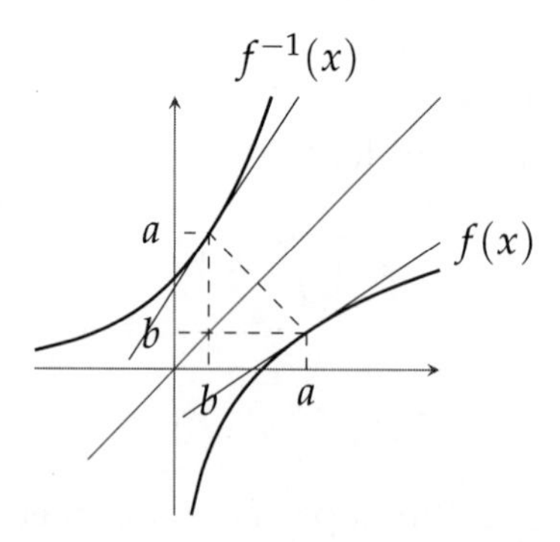

An Alternate Proof: First, we note that f^{-1} is a monotone function whose image is an interval, hence f^{-1} is continuous. Now define a function g by

$$g(y) = \begin{cases} \dfrac{f^{-1}(y) - f^{-1}(b)}{y - b} & \text{if } y \neq b, \\ 1/f'(a) & \text{if } y = b. \end{cases}$$

Substituting $y = f(x)$ and $b = f(a)$ gives

$$g(f(x)) = \begin{cases} \dfrac{x-a}{f(x)-f(a)} & \text{if } x \neq a, \\ 1/f'(a) & \text{if } x = a. \end{cases}$$

So, $g \circ f$ is continuous at a. Therefore, $g = g \circ f \circ f^{-1}$ is continuous at b. This gives the result. ■

Task 4.3.6

Differentiate the given functions and identify the points where the derivative exists:

(a) $g(x) = \sqrt{x}$.

(b) $h(x) = \sqrt{x + \sqrt{x}}$.

Inverse Trigonometric Functions

The sine function is neither one-one nor onto. We can make it onto simply by choosing the codomain to be $[-1,1]$ instead of $\mathbb{R}$. Now $\sin\colon \mathbb{R} \to [-1,1]$ is still not one-one, but we can choose a piece of the function which is one-one. A standard choice is to restrict the domain to $[-\pi/2, \pi/2]$. On this domain, sine is strictly increasing and hence one-one.

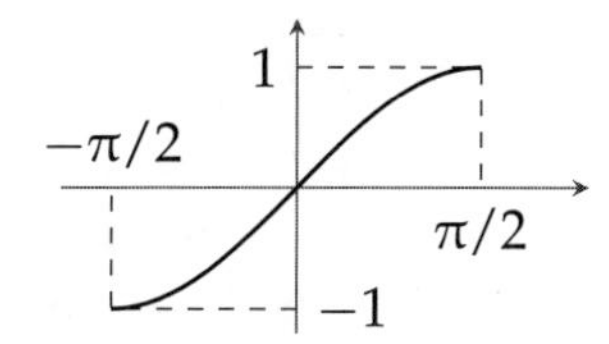

The graph of the bijection $\sin\colon [-\pi/2, \pi/2] \to [-1,1]$.

This restriction has an inverse function $\sin^{-1}\colon [-1,1] \to [-\pi/2, \pi/2]$. It is also called **arcsine** and its values are denoted by $\arcsin(x)$. It is continuous because it is monotone and its image is an interval. We can get its graph by reflecting the $y = \sin x$ graph in the $y = x$ line:

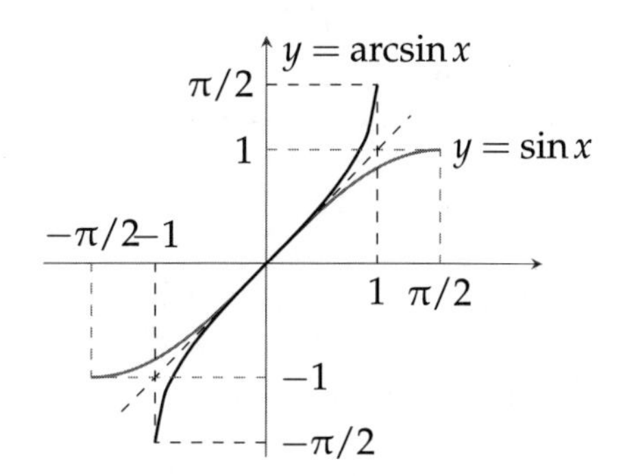

Similarly, the cosine function can be made a bijection by choosing the codomain to be $[-1,1]$ and restricting the domain to $[0,\pi]$. On this domain, cosine is strictly decreasing.

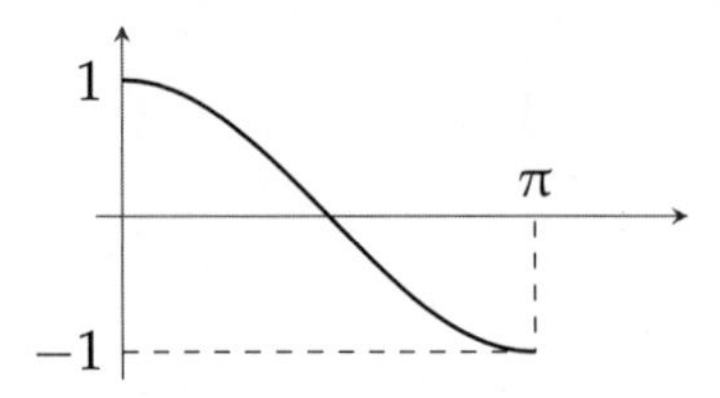

The graph of the bijection $\cos\colon [0,\pi] \to [-1,1]$.

This restriction has an inverse function $\cos^{-1}\colon [-1,1] \to [0,\pi]$. It is also denoted by arccos and is continuous because it is monotone and its image is an interval. We can get its graph by reflecting the $y = \cos x$ graph in the $y = x$ line.

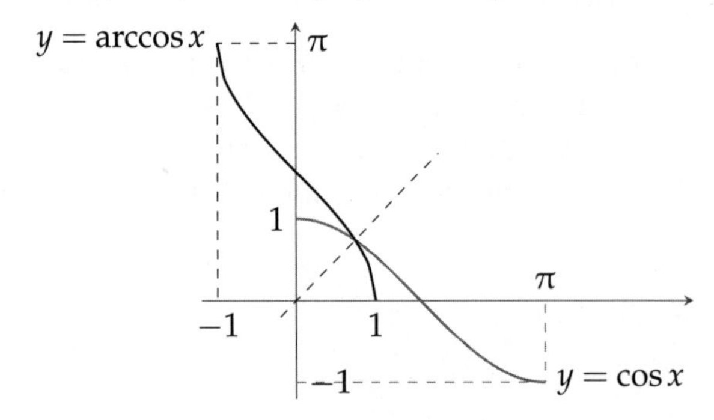

Comparing the graphs of $\sin^{-1} x$ and $\cos^{-1} x$ we see the first can be converted to the second by translating up by $\pi/2$ and then reflecting in the y-axis. This means that $\sin^{-1} x + \pi/2 = \cos^{-1}(-x)$. Similarly, we see that $-\sin^{-1} x + \pi/2 = \cos^{-1} x$. Which familiar identities do these reflect?

The portion of $\tan x$ defined on $(-\pi/2,\pi/2)$ is a bijection with $\mathbb{R}$. Its inverse is denoted by $\tan^{-1} x$ or $\arctan x$ and maps $\mathbb{R}$ to $(-\pi/2,\pi/2)$.

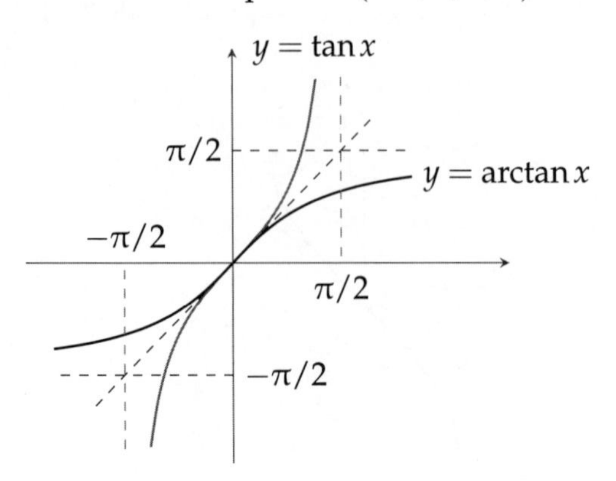

Theorem 4.3.7

The derivatives of the inverse trigonometric functions are:

$$\arcsin' x = \frac{1}{\sqrt{1-x^2}} \qquad \text{for } x \in (-1,1),$$

$$\arccos' x = \frac{-1}{\sqrt{1-x^2}} \qquad \text{for } x \in (-1,1),$$

$$\arctan' x = \frac{1}{1+x^2} \qquad \text{for } x \in \mathbb{R}.$$

Proof. We apply the formula for differentiating inverse functions to the arcsine function.

$$\arcsin' x = \frac{1}{\sin'(\arcsin x)} = \frac{1}{\cos(\arcsin x)}.$$

Now $\cos^2(\arcsin x) = 1 - \sin^2(\arcsin x) = 1 - x^2$. Since $\arcsin x \in [-\pi/2, \pi/2]$, we know that $\cos(\arcsin x) \geq 0$. Hence,

$$\arcsin' x = \frac{1}{\sqrt{1-x^2}}, \qquad \text{for } x \in (-1,1).$$

The calculation for arccosine is similar and is left to the reader. Finally,

$$\arctan' x = \frac{1}{\tan'(\arctan x)} = \frac{1}{\sec^2(\arctan x)} = \frac{1}{1+\tan^2(\arctan x)} = \frac{1}{1+x^2}.$$

■

The notation for inverse can be dangerous. Note the following:

- $\sin^{-1} x$ is the inverse sine function applied to x.
- $\sin x^{-1}$ is the sine function applied to $1/x$. The safer way to write it is $\sin(x^{-1})$.
- $(\sin x)^{-1}$ is $1/(\sin x)$. A common error is to mistake $\sin^{-1} x$ for $1/(\sin x)$.

Exponential Function

Theorem 4.3.8

The derivative of the exponential function is itself:

$$(e^x)' = e^x.$$

Proof. Consider $f(x) = \log x$. Its inverse function is $f^{-1}(x) = e^x$. Applying the formula for differentiating an inverse function, we get

$$(e^x)' = (f^{-1})'(x) = \frac{1}{f'(f^{-1}(x))} = \frac{1}{\log'(e^x)} = \frac{1}{1/e^x} = e^x.$$

■

It is a general principle in mathematics that objects or relations left unchanged by operations are especially important. The fact that the exponential function is unchanged by differentiation indicates that it will be a fundamental object in calculus.

Task 4.3.9

Let $a > 0$. Show that $(a^x)' = a^x \log a$.

The differentiation of the exponential function can be combined with the chain rule to differentiate arbitrary powers.

Theorem 4.3.10 (Power Rule)

If $r \in \mathbb{R}$, then $(x^r)' = rx^{r-1}$ for $x > 0$.

Proof. $(x^r)' = (e^{r \log x})' = \frac{r}{x} e^{r \log x} = \frac{r}{x} x^r = rx^{r-1}$. ∎

Example 4.3.11

We will differentiate the function $y = x^x$, with $x > 0$. We use the same technique as in the proof of the power rule.

$$(x^x)' = (e^{x \log x})' = e^{x \log x}(x \log x)' = x^x(1 + \log x).$$ □

At this point, we have a catalogue of basic functions whose derivatives are known, and techniques for dealing with both their algebraic combinations and their compositions. So we can differentiate everything that can be described by such combinations.

Task 4.3.12

Differentiate $\exp(\sqrt{x^2 + \arctan x})$.

The trick that we used to differentiate x^r and x^x can be formalized into a method called **logarithmic differentiation**. We start by noting that if f takes only positive values, then f is differentiable if and only if $g = \log \circ f$ is differentiable. (Since $g = \log \circ f \implies \exp \circ g = f$.) And then $g' = f'/f \implies f' = fg'$.

Example 4.3.13

Let us express the $(x^x)'$ calculation as logarithmic differentiation. Let $f(x) = x^x$ and $g(x) = \log f(x) = x \log x$. Then $g'(x) = 1 + \log x$. Hence, $f'(x) = f(x)g'(x) = x^x(1 + \log x)$. □

Task 4.3.14

Use logarithmic differentiation to find the derivative of the function $(\sin x)^{\cos x}$ for $x \in (0, \pi/2)$.

Example 4.3.15

We will show how differentiation can be used to discover the hyperbolic functions. Consider the hyperbola given by $x^2 - y^2 = 1$. We want to find functions $x(t), y(t)$ such that by varying t we can generate all the points on the branch of the hyperbola with $x > 0$.

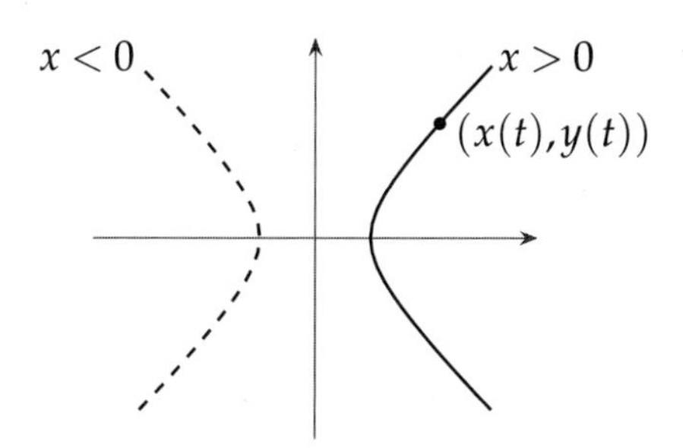

These functions have to satisfy $x(t)^2 - y(t)^2 = 1$. Differentiating both sides with respect to t, we get

$$x(t)x'(t) - y(t)y'(t) = 0 \quad \text{or} \quad \frac{x'(t)}{y'(t)} = \frac{y(t)}{x(t)}.$$

(In discovery mode, we do not worry about dividing by zero.) It is natural to try to arrange for $y'(t) = x(t)$ and $x'(t) = y(t)$. This leads to $y''(t) = y(t)$. We can easily check that functions of the form $y(t) = Ae^t + Be^{-t}$ satisfy this requirement. Now suppose we want the motion to start at $(1,0)$ when $t = 0$. This gives the equations $A + B = 0$ and $A - B = 1$, with solutions $A = 1/2$, $B = -1/2$. Hence,

$$y(t) = \frac{1}{2}e^t - \frac{1}{2}e^{-t} = \sinh t \quad \text{and} \quad x(t) = y'(t) = \frac{1}{2}e^t + \frac{1}{2}e^{-t} = \cosh t. \qquad \square$$

You may recall that we had already verified that the hyperbolic functions do trace the hyperbola.

Task 4.3.16

Prove that $\cosh' x = \sinh x$ *and* $\sinh' x = \cosh x$.

Inverse Hyperbolic Functions

Task 4.3.17

Show that $\sinh\colon \mathbb{R} \to \mathbb{R}$ *is a strictly increasing bijection.*

It follows that the hyperbolic sine function has an inverse that is strictly increasing as well as continuous. We denote it by $\sinh^{-1}$ or arsinh, following the same pattern as for inverse trigonometric functions.

The hyperbolic cosine function is even and hence not one-one. Therefore, we restrict the domain to $[0,\infty)$ and try again.

Task 4.3.18

Show that $\cosh\colon [0,\infty) \to [1,\infty)$ *is a strictly increasing bijection.*

The corresponding inverse function is called $\cosh^{-1}$ or arcosh. It is also strictly increasing and continuous.

Task 4.3.19

Prove that $(\sinh^{-1} x)' = \dfrac{1}{\sqrt{x^2+1}}$ *and* $(\cosh^{-1} x)' = \dfrac{1}{\sqrt{x^2-1}}$.

Exercises for § 4.3

1. Differentiate the following functions:

(a) $f(x) = \sqrt{x^2+1}$.

(b) $g(x) = \sin^2 x - \sin x^2$.

(c) $h(x) = \sin(\sin x)$.

(d) $k(x) = e^{-1/x^2}$.

(e) $\ell(x) = \log(1+x^2)$.

(f) $r(x) = \pi^x - x^\pi$.

2. Use implicit differentiation to find the tangent line at a point (x_0,y_0) on the hyperbola given by $\dfrac{x^2}{a^2} - \dfrac{y^2}{b^2} = 1$.

3. Consider the families of curves $y = cx^2$ and $x^2 + 2y^2 = k^2$, where c varies over all real numbers and k over all non-zero ones.

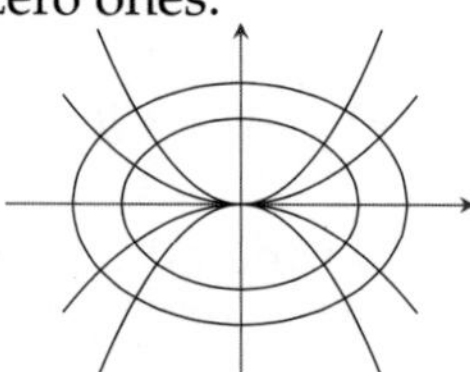

Show that whenever a curve from one family cuts a curve from the other family, their tangent lines are perpendicular to each other.

4. Graph the following functions (keep in mind that arcsine inverts only a part of sine):

(a) $\sin(\arcsin x)$.

(b) $\arcsin(\sin x)$.

5. Let $\cot^{-1}$ or arccot denote the inverse of $\cot\colon (0,\pi) \to \mathbb{R}$. Show that

$$\operatorname{arccot}'(x) = \frac{-1}{1+x^2}.$$

6. Let $\sec^{-1}$ or arcsec denote the inverse of $\sec\colon (0,\frac{\pi}{2}) \cup (\frac{\pi}{2},\pi) \to \mathbb{R} \setminus [-1,1]$. Show that

$$\operatorname{arcsec}' x = \frac{1}{|x|\sqrt{x^2-1}}.$$

7. Let $\csc^{-1}$ or arccsc denote the inverse of $\csc\colon (-\frac{\pi}{2},0) \cup (0,\frac{\pi}{2}) \to \mathbb{R} \setminus [-1,1]$. Show that

$$\operatorname{arccsc}' x = \frac{-1}{|x|\sqrt{x^2-1}}.$$

8. Use logarithmic differentiation to differentiate $y = x^{x^x}$, with $x > 0$.

9. Let f and g be differentiable functions, with $f(x) > 0$ for every x. Prove that $(f(x)^{g(x)})' = f(x)^{g(x)}\left(g'(x)\log f(x) + g(x)\dfrac{f'(x)}{f(x)}\right)$.

10. Prove the following formulas for the inverse hyperbolic functions:

(a) $\sinh^{-1} x = \log(x + \sqrt{x^2+1})$.

(b) $\cosh^{-1} x = \log(x + \sqrt{x^2-1})$.

4.4 The First Fundamental Theorem

As we have seen with the logarithm function, important functions may be defined not through an algebraic formula but as indefinite integrals. Such functions are automatically continuous. Are they differentiable? Can we calculate their derivatives? Let us consider the few simple examples we already know.

$f(x)$	a	$F(x) = \int_a^x f(t)\,dt$	$F'(x)$
$\mathrm{sgn}(x)$	0	$\|x\|$	$\mathrm{sgn}(x)$, for $x \neq 0$
x	0	$x^2/2$	x
x^2	0	$x^3/3$	x^2
$1/x$	1	$\log x$	$1/x$, for $x > 0$

It seems that on differentiation, F always reverts to f. Keeping the first example in mind, we should add the qualifier "where f is continuous".

This result is to be expected on physical grounds. Recall that one motivation for defining integration the way we did was to to obtain displacement from velocity. If we have done so correctly, then differentiating the integral that represents displacement should give us back the velocity.

The following rough argument gives geometric insight and also brings out the need for assuming continuity. The change $F(x+h) - F(x)$ is approximated by the area of the trapezium whose vertices are at x, $x+h$ on the x-axis and the corresponding points on the graph of f.

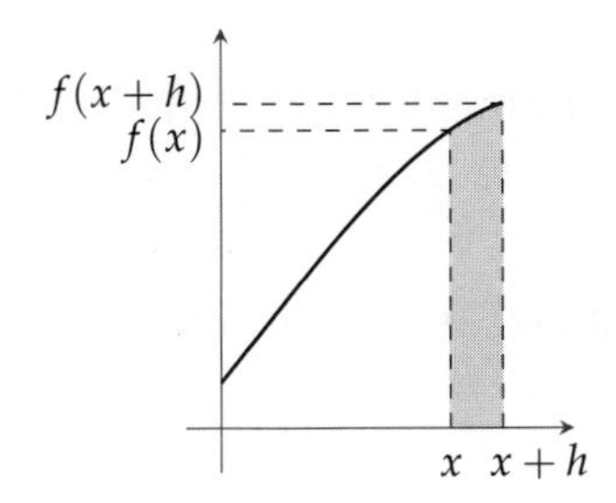

Hence, $\dfrac{F(x+h)-F(x)}{h} \approx \dfrac{1}{h}\dfrac{f(x)+f(x+h)}{2}h = \dfrac{f(x)+f(x+h)}{2} \to f(x)$, as $h \to 0$.

Theorem 4.4.1 (First Fundamental Theorem)

Let I be an interval and $f\colon I \to \mathbb{R}$ be integrable on each subinterval $[\alpha, \beta] \subseteq I$. Fix $a \in I$ and consider the indefinite integral $F\colon I \to \mathbb{R}$ defined by

$$F(x) = \int_a^x f(t)\,dt.$$

Then $F'(c) = f(c)$ if f is continuous at c. (If c is an end-point, use the appropriate one-sided notion of continuity and differentiability.)

Proof. For $h \neq 0$ we have

$$F(c+h) - F(c) = \int_a^{c+h} f(t)\,dt - \int_a^c f(t)\,dt = \int_c^{c+h} f(t)\,dt.$$

Hence,

$$\begin{aligned} F(c+h) - F(c) - hf(c) &= \int_c^{c+h} f(t)\,dt - \int_c^{c+h} f(c)\,dt \\ &= \int_c^{c+h} (f(t) - f(c))\,dt. \end{aligned}$$

Define $\varphi(h) = \frac{1}{h}\int_c^{c+h}(f(t) - f(c))\,dt$. Consider $\epsilon > 0$. If f is continuous at c, there is a $\delta > 0$ such that $|t - c| < \delta$ implies $|f(t) - f(c)| < \epsilon$. Therefore, if $0 < |h| < \delta$, we obtain

$$|\varphi(h)| = \frac{1}{|h|}\left|\int_c^{c+h} (f(t) - f(c))\,dt\right| \leq \frac{1}{|h|}|h|\epsilon = \epsilon.$$

Therefore, $\varphi(h) \to 0$ as $h \to 0$, and so $F'(c) = f(c)$. ■

Example 4.4.2

Suppose we have to differentiate $F(x) = \int_0^x \sin\sqrt{t}\,dt$. By the first fundamental theorem we know immediately that $F'(x) = \sin\sqrt{x}$. We do not have to first find a closed form expression for $F(x)$! □

Example 4.4.3

We shall combine the first fundamental theorem and the chain rule to differentiate $G(x) = \int_x^{x^2} \sin\sqrt{t}\,dt$, $x > 0$. First, let $F(x) = \int_0^x \sin\sqrt{t}\,dt$, as in the previous example. Then

$$G(x) = \int_0^{x^2} \sin\sqrt{t}\,dt - \int_0^x \sin\sqrt{t}\,dt = F(x^2) - F(x).$$

Hence, by the chain rule,

$$G'(x) = F'(x^2)2x - F'(x) = 2x\sin\sqrt{x^2} - \sin\sqrt{x} = 2x\sin x - \sin\sqrt{x}. \quad \square$$

At this point, the first fundamental theorem is another method of differentiation. However, it has also established a bridge between the realms of differentiation and integration. We shall use this bridge in the next chapter to supply integration with a rich supply of computational techniques.

Exercises for § 4.4

1. Compute and graph $F(x)$.

(a) $F(x) = \int_0^x tH(t)\,dt.$

(b) $F(x) = \int_0^x \operatorname{sgn}|t|\,dt.$

(c) $F(x) = \int_0^x [t]\,dt.$

(d) $F(x) = \int_0^x (-1)^{[t]}\,dt.$

(H is the Heaviside step function)

2. Find the derivative $F'(x)$.

(a) $F(x) = \int_0^{x^3} (1+t^2)^{-3}dt.$

(b) $F(x) = \int_x^{x^2} (1+t^2)^{-3}dt.$

3. We say that f is a C^k function if its kth derivative $f^{(k)}$ exists at every point of the domain of f and is continuous. Prove that if f is a C^k function then its indefinite integral $F(x) = \int_a^x f(t)\,dt$ is a C^{k+1} function.

4. Differentiate $g(x) = \int_0^{\int_0^x \sqrt{1+t^3}\,dt} \sqrt{1+t^3}\,dt.$

5. If f is continuous on an interval I and g,h are differentiable with range in I, differentiate $\int_{g(x)}^{h(x)} f(t)\,dt.$

4.5 Extreme Values and Monotonicity

We now make a start on using differentiation to explore different aspects of functions. The first question we consider is "How large or small a value can a certain function take?"

We say a function $f\colon D \to \mathbb{R}$ has a **global** or **absolute maximum** at a point c if $f(c) \geq f(x)$ for every $x \in D$. Similarly, it has a **global** or **absolute minimum** at a point d if $f(d) \leq f(x)$ for every $x \in D$.

Various situations are possible:

1. f may not have an absolute maximum or an absolute minimum. For example, $f(x) = x\colon \mathbb{R} \to \mathbb{R}$ and $f(x) = 1/x : (0,1) \to \mathbb{R}$.
2. f may have an absolute maximum but not an absolute minimum. For example, $f(x) = -x^2\colon \mathbb{R} \to \mathbb{R}$.
3. f may have an absolute minimum but not an absolute maximum. For example, $f(x) = x^2\colon \mathbb{R} \to \mathbb{R}$.

4. f has both an absolute minimum and an absolute maximum. And these may occur several times. For example, $\sin\colon \mathbb{R} \to \mathbb{R}$.

The extreme value theorem (Theorem 3.5.2) tells us that if the function is continuous and the domain is of the form $[a,b]$ then an absolute maximum and an absolute minimum will certainly be present.

Absolute maxima and minima are collectively known as **absolute extrema**.

Local Extrema and Fermat's Theorem

We do not always need to know the very largest (or smallest) value of a function. If we kick a ball into some rough ground, we know it will stop in a depression, but it need not stop in the deepest one. The stopping point will be lower than the immediately surrounding points, but perhaps not lower than further off ones.

We say a function $f\colon D \to \mathbb{R}$ has a **local** or **relative maximum** at a point c if there is an open interval I containing c such that $f(c) \geq f(x)$ for every $x \in I \cap D$. Similarly, it has a **local** or **relative minimum** at a point d if there is an open interval I containing d such that $f(d) \leq f(x)$ for every $x \in I \cap D$. Local maxima and minima are collectively known as **local extrema**.

Obviously, an absolute maximum will also be a local maximum, and an absolute minimum will be a local minimum. But local extrema need not be absolute extrema, and a function could well have local extrema without having any absolute extreme.

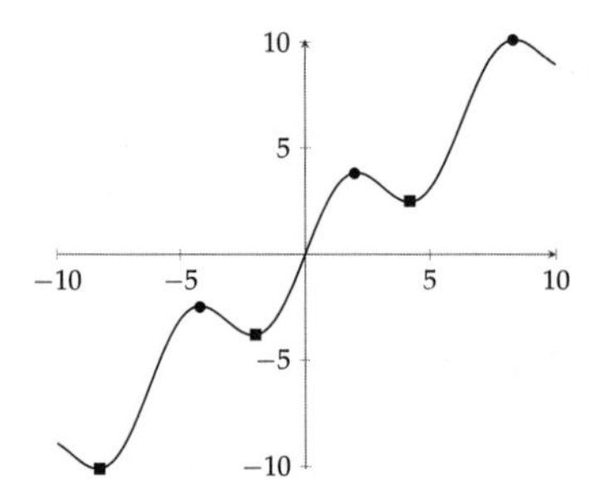

The function $x + 2\sin x$ has several local maxima (discs) and local minima (squares) that are not absolute extrema.

Theorem 4.5.1 (Fermat's Theorem)

Let $f(x)$ have a local extreme at an interior point c of an interval in its domain. Then either $f'(c)$ does not exist or $f'(c) = 0$.

Proof. Suppose $f'(c)$ exists. We have to show that $f'(c) = 0$. Suppose $f'(c) > 0$, that is,

$$\lim_{x \to c} \frac{f(x) - f(c)}{x - c} > 0.$$

Since the limit is positive, the values $\frac{f(x)-f(c)}{x-c}$ must themselves be positive once we are close to c. That is, there must be a $\delta > 0$ such that $0 < |x-c| < \delta \implies \frac{f(x)-f(c)}{x-c} > 0$. Then,

$$c-\delta < x < c \implies f(x) < f(c) \implies c \text{ is not a point of local minimum,}$$
$$c < x < c+\delta \implies f(x) > f(c) \implies c \text{ is not a point of local maximum.}$$

This rules out $f'(c) > 0$. We similarly rule out $f'(c) < 0$. ■

Here is an example of a local extreme that occurs at a point where f' does not exist.

Example 4.5.2

Consider $f(x) = |x|$. It has a local minimum at $x = 0$ but $f'(0)$ is not defined. □

In the next example we have a point where f' is zero but it is not a local extreme.

Example 4.5.3

Consider $f(x) = x^3$. Then $f'(0) = 0$ but there is no local extreme at $x = 0$. □

We call a point c a **critical point** or **critical number** of $f(x)$ if it is an interior point c of an interval in the domain and either $f'(c)$ does not exist or $f'(c) = 0$.

Let $f(x)$ have an interval $[a,b]$ as domain. Then, by fermat's theorem, the local extremes of f occur either at critical points or at the end-points of the domain.

Example 4.5.4

Consider $f(x) = x^3 - 3x + 1$ with domain $[0,3]$. We make the following calculations:

1. Calculate the function values at the endpoints: $f(0) = 1$ and $f(3) = 19$.
2. Find the critical points. Since f is differentiable we only have to look for $f'(c) = 0$. This gives $3c^2 - 3 = 0$ or $c = \pm 1$. Thus $c = 1$ is the only critical point (in the given domain).
3. Calculate the function values at the critical points: $f(1) = -1$.

Thus, the candidates for absolute extremes are only $f(0) = 1$, $f(1) = -1$ and $f(3) = 19$. So, the absolute maximum is at $x = 3$ and the absolute minimum is at $x = 1$. □

The common errors made by students in such problems are: ignoring the endpoints, and forgetting those critical points where f' does not exist.

Monotonicity

We have seen earlier (Theorem 4.1.11) that if f is monotonic and differentiable, then f' does not change sign. Now we shall prove the converse.

Theorem 4.5.5 (Monotonicity Theorem)

Suppose I is an interval and $f\colon I \to \mathbb{R}$ is differentiable on I.

1. *If $f'(x) > 0$ for every $x \in I$ then f is strictly increasing.*
2. *If $f'(x) \geq 0$ for every $x \in I$ then f is increasing.*

We also have the corresponding statements regarding negative derivatives and decreasing functions.

Proof. First, suppose $f'(x) > 0$ for every $x \in I$. Let $p,q \in I$ with $p < q$. We have to show that $f(p) < f(q)$.

Since f is continuous on $[p,q]$ it achieves its maximum and minimum over this interval. By fermat's theorem the points of maximum and minimum can only be the endpoints p,q.

If the maximum and minimum values are equal, then f is a constant function, and $f' = 0$. So, they are not equal and $f(p) \neq f(q)$. Suppose $f(q)$ is the minimum value over $[p,q]$. Then

$$f'(q) = \lim_{x \to q-} \frac{f(x) - f(q)}{x - q} \leq 0,$$

since $p < x < q$ implies $f(x) \geq f(q)$. This contradicts the positivity of f'. It follows that $f(q)$ is the maximum value over $[p,q]$ and hence $f(p) < f(q)$.

Now suppose we only have $f'(x) \geq 0$ for every $x \in I$. Let $p,q \in I$ with $p < q$. Take any $\epsilon > 0$ and consider the function $g(x) = f(x) + \epsilon x$. Then $g'(x) = f'(x) + \epsilon > 0$ and g is strictly increasing. Now

$$g(p) < g(q) \implies f(q) - f(p) > \epsilon(p - q).$$

Thus, $f(q) - f(p)$ is greater than every negative number and hence must be non-negative. ∎

Example 4.5.6

We will show that the equation $x^3 + 3x + 1 = 0$ has exactly one solution.

Let $f(x) = x^3 + 3x + 1$. Then $f(x)$ is a polynomial, hence continuous and differentiable everywhere.

We have $f(-1) = -3 < 0$ and $f(0) = 1 > 0$. So, by intermediate value theorem applied to $f\colon [-1,0] \to \mathbb{R}$, we have a $c \in (-1,0)$ such that $f(c) = 0$, that is, $c^3 + 3c + 1 = 0$. (We have not found c but we know it is somewhere in there. If required, we can use the bisection method to further narrow the range in which it lies.)

Now calculate the derivative: $f'(x) = 3x^2 + 3 \geq 3 > 0$, hence f is strictly increasing, therefore one-one. So, there can only be one c with $f(c) = 0$. □

Theorem 4.5.7

Let f,g be differentiable functions from an interval I to $\mathbb{R}$.

1. *If* $f'(x) = 0$ *for each* $x \in I$ *then* $f(x) =$*constant.*
2. *If* $f'(x) = g'(x)$ *for each* $x \in I$ *then* $f(x) = g(x)+$*constant.*

Proof. Exercise. ∎

These results often enable us to characterize a function by properties of its derivative. For example, suppose a function f satisfies $f'(x) = f(x)$ for every x. One such function is e^x. In fact, every function of the form Ae^x has this property. So we wonder whether they are the only ones to have this property. We have a positive answer:

Theorem 4.5.8
If $f'(x) = kf(x)$ *on an interval* I *then* $f(x) = Ae^{kx}$.

Proof. Consider $g(x) = f(x)e^{-kx}$. Then

$$g'(x) = f'(x)e^{-kx} - kf(x)e^{-kx} = kf(x)e^{-kx} - kf(x)e^{-kx} = 0.$$

Hence, $g(x) = A$, a constant, and $f(x) = Ae^{kx}$. ∎

Task 4.5.9
Suppose $f\colon \mathbb{R} \to \mathbb{R}$ *is differentiable,* $f' = f$ *and* $f(0) = 1$. *Show that* $f(x) = e^x$.

Now consider the sine and cosine functions. They satisfy the relation $f'' = -f$. More generally, every combination $a\cos x + b\sin x$ satisfies this relation. Again, we wonder if they are the only ones. We begin with a special case.

Task 4.5.10
Suppose $f\colon \mathbb{R} \to \mathbb{R}$ *is differentiable,* $f'' = -f$ *and* $f(0) = f'(0) = 0$. *Show that* $f(x) = 0$. (Hint: Differentiate the function $f^2 + (f')^2$).

Task 4.5.11
Suppose $f\colon \mathbb{R} \to \mathbb{R}$ *is differentiable and* $f'' = -f$. *Show that if* $f(0) = a$ *and* $f'(0) = b$, *then* $f(x) = a\cos x + b\sin x$.

An equation involving a function and its derivatives is called a **differential equation**. Mathematical modeling of physical and economic systems usually leads to differential equations and the goal is to find all the functions that satisfy them. These functions are called their **solutions**. We have seen that the solutions of $f' = kf$ have the form $f(x) = Ae^{kx}$ while solutions of $f'' = -f$ have the form $f(x) = a\cos x + b\sin x$. These are two simple but quite important examples of differential equations. We shall explore differential equations further in §5.6 and in the supplementary exercises of Chapter 5.

We close with a curious property of derivatives. The derivative of a function need not be continuous. Nevertheless, it always has the intermediate value property!

Theorem 4.5.12 (Darboux's Theorem)

Let $a < b$ and $f\colon [a,b] \to \mathbb{R}$ be differentiable. Suppose L is strictly between $f'(a)$ and $f'(b)$. Then there is $c \in (a,b)$ such that $f'(c) = L$.

Proof. First, suppose $L = 0$. Then $f'(a)f'(b) < 0$. Since f is continuous, it assumes its maximum and minimum values on $[a,b]$. If either occurs at a point $c \in (a,b)$, then by fermat's theorem we have $f'(c) = 0$. So, suppose they are assumed only on a,b. We may assume that the maximum is assumed on a and the minimum on b. Then $f'(a), f'(b) \leq 0$, a contradiction. This resolves the $L = 0$ case.

Now let $L \neq 0$. Consider $g(x) = f(x) - Lx$. Then $g'(a) = f'(a) - L < 0 < f'(b) - L = g'(b)$. So, there is $c \in (a,b)$ such that $g'(c) = 0$, hence $f'(c) = L$. ∎

Exercises for § 4.5

1. Consider the function $f(x) = xe^{-x^2}$ with domain $[-1/2, 2]$.

(a) Identify the critical points.

(b) Find the absolute maximum and minimum values of the function.

2. Consider the function $f(x) = 1 - x^{2/3}$ with domain $[-1,1]$.

(a) Identify the critical points.

(b) Find the absolute maximum and minimum values of the function.

3. Consider the rectangle inscribed inside a triangle as shown below. What is its maximum possible area?

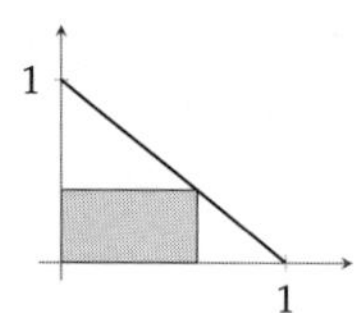

4. Prove that $f(x) = \dfrac{x^3}{3} + 2x - 2\cos x$ has exactly one zero.

5. Show that $x^2 = x\sin x + \cos x$ for exactly two values of x.

6. Suppose that $f\colon \mathbb{R} \to \mathbb{R}$ satisfies $f^{(n+1)} = 0$. Prove that f is a polynomial with degree n or less.

7. Find the equation $y(x)$ of a curve such that the tangent line at the point (x,y) intersects the x-axis at $x - 1$.

8. Suppose a function f satisfies the differential equation $f'(x) = k(M - f(x))$. Find the general form of f. (Hint: Consider $g(x) = M - f(x)$.)

9. Find all functions $f\colon \mathbb{R} \to \mathbb{R}$ with the property that $x \neq y$ implies $f(x) - f(y) \leq (x-y)^2$.

10. Let f be a differentiable function such that every tangent line to its graph passes through the origin. Show that the graph of f is a line through the origin.

11. Prove that there is no function f such that $f'(x) = \operatorname{sgn}(x)$ for every $x \in \mathbb{R}$.

12. Let I be an interval and $f\colon I \to \mathbb{R}$ a differentiable function such that $f'(x)$ is never zero. Show that f is strictly monotonic.

13. Prove that if a derivative f' is monotonic, then it is continuous.

4.6 Derivative Tests and Curve Sketching

We defined local extrema and critical points in §4.5. Fermat's theorem informs us that a local extreme in the interior of an interval can only occur at a critical point. At the same time, it is possible that a critical point fails to be a local extreme. A critical point of the last kind is called a **saddle point**. Naturally, we wish to be able to classify a given critical point as a local maximum, local minimum, or saddle point. One way to do this is to use the values of the derivative of the function on either side of the critical point. For example, consider the following graph:

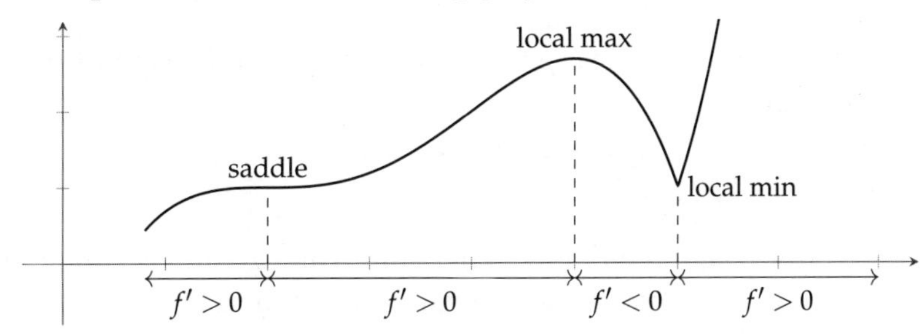

As we move from left to right and pass through the saddle point, the derivative changes from positive to zero and back to positive. Thus, it has the same sign on each side of the saddle point. As we pass through the local maximum the derivative changes from positive to negative, and at the local minimum it changes from negative to positive. These observations give a test for deciding whether a critical point is a local extreme and of what kind.

Theorem 4.6.1 (First Derivative Test)

Let a function f be continuous on an interval (a,b) and let $c \in (a,b)$ be a critical point of f. Suppose f is differentiable on (a,b) except perhaps at c. Then

1. *If $f'(x) \geq 0$ for $x \in (a,c)$ and $f'(x) \leq 0$ for $x \in (c,b)$, then f has a local maximum at c.*
2. *If $f'(x) \leq 0$ for $x \in (a,c)$ and $f'(x) \geq 0$ for $x \in (c,b)$, then f has a local minimum at c.*
3. *If f' has the same sign on each side of c, then f has a saddle point at c.*

Proof. Suppose $f'(x) \geq 0$ for $x \in (a,c)$ and $f'(x) \leq 0$ for $x \in (c,b)$. By the monotonicity theorem, f is increasing on (a,c) and decreasing on (c,b). The continuity of f then gives us that f is increasing on $(a,c]$ and decreasing on $[c,b)$. For, suppose there is $x_1 < c$ with $f(x_1) > f(c)$. By the intermediate value theorem, there is $x_2 \in (x_1,c)$ with $f(x_2) = \frac{1}{2}(f(x_1)+f(c)) < f(x_1)$, violating the fact that f is increasing on (a,c). This shows that f is increasing on $(a,c]$. Similarly, f is decreasing on $[c,b)$.

It follows that $f(c)$ is the largest value taken by $f(x)$ on (a,b) and hence there is a local maximum at c.

Similarly, if $f'(x) < 0$ for $x \in (a,c)$ and $f'(x) > 0$ for $x \in (c,b)$, there is a local minimum at c.

But if $f'(x)$ has the same sign on both sides of c, then values on one side are higher and on the other are lower. Hence, there is neither a local maximum nor a local minimum at c. ■

Example 4.6.2

Consider $f(x) = x^2e^x$. This is a differentiable function, so its critical points are given by the derivative being zero. We have $f'(x) = 2xe^x + x^2e^x = x(x+2)e^x$. Hence,

$$f'(c) = 0 \iff c(c+2) = 0 \iff c = 0, -2$$

To identify the nature of the critical points we have to find the sign of the derivative on either side of them:

	$x < -2$	$-2 < x < 0$	$x > 0$
$f'(x)$	$+$	$-$	$+$

By the first derivative test, there is a local maximum at -2 and a local minimum at 0. The function increases on $(-\infty,-2)$ to the value $4e^{-2} \approx 0.54$ at -2, then decreases to the value 0 at 0. Beyond 0 it increases again. Note that $\lim_{x\to\infty} x^2e^x = \infty$ and $\lim_{x\to-\infty} x^2e^x = \lim_{x\to\infty} \frac{x^2}{e^x} = 0$.

Here is the graph of $f(x)$ showing these features:

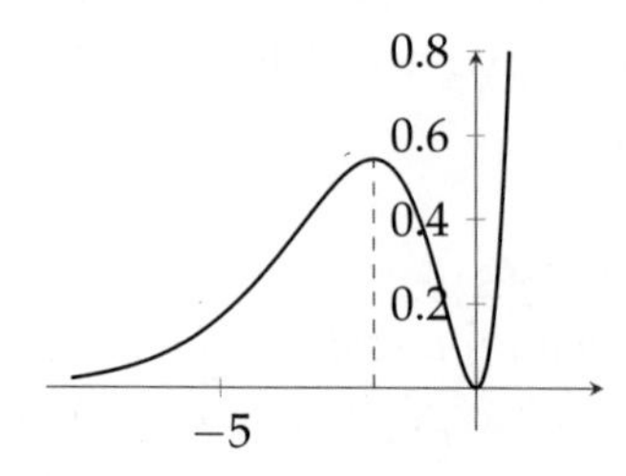

□

Example 4.6.3

Consider $f(x) = x + \sin x$ on $[0,2\pi]$. We have $f'(x) = 1 + \cos x$ and so $f'(c) = 0 \iff \cos c = -1$. Thus the only critical point in the given domain is $c = \pi$. We have

$f'(x) = 1 + \cos(x) \geq 0$ always and so $f(x)$ is monotonically increasing in the domain. In particular, although there is a horizontal tangent at $c = \pi$, it is not a local extreme but a saddle point.

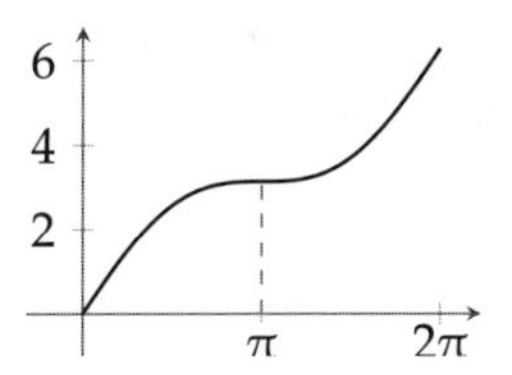

□

We have seen that the first derivative tells us whether a function is increasing or decreasing, and how fast. We can apply the same logic to get information from the second derivative. The sign of f'' will determine whether f' is rising or falling, and therefore whether the graph of f rises or falls with increasing or decreasing steepness.

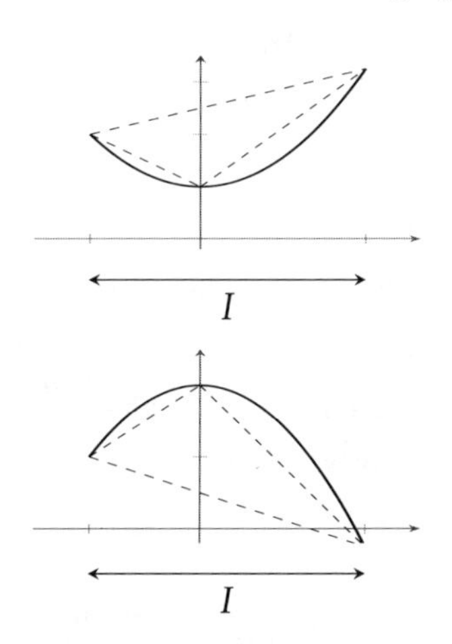

A function $f\colon I \to \mathbb{R}$ is said to be **convex** on I if its graph over *every* interval $[a,b]$ in I lies *below* the secant line through the endpoints of the graph over that interval.

The graph of a convex function turns upwards as we move from left to right. The special exercises at the end of this chapter provide more insight related to this comment.

Similarly, $f\colon I \to \mathbb{R}$ is said to be **concave** on I if its graph over *every* interval $[a,b]$ in I lies *above* the secant line through the endpoints of the graph over that interval. The graph of a concave function turns downwards as we move from left to right.

The formal definitions are as follows: Consider a function $f\colon I \to \mathbb{R}$ where I is an interval. Then

1. f is called **convex** on I if for every $a,x,b \in I$ with $a < x < b$, we have

$$f(x) \leq f(a) + \frac{f(b) - f(a)}{b - a}(x - a) \tag{4.1}$$

2. f is called **concave** on I if for every $a,x,b \in I$ with $a < x < b$, we have

$$f(x) \geq f(a) + \frac{f(b) - f(a)}{b - a}(x - a) \tag{4.2}$$

<u>**Task 4.6.4**</u>

Can a function be both convex and concave?

If the inequalities 4.1 and 4.2 are strict, we call f **strictly convex** and **strictly concave** respectively.

The convex or concave nature of a function is called its **convexity**. A function may be convex over one interval and concave over another. A point where the function is continuous and switches from strictly convex on one side to strictly concave on the other is called an **inflection point** of the function. For example, the sine function is strictly convex on $[-\pi, 0]$ and strictly concave on $[0, \pi]$. Hence, 0 is an inflection point.

Theorem 4.6.5 (Convexity Test)

Let f be twice differentiable on an interval I. Then

1. *$f'' \geq 0$ on I implies f is convex on I.*
2. *$f'' \leq 0$ on I implies f is concave on I.*
3. *If f'' is continuous at an inflection point c then $f''(c) = 0$.*

If the inequalities are strict, so is the convexity.

Proof. First, suppose $f'' \geq 0$ on I. Let $c, d \in I$ with $c < d$. The secant from $(c, f(c))$ to $(d, f(d))$ has the equation

$$y = f(c) + \frac{f(d) - f(c)}{d - c}(x - c).$$

Consider the difference $g(x) = f(c) + \dfrac{f(d) - f(c)}{d - c}(x - c) - f(x)$. Note that $g(c) = g(d) = 0$. Further, $g'' = -f'' \leq 0$ and so g' is a decreasing function.

We wish to show that for each $x \in (c, d)$, $g(x) \geq 0$. Suppose that $g(x) < 0$ at some point $x \in (c, d)$. By the monotonicity theorem, we obtain α, β as follows:

- $\alpha \in (c, x)$ and $g'(\alpha) < 0$,
- $\beta \in (x, d)$ and $g'(\beta) > 0$.

This contradicts g' being a decreasing function. Hence, $g(x) < 0$ is impossible, and f is convex.

If $f'' \leq 0$ on I, apply the first part to $-f$.

For the third part, suppose $f''(c) > 0$. Then, by continuity, $f'' > 0$ in an interval I centered at c. So, f is convex on I and c is not an inflection point. This rules out $f''(c) > 0$. We can similarly rule out $f''(c) < 0$. ■

Theorem 4.6.6 (Second Derivative Test)

Let f have a critical point at c and f'' be continuous in an open interval containing c. Then

1. *$f''(c) > 0$ implies there is a local minimum at c.*
2. *$f''(c) < 0$ implies there is a local maximum at c.*

Proof. Let $f''(c) > 0$. By continuity, $f'' > 0$ in an open interval containing c. Then f' is strictly increasing in that interval. Hence, f' changes from negative to positive at c, and there is a local minimum at c (by the first derivative test).

For the second part, apply the first part to $-f$. ■

Example 4.6.7

Let $f(x) = x^2e^x$. We saw earlier that this has a local maximum at -2 and a local (as well as absolute) minimum at 0. Now we identify the inflection points and convexity. First, we calculate the second derivative:

$$f'(x) = (x^2 + 2x)e^x \implies f''(x) = (x^2 + 4x + 2)e^x.$$

Then we identify the possible inflection points.

$$f''(c) = 0 \iff c^2 + 4c + 2 = 0 \iff c = -2 \pm \sqrt{2} \approx -3.4, -0.6.$$

	$x < -2-\sqrt{2}$	$-2-\sqrt{2} < x < -2+\sqrt{2}$	$x > -2+\sqrt{2}$
$f''(x)$	$+$	$-$	$+$
Convexity	Convex	Concave	Convex

Note that $f''(-2) = -2e^{-2} < 0$ confirms the local maximum at -2 and $f''(0) = 2 > 0$ confirms the local minimum at 0.

Here is the graph of $f(x)$ showing the convex parts as solid curves and the concave part as a dashed curve:

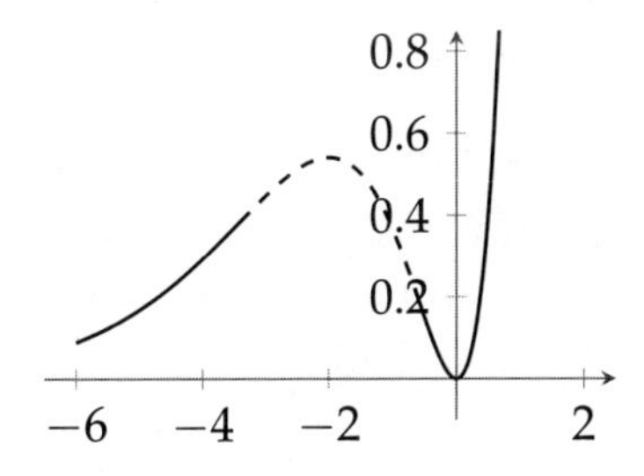

□

Example 4.6.8

Let $f(x) = x + \sin(x)$ on the interval $[0, 2\pi]$. We saw in Example 4.6.3 that the only critical point is at $x = \pi$ and this is not a local maximum or minimum. Now we calculate the second derivative.

$$f'(x) = 1 + \cos(x) \implies f''(x) = -\sin(x) \implies f''(c) = 0 \text{ at } x = 0, \pi, 2\pi.$$

	$0 < x < \pi$	$\pi < x < 2\pi$
$f''(x)$	$-$	$+$
Convexity	Concave	Convex

□

Remark: If c is a critical point of f and $f''(c) = 0$, the second derivative test is *inconclusive*. For example, each of the following functions has a critical point at $c = 0$ with $f''(c) = 0$ but the first has a local maximum there, the next has a local minimum, and the last has a saddle point: $-x^4$, x^4, x^3.

Curve Sketching

We have seen how first and second derivative calculations can give us key features of a graph. We can capture all the essential aspects of a function's behavior by supplementing these with the following: domain, axis-intercepts, points of discontinuity, symmetry (even, odd, periodic), asymptotes (vertical, horizontal, slant).

Example 4.6.9

$f(x) = \arctan\left(\dfrac{x-1}{x+1}\right)$.

Domain: Since the domain of arctan is $\mathbb{R}$, the only point where this expression is undefined is $x = -1$. Hence, the domain is $\mathbb{R} \setminus \{-1\}$. Note also that $f(x) \in (-\pi/2, \pi/2)$.

Intercepts: The function is zero at $x = 1$. It cuts the y-axis at $y = f(0) = \arctan(-1) = -\pi/4$.

Symmetry: We have $f(2) = \arctan(1/3)$ and $f(-2) = \arctan(3)$. They are positive and unequal (arctan is one-one), so $f(x)$ is neither even nor odd.

Vertical Asymptotes: As $f(x)$ is continuous on its domain, the only possibility of vertical asymptotes is at $x = -1$. So we calculate the one-sided limits there:

$$\lim_{x\to-1+} \frac{x-1}{x+1} = \lim_{t\to0+} \frac{t-2}{t} = -\infty \implies \lim_{x\to-1+} \arctan\left(\frac{x-1}{x+1}\right) = -\frac{\pi}{2},$$

$$\lim_{x\to-1-} \frac{x-1}{x+1} = \lim_{t\to0+} \frac{t+2}{t} = \infty \implies \lim_{x\to-1-} \arctan\left(\frac{x-1}{x+1}\right) = \frac{\pi}{2}.$$

Since the limits are finite there is no vertical asymptote at $x = -1$. They are still useful in plotting the graph.

Horizontal Asymptotes:

$$\lim_{x\to\infty} \left(\frac{x-1}{x+1}\right) = 1 \quad \implies \quad \lim_{x\to\infty} \arctan\left(\frac{x-1}{x+1}\right) = \arctan(1) = \frac{\pi}{4},$$

$$\lim_{x\to-\infty} \left(\frac{x-1}{x+1}\right) = 1 \quad \implies \quad \lim_{x\to-\infty} \arctan\left(\frac{x-1}{x+1}\right) = \arctan(1) = \frac{\pi}{4}.$$

Therefore, $y = \pi/4$ is a horizontal asymptote on both sides.

Critical Points:

$$\begin{aligned}
\frac{d}{dx}\arctan\left(\frac{x-1}{x+1}\right) &= \frac{1}{1+\left(\frac{x-1}{x+1}\right)^2}\frac{d}{dx}\left(\frac{x-1}{x+1}\right) \\
&= \frac{(x+1)^2}{(x+1)^2+(x-1)^2} \times \frac{(x+1)-(x-1)}{(x+1)^2} \\
&= \frac{2}{2x^2+2} = \frac{1}{x^2+1}.
\end{aligned}$$

The derivative $f'(x)$ always exists and is never zero, so there are no critical points. In fact, $f'(x) > 0$ and so f is strictly increasing on any interval in its domain. So, f is strictly increasing on $(-\infty, -1)$ and also on $(-1, \infty)$.

Convexity: $f''(x) = \frac{d}{dx}\frac{1}{x^2+1} = \frac{-2x}{(x^2+1)^2}$.

We have $f''(x) > 0$ for $x < 0$ and $f''(x) < 0$ for $x > 0$. So, f is convex on $(-\infty, -1)$ and on $(-1, 0)$. It is concave on $(0, \infty)$. The only inflection point is $x = 0$.

Here is the graph of f:

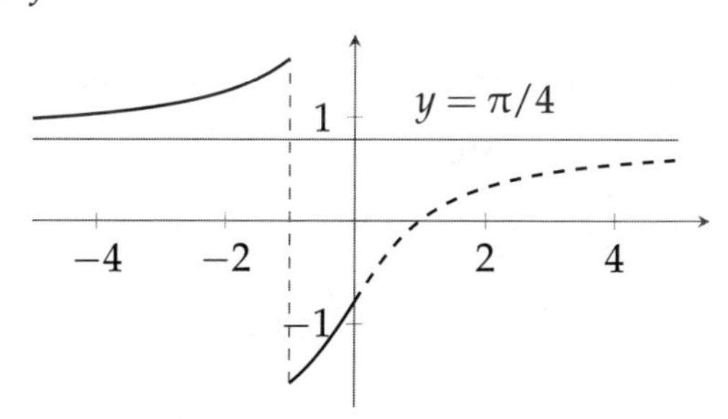

□

Example 4.6.10

Let $f(x) = \frac{x^2}{x^2+9}$.

Domain: Clearly, the domain is $\mathbb{R}$. And the image is in $[0, 1)$.

Intercepts: The function is zero at $x = 0$.

Symmetry: The function is even.

Vertical Asymptotes: As f is continuous on $\mathbb{R}$ it has no vertical asymptotes.

Horizontal Asymptotes:

$$\begin{aligned}
\lim_{x\to\infty} \frac{x^2}{x^2+9} &= \lim_{x\to\infty} \frac{1}{1+9/x^2} = 1, \\
\lim_{x\to-\infty} \frac{x^2}{x^2+9} &= \lim_{x\to-\infty} \frac{1}{1+9/x^2} = 1.
\end{aligned}$$

So $y = 1$ is a horizontal asymptote on each side.

Critical Points:

$$\frac{d}{dx}\left(\frac{x^2}{x^2+9}\right) = \frac{2x(x^2+9) - x^2(2x)}{(x^2+9)^2} = \frac{18x}{(x^2+9)^2}.$$

The only critical point is $x = 0$. We have $f'(x) < 0$ for $x < 0$ and $f'(x) > 0$ for $x > 0$. So, first derivative test implies that there is a local minimum at $x = 0$. Note that $f(0) = 0$.

Convexity:

$$\frac{d^2}{dx^2}\left(\frac{x^2}{x^2+9}\right) = -\frac{d}{dx}\left(\frac{18x}{(x^2+9)^2}\right) = 18\frac{(x^2+9) - 4x^2}{(x^2+9)^3} = 54\frac{3 - x^2}{(x^2+9)^3}.$$

The possible inflection points are $x = \pm\sqrt{3}$. Note that $f(\pm\sqrt{3}) = 0.25$.

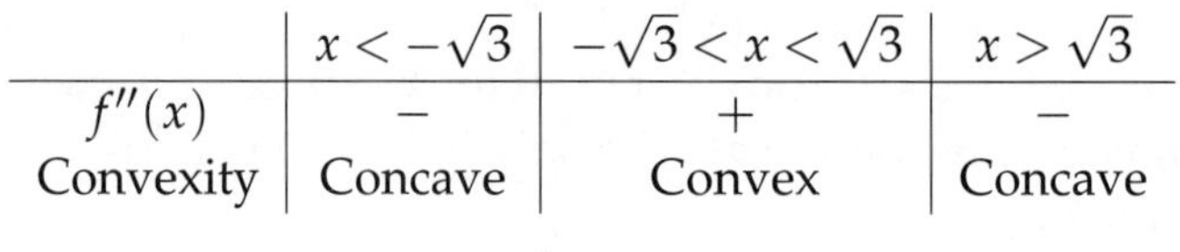

	$x < -\sqrt{3}$	$-\sqrt{3} < x < \sqrt{3}$	$x > \sqrt{3}$
$f''(x)$	$-$	$+$	$-$
Convexity	Concave	Convex	Concave

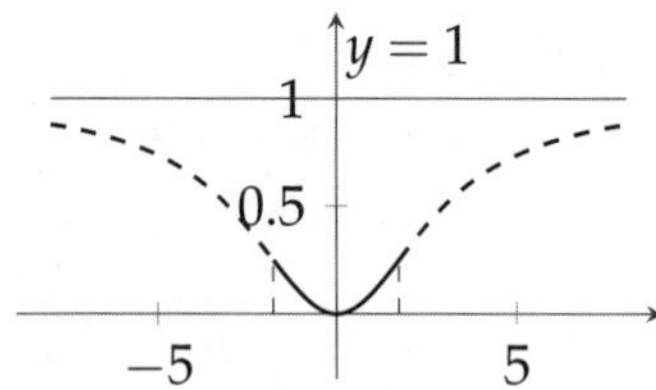

□

Example 4.6.11

Let $f(x) = (x - x^3)^{1/3}$. The computations are a little lengthy, so we just give the results. Verifying them will be an excellent test of your algebra and differentiation skills!

Domain: The domain is $\mathbb{R}$.

Intercepts: The function is zero at $x = 0, \pm 1$.

Symmetry: The function is odd.

Vertical Asymptotes: As f is continuous on $\mathbb{R}$, it has no vertical asymptotes.

Slant Asymptotes: It is evident that the function has infinite limits at $\pm\infty$ and so has no horizontal asymptotes. The form of the function suggests it should behave like $y = -x$ for large x. Let us look for slant asymptotes:

$$\begin{aligned}
a &= \lim_{x\to\infty} \frac{(x - x^3)^{1/3}}{x} = \lim_{x\to\infty}\left(\frac{1}{x^2} - 1\right)^{1/3} = -1,\\
b &= \lim_{x\to\infty}((x - x^3)^{1/3} - (-x))\\
&= \lim_{x\to\infty} \frac{x^{-1}}{(\frac{1}{x^2} - 1)^{2/3} - (\frac{1}{x^2} - 1)^{1/3} + 1} = 0.
\end{aligned}$$

This confirms that $y = ax + b = -x$ is a slant asymptote. By symmetry, it is a slant asymptote on both sides.

Critical Points:

$$\frac{df}{dx}(x) = \frac{1-3x^2}{(x-x^3)^{2/3}}.$$

The derivative is undefined for $x = 0, \pm 1$. It has infinite limit at these points, indicating a vertical slope. The derivative is zero at $x = \pm 1/\sqrt{3}$. Thus, there are five critical points. The intervals of increase and decrease are:

x	$(-\infty,-1)$	$(-1,\frac{-1}{\sqrt{3}})$	$(\frac{-1}{\sqrt{3}},0)$	$(0,\frac{1}{\sqrt{3}})$	$(\frac{1}{\sqrt{3}},1)$	$(1,\infty)$
$f'(x)$	$-$	$-$	$+$	$+$	$-$	$-$

Convexity:

$$\frac{d^2f}{dx^2}(x) = \frac{2+6x^2}{9(x-x^3)^{5/3}}.$$

The second derivative is never zero. But there are possible inflection points where it is undefined, that is, at $x = 0, \pm 1$.

	$x < -1$	$-1 < x < 0$	$0 < x < 1$	$1 < x$
$f''(x)$	$-$	$+$	$-$	$+$
Convexity	Concave	Convex	Concave	Convex

The resulting graph is drawn below:

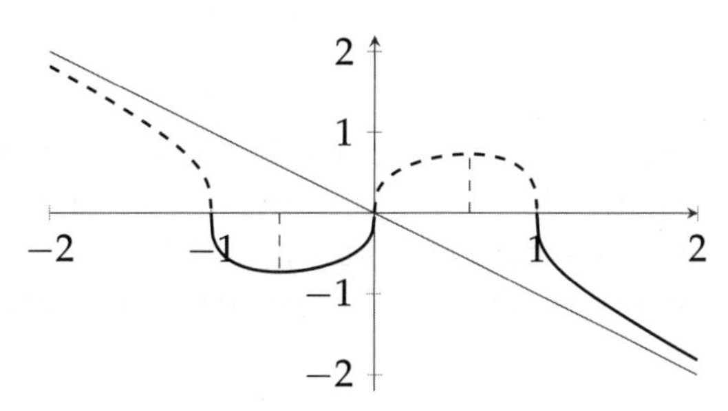

□

Exercises for § 4.6

1. Give an example of a function with a saddle point where the function is not differentiable.

2. For each of the following functions, use the first derivative to find and classify the critical points, and identify the intervals of decrease and increase:

(a) $f(x) = \sin^2 x,\ x \in [-\pi, \pi]$.

(b) $g(x) = e^{-x^2/2},\ x \in [-3,3]$.

(c) $h(x) = xe^{-x^2/2},\ x \in [-3,3]$.

(d) $k(x) = x|x-1|,\ x \in [-1,2]$.

3. For each function in the previous exercise, use the second derivative to find the inflection points, identify the intervals of convexity and concavity, and incorporate this information in their graphs.

4. Consider the function plotted in Example 4.6.9. Show that on each of the intervals $(-\infty,-1)$ and $(-1,\infty)$ it has the form $\arctan(x)+C$. Is this surprising?

5. Show that every cubic polynomial has exactly one inflection point.

6. Let $u\colon [0,1]\to\mathbb{R}$ be a twice continuously differentiable function that satisfies the differential equation $u''(x)=e^x u(x)$.

(a) Show that u does not have a positive local maximum or a negative local minimum in $(0,1)$.

(b) Suppose $u(0)=u(1)=0$. Show that $u=0$.

7. Suppose f satisfies $x^2f''(x)+4xf'(x)+2f(x)\geq 0$ on (a,b) and $f(a)=f(b)=0$. Show that $f(x)\leq 0$ on $[a,b]$. (Hint: Consider $g(x)=x^2f(x)$.)

8. Prove that $e^x > 1+x+x^2/2$ for $x>0$.

9. Consider a function f on an interval I. Prove the following:

(a) f is convex on I if and only if $f((1-t)x+ty)\leq(1-t)f(x)+tf(y)$ for any $x,y\in I$, and $t\in[0,1]$.

(b) f is concave on I if and only if $f((1-t)x+ty)\geq(1-t)f(x)+tf(y)$ for any $x,y\in I$, and $t\in[0,1]$.

Thematic Exercises

Convex Functions and Inequalities

In the following exercises, we explore what convex functions look like, and use them to obtain some famous inequalities. Corresponding results can be obtained for concave functions by noting that f is concave if and only if $-f$ is convex.

A1. Prove that f is convex on an interval I if and only if for each triple a,x,b in I with $a<x<b$, we have

$$\frac{f(x)-f(a)}{x-a}\leq\frac{f(b)-f(a)}{b-a}$$

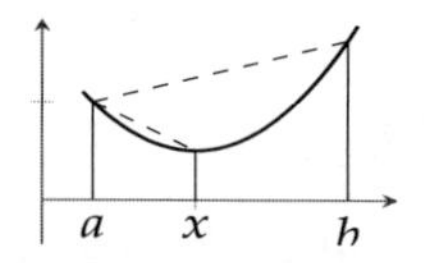

A2. Let $f\colon I\to\mathbb{R}$ be convex on an interval I, and $a,b,c,d\in I$ with $a<b\leq c<d$. Prove that

$$\frac{f(b)-f(a)}{b-a}\leq\frac{f(d)-f(c)}{d-c}$$

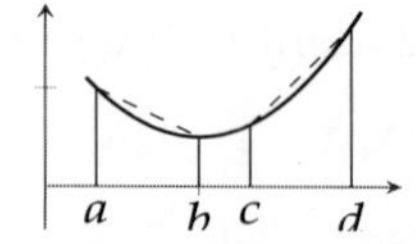

A3. Let $f\colon I\to\mathbb{R}$ be convex on an open interval I. Prove the following:

(a) The one-sided derivatives f'_+ and f'_- exist and satisfy $f'_+\geq f'_-$ at every point of I.

(b) The one-sided derivatives f'_+ and f'_- are increasing functions.

(c) f is continuous.

A4. Give an example of a discontinuous convex function on a closed interval.

A5. Let f be convex on an open interval I. A line $y = f(a) + m(x-a)$ through $(a, f(a))$ is called a **support line** if $f(x) \geq f(a) + m(x-a)$ for every $x \in I$. Prove the following.

(a) There is at least one support line through every point of the graph of f.

(b) The function f is differentiable at a if and only if there is a unique support line through $(a, f(a))$. This support line is also the tangent line to f at this point.

A6. (Bernoulli's Inequality) If $r \geq 1$ and $x > -1$ then $(1+x)^r \geq 1 + rx$.

Thus, differentiability of convex functions actually corresponds to the way tangent lines are usually described in school, as lines that touch the graph but do not cross it. This is, of course, false for general differentiable functions.

A7. Let $f\colon I \to \mathbb{R}$ be convex on an interval I. Prove that if f is differentiable at $a, b \in I$ with $a < b$, then $f'(a) \leq f'(b)$.

A8. Prove that a differentiable convex function will be continuously differentiable.

A9. Let a function f be defined on an interval I. Show the following:

(a) f is strictly convex on I if $f((1-t)x + ty) < (1-t)f(x) + tf(y)$ for any $x, y \in I$ and $t \in (0,1)$.

(b) If f is strictly convex, and is differentiable at a, then the tangent line through a intersects the graph of f only at $(a, f(a))$.

Tangent lines to graphs of strictly convex functions correspond exactly to the school description of "touching only at one point." In general, tangent lines may intersect the function graph at many points. In the most extreme case, if the function is linear, the function graph and the tangent line are the same!

A10. Consider a convex function f on an interval I. Prove the following:

(a) Any local minimum of f is a global minimum. The points where f has a local minimum form an interval, over which f is constant.

(b) If f is strictly convex, it has a unique minimum.

(c) If f is strictly convex and differentiable, then it has at most one critical point, and that critical point is a global minimum.

A11. (Jensen's Inequality) Let f be convex on I. Further, let $w_1, \ldots, w_n > 0$ such that $\sum_{i=1}^{n} w_i = 1$. Prove that for any $x_1, \ldots, x_n \in I$ we have the following:

(a) $\displaystyle\sum_{i=1}^{n} w_i x_i \in I$.

(b) $f(\sum_{i=1}^{n} w_i x_i) \leq \sum_{i=1}^{n} w_i f(x_i)$. If f is strictly convex then equality occurs if and only if $x_1 = \cdots = x_n$.

(c) If we drop the $\sum_{i=1}^{n} w_i = 1$ assumption, we still have $f\left(\frac{\sum_{i=1}^{n} w_i x_i}{\sum_{i=1}^{n} w_i}\right) \leq \frac{\sum_{i=1}^{n} w_i f(x_i)}{\sum_{i=1}^{n} w_i}$.

(d) If f is concave, the inequality reverses.

Given some positive numbers $x_1, \ldots, x_n$, we define the following kinds of "averages":

Root Mean Square: $RMS = \sqrt{\frac{x_1^2 + \cdots + x_n^2}{n}}$

Arithmetic Mean: $AM = \frac{x_1 + \cdots + x_n}{n}$

Geometric Mean: $GM = (x_1 \cdots x_n)^{1/n}$

Harmonic Mean: $HM = \frac{n}{\frac{1}{x_1} + \cdots + \frac{1}{x_n}}$

A12. Use Jensen's inequality to prove the following inequalities:

$$RMS \geq AM \geq GM \geq HM.$$

When will equality be possible? (Hint: For $RMS \geq AM$ use $f(x) = x^2$, for $AM \geq GM$, use $f(x) = -\log x$, and $GM \geq HM$ follows from $AM \geq GM$.)

A13. Let $p, q > 1$ with $\frac{1}{p} + \frac{1}{q} = 1$. Prove the following:

(a) (Young's Inequality) If $a, b \geq 0$ then $ab \leq \frac{a^p}{p} + \frac{b^q}{q}$.

(b) (Hölder's Inequality) If $a_1, \ldots, a_n, b_1, \ldots, b_n \geq 0$, then

$$\sum_{i=1}^{n} a_i b_i \leq \left(\sum_{i=1}^{n} a_i^p\right)^{1/p} \left(\sum_{i=1}^{n} b_i^q\right)^{1/q}.$$

(Hint: First prove the special case when $\sum_{i=1}^{n} a_i^p = \sum_{i=1}^{n} b_i^q = 1$.)

(c) (Minkowski's Inequality) If $a_1, \ldots, a_n, b_1, \ldots, b_n \in \mathbb{R}$, then

$$\left(\sum_{i=1}^{n} |a_i + b_i|^p\right)^{1/p} \leq \left(\sum_{i=1}^{n} |a_i|^p\right)^{1/p} + \left(\sum_{i=1}^{n} |b_i|^p\right)^{1/p}.$$

(Hint: $\sum |a_i + b_i|^p \leq \sum |a_i||a_i + b_i|^{p-1} + \sum |b_i||a_i + b_i|^{p-1}$ and apply Hölder's inequality.)

A14. (Jensen's Inequality for Integrals) Let F be convex on I. Let $g\colon [a,b] \to I$ be continuous. Prove the following:

(a) The point $x_0 = \frac{1}{b-a}\int_a^b g(x)\,dx$ belongs to I.

(b) There is an $m \in \mathbb{R}$ such that $F(x_0) + m(g(x) - x_0) \leq F(g(x))$ for every $x \in [a,b]$. (Hint: Exercise A5.)

(c) $F\left(\frac{1}{b-a}\int_a^b g(x)\,dx\right) \leq \frac{1}{b-a}\int_a^b F(g(x))\,dx$. If F is strictly convex, the inequality will be strict.

A15. Let $g\colon [0,1] \to (0,\infty)$ be a continuous function. Prove that

$$\exp\left(\int_0^1 \log g(x)\,dx\right) < \int_0^1 g(x)\,dx.$$

A16. (Jensen's Inequality for Weighted Integrals) Let F be convex on I. Let $g\colon [a,b] \to I$ be continuous. Let $p\colon [a,b] \to [0,\infty)$ satisfy $\int_a^b p(x)\,dx = 1$. Show that

$$F\left(\int_a^b g(x)p(x)\,dx\right) \leq \int_a^b F(g(x))p(x)\,dx.$$

A17. Let $g\colon [a,b] \to \mathbb{R}$ be continuous and $p\colon [a,b] \to [0,\infty)$ satisfy $\int_a^b p(x)\,dx = 1$. Show that

$$\left(\int_a^b g(x)p(x)\,dx\right)^2 \leq \int_a^b g(x)^2 p(x)\,dx.$$

A18. Let $g\colon [a,b] \to (0,\infty)$ be continuous and $p\colon [a,b] \to [0,\infty)$ satisfy $\int_a^b p(x)\,dx = 1$. Show that

$$\int_a^b g(x)p(x)\,dx \geq \left(\int_a^b g(x)^{-1}p(x)\,dx\right)^{-1}.$$

A19. (Hölder's Inequality) Let $f,g\colon [a,b] \to [0,\infty)$ be continuous, and $p,q > 1$ with $\frac{1}{p} + \frac{1}{q} = 1$. Show that

$$\int_a^b f(x)g(x)\,dx \leq \left(\int_a^b f(x)^p\,dx\right)^{1/p}\left(\int_a^b g(x)^q\,dx\right)^{1/q}.$$

A20. (Minkowski's Inequality) Let $f,g\colon [a,b] \to \mathbb{R}$ be continuous, and $p > 1$. Show that

$$\left(\int_a^b |f(x)+g(x)|^p\,dx\right)^{1/p} \leq \left(\int_a^b |f(x)|^p\,dx\right)^{1/p} + \left(\int_a^b |g(x)|^p\,dx\right)^{1/p}.$$

5 | Techniques of Integration

Our work on differentiation in the previous chapter has brought us to a fork in the road. We can pursue the implications of the first fundamental theorem to obtain techniques for computing integrals. Alternately, we can use the Fermat and monotonicity theorems to further develop the relationship between functions and their derivatives, leading to new techniques of calculating limits, approximation of functions by polynomials, use of integration to measure arc length, surface area and volume, and error estimates for numerical calculations of integrals. We have chosen to take up integration in this chapter. If you are more interested in the other applications of differentiation you can read Chapter 6 first.

5.1 The Second Fundamental Theorem

A function F is called an **anti-derivative** of f if $F' = f$. Let us make some observations regarding the existence and uniqueness of anti-derivatives:

1. Not every function has an anti-derivative. By Darboux's theorem (Theorem 4.5.12), if $f = F'$ then f has the intermediate value property. Thus, a function with a jump discontinuity, like the Heaviside step function or the greatest integer function, cannot have an anti-derivative.
2. On the other hand, the first fundamental theorem shows that every continuous function on an interval has an anti-derivative.
3. A function's anti-derivative is not unique. For example, both $\sin x$ and $1 + \sin x$ are anti-derivatives of $\cos x$.
4. On the other hand, two anti-derivatives of the same function over an interval can differ only by a constant. Theorem 4.5.7 states that if $F' = G'$ on an interval I, then $F - G$ is constant. Thus, every anti-derivative of $\cos x$ over an interval I has to have the form $\sin x + C$, where $C \in \mathbb{R}$.
5. Over non-overlapping intervals, two anti-derivatives of a function need not differ by the same constant. For example, the Heaviside step function and the zero function are anti-derivatives of the zero function over $(-\infty, 0) \cup (0, \infty)$.

The first fundamental theorem established a connection between integration and differentiation: if we are able to calculate the definite integrals of a continuous function, then the first fundamental theorem gives us its anti-derivative. The next theorem uses that connection to provide an approach for evaluating definite integrals by using anti-derivatives.

Theorem 5.1.1 (Second Fundamental Theorem)
Suppose that $f\colon [a,b] \to \mathbb{R}$ is a continuous function and $F\colon [a,b] \to \mathbb{R}$ satisfies $F' = f$. Then

$$\int_a^b f(t)\,dt = F(b) - F(a).$$

The difference $F(b) - F(a)$ is denoted by $F(x)\Big|_a^b$.

Proof. Define $G(x) = \int_a^x f(t)\,dt$. By the first fundamental theorem we know that $G'(x) = f(x) = F'(x)$ on $[a,b]$. Hence, by Theorem 4.5.7, $F(x) = G(x) + c$. Therefore,

$$F(b) - F(a) = G(b) - G(a) = \int_a^b f(t)\,dt - \int_a^a f(t)\,dt = \int_a^b f(t)\,dt. \qquad \blacksquare$$

The conclusion of the second fundamental theorem can also be written as

$$\int_a^b g'(t)\,dt = g(b) - g(a) \quad \text{or} \quad g(b) = g(a) + \int_a^b g'(t)\,dt.$$

This form is sometimes called the **net change theorem**: The total change in the function value is obtained by accumulating its instantaneous rate of change.

The second fundamental theorem is our main resource for integration problems. All techniques that we will encounter later are based on it. It says that to find the value of a definite integral of f we should first find an anti-derivative F of f. In this way, every derivative calculation is also the solution of an integration problem. For this reason, the process of finding an anti-derivative is also called integration.

Example 5.1.2

$$\sin' x = \cos x \implies \int_a^b \cos x\,dx = \sin x\Big|_a^b = \sin b - \sin a. \qquad \square$$

Example 5.1.3
For any integer $n \neq 0$,

$$(x^n)' = nx^{n-1} \implies \left(\frac{x^n}{n}\right)' = x^{n-1} \implies \int_a^b x^{n-1}\,dx = \frac{x^n}{n}\Big|_a^b = \frac{b^n - a^n}{n}.$$

We can allow n to be any non-zero real as well, of course restricting the domain of the integrand to $x > 0$. $\square$

Example 5.1.4

Consider $f(x) = \log|x|$. For $x \neq 0$,

$$f(x) = \begin{cases} \log x & \text{if } x > 0 \\ \log(-x) & \text{if } x < 0 \end{cases} \implies f'(x) = \begin{cases} 1/x & \text{if } x > 0 \\ -1/(-x) & \text{if } x < 0 \end{cases} = \frac{1}{x}.$$

Hence, if a and b have the same sign,

$$\int_a^b \frac{1}{x}\,dx = \log|b| - \log|a| = \log|b/a|. \qquad \square$$

Task 5.1.5

We are given that f satisfies $f'(x) = 1/x$ for $x \neq 0$ and $f(1) = 1$. Can we conclude that $f(x) = \log|x| + 1$?

We use the notation $\int f(x)\,dx$ for the collection of anti-derivatives of $f(x)$. From here on, we will assume that all anti-derivative calculations are over intervals unless stated otherwise. Hence, the anti-derivatives of a function will only differ by constants. We shall indicate this by writing statements like $\int \sin x\,dx = -\cos x + C$, with the C standing for an arbitrary real number.

The terms "integral" and "integration" are used for both definite integrals and anti-derivatives.

We have the following obvious facts, which follow from the corresponding properties of differentiation:

1. $\displaystyle\int c f(x)\,dx = c \int f(x)\,dx.$
2. $\displaystyle\int \big(f(x) + g(x)\big)\,dx = \int f(x)\,dx + \int g(x)\,dx.$

Let us work out some examples using this notation:

Example 5.1.6

Consider $f(x) = \log x$. Applying the product rule to $x \log x$ gives

$$\begin{aligned} (x\log x)' = \log x + 1 &\implies \int (\log x + 1)\,dx = x\log x + C \\ &\implies \int \log x\,dx + \int 1\,dx = x\log x + C \\ &\implies \int \log x\,dx + x = x\log x + C \\ &\implies \int \log x\,dx = x\log x - x + C. \end{aligned} \qquad \square$$

Example 5.1.7

Since $(\log|x|)' = 1/x$, we have $(\log|f(x)|)' = f'(x)/f(x)$. Therefore,

$$\int \frac{f'(x)}{f(x)}\,dx = \log|f(x)| + C.$$

Here are some applications:

$$\begin{aligned}
\int \frac{x}{x^2+1}\,dx &= \frac{1}{2}\int \frac{2x}{x^2+1}\,dx = \frac{1}{2}\log(x^2+1) + C,\\
\int \tan x\,dx &= \int \frac{\sin x}{\cos x}\,dx = -\int \frac{-\sin x}{\cos x}\,dx = -\log|\cos x| + C = \log|\sec x| + C,\\
\int \cot x\,dx &= \int \frac{\cos x}{\sin x}\,dx = \log|\sin x| + C = -\log|\csc x| + C,\\
\int \sec x\,dx &= \int \frac{\sec x(\sec x + \tan x)}{\sec x + \tan x}\,dx = \int \frac{\sec^2 x + \sec x \tan x}{\sec x + \tan x}\,dx\\
&= \log|\sec x + \tan x| + C,\\
\int \csc x\,dx &= \int \frac{\csc x(\csc x + \cot x)}{\csc x + \cot x}\,dx = \int \frac{\csc^2 x + \csc x \cot x}{\csc x + \cot x}\,dx\\
&= -\log|\csc x + \cot x| + C,\\
\int \frac{1}{x^2-1}\,dx &= \int \frac{1}{(x+1)(x-1)}\,dx = \frac{1}{2}\int \left(\frac{1}{x-1} - \frac{1}{x+1}\right)dx\\
&= \frac{1}{2}(\log|x-1| - \log|x+1|) + C = \frac{1}{2}\log\left|\frac{x-1}{x+1}\right| + C.
\end{aligned}$$

□

Task 5.1.8

Use the techniques of Examples 5.1.6 and 5.1.7 to evaluate the integral $\int \arctan x\,dx$.

Our results so far are collected below (we have not written the "$+C$").

$$\begin{aligned}
&\int 0\,dx = 0 && \int x^n\,dx = \frac{x^{n+1}}{n+1} \text{ if } n \neq 1\\
&\int \frac{1}{x}\,dx = \log|x| && \int \log x\,dx = x(\log x - 1)\\
&\int e^x\,dx = e^x && \int \sin x\,dx = -\cos x\\
&\int \cos x\,dx = \sin x && \int \tan x\,dx = \log|\sec x|\\
&\int \cot x\,dx = -\log|\csc x| && \int \sec x\,dx = \log|\sec x + \tan x|\\
&\int \csc x\,dx = -\log|\csc x + \cot x| && \int \frac{1}{\sqrt{1-x^2}}\,dx = \arcsin x\\
&\int \frac{1}{x^2+1}\,dx = \arctan x && \int \frac{x}{x^2+1}\,dx = \frac{1}{2}\log(x^2+1)
\end{aligned}$$

$$\int \frac{1}{\sqrt{x^2+1}}\,dx = \sinh^{-1} x \qquad \int \frac{1}{\sqrt{x^2-1}}\,dx = \cosh^{-1} x$$

$$\int \frac{1}{x^2-1}\,dx = \frac{1}{2}\log\left|\frac{x-1}{x+1}\right| \qquad \int \arctan x\,dx = x\arctan x - \frac{\log(1+x^2)}{2}$$

Exercises for § 5.1

1. Compute the following anti-derivatives:

(a) $\int (3x^2-1)\,dx$. (b) $\int \frac{x^2+3x+2}{x^2+x+1}\,dx$. (c) $\int \cos^2 t\,dt$.

2. Compute the following anti-derivatives:

(a) $\int \frac{x^2+5x-1}{\sqrt{x}}\,dx$. (b) $\int \cos 2t\,dt$. (c) $\int \frac{x+2}{x^2-1}\,dx$.

3. The fundamental theorems connect anti-derivatives and integrals. The two concepts are, however, distinct. We have already noted that an integrable function may not have an anti-derivative (for example, the Heaviside step function). Here is an example of a function f that has an anti-derivative but is not integrable.

(a) Let $F(x) = x^{3/2}\sin(1/x)$ when $x \neq 0$ and $F(0) = 0$. Show that F is differentiable for every x.

(b) Let $f = F'$. Show that f is unbounded on $[0,1]$ and hence not integrable on that interval.

4. Find $f\colon \mathbb{R}\setminus\{0\} \to \mathbb{R}$ such that $f'(x) = 1/x$, $f(1) = 1$ and $f(-1) = 2$.

5. Compute the following definite integrals:

(a) $\int_0^{\pi/4} \tan t\,dt$. (b) $\int_0^1 \frac{1}{x^2-4}\,dx$. (c) $\int_{-1}^1 \arctan u\,du$.

6. Compute the following definite integrals:

(a) $\int_0^{\pi/3} \sec x\,dx$. (b) $\int_0^{\pi/4} \tan^2 x\,dx$. (c) $\int_2^4 2^x\,dx$.

7. Find the area of the shaded region.

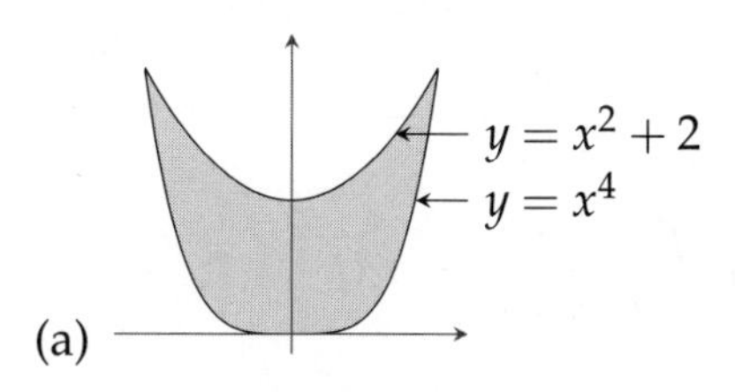

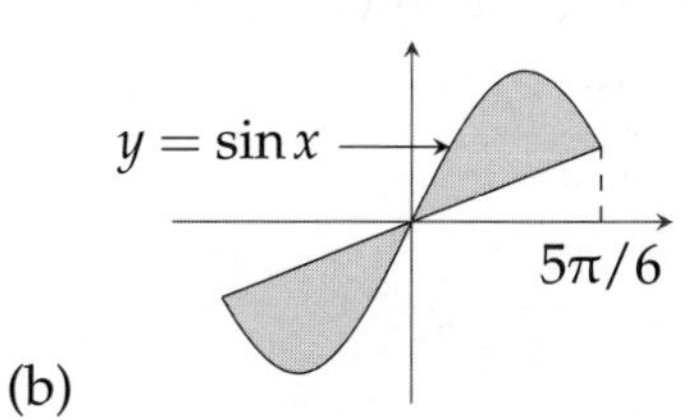

8. Show that $\int_a^{a+2\pi} \cos\theta\, d\theta = 0$. (Hint: Theorem 2.2.8.)

9. Verify that for any non-zero real number a, $\int \cos ax\, dx = \frac{1}{a}\sin ax + C$ and $\int \sin ax\, dx = -\frac{1}{a}\cos ax + C$. Use these and the trigonometric identities to carry out the following integrations:

(a) $\int_0^{2\pi} \cos 3x \cos 2x\, dx$.

(b) $\int_0^{\pi} \cos 3x \sin 2x\, dx$.

(c) $\int_0^{\pi/2} \cos^2 x\, dx$.

(d) $\int_0^{\pi/2} \sin^2 x\, dx$.

(e) $\int_{\pi/2}^{\pi} \cos^2 x \sin x\, dx$.

(f) $\int_{\pi/6}^{\pi/2} \cos^3 x\, dx$.

In the next three exercises, we take up the problem of integrating a rational function of the form $p(x)/q(x)$ where $p(x)$ is linear and $q(x) = x^2 + bx + c$ is quadratic. Integration of general rational functions is treated in §5.4.

10. Suppose $q(x)$ has distinct real roots α, β. Show the following:

(a) $\dfrac{p(x)}{q(x)} = \dfrac{Ax+B}{(x-\alpha)(x-\beta)} = \dfrac{C}{x-\alpha} + \dfrac{D}{x-\beta}$.

(b) $\int \dfrac{p(x)}{q(x)}\, dx = C\log|x-\alpha| + D\log|x-\beta| + E$.

11. Suppose $q(x)$ has a repeated real root α. Show the following:

(a) $\dfrac{p(x)}{q(x)} = \dfrac{Ax+B}{(x-\alpha)^2} = \dfrac{A}{x-\alpha} + \dfrac{C}{(x-\alpha)^2}$.

(b) $\int \dfrac{p(x)}{q(x)}\, dx = A\log|x-\alpha| - \dfrac{C}{x-\alpha} + E$.

12. Suppose $q(x)$ has no real roots. Show the following:

(a) $\dfrac{p(x)}{q(x)} = \dfrac{Ax+B}{(x-a)^2+b^2}$.

(b) $\int \dfrac{p(x)}{q(x)}\, dx = \dfrac{A}{2}\log((x-a)^2+b^2) + \dfrac{B+Aa}{b}\arctan\left(\dfrac{x-a}{b}\right) + E$.

13. Compute the following anti-derivatives:

(a) $\int \dfrac{3x-2}{x^2+2x+1}\, dx$.

(b) $\int \dfrac{7x}{x^2-3x+2}\, dx$.

(c) $\int \dfrac{7x+3}{x^2-2x+10}\, dx$.

(d) $\int \dfrac{x^3}{x^2-2x+1}\, dx$.

5.2 Integration by Substitution

The connection between differentiation and integration established by the second fundamental theorem enables the conversion of methods of differentiation to methods of integration. Let us take the chain rule first.

Theorem 5.2.1 (Substitution Method)

Suppose f is continuous on an interval I while $\varphi\colon [a,b] \to I$ is continuously differentiable. Then

$$\int_a^b f(\varphi(x))\,\varphi'(x)\,dx = \int_{\varphi(a)}^{\varphi(b)} f(u)\,du.$$

Proof. Since f is continuous it has an anti-derivative. Let $f = F'$. Then $f(\varphi(x))\,\varphi'(x) = F'(\varphi(x))\,\varphi'(x) = (F \circ \varphi)'(x)$. Therefore,

$$\begin{aligned}\int_a^b f(\varphi(x))\,\varphi'(x)\,dx &= \int_a^b (F \circ \varphi)'(x)\,dx \\ &= F(\varphi(b)) - F(\varphi(a)) = \int_{\varphi(a)}^{\varphi(b)} f(u)\,du.\end{aligned}$$ ■

This rule is easily remembered by means of the following convention. Consider the substitution $u = \varphi(x)$. Then $\dfrac{du}{dx} = \varphi'(x)$. The convention is that in an integration problem we are allowed to substitute $u = \varphi(x)$ together with $du = \varphi'(x)\,dx$. The substitution rule justifies the application of this convention, combined with a corresponding change of limits.

$$\int_a^b f(\underbrace{\varphi(x)}_{u})\,\underbrace{\varphi'(x)\,dx}_{du}$$

Example 5.2.2

Consider $\int_0^{\pi} x\sin(x^2)\,dx$. We look for a substitution that would create functions that are easier to integrate. For example, substituting $u = x^2$ converts $\sin(x^2)$ to $\sin u$. Let us try out this particular substitution. As discussed above, it leads to $du = 2x\,dx$ and so $\frac{1}{2}du = x\,dx$. Further, the limits change as follows: $x = 0 \implies u = 0$, $x = \pi \implies u = \pi^2$. Hence,

$$\int_0^{\pi} x\sin(x^2)\,dx = \int_0^{\pi^2} \frac{1}{2}\sin u\,du = -\frac{1}{2}\cos u\Big|_0^{\pi^2} = \frac{1}{2}(1 - \cos(\pi^2)).$$ □

The substitution method can also be used to find anti-derivatives. If $F' = f$, then

$$\int f(\varphi(x))\varphi'(x)\,dx = \int (F \circ \varphi)'(x)\,dx = (F \circ \varphi)(x) + C = F(\varphi(x)) + C.$$

We represent this calculation by the following abbreviation:

$$\int f(\varphi(x))\varphi'(x)\,dx = \int f(u)\,du, \quad \text{where } u = \varphi(x).$$

Example 5.2.3

To evaluate $\int \sin(2x - \pi)\,dx$ we make the substitution $u = 2x - \pi$, and so $du = 2dx$. Then

$$\int \sin(2x - \pi)\,dx = \frac{1}{2}\int \sin u\,du = -\frac{1}{2}\cos u + C = -\frac{1}{2}\cos(2x - \pi) + C.$$

□

Example 5.2.4

To evaluate $\int \frac{x}{\sqrt{1-x^2}}\,dx$ we substitute $y = 1 - x^2$, so that $dy = -2x\,dx$. Then

$$\int \frac{x}{\sqrt{1-x^2}}\,dx = -\frac{1}{2}\int \frac{1}{\sqrt{y}}\,dy = -\sqrt{y} + C = -\sqrt{1-x^2} + C.$$

□

Example 5.2.5

To evaluate $\int \frac{\log x}{x}\,dx$ we substitute $y = \log x$, so that $dy = \frac{1}{x}\,dx$. Then

$$\int \frac{\log x}{x}\,dx = \int y\,dy = \frac{y^2}{2} + C = \frac{(\log x)^2}{2} + C.$$

□

Example 5.2.6

To evaluate $\int \sin^3 x\,dx$ we first rearrange it as follows:

$$\int \sin^3 x\,dx = \int \sin^2 x \sin x\,dx = \int (1 - \cos^2 x)\sin x\,dx.$$

Now we substitute $y = \cos x$, so that $dy = -\sin x\,dx$.

$$\int (1 - \cos^2 x)\sin x\,dx = -\int (1 - y^2)\,dy = \frac{y^3}{3} - y + C = \frac{\cos^3 x}{3} - \cos x + C.$$

□

One way to carry out the substitution method for definite integrals is to first employ it for the corresponding anti-derivative calculation, and then incorporate the limits of integration. For example, to evaluate $\int_0^{\pi/2} \sin^3 x\,dx$, we use the earlier calculation of $\int \sin^3 x\,dx$:

$$\int \sin^3 x\,dx = \frac{\cos^3 x}{3} - \cos x + C \Longrightarrow \int_0^{\pi/2} \sin^3 x\,dx = \left(\frac{\cos^3 x}{3} - \cos x\right)\Bigg|_0^{\pi/2} = \frac{2}{3}.$$

Example 5.2.7

To evaluate $\int_0^\pi \sin^2 x\, dx$ we first use the half-angle formula to calculate the anti-derivative:

$$\int \sin^2 x\, dx = \int \frac{1-\cos 2x}{2}\, dx = \frac{x}{2} - \frac{\sin 2x}{4} + C.$$

Hence,

$$\int_0^\pi \sin^2 x\, dx = \left(\frac{x}{2} - \frac{\sin 2x}{4}\right)\Big|_0^\pi = \frac{\pi}{2}.$$ □

Task 5.2.8

Is the following calculation correct?

$$\int \frac{1}{x}\, dx = \log|x| \implies \int_{-1}^{2} \frac{1}{x}\, dx = \log|x|\Big|_{-1}^{2} = \log 2.$$

The trigonometric identities can be used to solve integrals of algebraic functions. For example, $\cos^2 x = 1 - \sin^2 x$ may help when the integrand contains $\sqrt{a^2 - x^2}$, while $\sec^2 x = 1 + \tan^2 x$ is a useful resource when the integrand contains $\sqrt{a^2 + x^2}$. These "trigonometric substitutions" are also examples where we reverse the manner in which we use the substitution method. We take the given integral as $\int f(x)\, dx$ and then substitute $x = \varphi(u)$ to convert it to $\int f(\varphi(u))\varphi'(u)\, du$, along with a corresponding change of limits in case of definite integrals.

Example 5.2.9

To evaluate $\int_0^2 x^3\sqrt{4-x^2}\, dx$ we substitute $x = 2\sin\theta$. Then $dx = 2\cos\theta\, d\theta$ and we get

$$\begin{aligned}
\int_0^2 x^3\sqrt{4-x^2}\, dx &= 32\int_0^{\pi/2} \sin^3\theta\sqrt{1-\sin^2\theta}\cos\theta\, d\theta \\
&= 32\int_0^{\pi/2} \sin^3\theta\cos^2\theta = 32\int_0^{\pi/2} (\cos^2\theta - \cos^4\theta)\sin\theta\, d\theta \\
&= -32\int_1^0 (u^2 - u^4)\, du = 32\left(\frac{u^3}{3} - \frac{u^5}{5}\right)\Big|_0^1 = \frac{64}{15}.
\end{aligned}$$ □

Example 5.2.10

To evaluate $\displaystyle\int_0^1 \frac{1}{\sqrt{1+x^2}}\, dx$, substitute $x = \tan\theta$. Then $dx = \sec^2\theta\, d\theta$, and

$$\int_0^1 \frac{1}{\sqrt{1+x^2}}\, dx = \int_0^{\pi/4} \sec\theta\, d\theta = \log(\sec\theta + \tan\theta)\Big|_0^{\pi/4} = \log(1+\sqrt{2}).$$ □

Example 5.2.11

We have noticed many similarities among the algebraic properties of the trigonometric and the hyperbolic functions. Let us see a geometric one. We defined the radian measure of an angle as being twice the area of the corresponding sector of a unit circle. This means that when a point on the unit circle is represented by an ordered pair $(\cos t, \sin t)$ then t is twice the area swept out by the point as it moves counter-clockwise from $(1,0)$ to its final position. We shall see that the same happens with a point $(\cosh t, \sinh t)$ on the hyperbola $x^2 - y^2 = 1$.

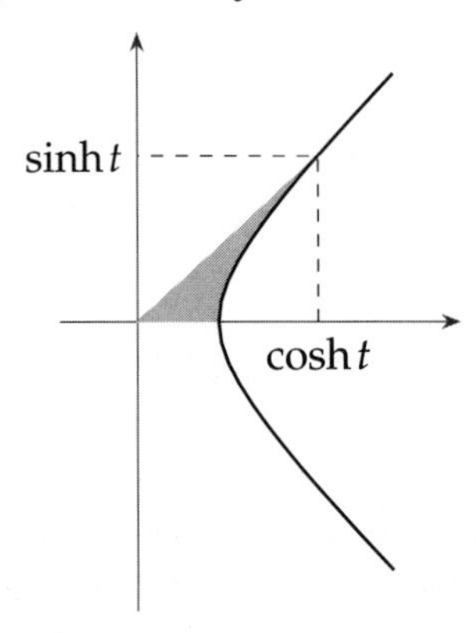

The swept out area is

$$A = \frac{1}{2}(\cosh t)(\sinh t) - \int_1^{\cosh t} \sqrt{x^2 - 1}\, dx.$$

To evaluate the integral, substitute $x = \cosh u$. Then we have $\sqrt{x^2 - 1} = \sqrt{\cosh^2 u - 1} = \sinh u$ and $dx = \sinh u\, du$. Therefore,

$$\int_1^{\cosh t} \sqrt{x^2 - 1}\, dx = \int_0^t \sinh^2 u\, du = \int_0^t \frac{\cosh(2u) - 1}{2}\, du = \frac{\sinh 2t}{4} - \frac{t}{2}.$$

Hence, $A = \frac{1}{2}(\cosh t)(\sinh t) - \frac{\sinh 2t}{4} + \frac{t}{2} = \frac{t}{2}$. □

Exercises for § 5.2

1. Apply the substitution method to find the following anti-derivatives:

(a) $\int x^2 \sqrt{x+1}\, dx$.

(b) $\int \frac{1}{\sqrt{x-1} + \sqrt{x+1}}\, dx$.

(c) $\int \frac{(\log x)^2}{x}\, dx$.

(d) $\int \frac{\log \sqrt{x}}{\sqrt{x}}\, dx$.

2. Apply the substitution method to find the following anti-derivatives:

(a) $\displaystyle\int \frac{1+e^x}{1-e^x}\,dx.$

(b) $\displaystyle\int \cos\theta\, e^{\sin\theta}\,dx.$

(c) $\displaystyle\int \sin\theta\cos^2\theta\,dx.$

(d) $\displaystyle\int \tan\theta\sec^2\theta\,d\theta.$

3. Apply the substitution method to calculate the following definite integrals:

(a) $\displaystyle\int_0^1 x\,e^{-x^2}\,dx.$

(b) $\displaystyle\int_4^5 (x^2+7)\sqrt{x-4}\,dx.$

(c) $\displaystyle\int_{-1}^1 \frac{u}{(u^2+2)^{10}}\,du.$

(d) $\displaystyle\int_0^{\pi/6} \tan^3\theta\,d\theta.$

4. Let $m,k\in\mathbb{N}$. Show that

$$\int \sin^m x\cos^{2k-1}x\,dx = \sum_{i=0}^{k-1}(-1)^i\binom{k-1}{i}\frac{(\sin x)^{m+2i+1}}{m+2i+1},$$

$$\int \cos^m x\sin^{2k-1}x\,dx = \sum_{i=0}^{k-1}(-1)^{i+1}\binom{k-1}{i}\frac{(\cos x)^{m+2i+1}}{m+2i+1}.$$

5. Let $m\in\mathbb{N}$. Show that

$$\int \cos^{2m}(ax)\,dx = \frac{1}{2^m}\sum_{k=0}^{m}\binom{m}{k}\int \cos^k(2ax)\,dx,$$

$$\int \sin^{2m}(ax)\,dx = \frac{1}{2^m}\sum_{k=0}^{m}(-1)^k\binom{m}{k}\int \cos^k(2ax)\,dx.$$

6. Compute the following definite integrals:

(a) $\displaystyle\int_0^{\pi/2} \sin^3 x\cos^3 x\,dx.$

(b) $\displaystyle\int_0^{\pi} \sin^4 x\,dx.$

(c) $\displaystyle\int_0^{\pi} \sin^5 x\,dx.$

(d) $\displaystyle\int_0^{\pi/2} \sin^4 x\cos^2 x\,dx.$

7. Use trigonometric substitutions to compute the following definite integrals:

(a) $\displaystyle\int_0^{1/2} \sqrt{1-x^2}\,dx.$

(b) $\displaystyle\int_2^3 \frac{1}{\sqrt{x^2-1}}\,dx.$

(c) $\displaystyle\int_2^3 \frac{1}{x\sqrt{x^2-1}}\,dx.$

(d) $\displaystyle\int_2^4 \frac{1}{x\sqrt{x-1}}\,dx.$

(e) $\displaystyle\int_1^2 \frac{1}{x\sqrt{4+x^2}}\,dx.$

(f) $\displaystyle\int_0^1 \frac{1}{x^2+x+1}\,dx.$

5.3 Integration by Parts

In the last section we used the chain rule to develop the method of substitution for solving integrals. Now let us consider the product rule:

$$\big(f(x)g(x)\big)' = f'(x)g(x) + f(x)g'(x).$$

The anti-derivative version of this is

$$f(x)g(x) = \int f'(x)g(x)\,dx + \int f(x)g'(x)\,dx.$$

We rearrange it to

$$\int f(x)g'(x)\,dx = f(x)g(x) - \int f'(x)g(x)\,dx.$$

This changes the function we have to integrate. When we are lucky (or clever), the new function is easier to integrate. Again, we have a version for definite integrals.

Theorem 5.3.1 (Integration by Parts)
If f' and g' are continuous, then

$$\int_a^b f(x)g'(x)\,dx = f(x)g(x)\Big|_a^b - \int_a^b f'(x)g(x)\,dx.$$

Proof. Since f' and g' are continuous, so are $f'g$, fg' and $(fg)'$. The second fundamental theorem gives

$$f(x)g(x)\Big|_a^b = \int_a^b (f(x)g(x))'\,dx = \int_a^b f(x)g'(x)\,dx + \int_a^b f'(x)g(x)\,dx. \qquad \blacksquare$$

A convenient way to remember this is to use the differential notation that we introduced for the substitution method. Writing $df(x) = f'(x)\,dx$ and $dg(x) = g'(x)\,dx$, we can express integration by parts as

$$\int_a^b f(x)\,dg(x) = f(x)g(x)\Big|_a^b - \int_a^b g(x)\,df(x).$$

Or, even more briefly, as

$$\int_a^b f\,dg = fg\Big|_a^b - \int_a^b g\,df.$$

Example 5.3.2
Consider $\int x\sin x\,dx$. Set $f(x) = x$ and $g'(x) = \sin x$. Then $f'(x) = 1$ and $g(x) = -\cos x$. Hence,

$$\int x\sin x\,dx = -x\cos x + \int \cos x\,dx = -x\cos x + \sin x + C.$$

We can use this to calculate definite integrals. For example,

$$\int_0^{\pi} x \sin x \, dx = (-x\cos x + \sin x)\Big|_0^{\pi} = \pi. \qquad \square$$

Example 5.3.3

Consider $\int xe^x \, dx$. Set $f(x) = x$ and $g'(x) = e^x$. Then $f'(x) = 1$ and $g(x) = e^x$. Hence,

$$\int xe^x \, dx = xe^x - \int e^x \, dx = xe^x - e^x + C. \qquad \square$$

Example 5.3.4

Consider $\int x \log x \, dx$. Set $f(x) = \log x$ and $g'(x) = x$. Then $f'(x) = 1/x$ and $g(x) = x^2/2$. Hence,

$$\int x \log x \, dx = \frac{x^2}{2} \log x - \int \frac{x}{2} \, dx = \frac{x^2}{2} \log x - \frac{x^2}{4} + C. \qquad \square$$

Task 5.3.5

Use integration by parts twice to evaluate the given integrals:

(a) $\displaystyle\int x^2 e^x \, dx.$ (b) $\displaystyle\int x^2 \sin x \, dx.$

Integration by parts can be useful even when the integrand is not in the form of a product. We just introduce a factor of 1. Here are two elementary examples where this pays off:

Example 5.3.6

Consider $\int \log x \, dx$. Set $f(x) = \log x$ and $g'(x) = 1$. Then $f'(x) = 1/x$ and $g(x) = x$. Hence,

$$\int \log x \, dx = x \log x - \int 1 \, dx = x \log x - x + C. \qquad \square$$

Example 5.3.7

Consider $\int \arctan x \, dx$. Set $f(x) = \arctan x$ and $g'(x) = 1$. Then $f'(x) = 1/(1+x^2)$ and $g(x) = x$. Hence,

$$\int \arctan x \, dx = x \arctan x - \int \frac{x}{1+x^2} \, dx = x \arctan x - \frac{1}{2} \log(1+x^2) + C.$$

$\square$

Example 5.3.8

The integral $\int \sec^3 x\,dx$ has a habit of cropping up in integrations of trigonometric functions or in trigonometric substitutions. We can tackle it by integration by parts as follows:

$$\begin{aligned}\int \sec^3 x\,dx = \int \sec x \sec^2 x\,dx &= \sec x \tan x - \int \sec x \tan^2 x\,dx \\ &= \sec x \tan x + \int \sec x\,dx - \int \sec^3 x\,dx.\end{aligned}$$

Hence, $\displaystyle\int \sec^3 x\,dx = \frac{1}{2}\Big(\sec x \tan x + \log|\sec x + \tan x|\Big) + C.$ □

Example 5.3.9

Sometimes it can be faster to use the product rule directly rather than go through integration by parts. Consider the problem of integrating $e^x \sin x$. The standard method is to apply integration by parts twice and then do a rearrangement. Here is an alternate approach:

$$\begin{aligned}(e^x \sin x)' &= e^x \cos x + e^x \sin x, \\ (e^x \cos x)' &= e^x \cos x - e^x \sin x.\end{aligned}$$

Hence, $$\begin{aligned}(e^x \sin x + e^x \cos x)' &= 2e^x \cos x, \\ (e^x \sin x - e^x \cos x)' &= 2e^x \sin x.\end{aligned}$$

Thus, we have killed two birds with one stone:

$$\begin{aligned}\int e^x \cos x\,dx &= \frac{1}{2}(e^x \sin x + e^x \cos x) + C, \\ \int e^x \sin x\,dx &= \frac{1}{2}(e^x \sin x - e^x \cos x) + C.\end{aligned}$$ □

Reduction Formulas

Using integration by parts, we can iteratively reduce powers in the integrand until we reach a situation where we have eliminated them. Formulas describing such iterations are called **reduction formulas**. We give an instance below:

$$\begin{aligned}\int \sin^n x\,dx &= \int \underbrace{\sin^{n-1} x}_{f}\,\underbrace{\sin x\,dx}_{dg} \\ &= \underbrace{(\sin^{n-1} x)}_{f}\underbrace{(-\cos x)}_{g} - \int \underbrace{(-\cos x)}_{g}\underbrace{(n-1)\sin^{n-2} x \cos x\,dx}_{df} \\ &= -\cos x \sin^{n-1} x + (n-1)\int \cos^2 x \sin^{n-2} x\,dx \\ &= -\cos x \sin^{n-1} x + (n-1)\int \sin^{n-2} x\,dx - (n-1)\int \sin^n x\,dx\end{aligned}$$

$$\Longrightarrow n\int \sin^n x\,dx = -\cos x\sin^{n-1}x + (n-1)\int \sin^{n-2}x\,dx$$
$$\Longrightarrow \int \sin^n x\,dx = -\frac{1}{n}\cos x\sin^{n-1}x + \frac{n-1}{n}\int \sin^{n-2}x\,dx.$$

This reduction formula allows us to obtain the integral of any $\sin^n x$ in terms of the integral of $\sin^{n-2}x$, then in terms of $\sin^{n-4}x$ and so on, until we reach a power of 0 or 1.

Example 5.3.10

$$\begin{aligned}\int \sin^4 x\,dx &= -\frac{\cos x\sin^3 x}{4} + \frac{3}{4}\int \sin^2 x\,dx\\ &= -\frac{\cos x\sin^3 x}{4} + \frac{3}{4}\left[-\frac{\cos x\sin x}{2} + \frac{1}{2}\int 1\,dx\right]\\ &= -\frac{\cos x\sin^3 x}{4} - \frac{3}{8}\cos x\sin x + \frac{3}{8}x + C.\end{aligned}$$

□

Example 5.3.11

Let us apply this reduction formula to definite integrals over $[0,\pi/2]$. We get

$$\begin{aligned}\int_0^{\pi/2} \sin^n x\,dx &= -\frac{1}{n}\cos x\sin^{n-1}x\Big|_0^{\pi/2} + \frac{n-1}{n}\int_0^{\pi/2}\sin^{n-2}x\,dx\\ &= \frac{n-1}{n}\int_0^{\pi/2}\sin^{n-2}x\,dx.\end{aligned}$$

For example,

$$\int_0^{\pi/2}\sin^4 x\,dx = \frac{3}{4}\int_0^{\pi/2}\sin^2 x\,dx = \frac{3}{4}\cdot\frac{1}{2}\int_0^{\pi/2}1\,dx = \frac{3\cdot 1}{4\cdot 2}\cdot\frac{\pi}{2}.$$

In general,

$$\begin{aligned}\int_0^{\pi/2}\sin^{2n}x\,dx &= \frac{2n-1}{2n}\cdot\frac{2n-3}{2n-2}\cdots\frac{1}{2}\int_0^{\pi/2}1\,dx\\ &= \frac{(2n-1)(2n-3)\cdots 1}{(2n)(2n-2)\cdots 2}\frac{\pi}{2} = \frac{(2n)!}{4^n(n!)^2}\frac{\pi}{2},\\ \int_0^{\pi/2}\sin^{2n+1}x\,dx &= \frac{2n}{2n+1}\cdot\frac{2n-2}{2n-1}\cdots\frac{2}{3}\int_0^{\pi/2}\sin x\,dx\\ &= \frac{(2n)(2n-2)\cdots 2}{(2n+1)(2n-1)\cdots 3} = \frac{4^n(n!)^2}{(2n+1)!}.\end{aligned}$$

□

The exercises for this section will introduce you to several other reduction formulas.

Exercises for § 5.3

1. Apply integration by parts, possibly in combination with other techniques like the substitution method, to find the following anti-derivatives:

(a) $\int \sin\sqrt{x}\, dx$.

(b) $\int \sin(\log x)\, dx$.

(c) $\int \arcsin x\, dx$.

(d) $\int x(\arctan x)^2\, dx$.

2. Calculate the following definite integrals:

(a) $\int_0^1 xe^x\, dx$.

(b) $\int_1^e x\log x\, dx$.

(c) $\int_{1/\sqrt{2}}^1 \frac{\arcsin x}{x^2}\, dx$.

(d) $\int_0^1 \arctan\sqrt{x}\, dx$.

3. Find the area of the shaded region.

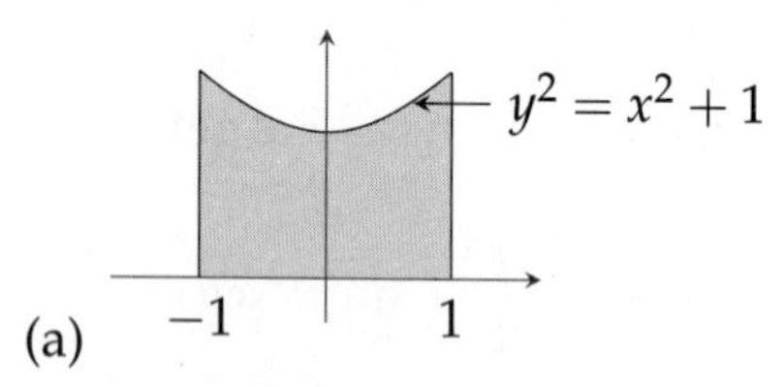

(a)

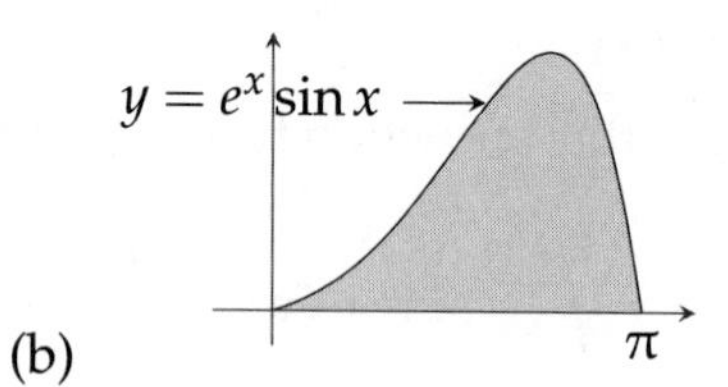

(b)

4. Prove the given reduction formula and apply it to compute the given integrals:

(a) $\int \cos^n x\, dx = \frac{\sin x \cos^{n-1} x}{n} + \frac{n-1}{n}\int \cos^{n-2} x\, dx$.

(b) $\int \cos^4 x\, dx$.

(c) $\int_0^{\pi/2} \cos^{2n} x\, dx$.

(d) $\int_0^{\pi/2} \cos^{2n+1} x\, dx$.

5. Prove the given reduction formula and its consequences:

(a) $\int (\log x)^n\, dx = x(\log x)^n - n\int (\log x)^{n-1}\, dx$.

(b) $\int (\log x)^n\, dx = x\sum_{k=0}^{n}(-1)^{n-k}\frac{n!}{k!}(\log x)^k + C$.

(c) $\int x^n e^x\, dx = e^x\sum_{k=0}^{n}(-1)^{n-k}\frac{n!}{k!}x^k + C$.

6. Prove the given reduction formulas:

(a) $$\int \frac{dx}{(x^2+a^2)^n} = \frac{x}{2a^2(n-1)(x^2+a^2)^{n-1}} + \frac{2n-3}{2a^2(n-1)} \int \frac{dx}{(x^2+a^2)^{n-1}}.$$

(b) $$\int \frac{dx}{(x^2-a^2)^n} = -\frac{x}{2a^2(n-1)(x^2-a^2)^{n-1}} + \frac{2n-3}{2a^2(n-1)} \int \frac{dx}{(x^2-a^2)^{n-1}}.$$

(c) $$\int \sec^n x\,dx = \frac{\sec^{n-2} x \tan x}{n-1} + \frac{n-2}{n-1} \int \sec^{n-2} x\,dx.$$

7. Compute the following integrals:

(a) $\displaystyle\int \frac{dx}{(x^2+2x+2)^2}$. (b) $\displaystyle\int \sec^4 x\,dx$.

8. (Second mean value theorem for integration) Let $f,g\colon [a,b] \to \mathbb{R}$ be continuous, with f also having a continuous derivative that does not change sign. Show that there is a $c \in [a,b]$ such that

$$\int_a^b f(x)g(x)\,dx = f(a) \int_a^c g(x)\,dx + f(b) \int_c^b g(x)\,dx.$$

(Hint: Let $G(x) = \int_a^x g(s)\,ds$ and apply integration by parts to $\int_a^b f(x)g(x)\,dx = \int_a^b f(x)G'(x)\,dx$.)

5.4 Partial Fractions

In this section we will learn how to integrate a rational function by breaking it down into a sum of rational functions with simple denominators, and then applying suitable substitutions or integration by parts. We have already seen several examples of integration of rational functions. For example, we have

$$\int \frac{dx}{x-a} = \log|x-a|,$$

$$\int \frac{dx}{x^2+a^2} = \frac{1}{a}\arctan\left(\frac{x}{a}\right),$$

$$\int \frac{x\,dx}{x^2+a^2} = \frac{1}{2}\log(x^2+a^2),$$

with the following holding when $n > 1$:

$$\int \frac{dx}{(x-a)^n} = \frac{-1}{n-1}\frac{1}{(x-a)^{n-1}},$$

$$\int \frac{dx}{(x^2+a^2)^n} = \frac{x}{2a^2(n-1)(x^2+a^2)^{n-1}} + \frac{2n-3}{2a^2(n-1)}\int \frac{dx}{(x^2+a^2)^{n-1}},$$

$$\int \frac{x\,dx}{(x^2+a^2)^n} = \frac{-1}{2(n-1)(x^2+a^2)^{n-1}}.$$

Our approach will be based on the fact that every real polynomial can be expressed as a product of factors, each of which is either linear or quadratic with no real roots. We will not prove this fact. In the practical world, what is more important is actually finding these factors, and that task is not eased by the knowledge that the factors exist. So, we shall concentrate on what can be done once the factors have been found, namely the reduction of the problem to the integrals already listed above.

We begin with the observation that we can focus on rational functions $p(x)/q(x)$ with $\deg p < \deg q$. For we can always divide p by q to get $p(x) = s(x)q(x) + r(x)$ with r either being 0 or satisfying $\deg r < \deg q$. Then

$$\frac{p(x)}{q(x)} = s(x) + \frac{r(x)}{q(x)}.$$

Now $s(x)$ is a polynomial and easy to integrate, so we need only analyse $r(x)/q(x)$.

A pair of polynomials q_1 and q_2 is called **coprime** if the only polynomials that are their common factors are the constants.

Theorem 5.4.1

Suppose $\deg p < \deg q$ and $q = q_1q_2$ is a factoring into non-constant coprime polynomials. Then there are polynomials r_1 and r_2 such that $\deg r_i < \deg q_i$ for $i = 1,2$, and

$$\frac{p(x)}{q(x)} = \frac{r_1(x)}{q_1(x)} + \frac{r_2(x)}{q_2(x)}.$$

Proof. Let $\mathcal{I}$ be the collection of all combinations of the form $p_1q_1 + p_2q_2$, where p_1, p_2 are arbitrary polynomials. Let $\ell \in \mathcal{I}$ be a non-zero polynomial of least degree. Then $q_i = t_i\ell + r_i$ where $\deg r_i < \deg \ell$ or $r_i = 0$. Now, $q_i = t_i(p_1q_1 + p_2q_2) + r_i$ implies $r_i \in \mathcal{I}$. So, we must have $r_i = 0$ and hence ℓ is a factor of each q_i. Therefore, ℓ is a non-zero constant and we may take it to be 1. Hence, $1 = p_1q_1 + p_2q_2$. It follows that every polynomial belongs to $\mathcal{I}$!

Therefore, $p = t_1q_1 + t_2q_2$. Now $t_1 = t_1'q_2 + r_2$ and $t_2 = t_2'q_1 + r_1$, with $r_i = 0$ or $\deg r_i < \deg q_i$. This gives

$$p = (t_1'q_2 + r_2)q_1 + (t_2'q_1 + r_1)q_2 = (t_1' + t_2')q + r_2q_1 + r_1q_2.$$

On matching degrees, we see we must have $t_1' + t_2' = 0$ and hence $p = r_2q_1 + r_1q_2$. Dividing by q gives the desired expression. ■

Example 5.4.2

We have noted that every polynomial is a product of linear and quadratic factors, with the quadratic factors having no real roots. For example, the factoring may look as follows:

$$q(x) = (x+1)^2(x^2+1)^2.$$

Further, let $p(x) = x^5 + 7x^4 + 5x^3 + 12x^2 + 4x + 7$. By application of the previous theorem, we see that

$$\frac{x^5 + 7x^4 + 5x^3 + 12x^2 + 4x + 7}{(x+1)^2(x^2+1)^2} = \frac{Ax+B}{(x+1)^2} + \frac{Cx^3 + Dx^2 + Ex + F}{(x^2+1)^2}.$$ □

From the preceding theorem and example, we see that we need to work on two fronts:

1. Methods to find the unknown constants on the right-hand side.
2. Methods to integrate rational functions of the form $p(x)/q(x)^n$ where $q(x)$ is either linear or quadratic.

Theorem 5.4.3

Consider a rational function $p(x)/q(x)^n$ with $\deg p < n(\deg q)$. It can be expressed as

$$\frac{p(x)}{q(x)^n} = \frac{r_1(x)}{q(x)^n} + \frac{r_2(x)}{q(x)^{n-1}} + \cdots + \frac{r_n(x)}{q(x)},$$

with each r_i satisfying either $r_i = 0$ or $\deg r_i < \deg q$.

Proof. Divide by q repeatedly. This gives

$$\begin{aligned} p(x) &= p_1(x)q(x) + r_1(x) \\ &= (p_2(x)q(x) + r_2(x))q(x) + r_1(x) \\ &= ((p_3(x)q(x) + r_3(x))q(x) + r_2(x))q(x) + r_1(x) \\ &\vdots \\ &= \sum_{i=1}^{n} r_i(x)q(x)^{i-1}, \end{aligned}$$

with each $r_i(x)$ satisfying either $r_i = 0$ or $\deg r_i < \deg q$. Now divide by $q(x)^n$ to get the result. ■

Let us write out the consequences of this theorem when $q(x)$ is linear or quadratic.

Corollary 5.4.4

Consider $p(x)/q(x)^n$ with $\deg p < n(\deg q)$.

1. *If $q(x) = x - a$, the function can be expressed as*

$$\frac{p(x)}{q(x)^n} = \frac{A_1}{(x-a)^n} + \frac{A_2}{(x-a)^{n-1}} + \cdots + \frac{A_n}{(x-a)}$$

with $A_i \in \mathbb{R}$.

2. *If $q(x) = x^2 + \alpha x + \beta$, the function can be expressed as*

$$\frac{p(x)}{q(x)^n} = \frac{B_1 x + C_1}{(x^2+\alpha x+\beta)^n} + \frac{B_2 x + C_2}{(x^2+\alpha x+\beta)^{n-1}} + \cdots + \frac{B_n x + C_n}{(x^2+\alpha x+\beta)}$$

with $B_i, C_i \in \mathbb{R}$. ■

On combining Theorem 5.4.1 with Corollary 5.4.4 we see that any rational function $p(x)/q(x)$ with $\deg p < \deg q$ can be expressed as a sum of terms of the form $A/(x-a)^k$ or $(Bx+C)/(x^2+\alpha x+\beta)^k$, which we shall call its **partial fraction decomposition**.

Example 5.4.5

If we return to Example 5.4.2, the decomposition can be further improved to

$$\frac{x^5+7x^4+5x^3+12x^2+4x+7}{(x+1)^2(x^2+1)^2} = \sum_{i=1}^{2} \frac{A_i}{(x+1)^i} + \sum_{i=1}^{2} \frac{B_i x + C_i}{(x^2+1)^i}. \qquad \square$$

We have established that every rational function can be expressed in a simpler form that is convenient for integration. However, we still need a way to compute the coefficients of the partial fractions. One way is to multiply through by the denominator, and then match the coefficients on both sides. Rather than doing the full expansion in one go, we demonstrate below how one can conveniently carry out the calculations in easy stages.

To warm up, we begin with examples where the denominator has only linear factors.

Example 5.4.6

Consider $\dfrac{x^2+2x+2}{(x-1)^3}$. Its partial fraction decomposition has the form

$$\frac{x^2+2x+2}{(x-1)^3} = \frac{A}{(x-1)^3} + \frac{B}{(x-1)^2} + \frac{C}{(x-1)}.$$

Multiply both sides by $(x-1)^3$:

$$x^2+2x+2 = A + B(x-1) + C(x-1)^2.$$

Put $x = 1$ to get $A = 5$. If we substitute this in the last expression and also move the A term to the left-hand side, we see that both sides must be divisible by $x-1$. Dividing by $x-1$ gives

$$x+3 = B + C(x-1).$$

Again, $x = 1$ gives $B = 4$, and then $C = 1$. So the partial fraction decomposition is

$$\frac{x^2+2x+2}{(x-1)^3} = \frac{5}{(x-1)^3} + \frac{4}{(x-1)^2} + \frac{1}{(x-1)}.$$

This is easy to integrate:

$$\int \frac{x^2+2x+2}{(x-1)^3}\,dx = -\frac{5/2}{(x-1)^2} - \frac{4}{(x-1)} + 2\log|x-1| + C. \qquad \square$$

Example 5.4.7

Consider $\dfrac{x^3+9x^2+8}{(x-1)^2(x+2)^2}$. Its partial fraction decomposition has the form

$$\frac{x^3+9x^2+8}{(x-1)^2(x+2)^2} = \frac{A}{(x-1)^2} + \frac{B}{(x-1)} + \frac{C}{(x+2)^2} + \frac{D}{(x+2)}.$$

We proceed as in the previous example. Multiplying by $(x-1)^2$ and then evaluating at $x = 1$ gives $A = 2$. Substitute this and simplify to get

$$\frac{x^2+8x}{(x-1)(x+2)^2} = \frac{B}{(x-1)} + \frac{C}{(x+2)^2} + \frac{D}{(x+2)}.$$

Now multiply by $x-1$ and evaluate at $x = 1$ to obtain $B = 1$. The decomposition becomes

$$\frac{4}{(x+2)^2} = \frac{C}{(x+2)^2} + \frac{D}{(x+2)}.$$

This immediately gives $D = 0$ and $C = 4$. Therefore,

$$\frac{x^3+9x^2+8}{(x-1)^2(x+2)^2} = \frac{2}{(x-1)^2} + \frac{1}{(x-1)} + \frac{4}{(x+2)^2}.$$

Hence,

$$\int \frac{x^3+9x^2+8}{(x-1)^2(x+2)^2}\,dx = \frac{-2}{x-1} + \log|x-1| - \frac{4}{x+2} + C. \qquad \square$$

Now we take up examples where the denominator has only quadratic factors. We will not be able to use roots to simplify calculations, but the division algorithm helps us cross this obstacle.

Example 5.4.8

Consider $\dfrac{x^5+4x^3-x^2+3x}{(x^2+1)^3}$. This is the situation of Theorem 5.4.3 and we apply the procedure used in its proof.

$$\begin{aligned} x^5+4x^3-x^2+3x &= (x^3+3x-1)(x^2+1)+1 \\ &= (x(x^2+1)+2x-1)(x^2+1)+1 \\ &= x(x^2+1)^2+(2x-1)(x^2+1)+1. \end{aligned}$$

Hence,

$$\frac{x^5+4x^3-x^2+3x}{(x^2+1)^3} = \frac{x}{x^2+1} + \frac{2x-1}{(x^2+1)^2} + \frac{1}{(x^2+1)^3}.$$

We carry out the integration using the results stated at the start of this section. The result is

$$\int \frac{x^5+4x^3-x^2+3x}{(x^2+1)^3}\,dx = \frac{1}{2}\log(x^2+1) - \frac{\arctan x}{8} - \frac{2+x}{2(x^2+1)} + \frac{3x^3+5x}{8(x^2+1)^2} + C.$$

□

Example 5.4.9

Next, we consider an example with two different quadratic factors: $\frac{1}{(x^2+1)^2(x^2+x+1)}$. Its decomposition has the form

$$\frac{1}{(x^2+1)^2(x^2+x+1)} = \frac{Ax+B}{x^2+1} + \frac{Cx+D}{(x^2+1)^2} + \frac{Ex+F}{x^2+x+1}.$$

Multiply through by $(x^2+1)^2(x^2+x+1)$ to create a polynomial equation:

$$1 = (Ax+B)(x^2+1)(x^2+x+1) + (Cx+D)(x^2+x+1) + (Ex+F)(x^2+1)^2.$$

Each side must have remainder 1 if we divide by x^2+1. This gives $1 = Dx - C$ and hence $C = -1$, $D = 0$. The decomposition becomes

$$\frac{x+1}{(x^2+1)(x^2+x+1)} = \frac{Ax+B}{x^2+1} + \frac{Ex+F}{x^2+x+1}.$$

This gives

$$x+1 = (Ax+B)(x^2+x+1) + (Ex+F)(x^2+1).$$

Comparing remainders on dividing by x^2+1 gives $x+1 = Bx - A$, so $A = -1$ and $B = 1$. The decomposition now reduces to

$$\frac{x}{x^2+x+1} = \frac{Ex+F}{x^2+x+1}.$$

Hence, $E = 1$ and $F = 0$, and the partial fraction decomposition is completely worked out. □

We have seen how to compute partial fractions that only involve linear factors or only quadratic factors. To finish off, we need to show how to work with a rational function where both are present. The strategy is to first work out the coefficients for the linear factors, eventually reducing to the case of quadratic factors.

Example 5.4.10

Consider the partial fraction decomposition

$$\frac{x^5+7x^4+5x^3+12x^2+4x+7}{(x+1)^2(x^2+1)^2} = \sum_{i=1}^{2} \frac{A_i}{(x+1)^i} + \sum_{i=1}^{2} \frac{B_i x + C_i}{(x^2+1)^i}.$$

Multiply both sides by $(x+1)^2$:

$$\frac{x^5+7x^4+5x^3+12x^2+4x+7}{(x^2+1)^2} = A_1(x+1) + A_2 + \sum_{i=1}^{2} \frac{B_i x + C_i}{(x^2+1)^i}(x+1)^2.$$

Put $x=-1$ to get $A_2 = 4$. Substitute this in the decomposition and cancel $x+1$ to get

$$\frac{x^4+2x^3+3x^2+x+3}{(x^2+1)^2} = A_1 + \sum_{i=1}^{2} \frac{B_i x + C_i}{(x^2+1)^i}(x+1).$$

Again put $x=-1$ and get $A_1 = 1$. Substitute and simplify:

$$\frac{2x^2-x+2}{(x^2+1)^2} = \sum_{i=1}^{2} \frac{B_i x + C_i}{(x^2+1)^i}.$$

Multiply both sides by $(x^2+1)^2$:

$$2x^2 - x + 2 = (B_1 x + C_1)(x^2+1) + (B_2 x + C_2).$$

At this stage, we only have a few terms to deal with, and we can read off the coefficients easily: $B_1 = 0$, $C_1 = 2$, $B_2 = -1$, $C_2 = 0$. We have finally reached our goal.

$$\frac{x^5+7x^4+5x^3+12x^2+4x+7}{(x+1)^2(x^2+1)^2} = \frac{1}{x+1} + \frac{4}{(x+1)^2} + \frac{2}{x^2+1} - \frac{x}{(x^2+1)^2}.$$

The integration is left to you! □

Exercises for § 5.4

1. Use the method of partial fractions to find the following anti-derivatives:

(a) $\displaystyle\int \frac{1}{x^2-1}\,dx.$

(b) $\displaystyle\int \frac{1}{x^3-1}\,dx.$

(c) $\displaystyle\int \frac{1}{(x^2+x+1)^2}\,dx.$

(d) $\displaystyle\int \frac{x}{(x-1)^2}\,dx.$

2. Consider a function of the form $p(x)/(x-a)^n$ with $\deg p < n$. Let its partial fraction decomposition be $\sum_{k=0}^{n-1} A_k/(x-a)^{n-k}$. Show that

$$A_k = \frac{p^{(k)}(a)}{k!}.$$

3. Consider a function of the form $p(x)/(x-a_1)\cdots(x-a_n)$ with $\deg p < n$. Let its partial fraction decomposition be $\sum_{k=1}^{n} A_k/(x-a_k)$. Show that

$$A_k = \frac{p(a_k)}{\prod_{i:i\neq k}(x_k - x_i)}.$$

4. Compute the following integrals:

(a) $\displaystyle\int \frac{x^2-5x+12}{x(x+2)(x-3)}\,dx.$

(b) $\displaystyle\int \frac{x}{(x-3)^2(x+2)^2}\,dx.$

(c) $\displaystyle\int \frac{x^3+x+1}{(x^2+1)(x^2+2)}\,dx.$

(d) $\displaystyle\int \frac{x^3}{(x-1)(x^2+4)^2}\,dx.$

(e) $\displaystyle\int \frac{dx}{x^4-1}.$

(f) $\displaystyle\int \frac{dx}{x^4+1}.$

5. Use substitution to convert the integrand to a rational function and then apply partial fractions.

(a) $\displaystyle\int \frac{\sqrt{x}}{x-9}\,dx.$

(b) $\displaystyle\int \frac{1}{e^{2x}-1}\,dx.$

6. The substitution $x = \tan t/2$ converts trigonometric integrands to rational functions. First, check that it leads to

$$\sin x = \frac{2t}{1+t^2}, \quad \cos x = \frac{1-t^2}{1+t^2}, \quad dx = \frac{2dt}{1+t^2}.$$

Now apply this substitution to the following integration problems:

(a) $\displaystyle\int \frac{dx}{\sin x + \cos x}.$

(b) $\displaystyle\int \frac{\sin^2 x}{1+\sin^2 x}\,dx.$

5.5 Improper Integrals

Our definition of definite integrals requires a bounded function f over a closed and bounded domain $[a,b]$. Applications of integration often involve situations where these requirements are not met, and either the function or the domain is unbounded. Such integrals are called **improper**. We shall evaluate them by considering them as limits of "proper" ones.

On the other hand, the requirement of taking a closed interval is not important. Suppose f is bounded on $[a,b)$. One can define $f(b)=0$ and consider $\int_a^b f(x)\,dx$. You can easily check that the result is independent of the number assigned to $f(b)$. Further, the result equals $\displaystyle\lim_{t\to b-}\int_a^t f(x)\,dx$.

Improper integrals of the first kind

If the integrand f is bounded but the interval of integration is not, we have an **improper integral of the first kind**. These integrals are defined via limits as follows:

$$\int_a^\infty f(x)\,dx = \lim_{b\to\infty}\int_a^b f(x)\,dx,$$
$$\int_{-\infty}^b f(x)\,dx = \lim_{a\to-\infty}\int_a^b f(x)\,dx,$$
$$\int_{-\infty}^\infty f(x)\,dx = \int_{-\infty}^a f(x)\,dx + \int_a^\infty f(x)\,dx.$$

Obviously, we first need f to be integrable on each interval of integration $[a,b]$ in these definitions. In particular, f needs to be bounded on each $[a,b]$, though it need not be bounded on the entire unbounded interval.

If the defining limit exists and is finite, we say the improper integral **converges**. Else, we say it **diverges**. In the definition of $\int_{-\infty}^\infty f(x)\,dx$ we can use any convenient a, and then **both** $\int_{-\infty}^a f(x)\,dx$ and $\int_a^\infty f(x)\,dx$ need to converge for $\int_{-\infty}^\infty f(x)\,dx$ to be defined.

Example 5.5.1

$$\int_0^\infty e^{-x}\,dx = \lim_{b\to\infty}\int_0^b e^{-x}\,dx = \lim_{b\to\infty} -e^{-x}\Big|_0^b = \lim_{b\to\infty}(1-e^{-b}) = 1. \qquad \square$$

Example 5.5.2

Let us consider the convergence of $\displaystyle\int_1^\infty \frac{1}{x^p}\,dx$ when $p>0$. If $p=1$, we have

$$\int_1^\infty \frac{1}{x}\,dx = \lim_{b\to\infty}\int_1^b \frac{1}{x}\,dx = \lim_{b\to\infty}\log b = \infty.$$

If $p\neq 1$ we have

$$\int_1^\infty \frac{1}{x^p}\,dx = \lim_{b\to\infty}\int_1^b \frac{1}{x^p}\,dx = \frac{1}{1-p}\lim_{b\to\infty}\frac{1}{x^{p-1}}\Big|_1^b$$
$$= \frac{1}{1-p}\lim_{b\to\infty}\left(\frac{1}{b^{p-1}}-1\right) = \begin{cases} \dfrac{1}{p-1} & \text{if } p>1, \\ \infty & \text{if } p<1. \end{cases}$$

Overall,

$$\int_1^\infty \frac{1}{x^p}\,dx = \begin{cases} \dfrac{1}{p-1} & \text{if } p>1, \\ \infty & \text{if } p\leq 1. \end{cases}$$

Thus the integral diverges when $p\leq 1$. $\square$

For non-negative functions the improper integral represents the total area under the curve, and convergence of the improper integral corresponds to this area being finite. This interpretation suggests the following:

Theorem 5.5.3 (Comparison Theorem)

Suppose $f, g\colon [a,\infty) \to \mathbb{R}$ are continuous functions and $0 \le f(x) \le g(x)$ for every $x \in [a,\infty)$. If $\int_a^\infty g(x)\,dx$ converges, then $\int_a^\infty f(x)\,dx$ converges and $\int_a^\infty f(x)\,dx \le \int_a^\infty g(x)\,dx$.

Proof. Let $F(t) = \int_a^t f(x)\,dx$ and $G(t) = \int_a^t g(x)\,dx$. Since $f, g \ge 0$, the functions F, G are increasing. By the monotone convergence theorem (Theorem 3.7.3),

$$\int_a^\infty g(x)\,dx = \lim_{b\to\infty} G(b) \ge G(t) \ge F(t) \quad \text{for every } t \ge a.$$

Hence, F is bounded, and by the monotone convergence theorem again, we have the convergence of $\lim_{b\to\infty} F(b) = \displaystyle\int_a^\infty f(x)\,dx$. ■

Perhaps the most famous improper integral is the **Gaussian integral** $\int_{-\infty}^\infty e^{-x^2}\,dx = \sqrt{\pi}$. The associated improper indefinite integral

$$\Phi(x) = \frac{1}{\sqrt{2\pi}} \int_{-\infty}^x e^{-t^2/2}\,dt$$

is the cumulative distribution function of the standard normal distribution in probability theory and an important tool in the analysis of diverse random processes such as errors in measurement and fluctuations in stock prices. In this section we will be able to establish its convergence and calculate approximate values. Determining the exact value $\sqrt{\pi}$ of the Gaussian integral requires more advanced techniques. It is usually established in courses of multivariable calculus, where it is one of the striking applications of double integrals in polar coordinates (for example, see §17.5 of Marsden and Weinstein [22]). It can also be obtained using infinite products (see Problems 39–41 following §19 of Spivak [30]). We offer a proof using convex functions, in the supplementary exercises on gamma and beta functions after Chapter 7, as it brings together many of the concepts studied by us at various stages.

Example 5.5.4

Consider the improper integral $\int_0^\infty e^{-x^2}\,dx$. We have

$$\int_0^\infty e^{-x^2}\,dx = \int_0^1 e^{-x^2}\,dx + \int_1^\infty e^{-x^2}\,dx.$$

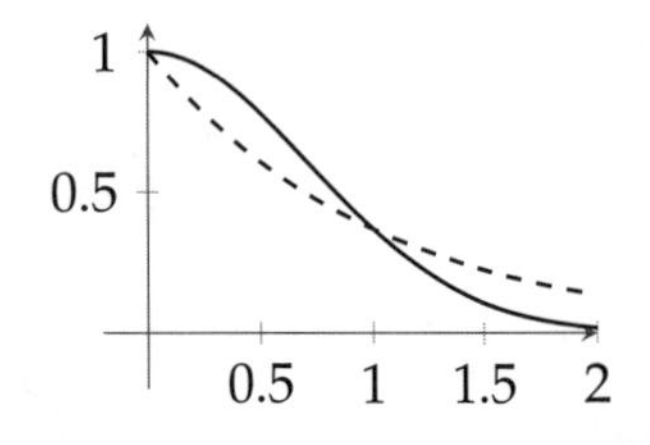

For $x \ge 1$ we have $x^2 \ge x$ and hence $0 \le e^{-x^2} \le e^{-x}$. Since $\int_1^\infty e^{-x}\,dx$ converges, so does $\int_1^\infty e^{-x^2}\,dx$. Therefore, $\int_0^\infty e^{-x^2}\,dx$ converges. Similarly, $\int_{-\infty}^0 e^{-x^2}\,dx$ converges.

Hence, $\int_{-\infty}^{\infty} e^{-x^2}\,dx = \int_{-\infty}^{0} e^{-x^2}\,dx + \int_{0}^{\infty} e^{-x^2}\,dx$ converges. □

Example 5.5.5

We can find approximations to $\int_0^\infty e^{-x^2}\,dx$ as follows: First, truncate the interval of integration to some $[0,b]$. Then use a step function to get an approximate value of $\int_0^b e^{-x^2}\,dx$.

As an example let us first set $b = 3$. Next we set $n = 6$ and partition $[0,3]$ into 6 equal subintervals. On the ith subinterval we approximate $f(x) = e^{-x^2}$ by its value at the midpoint c_i.

c_i	0.25	0.75	1.25	1.75	2.25	2.75
$f(c_i)$	0.9394	0.5698	0.2096	0.0468	0.0063	0.0005

Now

$$\int_0^\infty e^{-x^2}\,dx \approx \int_0^3 e^{-x^2}\,dx \approx \sum_{i=1}^{6} f(c_i) \times 0.5 = 0.886213\ldots.$$

The exact value of the integral is $\sqrt{\pi}/2 = 0.886226\ldots$. With these few calculations we already have accuracy to 4 decimal places! □

Improper integrals of the second kind

Improper integrals of the second kind occur when f has a vertical asymptote, such as when we try to integrate $1/\sqrt{x}$ over $[0,1]$. These are defined by taking limits at the points where f has a vertical asymptote.

Example 5.5.6

The function $1/\sqrt{x}$ is unbounded on $(0,1]$. On the other hand, it is continuous on $[a,1]$ for every $a \in (0,1)$. Therefore, we can define its improper integral on $[0,1]$ by integrating on $[a,1]$ and then letting $a \to 0+$:

$$\int_0^1 \frac{1}{\sqrt{x}}\,dx = \lim_{a\to 0+} \int_a^1 \frac{1}{\sqrt{x}}\,dx = \lim_{a\to 0+} 2\sqrt{x}\Big|_a^1 = 2\lim_{a\to 0+}(1-\sqrt{a}) = 2.$$ □

Task 5.5.7

Show that $\int_0^1 x^\alpha\,dx$ converges for $-1 < \alpha < 0$ and diverges for $\alpha \le -1$.

In the preceding Example and Task we were able to directly compute the integral and establish convergence. If we are unable to compute the integral, we may still be able to establish convergence by a comparison with a simpler integral.

Theorem 5.5.8 (Comparison Theorem)

Suppose $f, g\colon (a,b] \to \mathbb{R}$ are continuous functions that are unbounded on $(a,b]$ but bounded on every $[x,b]$ with $a < x < b$, and $0 \le f(x) \le g(x)$ for every $x \in (a,b]$. If $\int_a^b g(x)\,dx$ converges, then $\int_a^b f(x)\,dx$ converges and $\int_a^b f(x)\,dx \le \int_a^b g(x)\,dx$.

Proof. Exercise. ■

Example 5.5.9

Consider the improper integral $\int_0^1 e^{-x}x^{\alpha}\,dx$ with $\alpha < 0$. It is improper because $\lim\limits_{x\to 0+} e^{-x}x^{\alpha} = \infty$. We compute as follows for $0 < x \le 1$:

$$-1 < \alpha < 0 \implies 0 < e^{-x}x^{\alpha} \le x^{\alpha} \text{ and } \int_0^1 x^{\alpha}\,dx \text{ converges.}$$

$$\alpha \le -1 \implies e^{-x}x^{\alpha} \ge e^{-1}x^{\alpha} > 0 \text{ and } \int_0^1 x^{\alpha}\,dx \text{ diverges.}$$

Hence, by the comparison theorem, the integral converges for $-1 < \alpha < 0$ and diverges for $\alpha \le -1$. □

Gamma Function

The **gamma function** is an instance of an improper integral that involves both an unbounded interval as well as an unbounded function.

$$\Gamma(x) = \int_0^{\infty} e^{-t}t^{x-1}\,dt \qquad (x > 0).$$

Note that apart from the unbounded interval of integration, the integrand goes to infinity at zero when $0 < x < 1$. We split the integral as follows:

$$\Gamma(x) = \int_0^1 e^{-t}t^{x-1}\,dt + \int_1^{\infty} e^{-t}t^{x-1}\,dt.$$

The convergence of the integral is established by the following calculations:

1. The integral from 0 to 1 is improper when $0 < x < 1$ and we have already established its convergence in the previous Example.
2. For any fixed $x > 0$, $e^{-t/2}t^{x-1} \to 0$ as $t \to \infty$. Hence, there is an a such that $t \ge a$ implies $e^{-t/2}t^{x-1} \le 1$. Therefore, for $x \ge a$, $0 \le e^{-t}t^{x-1} \le e^{-t/2}$. Again, the comparison theorem gives the convergence of $\int_a^{\infty} e^{-t}t^{x-1}\,dt$ and hence of $\int_1^{\infty} e^{-t}t^{x-1}\,dt$.

We now apply integration by parts to obtain a relationship between different values of $\Gamma(x)$. Let $0 < a < b$. Then

$$\int_a^b e^{-t}t^x\,dt = -e^{-t}t^x\Big|_a^b + x\int_a^b e^{-t}t^{x-1}\,dt.$$

We have $\lim_{t\to\infty} e^{-t}t^x = \lim_{t\to 0+} e^{-t}t^x = 0$. Hence, letting $a \to 0+$ and $b \to \infty$, we get

$$\Gamma(x+1) = x\Gamma(x).$$

It is easy to compute that $\Gamma(1) = 1$. Hence,

$$\Gamma(2) = 1 \cdot \Gamma(1) = 1, \quad \Gamma(3) = 2 \cdot \Gamma(2) = 2 \cdot 1, \quad \Gamma(4) = 3 \cdot \Gamma(3) = 3 \cdot 2 \cdot 1, \ldots.$$

In general,

$$\Gamma(n+1) = n! \qquad n = 0,1,2,\ldots.$$

Task 5.5.10

Show that $\Gamma(1/2)$ equals the Gaussian integral $\int_{-\infty}^{\infty} e^{-x^2}\,dx$.

Exercises for § 5.5

1. Discuss the convergence of $\int_0^\infty x^\alpha\,dx$ for $\alpha \in \mathbb{R}$.

2. Prove that the given improper integrals converge and find their value:

(a) $\displaystyle\int_0^\infty \frac{dx}{1+x^2}$. (b) $\displaystyle\int_{-\infty}^\infty xe^{-x^2}\,dx$. (c) $\displaystyle\int_1^\infty \frac{\log x}{x^2}\,dx$.

3. Prove that the given improper integrals converge and find their value:

(a) $\displaystyle\int_0^1 \frac{1}{\sqrt{t(1-t)}}\,dt$. (b) $\displaystyle\int_0^1 \log x\,dx$. (c) $\displaystyle\int_0^1 \frac{\log x}{\sqrt{x}}\,dx$.

4. Prove that the given improper integrals diverge:

(a) $\displaystyle\int_0^{\pi/2} \sec x\,dx$. (c) $\displaystyle\int_1^2 \frac{1}{x\log x}\,dx$.

(b) $\displaystyle\int_2^\infty \frac{1}{x\log x}\,dx$. (d) $\displaystyle\int_0^\infty \frac{1}{\sqrt{9+x^2}}\,dx$.

5. Use the comparison test to decide whether the given improper integrals converge or diverge:

(a) $\displaystyle\int_1^\infty e^{-x}x^{3/2}\,dx$. (c) $\displaystyle\int_0^\infty \frac{2+\sin x}{1+x}\,dx$.

(b) $\displaystyle\int_1^\infty \frac{1}{\sqrt{1+x^4}}\,dx$. (d) $\displaystyle\int_1^\infty e^{-x}\log x\,dx$.

6. (Limit Comparison Test) Let f, g be continuous and satisfy $0 \le f(x) \le g(x)$ for every $x \ge a$. Let

$$\lim_{x\to\infty} \frac{f(x)}{g(x)} = L \in \mathbb{R},\ L \ne 0.$$

Prove that $\int_a^\infty f(x)\,dx$ converges if and only if $\int_a^\infty g(x)\,dx$ converges. (Hint: For large enough x, we have $L-1 \le f(x)/g(x) \le L+1$.)

7. Prove that $\displaystyle\int_2^\infty \frac{x^2-5x+11}{x^4+x^3+x^2-x-10}\,dx$ converges. (Hint: Where are the zeroes of the denominator?)

8. State and prove the limit comparison test for improper integrals of the second kind.

9. Let f be continuous on $\mathbb{R}$. Then $\lim_{R\to\infty}\int_{-R}^{R} f(x)\,dx$ is called the **Cauchy principal value** of the improper integral $\int_{-\infty}^{\infty} f(x)\,dx$. Prove the following:

(a) If $\int_{-\infty}^{\infty} f(x)\,dx$ converges, then its value equals its Cauchy principal value.

(b) If f is odd, then the Cauchy principal value is zero, regardless of the convergence or divergence of the improper integral.

(c) If f is even, then the Cauchy principal value exists if and only if the improper integral converges.

10. A function $f\colon [a,\infty) \to \mathbb{R}$ is given to be continuous.

(a) Suppose $f(x) \to L \neq 0$ as $x \to \infty$. Can $\int_a^\infty f(x)\,dx$ converge?

(b) Suppose $\int_a^\infty f(x)\,dx$ converges. Does this imply that $f(x) \to 0$ as $x \to \infty$?

5.6 Ordinary Differential Equations

An **ordinary differential equation**, or ODE, is an equation involving a function $f(x)$ and some of its derivatives, as well as the variable x. The task is to solve for f. The order of the highest derivative of f that occurs in the ODE is called the **order** of the ODE. Here are some ODEs for an unknown function $y = f(x)$:

1. $y' = \cos x$ (first-order).
2. $y' = 5xy$ (first-order).
3. $y'' = -3y + 1$ (second-order).
4. $(y''')^2 + y\sec(y'') + y'/y + \tan x = 0$ (third-order).

Of course, one may use different variable and function names. For example, the variable may be time t, the unknown function may be position $x(t)$, and the ODE may be $x'' = -5x$.

An ODE may have no solution. For example, $(y')^2 + 1 = 0$ has no solution. However, an ODE arising out of a practical problem is unlikely to be like this. On the contrary, a typical ODE will have multiple solutions. The reason is that a differential equation has information about how a quantity changes. The final value of the quantity depends on how it changes (described by the ODE) as well as its starting state. If we know the starting state, we may be able to narrow down to exactly one solution.

Example 5.6.1

We have seen that every solution of $y' = y$ has the form $y = Ae^x$. Each value of A leads to a different solution. If we know that $y(0) = 5$, we can solve for A and get a unique solution,

$$y(0) = 5 \implies 5 = Ae^0 \implies A = 5 \implies y(x) = 5e^x. \qquad \square$$

A collection of data of the form $y^{(k)}(a) = 0$, with $k = 0, 1, \ldots, n-1$, for an nth-order ODE is called its **initial conditions**.

In the rest of this section we will study first-order ODEs. The supplementary exercises for this chapter contain further material on a particular variety of second-order ODEs.

Separable First-order ODE

The typical form for a first-order ODE is $y' = h(x,y)$, since the only quantities that can be involved are x, y, and y'. In many important examples, it is possible to separate $h(x,y)$ into a factor involving only x and a factor involving only y,

$$y' = f(x)g(y) \tag{5.1}$$

When this can be done we say we have a **separable** first-order ODE. Let us solve (5.1). First, we rearrange it to

$$\frac{y'}{g(y)} = f(x).$$

Both sides are functions of x and we integrate them with respect to x.

$$\int \frac{y'}{g(y)}\,dx = \int f(x)\,dx.$$

According to the substitution method, we can replace $y'\,dx$ with dy, to get

$$\int \frac{dy}{g(y)} = \int f(x)\,dx, \tag{5.2}$$

provided that f, g, and y' are continuous. This gives an equation involving y. If we are fortunate, we can solve it to obtain an explicit formula for y.

Example 5.6.2 (Exponential Growth and Decay)

Consider the separable ODE $y' = ky$. Applying (5.2), we get

$$\int \frac{dy}{y} = \int k\,dx.$$

Hence,

$$\log|y| = kx + c.$$

Therefore,

$$|y| = e^{kx+c} = e^c e^{kx} \quad \text{and} \quad y = Ae^{kx}.$$

While this process gives a solution with $A \neq 0$, we see that $A = 0$ also gives a valid solution. When $k > 0$ we have exponential *growth*, and when $k < 0$ we have exponential *decay*. □

Task 5.6.3

Show that the solutions $y(t)$ of $y' = M - ky$ have the form $y = (M - Ae^{-kt})/k$ if $k \neq 0$.

Example 5.6.4 (Logistic Growth)

Consider a population $y(t)$ governed by the separable ODE $y' = ky(M - y)$ and with values in $[0, M]$. We have the following implications:

$$\begin{aligned}
y' = ky(M-y) &\implies \int \frac{dy}{y(M-y)} = \int k\,dt \\
&\implies \frac{1}{M}\int \left(\frac{1}{y} + \frac{1}{M-y}\right) dy = \int k\,dt \\
&\implies \log\left(\frac{y}{M-y}\right) = kMt + d \\
&\implies \frac{y}{M-y} = Ae^{kMt} \\
&\implies y = \frac{AMe^{kMt}}{1 + Ae^{kMt}} = \frac{AM}{e^{-kMt} + A}.
\end{aligned}$$

This model describes a population whose growth is initially exponential but then tapers off as it approaches a maximum sustainable value M. Here is a graph of the solution with $k = M = 1$ and $A = 0.01$:

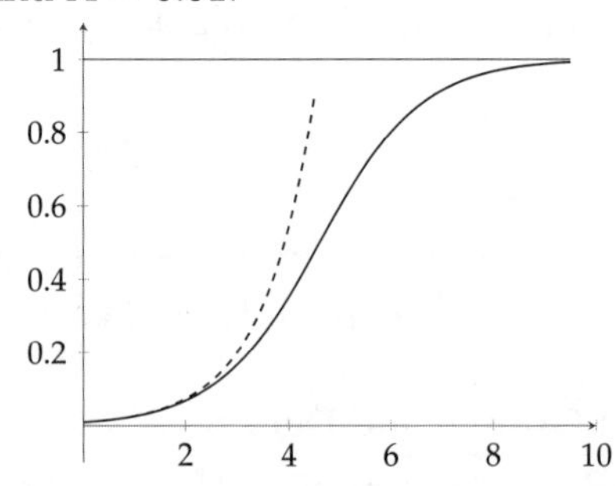

The solid curve shows the logistic growth. In its early stages it resembles the exponential growth corresponding to $y' = y$ and $y(0) = 0.01$. The parameter A in the logistic growth solution can be determined if we know the initial value $y(0)$.

$$y(0) = \frac{AM}{1+A} \implies (1+A)y(0) = AM \implies A = \frac{y(0)}{M - y(0)}.$$ □

When we solve a first-order ODE we typically get a family of solutions, generated by one parameter. The common formula for this family is called a **general solution** of the ODE.

ODE	General Solution
$y' = ky$	$y = Ae^{kt}$
$y' = M - ky$	$y = \frac{1}{k}(M - Ae^{-kt})$
$y' = ky(M - y)$	$y = \frac{AMe^{kMt}}{1 + Ae^{kMt}}$

When the parameter A is given a specific value, we get an individual solution, which is called a **particular solution**.

Example 5.6.5

Consider the equation $y' = -xy$. Separating variables gives $y'/y = -x$ and then $\log|y| = -\frac{x^2}{2} + C$. So the general solution is

$$y(x) = Ae^{-x^2/2}.$$

Knowledge of any $y(a)$ value will give a particular solution. For example, the initial condition $y(0) = 2$ gives $y(x) = 2e^{-x^2/2}$. Various particular solutions of $y' = -xy$ are shown below:

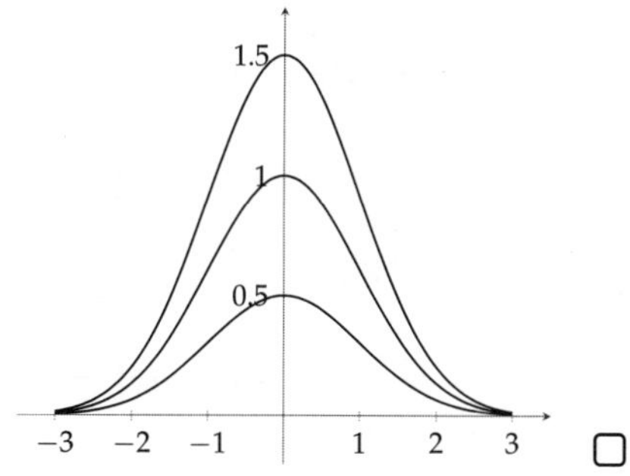

□

Task 5.6.6

Show that the solutions of $y' = y/x$ are straight lines. Plot the solutions satisfying the respective initial conditions $y(1) = -1, 0,$ and 1.

Example 5.6.7

Consider the initial value problem $y' = y^{1/3}, y(0) = 0$. We can obtain a general solution by separation of variables.

$$y' = y^{1/3} \implies \int y^{-1/3}\,dy = \int 1\,dx \implies \frac{3}{2}y^{2/3} = x + c \implies y = (\frac{2}{3}x + A)^{3/2}.$$

This gives the solution $y = (\frac{2}{3}x)^{3/2}$, for $x \geq 0$, of the given initial value problem. However, this is not the only solution, as $y = 0$ is another. This shows that a general solution may not catch *all* solutions, and an initial value problem may have multiple solutions. □

The next task illustrates that the integration may yield an equation between x and y that we may not find easy (or at all possible) to solve for an explicit formula giving y in terms of x.

Task 5.6.8

Show that a general solution of $y' = \dfrac{3x^2+1}{5y^4+1}$ *is* $y^5 + y = x^3 + x + C$.

Linear First-order ODE

A first-order ODE is called **linear** if it has the form

$$y' + P(x)y = Q(x). \tag{5.3}$$

For example, the ODE $y' = ky$ is linear as it can be arranged into $y' - ky = 0$, with $P(x) = -k$ and $Q(x) = 0$. A first-order linear ODE is called **homogeneous** if it has the form

$$y' + P(x)y = 0. \tag{5.4}$$

This is a separable ODE and we have learned how to solve it. We have

$$\begin{aligned} \frac{y'}{y} = -P(x) &\Longrightarrow \int \frac{dy}{y} = -\int P(x)\,dx \Longrightarrow \log|y(x)| = -R(x) + C \\ &\Longrightarrow y(x) = Ae^{-R(x)}. \end{aligned}$$

However, due to Example 5.6.7 we are concerned whether we have really found all solutions. The next theorem gives a positive answer.

Theorem 5.6.9

Suppose that the function $P(x)$ *in (5.4) is continuous on an interval* I. *Then every solution of (5.4) has the form* $y(x) = Ae^{-R(x)}$, *with* $A \in \mathbb{R}$ *and* $R'(x) = P(x)$.

Proof. Since $P(x)$ is continuous, it has an anti-derivative $R(x)$, and we can easily verify that $y(x) = Ae^{-R(x)}$ is a solution. Conversely, let y be a solution. Then we consider the ratio of y and $e^{-R(x)}$.

$$\left(\frac{y}{e^{-R(x)}}\right)' = (ye^{R(x)})' = (y' + P(x)y)e^{R(x)} = 0 \Longrightarrow \frac{y}{e^{-R(x)}} = A. \qquad \blacksquare$$

Example 5.6.10

Consider the ODE $xy' + (1-x)y = 0$. First, we put it in the standard form for a linear ODE,

$$y' + \underbrace{\left(\frac{1}{x} - 1\right)}_{P(x)} y = 0.$$

Now, $\int \left(\frac{1}{x} - 1\right) dx = \log(x) - x$. So the general solution is

$$y = Ae^{x-\log(x)} = A\frac{e^x}{x}.$$ □

The ODE (5.3) is called **non-homogeneous** if $Q(x)$ is not identically zero.

Theorem 5.6.11

Consider a non-homogeneous linear ODE of the form (5.3). Let y_p be a particular solution of this ODE and let y_h be the general solution of the corresponding homogeneous ODE (5.4). Then $y_h + y_p$ is the general solution of (5.3).

Proof. It is trivial to check that $y_h + y_p$ is a solution of $y' + P(x)y = Q(x)$. Now let y be any solution of $y' + P(x)y = Q(x)$. Then

$$(y - y_p)' + P(x)(y - y_p) = Q(x) - Q(x) = 0,$$

hence $y - y_p$ solves the homogeneous ODE and equals one of the members of the family y_h. ■

We now obtain a particular solution y_p of the non-homogeneous linear ODE $y' + P(x)y = Q(x)$ by using a powerful technique called **variation of parameters**, which was introduced by Euler and Lagrange in the eighteenth century.

Theorem 5.6.12

Suppose that the functions $P(x)$ and $Q(x)$ in (5.3) are continuous. Then a particular solution y_p of (5.3) can be obtained by

$$y_p = \left(\int Q(x)\, e^{R(x)}\, dx\right) e^{-R(x)},$$

where $R'(x) = P(x)$.

Proof. Let $R'(x) = P(x)$. We have seen that the general solution of (5.4) is $y_h = Ae^{-R(x)}$. We substitute a function $h(x)$ for the parameter A to obtain a candidate solution of (5.3),

$$y = h(x)\, e^{-R(x)}.$$

Then $y' = h'(x)\, e^{-R(x)} - h(x)P(x)\, e^{-R(x)} = h'(x)\, e^{-R(x)} - P(x)y$, hence $y' + P(x)y = h'(x)\, e^{-R(x)}$. Therefore, we need $h'(x)\, e^{-R(x)} = Q(x)$, or $h(x) = \int Q(x)\, e^{R(x)}\, dx$. ■

Example 5.6.13

Consider $xy' + (1-x)y = e^{2x}$. First, we put it in the standard form,

$$y' + \left(\frac{1}{x} - 1\right)y = \frac{e^{2x}}{x}.$$

We have already worked out the general solution of the homogeneous part as $y_h = A\frac{e^x}{x}$. By variation of parameters, a particular solution is calculated as follows:

$$h(x) = \int \frac{e^{2x}}{x} e^{R(x)}\, dx = \int \frac{e^{2x}}{x} e^{\log(x)-x}\, dx = e^x \implies y_p = \frac{e^{2x}}{x}.$$

Therefore, the general solution of this non-homogeneous equation is

$$y = y_h + y_p = A\frac{e^x}{x} + \frac{e^{2x}}{x}.$$

□

Theorem 5.6.14

Suppose that the functions $P(x)$ and $Q(x)$ in (5.3) are continuous on an interval I. Consider an initial condition $y(x_0) = y_0$ with $x_0 \in I$ and $y_0 \in \mathbb{R}$. Then (5.3) has a unique solution that satisfies this initial condition.

Proof. We already know that the general solution of (5.3) is

$$y = \left(\int Q(x)\, e^{R(x)}\, dx\right) e^{-R(x)} + Ae^{-R(x)},$$

with $R'(x) = P(x)$. We can take any choice of anti-derivative for $\int Q(x)\, e^{R(x)}\, dx$. Let us take $\int_{x_0}^{x} Q(t)\, e^{R(t)}\, dt$. Then the initial condition gives us

$$y_0 = \left(\int_{x_0}^{x_0} Q(t)\, e^{R(t)}\, dt\right) e^{-R(x_0)} + Ae^{-R(x_0)} = Ae^{-R(x_0)}.$$

Next, let us choose $R(x) = \int_{x_0}^{x} P(t)\, dt$. Then $R(x_0) = 0$ and we get $A = y_0$. We have reached the following solution that also satisfies the initial condition:

$$y = \left(\int_{x_0}^{x} Q(t)\, e^{R(t)}\, dt\right) e^{-R(x)} + y_0 e^{-R(x)}, \text{ with } R(x) = \int_{x_0}^{x} P(t)\, dt.$$

As for uniqueness, let y_1 be another solution of (5.3). Then $y_1 - y$ solves (5.4), hence we have $y_1 - y = Ae^{-R(x)}$. The common initial condition then gives $0 = Ae^{-R(x_0)}$, so $A = 0$ and $y_1 - y = 0$. ■

Task 5.6.15

Find a solution of the initial value problem $xy' + (1-x)y = e^{2x}$ and $y(1) = 0$.

Autonomous Differential Equations

A first-order ODE $y' = F(x,y)$ is **autonomous** if the variable x does not explicitly appear in it. That is, it has the form $y' = f(y)$. An autonomous ODE is separable and we can solve it as follows:

$$y' = f(y) \implies \int \frac{dy}{f(y)} = x + c \quad \text{if } f(y) \neq 0.$$

In principle, we have solved the ODE. Practically, we may find it difficult to carry out the integration or solve the resulting equation for y. In this section we shall see that we can obtain a qualitative description of the solutions of an autonomous ODE without actually solving it. We begin with two key observations.

First, since the solution by separation of variables requires $f(y) \neq 0$, let us consider what happens if $f(y_0) = 0$ for some y_0. We see that the constant function $y = y_0$ is a solution. Such a constant solution is called an **equilibrium solution** and the value y_0 is called a **critical value**.

Second, if $y(x)$ is a solution, then so is the shift $y_c(x) = y(x+c)$:

$$y_c'(x) = y'(x+c) = f(y(x+c)) = f(y_c(x)).$$

Example 5.6.16

The logistic equation $y' = ky(M-y)$ is an autonomous equation with critical values 0 and M. Its equilibrium solutions are $y = 0$ and $y = M$. While we have already solved this equation, let us see how much we could have found out about its solutions just from the form of the equation and without applying integration.

First, we plot the function $f(y) = ky(M-y)$ and the two equilibrium solutions. We have also marked an initial value $y(0) = y_0$ for a particular solution.

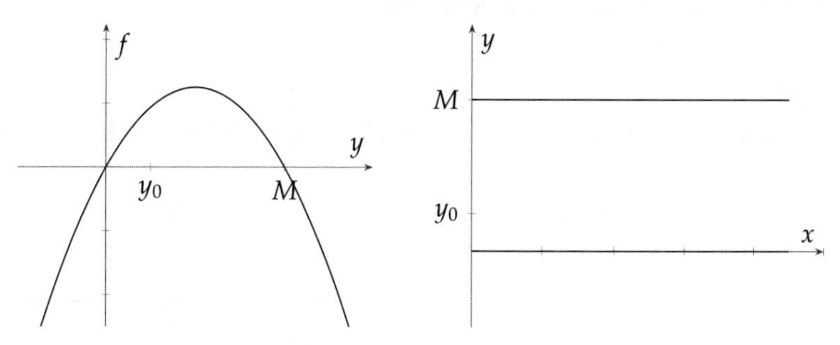

We will use the graph of $f(y)$ to read off the shapes of the solutions $y(x)$. Let us begin with the given $y(0) = y_0$. We see that $y'(0) = f(y_0)$ is positive, so the solution is initially an increasing one. Now, as y increases from y_0, so does $y' = f(y)$ and so the graph of $y(x)$ is initially convex. It stays convex until y reaches $M/2$. At this stage we have the following picture:

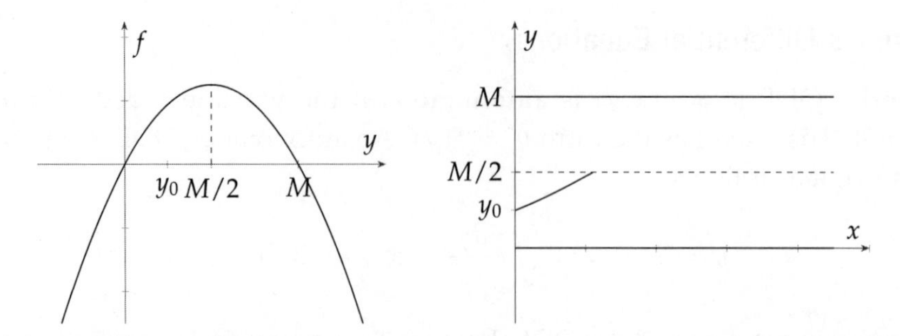

As y increases past $M/2$, y' becomes decreasing. Hence, the graph of y becomes concave and flattens out. The completed graph is shown below, along with a few shifts corresponding to different initial conditions between 0 and M. We have also shown examples of solutions with initial conditions that are either negative or more than 1. For these solutions, y' is always negative and so they are decreasing.

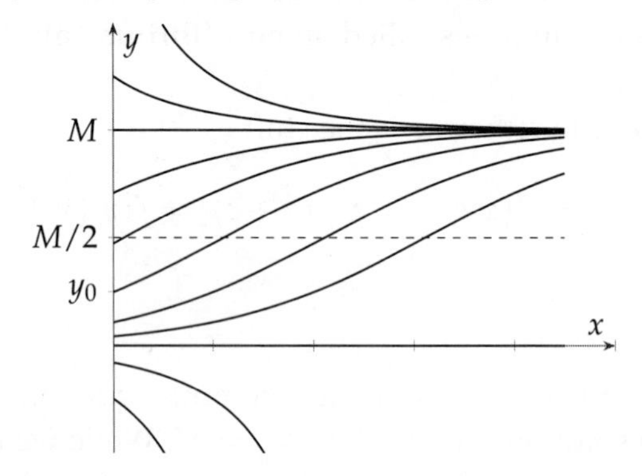

□

In the logistic equation, solutions starting near $y = 0$ move away from it and so $y = 0$ is called an **unstable equilibrium**. On the other hand, solutions starting near $y = M$ approach it asymptotically, hence $y = M$ is called a **stable equilibrium**. We may also have equilibrium points with mixed behavior.

Example 5.6.17

Consider the autonomous ODE $y' = y^2(1 - y)$. The equilibrium solutions are $y = 0$ and $y = 1$.

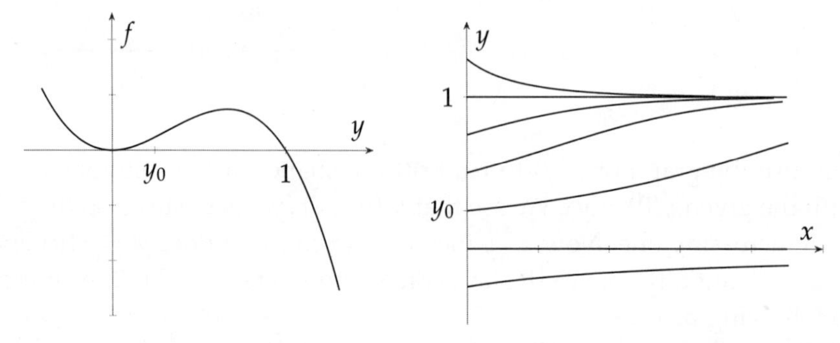

For positive y_0 the analysis is similar to the logistic equation. In particular, there is a stable equilibrium at $y = 1$. On the other hand, if $y(0)$ is negative, then y' is positive and so the solution increases up towards 0. □

In the last example, solutions starting just above 0 move away from $y = 0$, while those starting below are asymptotic to it. Such an equilibrium is called **semistable**.

The next few results give rigorous support to our visual observations.

Theorem 5.6.18

Consider an initial value problem $y' = f(y)$, $y(0) = y_0$, where $f: (a,b) \to \mathbb{R}$ is positive and continuous, and $y_0 \in (a,b)$. This initial value problem has a unique solution $y\colon (\alpha,\beta) \to (a,b)$, which is a strictly increasing bijection. In particular, $\lim_{x\to\beta-} y(x) = b$. (We may have $\beta = \infty$.)

Proof. Since f is positive and continuous on (a,b), so is $1/f$. Therefore, $1/f$ has a strictly increasing and surjective anti-derivative $F\colon (a,b) \to (\alpha,\beta)$. We may assume $F(y_0) = 0$. Now define $y(x) = F^{-1}(x)$. Then $y\colon (\alpha,\beta) \to (a,b)$ is a strictly increasing bijection, such that

$$y'(x) = \frac{1}{F'(F^{-1}(x))} = f(F^{-1}(x)) = f(y(x)) \quad \text{and} \quad y'(0) = F^{-1}(0) = y_0.$$

We have exhibited a solution $y(x)$. If $z(x)$ is any solution, then integrating both sides of $\dfrac{z'(x)}{f(z(x))} = 1$ gives $F(z(x)) = x + c$ and hence $0 = F(y_0) = F(z(0)) = c$. Therefore, $z(x) = F^{-1}(x)$, establishing uniqueness as well. ■

Task 5.6.19

State and prove a version of Theorem 5.6.18 in which the hypothesis $f > 0$ is replaced by $f < 0$.

Task 5.6.20

Consider an initial value problem $y' = f(y)$, $y(0) = y_0$, where f is continuous and y_0 belongs to the domain of f. Will there be a solution? Will it be unique?

Theorem 5.6.21

Consider an autonomous ODE $y' = f(y)$, with f being differentiable. Let c be a critical value.

1. *If $f'(c) < 0$ then $y = c$ is a stable equilibrium solution.*
2. *If $f'(c) > 0$ then $y = c$ is an unstable equilibrium solution.*

Proof. The main effort in proving the first part is in showing that for y_0 close to c the solution's domain will include $[0,\infty)$. We have $\delta > 0$ such that $0 < |y - c| < \delta$ implies

$$-\epsilon = \frac{3}{2}f'(c) < \frac{f(y)}{y - c} < \frac{1}{2}f'(c) = -\epsilon'.$$

Note that $c < y < c+\delta \implies f(y) < 0$ and $c-\delta < y < c \implies f(y) > 0$.

Now consider the case $c-\delta < y_0 < c$. From Theorem 5.6.18 we know there is a unique strictly increasing solution $y(x)$ with $y(0) = y_0$. Take any y in (y_0, c). Then

$$-\epsilon < -\frac{f(y)}{c-y} \implies \epsilon(c-y) > f(y) \implies \frac{1}{f(y)} > \frac{1}{\epsilon(c-y)}.$$

Using the notation of the proof of Theorem 5.6.18, we have

$$F(y) = \int_{y_0}^{y} \frac{dt}{f(t)} \geq \int_{y_0}^{y} \frac{dt}{\epsilon(c-t)} = -\frac{1}{\epsilon}\log\left(\frac{c-y}{c-y_0}\right) \to \infty \text{ as } y \to c-.$$

Hence, $y(x) = F^{-1}(x) \to c$ as $x \to \infty$.

The proof for the $c < y_0 < c+\delta$ case is similar and uses $\dfrac{f(y)}{y-c} < -\epsilon'$.

For the second part, since $f'(c) > 0$, there is a $\delta > 0$ such that $0 < |y-c| < \delta$ implies $\dfrac{f(y)}{y-c} > 0$. Hence, $f(y) > 0$ if $c < y < c+\delta$ and $f(y) < 0$ if $c-\delta < y < c$. Now apply Theorem 5.6.18 and Task 5.6.19 to $(c, c+\delta)$ and $(c-\delta, c)$. ■

Exercises for § 5.6

1. Use separation of variables to find the general solution.

(a) $y' + 2xy^2 = 0$.

(b) $y' \sin x = y \cos x$.

(c) $\sqrt{1+y}\,y' + \sqrt{1-x^2} = 0$.

(d) $(1-x)y' + xy^2 = 0$.

2. Solve $2xyy' = y^2 - x^2$ by substituting $u = y/x$.

3. Consider the logistic equation with an initial condition $y(0) = y_0 \in \mathbb{R}$. Does the domain of the corresponding solution depend on y_0?

4. Find the general solutions of these linear ODEs:

(a) $y' - y = e^x$.

(b) $xy' + 3y = x^2$.

(c) $xy' + 3y = \dfrac{\sin x}{x^2}$.

(d) $y' + y\cos x = e^{2x}$.

5. Solve the given initial value problem:

(a) $(1-x)y' = x(1+y)$, $y(0) = 0$.

(b) $\cos x \sin 2y\, y' = -\sin x \cos 2y$, $y(0) = \pi/2$.

(c) $y' - \dfrac{y}{x} + \csc\dfrac{y}{x} = 0$, $y(1) = 0$.

(d) $\tan\theta \dfrac{dr}{d\theta} - r = \tan^2\theta$, $r(\pi/4) = 0$.

6. Make rough plots of the solutions of the given autonomous ODEs without actually solving them, and classify the equilibrium solutions.

(a) $y' = \sin y$.

(b) $y' = \sin^2 y$.

(c) $y' = -y \log y$.

(d) $y' = 1/y$.

7. A **Bernoulli differential equation** has the form

$$y' + p(x)y = q(x)y^n, \qquad n \neq 0,1.$$

Jacob Bernoulli created this equation in 1719 while modeling projectile motion with drag proportional to a power of the velocity. The logistic equation is a Bernoulli ODE with $n = 2$.

(a) Show that the substitution $u = y^{1-n}$ converts the Bernoulli ODE to a linear first-order ODE.

(b) Find the general solution of $y' - y = xy^2$.

(c) Find the general solution of $y' - \dfrac{2y}{x} = -x^2y^2$.

8. Consider the family of parabolas given by $y = cx^2$, where c varies over all real numbers. Our task is to find curves that meet each of these parabolas orthogonally (that is, their tangents are perpendicular at every crossing point). Such a curve is called an **orthogonal trajectory** to the given family of curves.

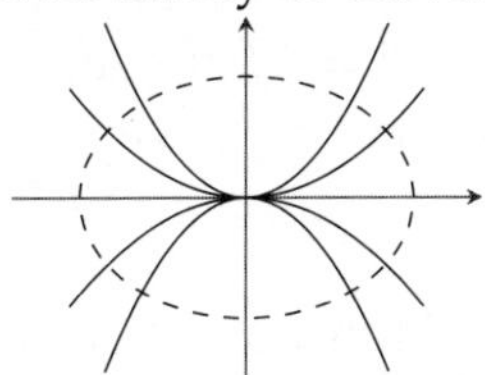

(a) Show that the orthogonal trajectories are governed by the differential equation $y' = -\dfrac{x}{2y}$.

(b) Show that the general solution of (a) is given by $x^2 + 2y^2 = k^2$.

9. Find the orthogonal trajectories to the given family of curves. Sketch some representatives of the family and the orthogonal trajectories.

(a) $x^2 - y^2 = k^2, k \geq 0$.

(b) $(x-r)^2 + y^2 = r^2, r > 0$.

10. Let $y(x)$ be a solution of an initial value problem $y' = f(y)$ and $y(0) = y_0$, with f continuous. Prove the following:

(a) If f is positive on (a, ∞) and $y_0 > a$ then $y(x) \to \infty$ as x increases.

(b) If f is negative on $(-\infty, a)$ and $y_0 < a$ then $y(x) \to -\infty$ as x increases.

Thematic Exercises

Second-order Linear ODE

We will solve an important class of second-order differential equations, called "linear ODE with constant coefficients." Such differential equations crop up in the study of mechanics, waves, electrical circuits, and market equilibrium. They have the form

$$f'' + af' + bf = g,$$

where $a, b \in \mathbb{R}$, and g is some given function. These generalize $f'' = -f$, which we have already solved in §4.5. We shall show that we can go from the special case to the general by a completion of squares, analogous to how we go from finding square roots to solving a general quadratic equation.

First, we consider another special case: $f'' = f$.

A1. Suppose $f\colon \mathbb{R} \to \mathbb{R}$ satisfies $f'' = f$. Show that $f(x) = Ae^x + Be^{-x}$. (Hint: Differentiate the functions $f + f'$ and $f - f'$.)

A2. Suppose $f\colon \mathbb{R} \to \mathbb{R}$ satisfies $f'' = f$. Show the following:

(a) If $f(0) = 0$ and $f'(0) = 1$ then $f(x) = \sinh x$.

(b) If $f(0) = 1$ and $f'(0) = 0$ then $f(x) = \cosh x$.

With the help of the cases $f'' = \pm f$ we can solve a general equation $f'' = kf$.

A3. Suppose $f\colon \mathbb{R} \to \mathbb{R}$ satisfies $f'' = kf$ for some $k \in \mathbb{R}$, $f(0) = A$ and $f'(0) = B$. Prove the following:

(a) If $k = 0$, then $f(x) = Ax + B$.

(b) If $k = -w^2 < 0$, then $f(x) = A \sin wx + B \cos wx$.

(c) If $k = w^2 > 0$, then $f(x) = A \sinh wx + B \cosh wx$.

(Hint: For (b) and (c), consider $g(x) = f(x/w)$.)

The differential equation $f'' + af' + bf = g$ is called "homogeneous" if $g = 0$ and "inhomogeneous" otherwise. The homogeneous case can be completely solved as follows:

A4. Consider the homogeneous differential equation $f'' + af' + bf = 0$. The corresponding "characteristic equation" is $\lambda^2 + a\lambda + b = 0$, with discriminant $d = a^2 - 4b$.

(a) Prove that f is a solution of this differential equation if and only if $f(x) = e^{-ax/2} v(x)$ where v satisfies $v'' + (b - \frac{a^2}{4})v = 0$.

(b) If $d = 0$ and the characteristic equation has repeated root λ, show that any solution has the form $f(x) = Ae^{\lambda x} + Bxe^{\lambda x}$.

(c) If $d > 0$ and the characteristic equation has distinct real roots λ_1, λ_2, show that any solution has the form $f(x) = Ae^{\lambda_1 x} + Be^{\lambda_2 x}$.

(d) If $d < 0$ and the characteristic equation has complex roots $r \pm wi$, show that any solution has the form $f(x) = Ae^{rx} \cos wx + Be^{rx} \sin wx$.

A5. Suppose the roots of the characteristic equation of $f'' + af' + bf = 0$ have negative real parts. Then every solution f satisfies $\lim_{x \to \infty} f(x) = 0$.

We have found a general form for the solutions of $f'' + af' + bf = 0$. It has two parameters A, B. By varying them we can generate all solutions. This form will be called the "general solution" of the homogeneous differential equation and we shall denote it by f_h. For example, if we consider $f'' + 2f' + f = 0$, then $f_h(x) = Ae^{-x} + Bxe^{-x}$.

Now consider an inhomogeneous differential equation $f'' + af' + bf = g$. The first important observation is that if we know one solution of this equation then we know all its solutions!

A6. Suppose $f_p(x)$ is a solution of $f'' + af' + bf = g$. Let f_h be the general solution of $f'' + af' + bf = 0$. Then the general solution of $f'' + af' + bf = g$ is $f = f_p + f_h$.

Here too, "variation of parameters" provides a general approach for finding f_p. We leave that for a course in differential equations. We give below an example of how f_p can often be found quite quickly by recognizing patterns of differentiation.

A7. Consider the differential equation $f''(x) - 3f'(x) + 2f(x) = \cos x$. Since cosine can be obtained by differentiating sine once or cosine twice, we try $f_p(x) = \alpha \cos x + \beta \sin x$.

(a) Substitute f_p in the given differential equation and show $\alpha = 0.1$, $\beta = -0.3$.

(b) Show the general solution is $f(x) = 0.1 \cos x - 0.3 \sin x + Ae^x + Be^{2x}$.

6 | Mean Value Theorems and Applications

When calculus is applied to problems of subjects like physics and economics, it usually leads to equations involving the first and second derivatives of functions, and the task is to recover the original function from these equations. If the derivative is completely known and is continuous, we can use the second fundamental theorem

$$f(x) = f(a) + \int_a^x f'(t)\,dt.$$

However, we usually have only a relation between the function and its derivatives rather than a full knowledge of the derivative. For example, we may know that $f'(x) = f(x)^2$ for every x. So, we need to find more ways of relating information about f' with information about f. We already have two important instances: Fermat's theorem and the monotonicity theorem. In this chapter we will explore several consequences of these results. The payoffs will be new techniques of calculating limits (§6.2), approximation of functions by polynomials (§6.3), use of integration to measure arc length, surface area, and volume (§6.4), and error estimates for numerical calculations of integrals (§6.5). Sections 6.2 and 6.3 are required for the final two chapters on sequences and series, while sections 6.4 and 6.5 are important for the applications of calculus.

Darboux's theorem says that if a function f is differentiable on an interval then f' will have the intermediate value property on that interval. Thus, f' cannot have a jump discontinuity and it behaves like a continuous function in some ways. Nevertheless, it need not be continuous or even bounded.

As an example, consider the function defined by $f(0) = 0$ and $f(x) = x^2 \sin(1/x)$ when $x \neq 0$. This function is differentiable at non-zero points by the chain rule. It is also differentiable at zero by a direct calculation:

$$f'(0) = \lim_{x \to 0} \frac{x^2 \sin(1/x) - 0}{x} = \lim_{x \to 0} x \sin \frac{1}{x} = 0.$$

Thus, f is differentiable at every point. However, f' is not continuous at zero:

$$\lim_{x \to 0} f'(x) = \lim_{x \to 0} \left(2x \sin \frac{1}{x} - \cos \frac{1}{x} \right) = -\lim_{x \to 0} \cos \frac{1}{x},$$

which does not exist.

We can modify the above example slightly to get a function that is differentiable but whose derivative is not bounded. Define $g(0) = 0$ and $g(x) = x^{3/2}\sin(1/x)$ when $x \neq 0$. We have

$$g'(x) = \begin{cases} 0 & \text{if } x = 0, \\ \dfrac{3}{2}\sqrt{x}\sin\dfrac{1}{x} - \dfrac{1}{\sqrt{x}}\cos\dfrac{1}{x} & \text{if } x \neq 0. \end{cases}$$

We say f is **continuously differentiable** on an interval I if it is differentiable on I and f' is continuous on I.

6.1 Mean Value Theorems

A "mean" value is one that can represent the entire set of values. One example is provided by the mean value theorem for integration. It guarantees a special value of a continuous function with domain $[a,b]$, which can be taken as its mean height in the sense that this value multiplied by the width $b-a$ equals the area under the graph of f. One may wonder if similar "mean" values are available for derivatives. This would be a single value of the derivative that would match the average rate of change of the function.

We begin our study of such mean values with some simple observations about what the derivative can tell us about absolute fluctuations of a function's values.

Theorem 6.1.1

Let f,g be differentiable functions from an interval I to $\mathbb{R}$.

(a) *(Racetrack Inequality) If $a,b \in I$, $f(a) \leq g(a)$ and $f'(x) \leq g'(x)$ on $[a,b]$, then $f(x) \leq g(x)$ on $[a,b]$.*
(b) *(Mean Value Inequality) If $a,b \in I$ and $m \leq f'(x) \leq M$ on $[a,b]$, then $m(x-a) \leq f(x) - f(a) \leq M(x-a)$ on $[a,b]$.*
(c) *If $a \in I$ and $|f'(x)| \leq M$ on I, then $|f(x) - f(a)| \leq M|x-a|$ on I.*

If any of the inequalities involving f' is strict, so is the corresponding inequality for f.

Proof. Part (a) is a consequence of the monotonicity theorem. Part (b) follows by applying (a) to the functions $m(x-a)$, $f(x) = f(a)$, and $M(x-a)$. The $x \geq a$ case of (c) is covered by (b). The $x < a$ case is obtained by symmetry. ∎

One application of these results is to put bounds on how much a function can change over an interval:

Example 6.1.2

Consider the function $f(x) = \sin x$ on an interval $[a,b]$. We have $|\sin' x| = |\cos x| \leq 1$ and hence $|\sin b - \sin a| \leq |b-a|$. □

Suppose you set out walking down a straight road. Sometimes you walk slowly, at other times you speed up. Once, you even double back thinking you have dropped your keys. After an hour of this you find yourself 2 km from your starting point. Was there some point when your velocity was exactly 2 km/hour?

Framed like this, the question may make one's head spin. However, it has a simple resolution. Assuming that the motion (position as a function of time) is always differentiable, so that velocity is defined, we see by Darboux's theorem that if our speed is never 2 km/hour then it is either always strictly more than 2 km/hour or always strictly less. By the strict form of the mean value inequality the distance travelled cannot be exactly 2 km, in contradiction to what is given.

To view the same problem from another perspective, we draw a possible graph of $f(t)$, your position at time t. The average velocity of 2 km/hour is the slope of the secant line joining the end points of the graph (we have $b = a + 1$ and $f(b) = f(a) + 2$). Imagine sliding this secant up and down. If we slide it far enough it will lose contact with the curve. At the last instant at which it has contact, it should touch the graph tangentially (points P and Q in our example). At the corresponding points t_P and t_Q, the instantaneous velocity matches the average one. This viewpoint enables a result with slightly weaker hypotheses.

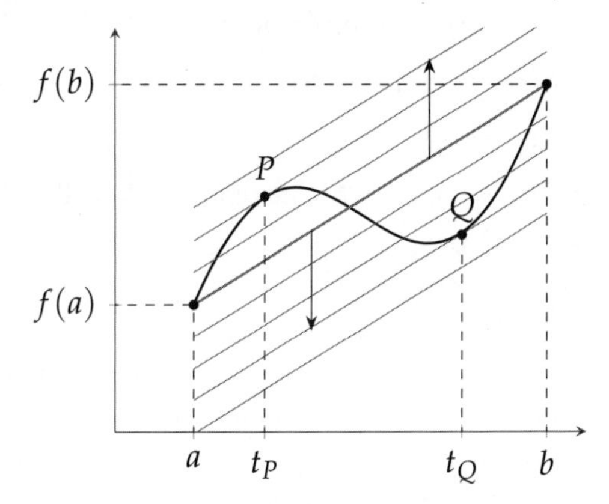

Theorem 6.1.3 (Mean Value Theorem)
Let $f\colon [a,b] \to \mathbb{R}$ satisfy the following:

1. *It is continuous on $[a,b]$.*
2. *It is differentiable on (a,b).*

Then there is a $c \in (a,b)$ such that $f'(c) = \dfrac{f(b) - f(a)}{b - a}$.

Proof. We shall follow the visual argument given above. It suggests that the desired points can be obtained by locating the extremes of the gap between the graph of f and the secant line joining its endpoints. Therefore, we start by defining the function

$$g(x) = f(x) - f(a) - \frac{f(b) - f(a)}{b - a}(x - a).$$

First, we observe that $g(a) = g(b) = 0$. Further, g is continuous on $[a,b]$ and differentiable on (a,b).

By the extreme value theorem, g achieves a maximum value M and a minimum value m on $[a,b]$. If $M = m$, then g is constant and hence zero. In this case, the graph of f is a line and every $c \in (a,b)$ has the desired property.

If $M \neq m$ then at least one of the maximum and minimum values of g is achieved at an interior point $c \in (a,b)$. By Fermat's theorem, $g'(c) = 0$. Hence $f'(c) = \dfrac{f(b) - f(a)}{b - a}$.

■

The above result is sometimes called **Lagrange's mean value theorem** to distinguish it from other versions. Many applications of the mean value theorem involve the following corollary:

Theorem 6.1.4 (Rolle's Theorem)

Let $f\colon [a,b] \to \mathbb{R}$ satisfy the following:

1. *It is continuous on $[a,b]$.*
2. *It is differentiable on (a,b).*
3. $f(a) = f(b)$.

Then there is a $c \in (a,b)$ such that $f'(c) = 0$. ■

Task 6.1.5

Suppose $a < b < c$, $f\colon [a,c] \to \mathbb{R}$ is twice continuously differentiable (that is, the function $f'' = (f')'$ is continuous), and $f(a) = f(b) = f(c) = 0$. Show that there is $\alpha \in (a,c)$ such that $f''(\alpha) = 0$.

We have studied the mean value theorem and Rolle's theorem, which yield certain special values of the derivative of a function. Our next result takes this further by relating special values of the derivatives of two functions.

Theorem 6.1.6 (Cauchy's Mean Value Theorem)

Let $f,g\colon [a,b] \to \mathbb{R}$ be continuous on $[a,b]$ and differentiable on (a,b). Suppose that $g'(x) \neq 0$ for every $x \in (a,b)$. Then there is a $c \in (a,b)$ such that

$$\frac{f'(c)}{g'(c)} = \frac{f(b) - f(a)}{g(b) - g(a)}.$$

Proof. First note that $g(a) \neq g(b)$. If they were equal, Rolle's theorem would give a $c \in (a,b)$ such that $g'(c) = 0$, in contradiction to the assumptions. Thus, both ratios in the theorem's conclusion are defined.

Let us first scale g so that its net change over $[a,b]$ matches the net change in f over $[a,b]$:

$$h(x) = g(x)\frac{f(b) - f(a)}{g(b) - g(a)}.$$

(You may want to check for yourself that $h(b) - h(a) = f(b) - f(a)$.) Now consider the difference $f(x) - h(x)$. Its net change over $[a,b]$ is zero. So, Rolle's theorem gives a

$c \in (a,b)$ such that $f'(c) - h'(c) = 0$. Then

$$0 = f'(c) - h'(c) = f'(c) - g'(c)\frac{f(b)-f(a)}{g(b)-g(a)} \implies \frac{f'(c)}{g'(c)} = \frac{f(b)-f(a)}{g(b)-g(a)}.$$ ■

Remark 1: Taking $g(x) = x$ in Cauchy's mean value theorem gives Lagrange's mean value theorem. So, this theorem is also called the **extended mean value theorem** or the **generalized mean value theorem.**

Remark 2: The motivation for this theorem comes from the motion of a particle in a plane. Suppose its location at any time t, $a \le t \le b$, is given by $(g(t), f(t))$. Then its total displacement during the time interval $[a,b]$ is $(g(b)-g(a), f(b)-f(a))$, while its velocity vector at any time t is $(g'(t), f'(t))$. Thus Cauchy's mean value theorem says that there is a time instant c when the velocity vector is parallel to the total displacement.

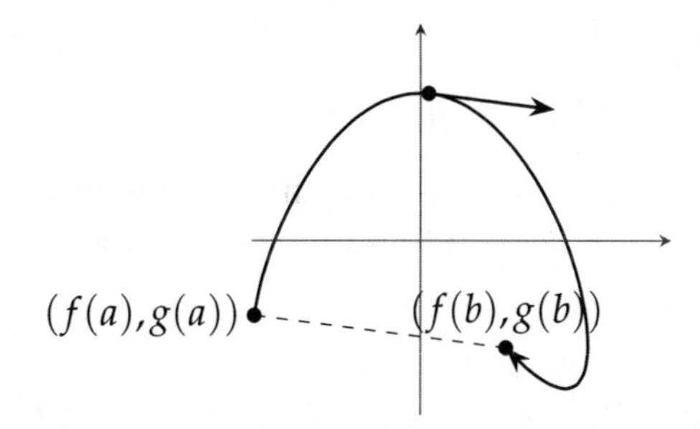

Exercises for § 6.1

1. Let f, g be differentiable functions on an interval I. Suppose that $a, b \in I$ with $f(b) \ge g(b)$ and $f'(x) \le g'(x)$ on $[a,b]$. Show that $f(x) \ge g(x)$ on $[a,b]$.

2. Let $y(t)$ be a positive function satisfying $y'(t) \le Ky(t)$ for $t \ge 0$. Show that $y(t) \le e^{Kt}y(0)$ for $t \ge 0$.

3. Prove that for any whole number n,

$$x > 0 \implies e^x > 1 + x + \frac{x^2}{2!} + \cdots + \frac{x^n}{n!}.$$

4. Verify that the given $f\colon [a,b] \to \mathbb{R}$ and $c \in \mathbb{R}$ satisfy $f'(c) = \dfrac{f(b)-f(a)}{b-a}$.

(a) $f(x) = x^2$ and $c = \dfrac{a+b}{2}$ (the arithmetic mean).

(b) $f(x) = 1/x$ and $c = \sqrt{ab}$ (the geometric mean), with $a, b > 0$.

5. The mean value theorem can be viewed as a unifying result from which several useful facts can be obtained. We have already proved the following, but now you should show they are also corollaries of the mean value theorem.

(a) If $f' = 0$ on an interval then f is constant.

(b) If $f' = g'$ on an interval then f and g differ by a constant.

(c) If $f' \geq 0$ on an interval then f is an increasing function.

(d) If $f' > 0$ on an interval then f is a strictly increasing function.

(e) If $m \leq f'(x) \leq M$ for every $x \geq a$, then $m(x-a) \leq f(x) - f(a) \leq M(x-a)$ for every $x \geq a$.

(f) If $|f'| \leq M$ on an interval I, and $a \in I$, then $|f(x) - f(a)| \leq M|x-a|$ for every $x \in I$.

6. Suppose f,g are continuous on $[a,b]$ and differentiable on (a,b) with $f(a) = g(a)$ and $f(b) = g(b)$. Prove that there is a $c \in (a,b)$ such that $f'(c) = g'(c)$.

7. Let f be an $n+1$ times differentiable function on $[0,1]$ such that

$$f(1) = f(0) = f'(0) = f''(0) = \cdots = f^{(n)}(0) = 0.$$

Prove that there is a $\xi \in (0,1)$ such that $f^{(n+1)}(\xi) = 0$.

8. Suppose f is continuous on $[1,2]$ and differentiable on $(1,2)$ with $f(2) = 2f(1)$. Prove that there is a $c \in (1,2)$ such that the tangent line to the graph of f at $(c, f(c))$ passes through the origin.

9. Suppose f is differentiable on (a,b) and continuous on $[a,b]$ with $f(a) = f(b) = 0$. Prove that for every $\lambda \in \mathbb{R}$, there exists $c \in (a,b)$ such that $f'(c) = \lambda f(c)$. (Hint: Create a function whose derivative has $f' - \lambda f$ as a factor.)

10. Consider $f\colon [0,\infty) \to \mathbb{R}$ defined by $f(0) = 0$ and $f(x) = x\sin(1/x)$ for $x > 0$. Show that for any $c > 0$ there are infinitely many points $t > 0$ such that $f'(t) = \sin(1/c)$.

11. Let f be differentiable on an interval I such that f' is an increasing function. Show that f is convex.

12. Consider a twice differentiable function f on an interval, which takes the values y_1, y_2 at distinct points x_1, x_2 respectively. The deviation of f from the interpolating line passing through (x_1, y_1) and (x_2, y_2) should be measured by its second derivative. The following steps will justify this intuition:

(a) Let $y = L(x)$ be the straight line through (x_1, y_1) and (x_2, y_2). Fix any x in (x_1, x_2). Define M by $f(x) - L(x) = M(x - x_1)(x - x_2)$, and then a function $G(t) = f(t) - L(t) - M(t - x_1)(t - x_2)$.

(b) Verify that $G(x_1) = G(x_2) = G(x) = 0$. Apply Rolle's theorem to obtain a $\xi \in (x_1, x_2)$ such that $G''(\xi) = 0$.

(c) Show that $f(x) - L(x) = \dfrac{f''(\xi)}{2}(x - x_1)(x - x_2)$.

13. Use the Cauchy mean value theorem to show that $1 - \dfrac{x^2}{2} < \cos x$ for $x \neq 0$.

6.2 L'Hôpital's rule

Many important limits have the form $\lim_{x\to a}\frac{f(x)}{g(x)}$ with $\lim_{x\to a} f(x) = \lim_{x\to a} g(x) = 0$. For example, all derivatives are of this type. They often cannot be handled by the algebra of limits, due to the vanishing denominator. We will give a powerful method for solving such limits.

A limit of the type $\lim_{x\to a}\frac{f(x)}{g(x)}$ with $\lim_{x\to a} f(x) = \lim_{x\to a} g(x) = 0$ is said to be an **indeterminate form** of type $\frac{0}{0}$. (The limits could also be one-sided.)

If f, g approach zero value at a with respective slopes m, n, then, near $x = a$, they resemble $m(x - a)$ and $n(x - a)$ respectively. Therefore, their ratio resembles m/n. This suggests that the limit of f/g should equal $f'(a)/g'(a)$.

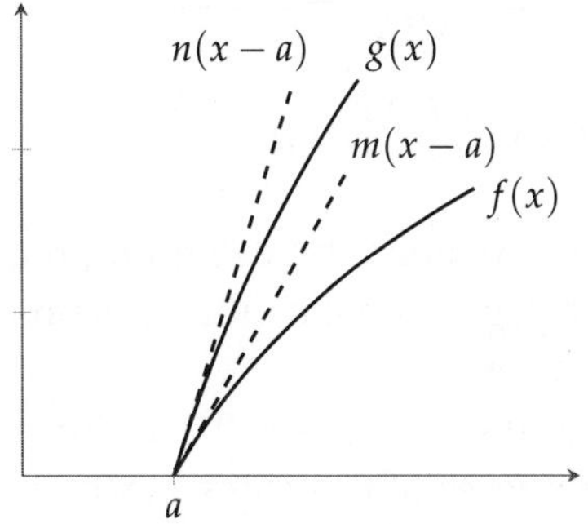

This insight gives the initial version of our method.

Theorem 6.2.1

Suppose $\lim_{x\to a} f(x) = \lim_{x\to a} g(x) = 0$, *$f$ and g are differentiable at a, and $g'(a) \neq 0$. Then* $\lim_{x\to a}\frac{f(x)}{g(x)} = \frac{f'(a)}{g'(a)}$.

Proof. Since f, g are differentiable at a, they are continuous there, and so $f(a) = g(a) = 0$. Further, $g'(a) \neq 0$ implies that there is an interval centered at a in which $g(x)$ is never zero.

$$\lim_{x\to a}\frac{f(x)}{g(x)} = \lim_{x\to a}\frac{(f(x) - f(a))/(x-a)}{(g(x) - g(a))/(x-a)} = \frac{f'(a)}{g'(a)}.$$

■

Example 6.2.2

Suppose that we have to evaluate $\lim_{x\to 0}\frac{x + \sin x}{\log(1-x)}$. The functions $f(x) = x + \sin x$, $g(x) = \log(1-x)$ satisfy the hypotheses of the above theorem. In fact, we have:

$$f'(x) = 1 + \cos x \implies f'(0) = 2,$$

$$g'(x) = \frac{-1}{1-x} \implies g'(0) = -1.$$

Therefore, $\lim_{x\to 0}\frac{x + \sin x}{\log(1-x)} = \frac{f'(0)}{g'(0)} = -2.$ □

This initial result turns out to be of limited utility. All too often, the expression f'/g' also turns out to be a $0/0$ form. We now give a version that can handle this situation, provided the functions are differentiable near the concerned point.

Theorem 6.2.3 (L'Hôpital's rule)

Let $f,g\colon (a,b) \to \mathbb{R}$ be differentiable functions that satisfy the following:

1. $\lim_{x\to a+} f(x) = \lim_{x\to a+} g(x) = 0$,
2. $g'(x) \neq 0$ *for every* $x \in (a,b)$,
3. $\lim_{x\to a+} \frac{f'(x)}{g'(x)} = L \in \mathbb{R}$.

Then $\lim_{x\to a+} \frac{f(x)}{g(x)} = L$.

Proof. We begin by extending the domain of f,g to $[a,b)$ by defining $f(a) = g(a) = 0$. Then f,g become continuous on $[a,b)$. Further, by Rolle's theorem, $g(x) \neq 0$ for every $x \in (a,b)$.

Let $0 < h < b - a$. Then, for each such h, the functions $f(x)$ and $g(x)$ satisfy the hypotheses of Cauchy's mean value theorem on the interval $[a,a+h]$. Hence, there is a $c_h \in (a,a+h)$ such that

$$\frac{f'(c_h)}{g'(c_h)} = \frac{f(a+h) - f(a)}{g(a+h) - g(a)} = \frac{f(a+h)}{g(a+h)}.$$

We have $a < c_h < a + h$. So, the sandwich theorem implies that $c_h \to a+$ as $h \to 0+$. Hence,

$$\lim_{x\to a+} \frac{f(x)}{g(x)} = \lim_{h\to 0+} \frac{f(a+h)}{g(a+h)} = \lim_{h\to 0+} \frac{f'(c_h)}{g'(c_h)} = \lim_{x\to a+} \frac{f'(x)}{g'(x)} = L.$$ ■

This rule can also be applied to left-hand limits and two-sided limits. We list these rules below. The proofs are minor modifications of the one for right-hand limits.

Theorem 6.2.4 (L'Hôpital's rule)

1. *(Left-hand limit) Suppose f,g are differentiable on (a,b), $g'(x) \neq 0$ for every $x \in (a,b)$, $\lim_{x\to b-} f(x) = \lim_{x\to b-} g(x) = 0$, and $\lim_{x\to b-} \frac{f'(x)}{g'(x)} = L$. Then $\lim_{x\to b-} \frac{f(x)}{g(x)} = L$.*
2. *(Two-sided limit) Suppose $a < b < c$, f,g are differentiable on $I = (a,b) \cup (b,c)$, $g'(x) \neq 0$ for every $x \in I$, $\lim_{x\to b} f(x) = \lim_{x\to b} g(x) = 0$, and $\lim_{x\to b} \frac{f'(x)}{g'(x)} = L \in \mathbb{R}$. Then $\lim_{x\to b} \frac{f(x)}{g(x)} = L$.* ■

Example 6.2.5

Consider $\lim_{x \to 1} \frac{\log x}{x-1}$. This is an indeterminate form of the type $0/0$ and $f(x) = \log x$, $g(x) = x - 1$ satisfy the first three hypotheses of L'Hôpital's rule for two-sided limits with $a = 0, b = 1, c = 2$. Further,

$$\lim_{x \to 1} \frac{f'(x)}{g'(x)} = \lim_{x \to 1} \frac{1/x}{1} = 1$$

Hence, $\lim_{x \to 1} \frac{\log x}{x-1} = 1$. (We could also have used Theorem 6.2.1.) □

Example 6.2.6

We know that $\lim_{x \to 0} \frac{\sin x}{x} = 1$. This means that for small x, $\sin x \approx x$. To improve this approximation we use L'Hôpital's rule to compare $\sin x - x$ with higher powers of x. First with x^2:

$$\begin{aligned} & \lim_{x \to 0} \frac{\sin x - x}{x^2} && (\frac{0}{0} \text{ form}) \\ = & \lim_{x \to 0} \frac{\cos x - 1}{2x} && (\frac{0}{0} \text{ form}) \\ = & \lim_{x \to 0} \frac{-\sin x}{2} = 0. \end{aligned}$$

So, the gap $\sin x - x$ is much smaller than x^2. Let us compare with x^3:

$$\begin{aligned} & \lim_{x \to 0} \frac{\sin x - x}{x^3} && (\frac{0}{0} \text{ form}) \\ = & \lim_{x \to 0} \frac{\cos x - 1}{3x^2} && (\frac{0}{0} \text{ form}) \\ = & \lim_{x \to 0} \frac{-\sin x}{3!\,x} = -\frac{1}{3!}. \end{aligned}$$

Thus, $\frac{\sin x - x}{x^3} \approx -\frac{1}{3!}$, or $\sin x \approx x - \frac{x^3}{3!}$ for small x. This process can be continued to get better and better polynomial approximations for $\sin x$. □

Task 6.2.7

Use L'Hôpital's rule to compare $\sin x - x + x^3/3!$ *with* x^5 *near zero and obtain the approximation* $\sin x \approx x - \frac{x^3}{3!} + \frac{x^5}{5!}$ *for small* x.

The graph, below, shows the progressive improvements in these approximations to $\sin x$:

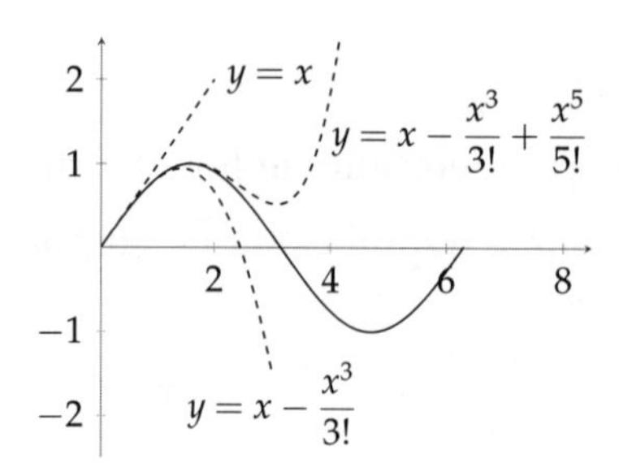

Now we shall see that a derivative cannot have a removable discontinuity. This supplements Darboux's result that a derivative cannot have a jump discontinuity. So, if a derivative has a discontinuity, it must be because one of its one-sided limits fails to exist.

Theorem 6.2.8

Suppose f is continuous at a and is differentiable for all $x \neq a$ in some open interval around a. If $\lim_{x \to a} f'(x)$ exists, then f is differentiable at a and $f'(a) = \lim_{x \to a} f'(x)$.

Proof. We notice that the expression that defines $f'(a)$ is a $0/0$ form and so we apply L'Hôpital's rule:

$$f'(a) = \lim_{x \to a} \frac{f(x) - f(a)}{x - a} = \lim_{x \to a} f'(x). \qquad \blacksquare$$

Limits Involving Infinity

We shall extend L'Hôpital's rule in various ways. Our first observation is that it is valid for limits at infinity.

Theorem 6.2.9 (L'Hôpital's rule)

Suppose f and g are differentiable on (a, ∞), $g'(x) \neq 0$ for every $x \in (a, \infty)$, $\lim_{x \to \infty} f(x) = \lim_{x \to \infty} g(x) = 0$, and $\lim_{x \to \infty} \frac{f'(x)}{g'(x)} = L \in \mathbb{R}$. Then $\lim_{x \to \infty} \frac{f(x)}{g(x)} = L$.

Proof. We begin by recalling that $\lim_{x \to \infty} f(x) = \lim_{t \to 0+} f(1/t)$. Hence,

$$\begin{aligned} \lim_{x \to \infty} \frac{f(x)}{g(x)} &= \lim_{t \to 0+} \frac{f(1/t)}{g(1/t)} = \lim_{t \to 0+} \frac{(f(1/t))'}{(g(1/t))'} = \lim_{t \to 0+} \frac{-f'(1/t)/t^2}{-g'(1/t)/t^2} \\ &= \lim_{t \to 0+} \frac{f'(1/t)}{g'(1/t)} = \lim_{x \to \infty} \frac{f'(x)}{g'(x)}. \qquad \blacksquare \end{aligned}$$

The ease of this proof is also an indication that this extension is not too significant. The next result is much more useful and its proof also needs a fresh approach.

Theorem 6.2.10 (L'Hôpital's rule)

1. *Each version of L'Hôpital's rule also holds if* $f, g \to \pm\infty$.
2. *Each version of L'Hôpital's rule also holds if* $L = \pm\infty$.

Proof. We start by recalling the standing assumption $g'(x) \neq 0$ and the implication that g is one-one.

1. We shall take up the case of the left-hand limit at $c \in \mathbb{R}$. Let $\lim_{x \to c-} \frac{f'(x)}{g'(x)} = L$. For any $\epsilon > 0$ there is an x_0 such that $x_0 \leq x < c$ implies

$$L - \epsilon < \frac{f'(x)}{g'(x)} < L + \epsilon.$$

Now take an $x \in (x_0, c)$. By Cauchy's mean value theorem there is a $\xi \in (x_0, x)$ such that

$$\frac{f(x) - f(x_0)}{g(x) - g(x_0)} = \frac{f'(\xi)}{g'(\xi)}.$$

Hence $L - \epsilon < \frac{f(x) - f(x_0)}{g(x) - g(x_0)} < L + \epsilon$, which we rearrange to

$$L - \epsilon < \frac{f(x)/g(x) - f(x_0)/g(x)}{1 - g(x_0)/g(x)} < L + \epsilon,$$

and then

$$(L - \epsilon)\left(1 - \frac{g(x_0)}{g(x)}\right) + \frac{f(x_0)}{g(x)} < \frac{f(x)}{g(x)} < (L + \epsilon)\left(1 - \frac{g(x_0)}{g(x)}\right) + \frac{f(x_0)}{g(x)}.$$

As $x \to c-$, $g(x_0)/g(x) \to 0$ and $f(x_0)/g(x) \to 0$. Hence, by taking x close to c we get

$$L - 2\epsilon < \frac{f(x)}{g(x)} < L + 2\epsilon.$$

This gives $\lim_{x \to c-} \frac{f(x)}{g(x)} = L$.

2. We show how to modify the proof of the first part of this theorem for the $L = \infty$ case. Given any $M \in \mathbb{R}$, there is an x_0 such that $x_0 \leq x < c$ implies

$$M + 1 < \frac{f'(x)}{g'(x)}.$$

Applying Cauchy's mean value theorem and proceeding as before we reach

$$(M + 1)\left(1 - \frac{g(x_0)}{g(x)}\right) + \frac{f(x_0)}{g(x)} < \frac{f(x)}{g(x)}, \quad \text{for } x \in (x_0, c).$$

By taking x close to c we get:

$$M < \frac{f(x)}{g(x)}.$$

■

Example 6.2.11

Consider $\lim_{x\to\infty}\frac{x}{e^x}$. This is an $\frac{\infty}{\infty}$ form, the numerator and denominator are differentiable on $(0,\infty)$, and the derivative of the denominator is always non-zero. So, L'Hôpital's rule can be applied:

$$\lim_{x\to\infty}\frac{x'}{(e^x)'}=\lim_{x\to\infty}\frac{1}{e^x}=0\implies\lim_{x\to\infty}\frac{x}{e^x}=0.$$ □

Example 6.2.12

On occasion we may need to apply L'Hôpital's rule repeatedly. Consider $\lim_{x\to\infty}\frac{x^2}{e^x}$. This is again an $\frac{\infty}{\infty}$ form, the numerator and denominator are continuously differentiable on $(0,\infty)$, and the derivative of the denominator is always non-zero. Further,

$$\lim_{x\to\infty}\frac{(x^2)'}{(e^x)'}=2\lim_{x\to\infty}\frac{x}{e^x}.$$

The second limit, by another application of L'Hôpital's rule (previous example), is 0. Hence, $\lim_{x\to\infty}\frac{x^2}{e^x}=0$.

We can repeat this argument to show that $\lim_{x\to\infty}\frac{x^n}{e^x}=0$ for any $n\in\mathbb{N}$. Thus, the exponential function grows faster than any power of x. □

Example 6.2.13

Consider $\lim_{x\to\infty}\frac{x^p}{\log x}$, with $p>0$. This is an $\frac{\infty}{\infty}$ form, the numerator and denominator are continuously differentiable on $(0,\infty)$ and the derivative of the denominator is always non-zero. Now

$$\lim_{x\to\infty}\frac{(x^p)'}{(\log x)'}=\lim_{x\to\infty}\frac{p\,x^{p-1}}{1/x}=\lim_{x\to\infty}p\,x^p=\infty.$$

Hence $\lim_{x\to\infty}\frac{x^p}{\log x}=\infty$. □

Other Indeterminate Forms

So far in this section we have considered indeterminate forms of the type $0/0$ or ∞/∞. These are limits of ratios $f(x)/g(x)$ where $f(x),g(x)\to 0$ or $f(x),g(x)\to\infty$, respectively. They are called indeterminate because the result is not determined only by the limits of the parts $f(x)$ and $g(x)$.

There are various other combinations as well whose limits are not determined by only the limits of the parts. We list them below.

Type $\infty - \infty$**:** These have the form $f(x) - g(x)$ where $f(x), g(x) \to \infty$. Depending on which term dominates, the result could be anything. For example, $\lim_{x\to\infty}(x^2 - x) = \lim_{x\to\infty} x(x-1) = \infty$ and $\lim_{x\to\infty}(x - x) = 0$. The typical approach is to rearrange the expression, perhaps by rationalizing, to make it a ratio. (See Examples 3.7.8 and 3.7.15.)

Type $0 \cdot \infty$**:** These have the form $f(x)g(x)$ where $f(x) \to 0$ and $g(x) \to \infty$. We take either f or g to the denominator: $\dfrac{g(x)}{1/f(x)}$ is an $\dfrac{\infty}{\infty}$ form (if $f > 0$), while $\dfrac{f(x)}{1/g(x)}$ is a $\dfrac{0}{0}$ form.

Type 1^∞**:** These have the form $f(x)^{g(x)}$ where $f(x) \to 1$ and $g(x) \to \infty$. A familiar example is $\lim_{x\to 0}(1+x)^{1/x} = e$. Applying the log function converts this to a $0 \cdot \infty$ form.

Type ∞^0**:** These have the form $f(x)^{g(x)}$ where $f(x) \to \infty$ and $g(x) \to 0$. Applying log converts this to a $0 \cdot \infty$ form.

Type 0^0**:** These have the form $f(x)^{g(x)}$ where $f(x) \to 0$ and $g(x) \to 0$. Applying log converts this to a $0 \cdot \infty$ form.

Example 6.2.14

The right-hand side limit of $x \log x$ at 0 is a $0 \cdot \infty$ form. We can convert it to a ratio and then apply L'Hôpital's rule.

$$\lim_{x\to 0+} x\log x = \lim_{t\to\infty} \frac{\log(1/t)}{t} = -\lim_{t\to\infty} \frac{\log t}{t} = -\lim_{t\to\infty} \frac{1/t}{1} = 0. \qquad \square$$

Example 6.2.15

Consider $\lim_{x\to\pi/2-}(\sin x)^{\tan x}$. This is a 1^∞ form. Let $y = (\sin x)^{\tan x}$. Then $\log y = (\tan x)\log(\sin x) = \dfrac{\log(\sin x)}{\cot x}$ and $\lim_{x\to\pi/2-} \dfrac{\log(\sin x)}{\cot x}$ is an $\dfrac{\infty}{\infty}$ form. Apply L'Hôpital's rule:

$$\lim_{x\to\frac{\pi}{2}-} \log y = \lim_{x\to\frac{\pi}{2}-} \frac{\log(\sin x)}{\cot x} = \lim_{x\to\frac{\pi}{2}-} \frac{\cot x}{-\csc^2 x} = -\lim_{x\to\frac{\pi}{2}-}(\cos x)(\sin x) = 0.$$

Finally, $\log y \to 0$ implies $y \to 1$. $\square$

Example 6.2.16

Consider $\lim_{x\to 0+} x^x$. This is a 0^0 form. Let $y = x^x$. Then $\log y = x\log x = \dfrac{\log x}{1/x}$, and $\lim_{x\to 0+} \dfrac{\log x}{1/x}$ is an $\dfrac{\infty}{\infty}$ form. Apply L'Hôpital's rule:

$$\lim_{x\to 0+} \frac{(\log x)'}{(1/x)'} = \lim_{x\to 0+} \frac{1/x}{-1/x^2} = -\lim_{x\to 0+} x = 0 \implies \lim_{x\to 0+} \log y = 0.$$

Hence $y \to 1$. □

A Red Flag

While L'Hôpital's rule is a powerful tool, it can get stuck in blind alleys and if used carelessly may also lead to wrong conclusions (which is the fault of the user and not the tool).

Example 6.2.17

Suppose $f(x) = e^{-x}$, $g(x) = 1/x$, and we have to calculate $\lim_{x\to\infty} \frac{f(x)}{g(x)}$. This is an $\frac{\infty}{\infty}$ form and we are allowed to apply L'Hôpital's rule. If we do, we get

$$\lim_{x\to\infty} \frac{f'(x)}{g'(x)} = \lim_{x\to\infty} \frac{-e^{-x}}{-1/x^2},$$

which is more complicated than the original limit! Of course, we can easily resolve this by first rearranging the expression to x/e^x. □

Example 6.2.18

Consider $f(x) = x + \sin x$ and $g(x) = 2x$. Then $\lim_{x\to\infty} \frac{f(x)}{g(x)}$ is an $\frac{\infty}{\infty}$ form and we may be tempted to calculate as follows:

$$\lim_{x\to\infty} \frac{f(x)}{g(x)} = \lim_{x\to\infty} \frac{f'(x)}{g'(x)} = \lim_{x\to\infty} \frac{1+\cos x}{2}, \text{ hence does not exist.}$$

However, this conclusion is not justified. If $\lim_{x\to\infty} \frac{f'(x)}{g'(x)}$ does not exist, then L'Hôpital's rule *fails to imply anything* about the original limit. In fact, we can apply the sandwich theorem to conclude that the original limit equals $1/2$. □

Exercises for § 6.2

1. Compute the following limits, where $a, b > 0$:

(a) $\lim_{x\to 1-} \frac{\log x}{\sqrt{1-x}}$.

(b) $\lim_{x\to a} \frac{x^n - a^n}{\log x - \log a}$.

(c) $\lim_{t\to 0} \frac{b - \sqrt{b^2 - t^2}}{t^2}$.

(d) $\lim_{x\to a} \frac{a - x - a\log a + a\log x}{a - \sqrt{2ax - x^2}}$.

(e) $\lim_{x\to 0} \frac{e^x - e^{-x}}{\log(1+x)}$.

(f) $\lim_{x\to \pi/2} \frac{1 - \sin x + \cos x}{\sin x + \cos x - 1}$.

(g) $\lim_{x \to 0} \dfrac{e^x - 1 - \log(1+x)}{x^2}$.

(h) $\lim_{x \to 1} \left(\dfrac{x}{x-1} - \dfrac{1}{\log x} \right)$.

(i) $\lim_{x \to 0} \left(\dfrac{1}{2x^2} - \dfrac{\pi}{2x \tan \pi x} \right)$.

(j) $\lim_{x \to 1} \dfrac{x^x - x}{1 - x + \log x}$.

(The exercises given above are from Leonhard Euler's 1755 calculus textbook [40]! Although it is in Latin, much of the mathematics is readily understood, as Euler introduced key parts of modern notation like $f(x)$ for the value of a function. Also see the article by Dunham [52].)

2. L'Hôpital's rule fails to help if applied directly to the given limits. Check this, and then evaluate them by some other method.

(a) $\lim_{x \to 0+} \dfrac{\sqrt{\sin x}}{\sqrt{x}}$.

(b) $\lim_{x \to 0+} \dfrac{x}{e^{-1/x}}$.

3. Mimic the approach of Example 6.2.6 to obtain the following approximations for x close to zero:

(a) $\cos x \approx 1 - \dfrac{x^2}{2!} + \dfrac{x^4}{4!}$.

(b) $e^x \approx 1 + x + \dfrac{x^2}{2!} + \dfrac{x^3}{3!}$.

4. Suppose that f is differentiable $n+1$ times at a. Prove the following:

(a) $\lim_{x \to a} \dfrac{f(x) - f(a) - f'(a)(x-a)}{(x-a)^2} = \dfrac{f''(a)}{2}$.

(b) $\lim_{x \to a} \dfrac{f(x) - \sum_{k=0}^{n} \frac{f^{(k)}(a)}{k!}(x-a)^k}{(x-a)^{n+1}} = \dfrac{f^{(n+1)}(a)}{(n+1)!}$.

5. Let $f(x) = x^2 \cos(1/x)$, $g(x) = \sin x$ for $x \neq 0$. Observe that $\lim_{x \to 0} f(x) = \lim_{x \to 0} g(x) = 0$ but $\lim_{x \to 0} \dfrac{f(x)}{g(x)} \neq \lim_{x \to 0} \dfrac{f'(x)}{g'(x)}$. What has gone wrong?

6. Following Euler, compute the sum $1 + 2 + \cdots + n$ as follows:

(a) Use the sum of a geometric progression to obtain

$$1 + 2x + 3x^2 + \cdots + nx^{n-1} = \frac{1 - (n+1)x^n + nx^{n+1}}{(1-x)^2}.$$

(b) Apply L'Hôpital's rule to the above identity to calculate the limit as $x \to 1$.

7. Suppose f is differentiable in an open interval containing a, and twice differentiable at a. Show that

$$f''(a) = \lim_{h \to 0} \frac{f(a+2h) - 2f(a+h) + f(a)}{h^2}.$$

8. Compute the following limits:

(a) $\lim_{x\to\infty} \frac{x}{\log(1+e^x)}$.

(b) $\lim_{x\to-\infty} xe^x$.

(c) $\lim_{t\to 0+} \sin t \log t$.

(d) $\lim_{x\to 0+} (\cos x)^{1/x}$.

(e) $\lim_{x\to 0+} (\log(1/x))^x$.

(f) $\lim_{x\to 0+} x^{1/\log x}$.

6.3 Taylor Polynomials

Polynomials are the simplest real functions, in that we only have to add and multiply to find their values. So, it is useful to approximate functions by polynomials to enable easy computation. We have already had a glimpse of this in the previous section, when we successively approximated $\sin x$ near $x = 0$ by x, $x - x^3/3!$ and $x - x^3/3! + x^5/5!$.

To approximate $f(x)$ by an n-degree polynomial near $x = a$, we shall use polynomials of the form

$$P_n(x) = a_0 + a_1(x-a) + a_2(x-a)^2 + \cdots + a_n(x-a)^n,$$

so that calculations at $x = a$ are simplified.

Let us suppose that f is differentiable $n+1$ times on an open interval I, and that $f^{(n+1)}(x) \leq M$ on I. Suppose $a \in I$. Apply the mean value inequality to get

$$f^{(n)}(x) - f^{(n)}(a) \leq M(x-a) \qquad \text{for } x > a.$$

Now integrate both sides over $[a, x]$ to get

$$f^{(n-1)}(x) - f^{(n-1)}(a) - f^{(n)}(a)(x-a) \leq \frac{M}{2}(x-a)^2.$$

At the next iteration, we have

$$f^{(n-2)}(x) - f^{(n-2)}(a) - f^{(n-1)}(a)(x-a) - \frac{f^{(n)}(a)}{2}(x-a)^2 \leq \frac{M}{3\cdot 2}(x-a)^3.$$

Continuing in this fashion, we finally obtain

$$f(x) - \sum_{k=0}^{n} \frac{f^{(k)}(a)}{k!}(x-a)^k \leq \frac{M}{(n+1)!}(x-a)^{n+1} \qquad \text{for } x > a. \tag{6.1}$$

Similarly, if we have $m \leq f^{(n+1)}(x)$ on I, we get

$$\frac{m}{(n+1)!}(x-a)^{n+1} \leq f(x) - \sum_{k=0}^{n} \frac{f^{(k)}(a)}{k!}(x-a)^k \qquad \text{for } x > a. \tag{6.2}$$

The polynomial defined by

$$T_n(x) = \sum_{k=0}^{n} \frac{f^{(k)}(a)}{k!}(x-a)^k$$
$$= f(a) + f^{(1)}(a)(x-a) + \frac{f^{(2)}(a)}{2!}(x-a)^2 + \cdots + \frac{f^{(n)}(a)}{n!}(x-a)^n$$

is called the nth **Taylor polynomial** of $f(x)$ centered at $x = a$. When $a = 0$ the Taylor polynomials are also called the **Maclaurin polynomials**.

We could not use integration to start this iterative process as $f^{(n+1)}$ may not have been integrable. Luckily, the mean value inequality helped us out. We *were* able to use integration in subsequent stages as the functions $f',\ldots,f^{(n)}$ were differentiable, hence continuous, hence integrable.

Task 6.3.1

If T_n is the nth Taylor polynomial of f centered at a, show that

$$T_n^{(k)}(a) = f^{(k)}(a) \text{ for } k = 0,1,\ldots,n.$$

Example 6.3.2

Let us calculate the Taylor polynomials of $\sin x$ centered at $a = 0$:

$$\begin{aligned}
f(x) &= \sin x & &\Longrightarrow & a_0 &= f(0) = 0,\\
f'(x) &= \cos x & &\Longrightarrow & a_1 &= f'(0) = 1,\\
f''(x) &= -\sin x & &\Longrightarrow & a_2 &= \frac{f''(0)}{2!} = 0,\\
f'''(x) &= -\cos x & &\Longrightarrow & a_3 &= \frac{f'''(0)}{3!} = -\frac{1}{3!}.
\end{aligned}$$

We see that $a_k = 0$ when k is even. And for odd $k = 2\ell + 1$ we have $a_k = \dfrac{(-1)^\ell}{(2\ell+1)!}$. Thus, the $(2n+1)^{\text{th}}$ Taylor polynomial has the form

$$T_{2n+1}(x) = \sum_{\ell=0}^{n} \frac{(-1)^\ell}{(2\ell+1)!} x^{2\ell+1} = x - \frac{x^3}{3!} + \frac{x^5}{5!} \cdots + (-1)^{2n+1}\frac{x^{2n+1}}{(2n+1)!}.$$

These are the same polynomials that we discovered earlier using L'Hôpital's rule. □

Example 6.3.3

The Taylor polynomials of $\cos x$ centered at $a = 0$ can be found similarly.

$$T_{2n}(x) = \sum_{\ell=0}^{n} \frac{(-1)^\ell}{(2\ell)!} x^{2\ell} = 1 - \frac{x^2}{2!} + \frac{x^4}{4!} \cdots + (-1)^{2n}\frac{x^{2n}}{(2n)!}.$$

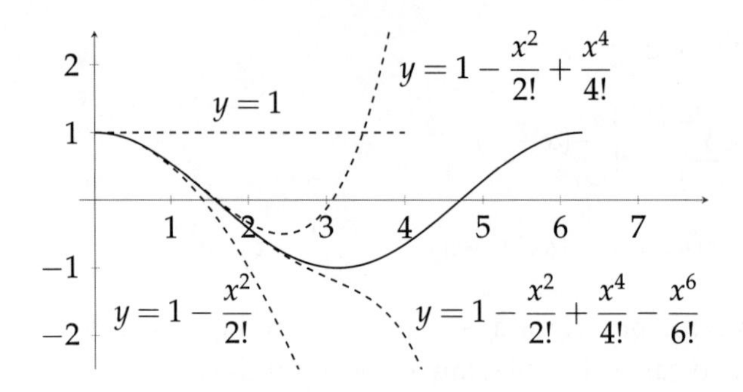

□

Example 6.3.4

The Taylor polynomials of e^x centered at $a=0$ are

$$T_n(x)=\sum_{k=0}^{n}\frac{x^k}{k!}=1+x+\frac{x^2}{2!}+\frac{x^3}{3!}+\cdots+\frac{x^n}{n!}.$$

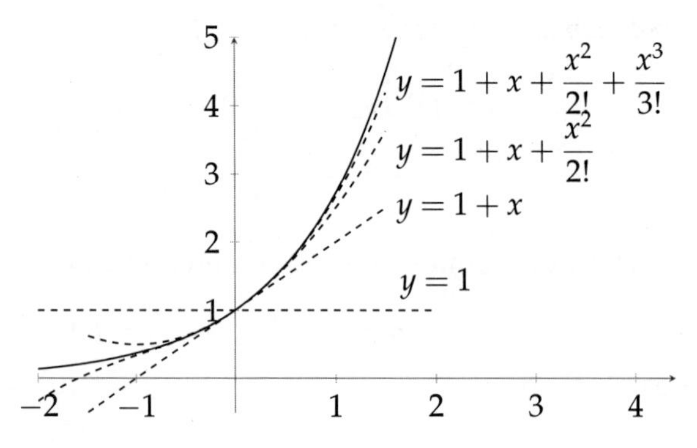

Putting $x=1$ gives $e\approx\sum_{k=0}^{n}\frac{1}{k!}=1+1+\frac{1}{2!}+\frac{1}{3!}+\cdots+\frac{1}{n!}$. □

Example 6.3.5

The Taylor polynomials of $f(x)=\log x$ centered at $a=1$ are:

$$\begin{aligned}T_n(x)&=\sum_{k=1}^{n}\frac{(-1)^{k-1}}{k}(x-1)^k\\&=(x-1)-\frac{(x-1)^2}{2}+\frac{(x-1)^3}{3}+\cdots+(-1)^{n-1}\frac{(x-1)^n}{n}.\end{aligned}$$

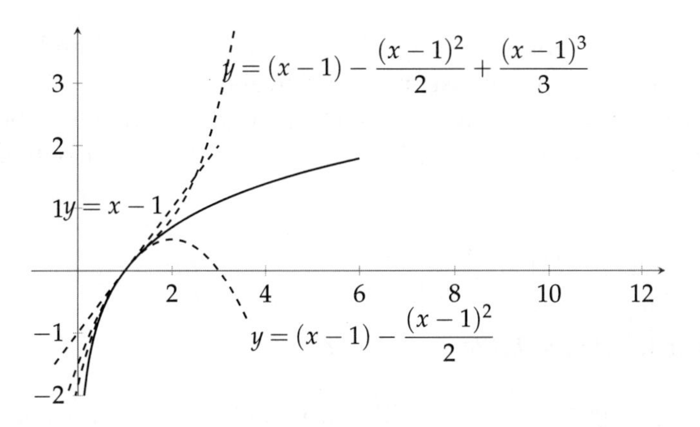

□

Task 6.3.6

Check that the Taylor polynomials of $\log(1+x)$ *centered at* $a = 0$ *are* $x - \frac{x^2}{2} + \frac{x^3}{3} \cdots + (-1)^{n+1}\frac{x^n}{n}$.

These examples may give the impression that higher degree Taylor polynomials always give better approximations, provided the function is repeatedly differentiable. This is not true. A rather drastic example is the function defined to be 0 when $x = 0$ and $e^{-1/|x|}$ when $x \neq 0$. It takes some work, but it turns out that at $a = 0$ all the derivatives of this function are zero. (See Exercise 7 for assistance on these calculations.) Therefore, all its Taylor polynomials centered at $a = 0$ are just 0 too! The graph below shows that this function is extremely flat near the origin.

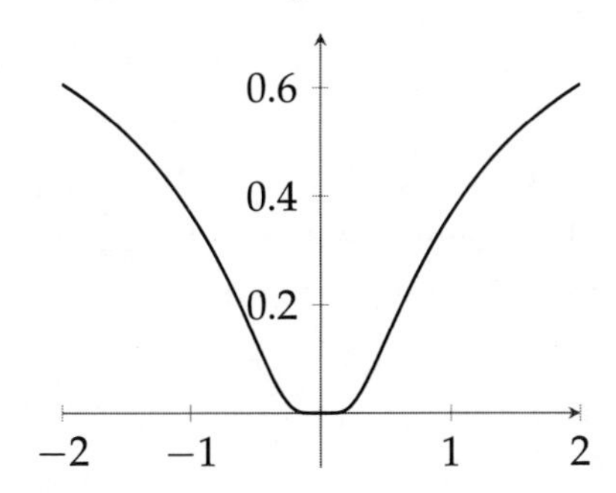

Theorem 6.3.7 (Taylor's Theorem)

Let I *be an interval,* $f\colon I \to \mathbb{R}$, *and* $a \in I$.

1. *Let* $f(x)$ *be differentiable* $n+1$ *times on* I, *and suppose* $|f^{(n+1)}(x)| \leq M$ *on* I.
2. *Let* $T_n(x)$ *be the* n^{th}*-degree Taylor polynomial of* $f(x)$ *centered at* a.

Then, for each $x \in I$,

$$|f(x) - T_n(x)| \leq \frac{M}{(n+1)!}|x-a|^{n+1}.$$

Proof. We have already established this for $x \geq a$ in Equations 6.1 and 6.2. The $x < a$ case can be converted to the $x > a$ case by reflection about $x = a$:

First, note that $x > a \iff y = 2a - x < a$. Define $g(x) = f(2a - x)$. Then $g^{(k)}(x) = (-1)^k f^{(k)}(x)$. This implies $|g^{(k)}(x)| \leq M$. Let T_n^g denote the Taylor polynomials of g. The bound on $|g^{(k)}|$ gives

$$|g(x) - T_n^g(x)| \leq \frac{M}{(n+1)!}|x - a|^{n+1} \quad \text{for } x > a.$$

Further, observe that $T_n^g(x) = T_n(2a - x)$. Hence,

$$|f(2a - x) - T_n(2a - x)| \leq \frac{M}{(n+1)!}|(2a - x) - a|^{n+1} \quad \text{for } x > a.$$

Substitute $y = 2a - x$ to get

$$|f(y) - T_n(y)| \leq \frac{M}{(n+1)!}|y - a|^{n+1} \quad \text{for } y < a.$$ ■

Example 6.3.8

Suppose we need to approximate $\sin 1.2$ to 4 decimal places. Applying Taylor's theorem to $\sin x$ with $a = 0$ and $x = 1.2$, we find that $M = 1$ and

$$|\sin 1.2 - T_n(1.2)| \leq \frac{1.2^{n+1}}{(n+1)!}.$$

To ensure $T_n(1.2)$ is sufficiently accurate, we need to choose n such that $\frac{1.2^{n+1}}{(n+1)!} \leq 5 \times 10^{-5}$. If we take $n = 8$ we get $\frac{1.2^9}{9!} = 1.4 \times 10^{-5} < 5 \times 10^{-5}$. So, the 8th degree Taylor polynomial suffices. However, the degree 8 term is zero in the Taylor expansion of $\sin x$ and so we only need the terms up to degree 7.

$$\sin 1.2 \approx 1.2 - \frac{1.2^3}{3!} + \frac{1.2^5}{5!} - \frac{1.2^7}{7!} \approx 0.932025.$$

(The exact value is $0.932039\ldots$, so our answer is satisfactory!) □

Example 6.3.9

Now let us approximate Euler's number e to 4 decimal places. Recall that we already know $e < 4$ and so the function e^x is bounded by 4 on $[0,1]$. Therefore, applying Taylor's theorem to e^x with $a = 0$ and $x = 1$, we find that

$$|e - T_n(1)| \leq \frac{4}{(n+1)!}.$$

To ensure $T_n(1)$ is sufficiently accurate, we need to choose n such that $\frac{4}{(n+1)!} \leq 5 \times 10^{-5}$. Again, $n = 8$ does the job. Therefore,

$$e \approx 1 + 1 + \frac{1}{2!} + \frac{1}{3!} + \cdots + \frac{1}{8!} = 2.718278\ldots.$$

(The exact value is $2.718281\ldots$) □

Taylor's theorem can be slightly refined if the final derivative is assumed to be continuous.

Theorem 6.3.10 (Remainder Theorem)

Let I be an interval, $f\colon I \to \mathbb{R}$, and $a \in I$. Further,

1. *Let $f(x)$ be continuously differentiable $n+1$ times on I.*
2. *Let $T_n(x)$ be the nth-degree Taylor polynomial of $f(x)$ centered at a.*

Then, for each $x \in I$, there is a ξ between a and x such that

$$f(x) - T_n(x) = \frac{f^{(n+1)}(\xi)}{(n+1)!}(x-a)^{n+1}.$$

Proof. We give the proof for $x > a$. By the extreme value theorem, $f^{(n+1)}$ achieves a minimum value m and a maximum value M on $[a, x]$. Then

$$\frac{m}{(n+1)!}(x-a)^{n+1} \leq f(x) - T_n(x) \leq \frac{M}{(n+1)!}(x-a)^{n+1}.$$

Hence,

$$m \leq \Big(f(x) - T_n(x)\Big)\frac{(n+1)!}{(x-a)^{n+1}} \leq M.$$

Now the intermediate value theorem gives a $\xi \in (a, x)$ such that

$$f^{(n+1)}(\xi) = \Big(f(x) - T_n(x)\Big)\frac{(n+1)!}{(x-a)^{n+1}}.$$ ■

In fact, the remainder theorem holds even if we only assume the existence of $f^{(n+1)}$ and not its continuity. See Exercise 12.

One application of the remainder theorem is to improve our ability to classify critical points.

Theorem 6.3.11

Let f have a critical point at a and be n times continuously differentiable at a. Suppose

$$f'(a) = f''(a) = \cdots = f^{(n-1)}(a) = 0 \quad \textit{and} \quad f^{(n)}(a) \neq 0.$$

The nature of the critical point is as follows:

1. *If n is even and $f^{(n)}(a) > 0$ then f has a local minimum at a.*
2. *If n is even and $f^{(n)}(a) < 0$ then f has a local maximum at a.*
3. *If n is odd then f has a saddle point at a.*

Proof. By continuity, there is an open interval I containing a such that $f^{(n)}$ does not change sign in I. For each $x \in I$ there is a $\xi \in I$ such that

$$f(x) = f(a) + \frac{f^{(n)}(\xi)}{n!}(x-a)^n.$$

If n is even and $f^{(n)}(a) > 0$ then we have $f^{(n)}(\xi) > 0$ for every $\xi \in I \setminus \{a\}$. It follows that $f(x) > f(a)$ for every $x \in I$ and hence there is a local minimum at a. The other cases are similar. ■

Exercises for § 6.3

1. Let $T_{f,n}$ denote the nth Taylor polynomial of f and $T_{Df,n}$ the nth Taylor polynomial of f', both centered at a. Verify that $T_{Df,n} = T'_{f,n+1}$.

2. Let $T_{f,n}$ denote the nth Taylor polynomial of f and $T_{\int f,n}$ the nth Taylor polynomial of $g(x) = \int_a^x f(t)\,dt$, both centered at a. Verify that $T_{\int f,n}(x) = \int_a^x T_{\int f,n-1}(t)\,dt$.

3. Find the Maclaurin polynomials of the given function.

(a) $\dfrac{1}{1-x}$.

(b) $\dfrac{1}{1-x^2}$.

(c) $\dfrac{1}{(1-x)^2}$.

(d) $\log(1-x)$.

(e) $\dfrac{1}{1+x^2}$.

(f) $\arctan x$.

4. Find the Taylor polynomials of the given function at the given center a.

(a) $\dfrac{1}{1-x}$, $a = -1$.

(b) e^x, $a = 1$.

5. Use the second degree Maclaurin polynomial of cosine to get an approximate solution of $x = \cos x$.

6. Suppose we are looking for a function that satisfies $f'' = f$, $f(0) = 0$ and $f'(0) = 1$. Can we at least identify its Maclaurin polynomials?

7. Consider the function $f\colon \mathbb{R} \to \mathbb{R}$ defined by $f(x) = e^{-1/|x|}$ if $x \neq 0$ and $f(0) = 0$. Use mathematical induction to prove that for each $n \in \mathbb{N}$,

(a) $f_+^{(n)}(0) = 0$,

(b) For $x > 0$, $f^{(n)}(x) = \dfrac{p(x)}{x^{2n}} e^{-1/x}$ where p is a polynomial with degree at most $n-1$.

Hence, all the Maclaurin polynomials of f are identically zero. (Hint: Prove the two statements simultaneously.)

8. Let $f\colon \mathbb{R} \to \mathbb{R}$ satisfy $|f(x)| \le 1$ and $|f''(x)| \le 1$ for every x. Prove that $|f'(x)| \le 2$ for every x. (Hint: Show that if $f'(a) > 2$ then $f(a+2) - f(a) > 2$.)

9. Complete the proofs of the remaining cases of Theorem 6.3.11.

10. Let p,q be polynomials such that that for some a, $p^{(n)}(a) = q^{(n)}(a)$ for every $n \ge 0$. Show that $p = q$.

11. Consider a polynomial function of the form $p(x) = (x-a)^n q(x)$ where $n \ge 2$ and q is a polynomial with $q(a) \ne 0$. Show that p has a critical point at a and classify it.

12. The following steps provide a proof of the remainder theorem that does not require continuity of $f^{(n+1)}$. We fix a point x and then find the remainder at x:

(a) Define a real number R by $f(x) = T_n(x) + R(x-a)^{n+1}$.

(b) Define a function $g(t) = f(t) - T_n(t) - R(t-a)^{n+1}$.

(c) Observe that $g(x) = g(a) = g'(a) = \cdots = g^{(n)}(a) = 0$.

(d) Apply Rolle's theorem to obtain ξ between a and x with $g^{(n+1)}(\xi) = 0$.

(e) Show that $R = f^{(n+1)}(\xi)/(n+1)!$.

13. Let $f\colon (a-R, a+R) \to \mathbb{R}$ be $n+1$ times continuously differentiable. Let $T_n(x)$ be the nth Taylor polynomial of f. Show the following:

(a) $\displaystyle\lim_{x\to a} \frac{f(x) - T_n(x)}{(x-a)^{n+1}} = \frac{f^{(n+1)}(a)}{(n+1)!}$.

(b) $\displaystyle\lim_{x\to a} \frac{f(x) - T_n(x)}{(x-a)^{n}} = 0$.

(c) If $P(x)$ is any polynomial of degree n or less and $\displaystyle\lim_{x\to a} \frac{f(x) - P(x)}{(x-a)^{n}} = 0$ for every x, then $P(x) = T_n(x)$.

14. Use the degree 5 Maclaurin polynomial of the sine function to estimate $\displaystyle\int_0^1 \frac{\sin x}{x}\,dx$, and give a bound for the error.

6.4 Riemann Sums and Mensuration

Our approach to definite integrals via upper and lower sums was created by Gaston Darboux in 1875. As we have seen, it is quite convenient for developing the basic theory of integral calculus. Yet it is not so convenient for direct calculations of integrals. The reason is that it requires us to discover bounds for the integrand over every interval of a partition, and this could be laborious. We will now consider a different approach that uses selected values of the function in place of bounds. This approach lets us generate estimates of integrals with ease, and also makes it easier to use the mean value theorem. It is in fact an earlier description of integration, created by Bernhard Riemann in 1854. In this section we explore the relationship of the Darboux and Riemann approaches, and use the greater flexibility of Riemann's approach to apply integration to calculations of arc length, surface area, and volume.

Riemann Sums

Given a bounded function f over an interval $[a,b]$, we again begin with a partition $P = \{x_1, \ldots, x_n\}$ of $[a,b]$. A **tag** is a choice of points $x_i^* \in (x_{i-1}, x_i)$ for each $i = 1, \ldots, n$. A **tagged partition** P^* is a partition P along with a tag. Each tagged partition creates a **Riemann sum**

$$R(f, P^*) = \sum_{i=1}^{n} f(x_i^*)(x_i - x_{i-1}),$$

which we view as an approximation to the integral of f over $[a,b]$.

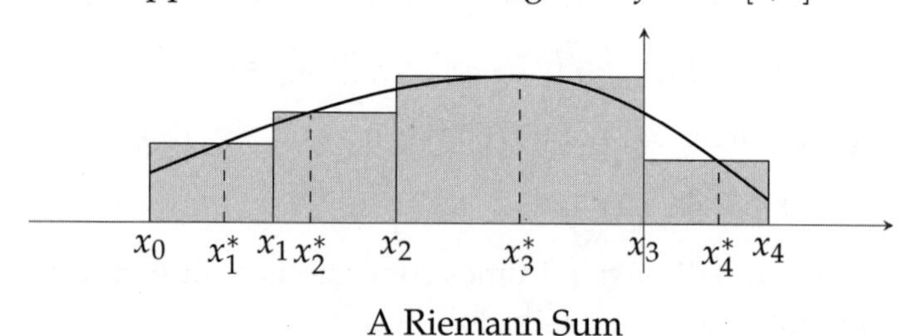

A Riemann Sum

When f is integrable, the Riemann sums provide easily computed estimates of the integral.

Theorem 6.4.1

Let f be integrable on $[a,b]$. Then for every $\epsilon > 0$ there is a partition P_ϵ such that for every refinement Q of P_ϵ and for every tag of Q we have $|\int_a^b f(x)\,dx - R(f, Q^*)| < \epsilon$.

Proof. By the Riemann condition, we have step functions s, t that satisfy $s \leq f \leq t$ on $[a,b]$ and $\int_a^b t(x)\,dx - \int_a^b s(x)\,dx < \epsilon$. Let P_ϵ be a partition that is adapted to s and t. ■

Moreover, we need not look at all possible tags to identify the integral.

Theorem 6.4.2

Suppose $f\colon [a,b] \to \mathbb{R}$ is integrable, $I \in \mathbb{R}$, and every partition P of $[a,b]$ has been allotted a tag such that the tagged partitions P^* satisfy the following: For every $\epsilon > 0$ there is a partition P'_ϵ such that for every refinement P of P'_ϵ we have $|I - R(f, P^*)| < \epsilon$. Then $I = \int_a^b f(x)\,dx$.

Proof. Fix $\epsilon > 0$. Let P_ϵ be the corresponding partition obtained from the previous theorem. Let $P = P_\epsilon \cup P'_\epsilon$. Let P^* be the tagged partition obtained from the current hypothesis. Then we have $|I - R(f,P^*)| < \epsilon$ and $|\int_a^b f(x)\,dx - R(f,P^*)| < \epsilon$. It follows that $|I - \int_a^b f(x)\,dx| < 2\epsilon$. Since this is true for every $\epsilon > 0$, we must have $I = \int_a^b f(x)\,dx$. ■

The following result is useful in dealing with products. We will need it when we discuss surface area later in this section.

Theorem 6.4.3

Consider continuous functions $f,g\colon [a,b] \to \mathbb{R}$. For any $\epsilon > 0$ we can choose a partition $P = \{x_1,\dots,x_n\}$ of $[a,b]$ such that for every choice of points $u_i, v_i \in [x_{i-1}, x_i]$ we have

$$\left| \int_a^b f(x)g(x)\,dx - \sum_{i=1}^n f(u_i)g(v_i)(x_i - x_{i-1}) \right| < \epsilon.$$

Proof. Choose a partition $P_{\epsilon/2}$ such that every tag of every refinement Q satisfies

$$\left| \int_a^b f(x)g(x)\,dx - R(fg,Q^*) \right| < \epsilon/2.$$

Let M be an upper bound of $|g|$ over $[a,b]$. By the small span theorem, there is a refinement $P = \{x_1,\dots,x_n\}$ of $P_{\epsilon/2}$ such that $|f(u_i) - f(v_i)| < \epsilon/2M(b-a)$ whenever $u_i, v_i \in [x_{i-1}, x_i]$. Let the v_i be considered as a tag of P. Then

$$\begin{aligned}
&\left| R(fg,P^*) - \sum_{i=1}^n f(u_i)g(v_i)(x_i - x_{i-1}) \right| \\
&\quad = \left| \sum_{i=1}^n f(v_i)g(v_i)(x_i - x_{i-1}) - \sum_{i=1}^n f(u_i)g(v_i)(x_i - x_{i-1}) \right| \\
&\quad = \left| \sum_{i=1}^n (f(v_i) - f(u_i))g(v_i)(x_i - x_{i-1}) \right| \\
&\quad \le \sum_{i=1}^n \frac{\epsilon}{2M(b-a)} M(x_i - x_{i-1}) = \epsilon/2.
\end{aligned}$$

■

Arc Length

We shall define the length of the curve formed by the graph of a function $f\colon [a,b] \to \mathbb{R}$. We take line segments as our basic shapes, with the length of the line segment joining (x_1,y_1) and (x_2,y_2) being $((x_2 - x_1)^2 + (y_2 - y_1)^2)^{1/2}$. Now let $P = \{x_0,\dots,x_n\}$ be a partition of $[a,b]$, and set $y_i = f(x_i)$. The total length of the line segments between successive points (x_i,y_i) on the graph is

$$S(f,P) = \sum_{i=1}^n ((x_i - x_{i-1})^2 + (y_i - y_{i-1})^2)^{1/2}.$$

If Q is a refinement of P, the triangle inequality in the plane implies that $S(f,P) \leq S(f,Q)$. Therefore, it is reasonable to define the **arc length** of the graph of f by

$$S(f) = \sup\{\, S(f,P) \mid P \text{ is a partition of } [a,b]\,\}.$$

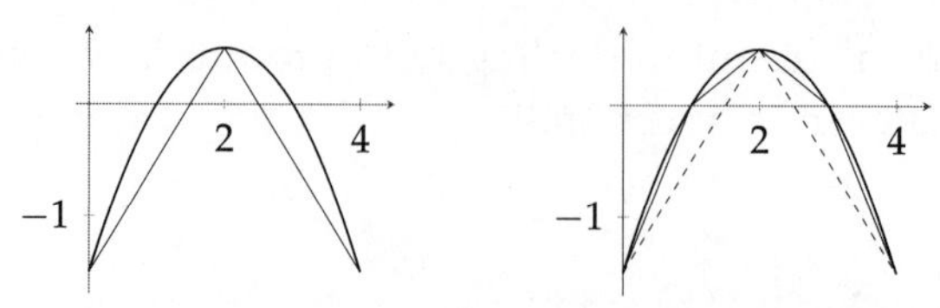

The existence of $S(f)$ is not guaranteed. It is possible, even with f continuous, that the collection $S(f,P)$ is unbounded. An example is provided by the function $x\cos(\pi/2x)$. (See Exercise 6) We call f **rectifiable** if $S(f)$ exists. The following observation is useful in arranging calculations:

Theorem 6.4.4 (Additivity of Arc Length)

Suppose $f\colon [a,b] \to \mathbb{R}$ and $g\colon [b,c] \to \mathbb{R}$ are rectifiable and $f(b) = g(b)$. Define $h\colon [a,c] \to \mathbb{R}$ by $h(x) = f(x)$ if $x \leq b$ and $h(x) = g(x)$ if $x \geq b$. Then h is rectifiable and $S(h) = S(f) + S(g)$.

Proof. Similar to the additivity of integrals over intervals. ∎

Theorem 6.4.5

Let $f\colon [a,b] \to \mathbb{R}$ be continuously differentiable. Then f is rectifiable and the arc length of its graph is given by

$$S(f) = \int_a^b \sqrt{1 + f'(x)^2}\, dx.$$

Proof. Given a partition $P = \{x_0, \ldots, x_n\}$ of $[a,b]$, let us write $y_i = f(x_i)$, $\triangle x_i = x_i - x_{i-1}$ and $\triangle y_i = y_i - y_{i-1}$. Then we have

$$S(f,P) = \sum_{i=1}^{n} ((\triangle x_i)^2 + (\triangle y_i)^2)^{1/2} = \sum_{i=1}^{n} \left(1 + \left(\frac{\triangle y_i}{\triangle x_i}\right)^2\right)^{1/2} \triangle x_i.$$

The mean value theorem gives $x_i^* \in (x_{i-1}, x_i)$ such that $f'(x_i^*) = \triangle y_i / \triangle x_i$. Taking these x_i^* as tags for the partition P, the $S(f,P)$ become Riemann sums:

$$S(f,P) = \sum_{i=1}^{n} \sqrt{1 + f'(x_i^*)^2}\, \triangle x_i = R(\sqrt{1 + (f')^2}, P^*).$$

The function $\sqrt{1 + (f')^2}$ is continuous and hence integrable. By Theorem 6.4.1 there is a partition P_1 such that every refinement Q of P_1 satisfies $|\int_a^b \sqrt{1 + f'(x)^2}\, dx - S(f,Q)| = |\int_a^b \sqrt{1 + f'(x)^2}\, dx - R(\sqrt{1 + (f')^2}, Q^*)| < 1$. Hence, for an arbitrary partition P,

$$S(f,P) \leq S(f, P \cup P_1) < \int_a^b \sqrt{1 + f'(x)^2}\, dx + 1.$$

It follows that the collection of all $S(f,P)$ is bounded above and f is rectifiable.

Given any $\epsilon > 0$, there is a partition P'_ϵ such that $S(f) - S(f,P'_\epsilon) < \epsilon$. If P is a refinement of P'_ϵ, then

$$S(f) - R(\sqrt{1+(f')^2}, P^*) = S(f) - S(f,P) \leq S(f) - S(f,P'_\epsilon) < \epsilon.$$

By Theorem 6.4.2, $S(f) = \int_a^b \sqrt{1+f'(x)^2}\,dx$. ■

Example 6.4.6

(Catenary) Consider the hyperbolic cosine function $\cosh x = \dfrac{e^x + e^{-x}}{2}$. We have $1 + (\cosh' x)^2 = 1 + (\sinh x)^2 = \cosh^2 x$. Therefore, the length of the graph over an interval $[0,a]$ is

$$\int_0^a \sqrt{1+(\cosh' x)^2}\,dx = \int_0^a \cosh x\,dx = \sinh a.$$ □

Example 6.4.7

(Parabola) We will compute the length of a parabola. Let $f(x) = x^2$. The length of the graph over $[0,a]$ is

$$S = \int_0^a \sqrt{1+(2x)^2}\,dx = \frac{1}{2}\int_0^{2a} \sqrt{1+x^2}\,dx.$$

One option is to substitute $x = \tan\theta$. This leads to the integral of $\sec^3\theta$, which we have carried out earlier and is quite complicated. More pleasant results are obtained by substituting $x = \sinh t$. Then we have $\sqrt{1+x^2} = \cosh t$ and $dx = \cosh t\,dt$. This gives

$$\begin{aligned} S &= \frac{1}{2}\int_0^b \cosh^2 t\,dt = \frac{1}{2}\int_0^b \frac{1+\cosh 2t}{2}\,dt \qquad (b = \sinh^{-1} 2a) \\ &= \frac{1}{4}\left(b + \frac{\sinh 2b}{2}\right) = \frac{b + \sinh b \cosh b}{4} = \frac{\sinh^{-1} 2a + 2a\sqrt{1+4a^2}}{4}. \end{aligned}$$ □

Example 6.4.8

(Circle) For our next application, we consider the lengths of circular arcs. A semicircle of radius R is obtained as the graph of $f(x) = \sqrt{R^2 - x^2}$ with $x \in [-R,R]$. We have

$$f'(x) = \frac{-x}{\sqrt{R^2-x^2}} \quad \text{and} \quad \sqrt{1+f'(x)^2} = \frac{R}{\sqrt{R^2-x^2}}.$$

Note that f is not differentiable at $x = \pm R$ so we cannot apply the integral formula in one go for the length of the semicircle.

Let us begin with the length of an arc whose central angle θ is less than $\pi/2$. The arc is obtained as the graph of f restricted to $[0, R\sin\theta]$. Hence, its length is

$$\int_0^{R\sin\theta} \sqrt{1+f'(x)^2}\,dx = \int_0^{R\sin\theta} \frac{R}{\sqrt{R^2-x^2}}\,dx = R\int_0^{\sin\theta} \frac{1}{\sqrt{1-x^2}}\,dx$$
$$= R\arcsin x\Big|_0^{\sin\theta} = R\theta.$$

Any circular arc can be cut into congruent pieces each of which has central angle less than $\pi/2$. Combined with the additivity of arc length, this extends the formula $R\theta$ to arbitrary θ. In particular, we recover the description of π as the ratio of a circle's circumference to its diameter. (Recall that we *defined* π as the ratio of a circle's area to its squared radius.) □

The square root in the integrand makes for difficult integration problems. Mostly, we can only hope to express the length as an integral, without being able to carry out the integration formally. This is progress too, as integrals can be systematically estimated by Riemann sums, a topic we explore in the next section.

As an example, let us consider the the problem of finding the length of part of an ellipse, given by the equation

$$\frac{x^2}{a^2} + \frac{y^2}{b^2} = 1 \qquad (a > b > 0)$$

An arc of the ellipse can be viewed as the graph of the function

$$f(x) = \frac{b}{a}\sqrt{a^2-x^2}.$$

This leads to the integral

$$\int \sqrt{1+f'(x)^2}\,dx = \int \sqrt{\frac{a^2+(b/a^2-1)x^2}{a^2-x^2}}\,dx$$
$$= a\int \sqrt{\frac{1+e^2w^2}{1-w^2}}\,dw \qquad (w = x/a)$$

where $e = (a^2-b^2)^{1/2}/a$ is called the **eccentricity** of the ellipse and measures its deviation from a circle. This integral turns out to be inexpressible in terms of any combination of our standard functions. We have to treat it as a new function!

Surface Area

A surface is a two dimensional object in a three dimensional world. (This is not a definition!) So, the general study of surfaces and their properties like area requires the use of functions of two variables. However, we can study certain surfaces whose symmetry enables us to make do with the calculus of functions of one variable.

A **surface of revolution** is obtained by taking a curve in the xy-plane and rotating it about the x-axis. The diagram given below shows the result of rotating the graph of $y = x^2$.

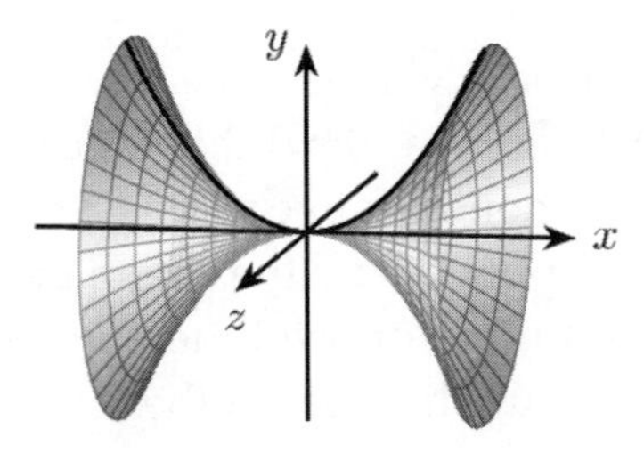

The surfaces of revolution include many of our familiar shapes such as cylinders, cones, and spheres.

Just as we used line segments to build the concept of arc length, we shall use frustums of cones to describe the surface area of a surface of revolution. Given the graph of a function $y = f(x)$, we first approximate it by line segments. Then we rotate these line segments around the x-axis to create frustums of cones that approximate the surface of revolution.

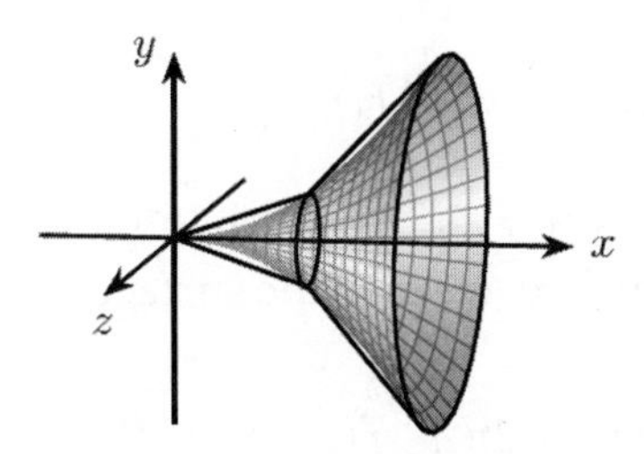

You may recall that the surface area of a right circular cone with radius r and slant height ℓ is $\pi r \ell$. This formula is obtained by cutting the cone along a generator and flattening it into a sector of a circle. From this, we can obtain the surface area of the frustum of a cone with end radii r_1, r_2, and slant height ℓ. First, visualize the full cone of which the frustum is a part.

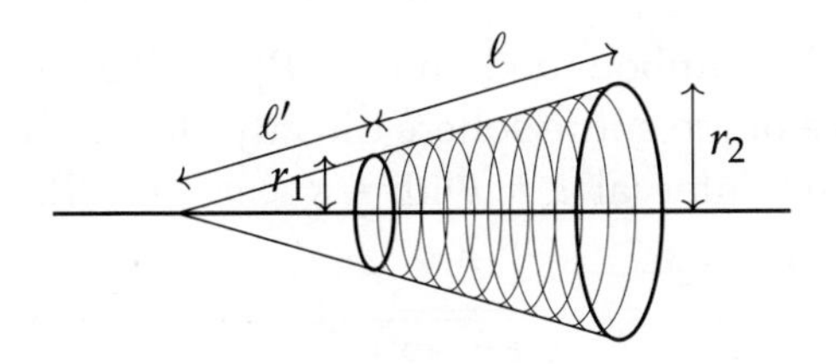

Let ℓ' be the slant height of the conical cap that completes the frustum. Using similar triangles, we get

$$\frac{\ell+\ell'}{\ell'} = \frac{r_2}{r_1}, \text{ hence } \ell' = \frac{r_1}{r_2 - r_1}\ell.$$

And now we compute the surface area F of the frustum:

$$F = \pi r_2(\ell + \ell') - \pi r_1 \ell' = \pi(r_1 + r_2)\ell.$$

This includes the limiting case of a cylinder, when $r_1 = r_2$.

Now consider a function $f\colon [a,b] \to [0,\infty)$ that is continuously differentiable. Note that we have required it to take non-negative values. Let $P = \{x_0, \ldots, x_n\}$ be a partition of $[a,b]$. On each $[x_{i-1}, x_i]$ we approximate the graph of $y = f(x)$ by the line segment joining its endpoints, and the surface of revolution by the frustum of a cone with end radii $f(x_{i-1})$ and $f(x_i)$. We add up the areas of these frustums to get

$$A(f,P) = \sum_{i=1}^{n} \pi(f(x_{i-1}) + f(x_i))\sqrt{(\triangle x_i)^2 + (\triangle y_i)^2},$$

where $\triangle x_i = x_i - x_{i-1}$ and $\triangle y_i = f(x_i) - f(x_{i-1})$. Now, we do the following in each $[x_{i-1}, x_i]$:

- Use the intermediate value theorem to get a $u_i \in (x_{i-1}, x_i)$ such that $f(u_i) = (f(x_{i-1}) + f(x_i))/2$.
- Use the mean value theorem to get a $v_i \in (x_{i-1}, x_i)$ such that $f'(v_i) = \triangle y_i / \triangle x_i$.

This gives

$$A(f,P) = 2\pi \sum_{i=1}^{n} f(u_i)\sqrt{1 + f'(v_i)^2}\,\triangle x_i.$$

By Theorem 6.4.3 the numbers $A(f,P)$ approach the corresponding definite integral as we take finer partitions. Therefore, we take the surface area of our surface of revolution to be

$$A(f) = 2\pi \int_a^b f(x)\sqrt{1 + f'(x)^2}\,dx.$$

Example 6.4.9

(Sphere) A sphere of radius R can be obtained by rotating the graph of the function $f\colon [-R,R] \to [0,R]$ defined by $f(x) = \sqrt{R^2 - x^2}$. We have

$$1 + f'(x)^2 = 1 + \frac{x^2}{R^2 - x^2} = \frac{R^2}{R^2 - x^2}.$$

We see that $\sqrt{1 + f'(x)^2}$ is unbounded on $[-R,R]$ and so we cannot use the integral formula to find the area of the entire sphere directly. But we can use it to find the area of the part lying over any interval $[a,b]$ with $-R < a < b < R$:

$$2\pi \int_a^b \sqrt{R^2 - x^2}\frac{R}{\sqrt{R^2 - x^2}}\,dx = 2\pi(b-a)R.$$

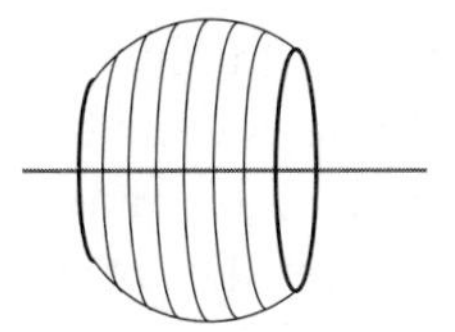

To get the area of the entire sphere we now let $a \to -R$ and $b \to R$, giving $4\pi R^2$. □

Example 6.4.10

(Catenoid) The surface created by rotating a catenary is called a **catenoid**. We can compute its surface area:

$$2\pi \int_0^a \cosh x \sqrt{1 + (\cosh' x)^2}\, dx = 2\pi \int_0^a \cosh^2 x\, dx = \pi \int_0^a (1 + \cosh 2x)\, dx$$
$$= \pi(a + \frac{\sinh 2a}{2}).$$

The catenoid turns out to be the surface of minimal area between two rings centered on the same axis. Let us at least see an example where a catenoid has less area than the frustum of a cone between two rings. Consider the catenoid obtained by rotating the catenary over $[0,1]$. Its area is $\pi(1 + \sinh(2)/2) \approx 8.8$. Now consider the frustum of a cone between the same circles. Its area is $\pi(1 + \cosh 1)\sqrt{1 + (\cosh 1 - 1)^2} \approx 9.1$. □

The fact that line segments have minimal length among curves, but cones do not have minimal area among surfaces of revolution, is the reason why our approach to surface area could not be as direct as for arc length.

Volume

We shall now take up the volume of the region enclosed by a surface of revolution. We call this region a **solid of revolution**. We shall obtain its volume by approximating it by a bunch of cylinders.

Let us first explore the volume of a cylinder whose height is h and base radius is r. We take cuboids to be our basic shapes, and define the volume of a cuboid to be the product of its three dimensions, which we can also express as base area times height. Let us fit a cuboid inside the cylinder.

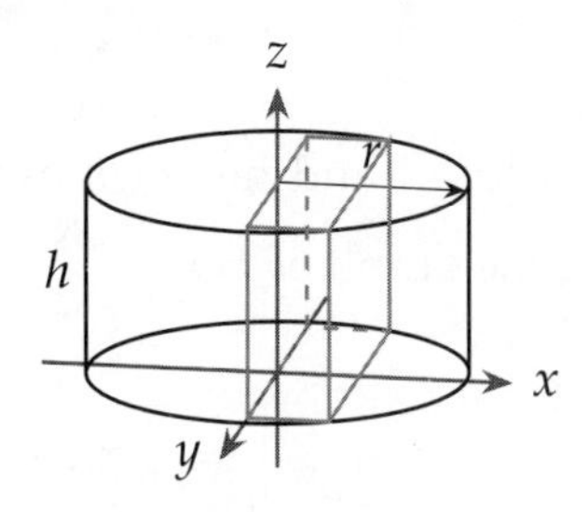

Suppose we stack up some cuboids in this way inside the cylinder. They have the common height h so their total volume is their total base area times the height h. As we fill the cylinder with thinner cuboids, the total base area of the cuboids approaches the area of the base circle, and therefore the total volume of the cuboids approaches $\pi r^2 h$. This gives $\pi r^2 h$ as the volume of the cylinder.

Now consider a continuous function $f\colon [a,b] \to \mathbb{R}$ with $f \geq 0$. Let $s\colon [a,b] \to \mathbb{R}$ be a step function such that $0 \leq s(x) \leq f(x)$ on $[a,b]$. Let $P = \{x_0, \ldots, x_n\}$ be a partition adapted to s. Rotating the graph of f around the x-axis creates a solid of revolution. Rotating the graph of s creates a collection of coaxial cylinders, contained inside the solid of revolution. If s_i is the vale of s on (x_{i-1}, x_i), then the total volume of the cylinders is $\sum_{i=1}^{n} \pi s_i^2 \triangle x_i$ and this number serves as an underestimate for the volume.

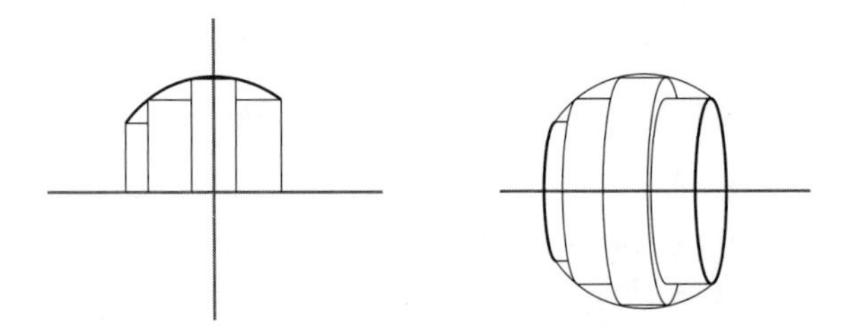

Similarly, a step function t that lies above f, creates cylinders with total volume $\sum_{i=1}^{n} \pi t_i^2 \triangle x_i$ and this is an overestimate for the volume of the solid of revolution. Now the function πf^2 is continuous, hence integrable. Therefore, the integral $\int_a^b \pi f(x)^2\, dx$ is the unique number lying above all the underestimates and below all the overestimates. We take it as the volume of the solid of revolution. This approach is called the **discs method** as it visualizes the solid as made of thin coaxial discs.

Example 6.4.11

(Sphere) A solid sphere of radius R can be obtained as the solid of revolution obtained by rotating the graph of $f(x) = \sqrt{R^2 - x^2}$, $-R \leq x \leq R$. Its volume is especially easy to calculate:

$$\int_{-R}^{R} \pi(R^2 - x^2)\, dx = \left(R^2 x - \frac{x^3}{3}\right)\Bigg|_{-R}^{R} = \frac{4}{3}\pi R^3. \qquad \square$$

Example 6.4.12

(Cone) A solid cone of base radius R and height h can be obtained as the solid of revolution obtained by rotating the graph of $f(x) = \frac{R}{h}x$, $0 \leq x \leq h$. Its volume is

$$\int_0^h \pi \left(\frac{R}{h}x\right)^2 dx = \pi \frac{R^2 x^3}{3h^2}\Bigg|_0^h = \frac{1}{3}\pi R^2 h. \qquad \square$$

It is good to verify that the basic formulas taught to us in school were correct, but it is not terribly exciting. Let us take on a more exotic shape. A **torus** is obtained by rotating a disc of radius r around a circle of larger radius R as shown below.

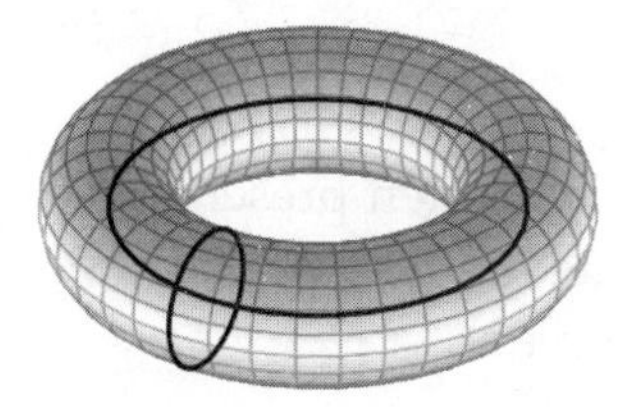

Example 6.4.13

(Torus) A torus can be generated by taking the region lying between the graphs of $y = R + \sqrt{r^2 - x^2}$ and $y = R - \sqrt{r^2 - x^2}$, $-r \le x \le r$, and rotating it around the x-axis. The volume is given by

$$\begin{aligned}\int_{-r}^{r} \pi(R + \sqrt{r^2 - x^2})^2\,dx &- \int_{-r}^{r} \pi(R - \sqrt{r^2 - x^2})^2\,dx \\ &= \pi \int_{-r}^{r} \left((R + \sqrt{r^2 - x^2})^2 - (R - \sqrt{r^2 - x^2})^2\right) dx \\ &= 4\pi R \int_{-r}^{r} \sqrt{r^2 - x^2}\,dx = 2\pi^2 R r^2.\end{aligned}$$

The volume can be expressed as $(2\pi R)(\pi r^2)$, the same as a cylinder of radius r and axial length $2\pi R$, as if the torus were obtained by bending such a cylinder until its two circular ends came together. ▢

With the help of improper integrals, surface area and volume can be calculated for unbounded shapes as well.

Example 6.4.14

(Gabriel's Horn) Consider the function $f(x) = 1/x$ with $x \ge 1$. Rotate its graph about the x-axis.

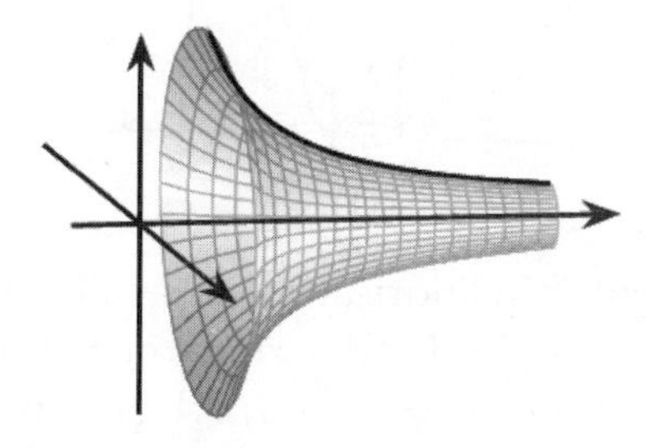

The volume enclosed by the horn is

$$V = \int_{1}^{\infty} \pi \frac{1}{x^2}\,dx = -\pi \lim_{b\to\infty} \frac{1}{x}\Big|_{1}^{b} = \pi.$$

Even though the horn has infinite extent, it encloses a finite volume. Now consider its surface area,

$$S = \int_1^\infty 2\pi\frac{1}{x}\sqrt{1+(-1/x^2)^2}dx \geq \int_1^\infty 2\pi\frac{1}{x}dx = 2\pi\lim_{b\to\infty}\log x\Big|_1^b = \infty.$$

The surface area is infinite. This is often presented as a paradox. Since the horn has infinite area, we need an infinite amount of paint to paint it. But we can fill it with only a finite amount! Do you agree? ▢

A solid of revolution can also be created by rotating a region in the xy-plane about the y-axis.

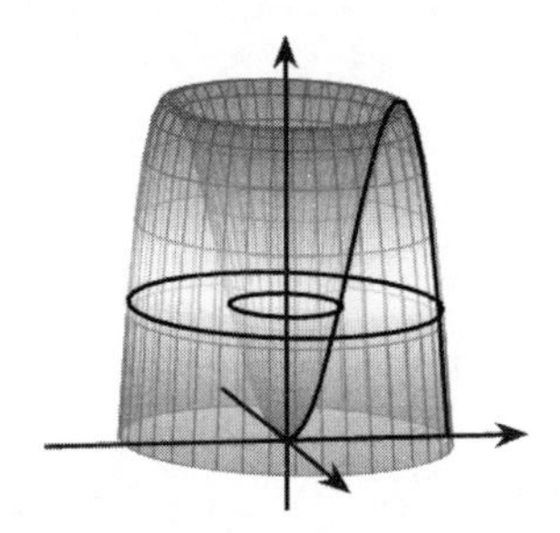

If we attempt to find the volume of such a solid by the discs method, we have to use horizontal discs, and finding their radii involves inverting the function $y = f(x)$ whose graph bounds the rotated region. This may be feasible, but an alternate approach is available, which is usually easier to implement. Instead of viewing the solid as made of coaxial discs, we view it as made of concentric cylindrical shells.

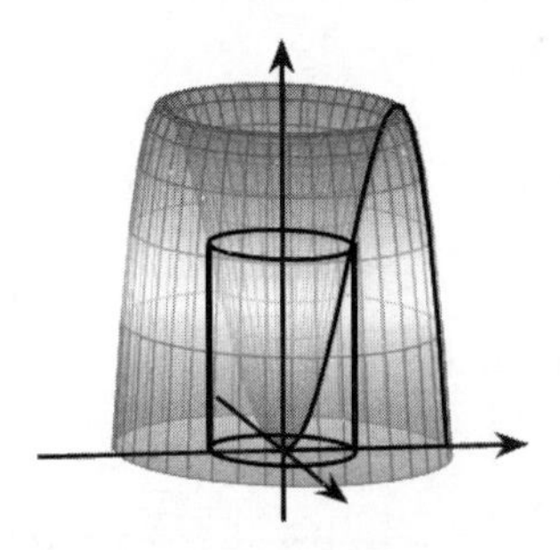

This method of using concentric cylindrical shells is called the **shell method**. Suppose we are rotating the region lying under the graph of the function $y = f(x)$ over the interval $[a,b]$ with $0 \leq a < b$. Let $P = \{x_0, \ldots, x_n\}$ be a partition of $[a,b]$ with a tag $x_1*, \ldots, x_n^*$. Consider the cylindrical shell with height $f(x_i^*)$, inner radius x_{i-1} and outer radius x_i. The volume of this shell is

$$\pi x_i^2 f(x_i^*) - \pi x_{i-1}^2 f(x_i^*) = 2\pi x_i^{**} f(x_i^*)\triangle x_i, \quad \text{where } x_i^{**} = \frac{x_{i-1}+x_i}{2}.$$

The total volume of the cylindrical shells is

$$\sum_{i=1}^{n} 2\pi x_i^{**} f(x_i^*) \triangle x_i.$$

By Theorem 6.4.3, these approach $\int_a^b 2\pi x f(x)\,dx$ as we take finer partitions.

Example 6.4.15

The volcano shaped solid we have been using to illustrate the shell method is generated by the region lying under the graph of $y = x^2 - x^4$ with $0 \le x \le 1$. Therefore, its volume is

$$\int_0^1 2\pi x(x^2 - x^4)\,dx = 2\pi\left(\frac{x^4}{4} - \frac{x^6}{6}\right)\Big|_0^1 = \frac{\pi}{6}. \qquad \square$$

Exercises for § 6.4

1. With the help of Riemann sums we can weaken the continuity assumption in the second fundamental theorem. Suppose F is differentiable on $[a,b]$ and $f = F'$ is integrable.

(a) Show that for every partition P of $[a,b]$ there is a tag such that $F(b) - F(a) = R(f,P^*)$.

(b) Conclude that $F(b) - F(a) = \int_a^b f(x)\,dx$.

2. Show that Theorem 6.4.2 becomes false if we drop the integrability assumption.

3. Prove the converse of Theorem 6.4.1: Suppose f is a function on $[a,b]$ and I is a real number such that for every $\epsilon > 0$ there is a partition P_ϵ of $[a,b]$ with the property that every tag of P_ϵ satisfies $|I - R(f,P^*)| < \epsilon$. Then $I = \int_a^b f(x)\,dx$.

4. Find the arc lengths of the graphs of the following functions:

(a) $y = \log(\sec x)$ with x varying from 0 to $\pi/4$.

(b) $y = (1 - x)^{3/2}$ with x varying from 1 to 4.

(c) $y = x^3$ with x varying from 0 to 1.

5. Plot the astroid $\left(\frac{x}{a}\right)^{2/3} + \left(\frac{y}{a}\right)^{2/3} = 1$ and find its length.

6. Consider the continuous function defined by $f(x) = 0$ if $x = 0$ and $f(x) = x\cos(\pi/2x)$ if $x \in (0,1/2]$. The following will show that the arc length of the graph of f is not defined:

(a) Consider the partition $P_n = \{\frac{1}{2n+1}, \frac{1}{2n}, \ldots, \frac{1}{3}, \frac{1}{2}\}$. Show that

$$L(f,P_n) \ge \frac{1}{2} + \frac{1}{3} + \cdots + \frac{1}{2n+1}.$$

(b) Show that $L(f, P_n) \geq \int_2^{2n+2} (1/x)\, dx = \log(n+1)$.

7. Find the surface area of a torus.

8. Use the shell method to find the volume of a torus.

9. An ellipsoid is obtained by rotating the ellipse $x^2/a^2 + y^2/b^2 = 1$ about the x-axis. Find its surface area and volume.

10. Find the volume of the bead created by drilling a cylindrical hole of radius r through a solid sphere of radius R.

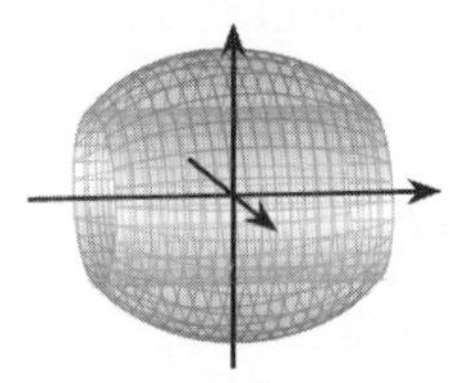

11. Consider the region A lying between the graph of $y = \sin x$ and the x-axis, with $0 \leq x \leq \pi$. Find the following:

(a) Surface area of the solid obtained by rotating A about the x-axis.

(b) Volume of the solid obtained by rotating A about the x-axis.

(c) Volume of the solid obtained by rotating A about the y-axis.

12. Consider a region lying between the graphs of $y = f(x)$ and $y = g(x)$ as shown. Suppose the region is symmetric about the line $y = \ell$, and has perimeter P and area A.

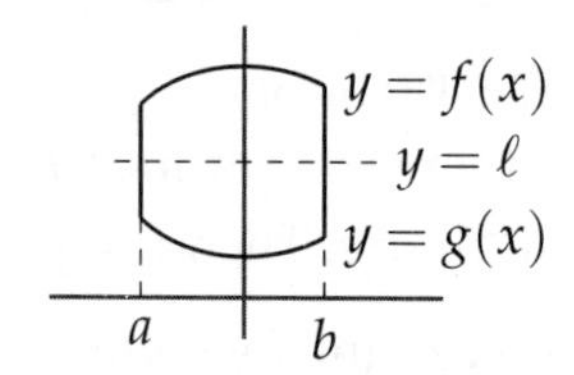

Consider the solid of revolution obtained by rotating the region around the x-axis.

(a) Show that its surface area is $2\pi\ell P$.

(b) Show that its volume is $2\pi\ell A$.

13. Find the volume of the infinitely long spindle-shaped solid created by rotating the graph of $y = \dfrac{1}{1+x^2}$ around the x-axis.

6.5 Numerical Integration

Integration can be difficult to carry out and some functions do not have conveniently expressible integrals. Our very first example of this is the log function. To get its values, we have to estimate the integrals of $1/x$. Once we become familiar with log and exp, we start thinking of them as basic functions, but then new functions crop up whose integrals cannot be expressed as finite combinations of the basic functions, such as polynomials, roots, trigonometric functions, logarithms, and exponentials. These can be as innocuous looking as e^{-x^2} and $\sqrt{1+\sin^2 x}$.

Therefore, we need to develop ways of estimating integrals and, more importantly, methods of estimating their accuracy. Estimates can be made by common sense but understanding the error requires theory.

Midpoint Rule

To approximate $\int_a^b f(x)\,dx$ we shall partition $[a,b]$ into n equal parts and let c_i be the midpoint of the ith subinterval. The **midpoint rule** approximates $\int_a^b f(x)\,dx$ by the Riemann sum

$$M_a^b(f) = \sum_{i=1}^{n} f(c_i)\triangle x, \quad \text{where } \triangle x = \frac{b-a}{n}.$$

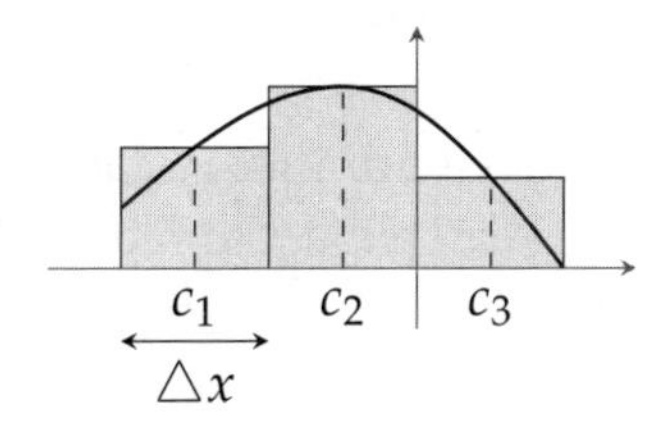

Any estimate needs error bounds if it is to be useful. Clearly, there cannot be any error bounds that will hold for *all* integrable functions or even for all continuous functions. However, if we have some control over how much f deviates from being straight, we may hope to bound the error due to approximating it by straight lines. Therefore, let f be twice continuously differentiable on $[a,b]$.

Let us consider the accuracy of each rectangle as an approximation to the corresponding portion of the integral. For simplicity, let the rectangle have base $[-h,h]$.

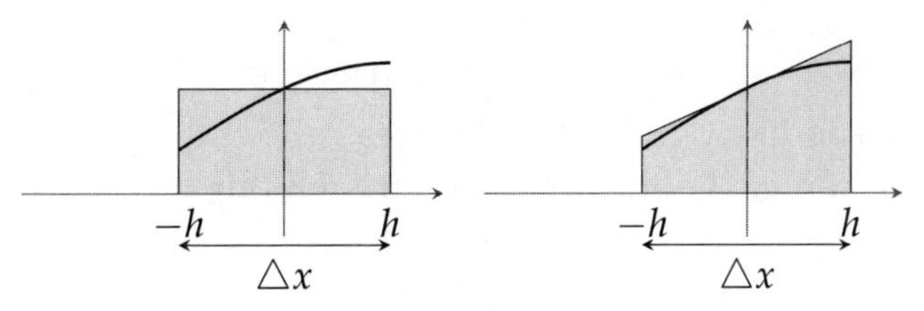

Rotating the top edge of the rectangle through its midpoint gives a trapezium with the same area. So, we rotate it until it becomes tangent to the graph of f. Now the top edge has become the graph of the first degree Taylor approximation $\ell(x)$ to the function $f(x)$. By the remainder theorem, the gap between f and ℓ at any point is

$$f(x) - \ell(x) = \frac{f''(t_x)}{2}x^2,$$

for some $t_x \in (-h,h)$. Note that the value $f''(t_x)$ must change continuously with x. By the mean value theorem for weighted integration (Theorem 3.6.7), we get

$$\int_{-h}^{h} f(x)\,dx - \int_{-h}^{h} \ell(x)\,dx = \int_{-h}^{h} \frac{f''(t_x)}{2} x^2\,dx$$
$$= f''(t) \int_{-h}^{h} \frac{x^2}{2}\,dx = \frac{h^3}{3} f''(t) = \frac{(\triangle x)^3}{24} f''(t),$$

for some $t \in [-h,h]$. We apply this result to each rectangle in the midpoint method to obtain:

$$\int_a^b f(x)\,dx - M_a^b(f) = \frac{(\triangle x)^3}{24} \sum_{i=1}^{n} f''(t_i)$$
$$= \frac{(\triangle x)^3}{24} n f''(t) = \frac{(\triangle x)^2}{24} (b-a) f''(t),$$

for some $t \in [a,b]$. (We have used the intermediate value theorem as well. Can you see where?)

The key part of this estimate is the quadratic $\triangle x$ factor. It indicates that if we double the calculations by halving $\triangle x$, we can expect to reduce the error by a factor of 4.

Example 6.5.1

In Example 5.5.5 we used the midpoint rule, without naming it, to estimate $\int_0^3 e^{-x^2}\,dx$. We set $n = 6$ and obtained the following values of f at the midpoints c_i.

c_i	0.25	0.75	1.25	1.75	2.25	2.75
$f(c_i)$	0.9394	0.5698	0.2096	0.0468	0.0063	0.0005

This gave

$$\int_0^3 e^{-x^2}\,dx \approx \sum_{i=1}^{6} f(c_i) \times 0.5 = 0.886213\ldots$$

while the precise value of this integral is 0.88620734.... Thus, with these few calculations we had accuracy to 4 decimal places.

Let us see if we can gauge the accuracy through our own calculations. We have $f'(x) = -2xe^{-x^2}$ and $f''(x) = (-2+4x^2)e^{-x^2}$. This gives the bound $M = 2$, and hence

$$\left| \int_0^3 e^{-x^2}\,dx - 0.886213\ldots \right| \le \frac{3 \times 2}{24}(0.5)^2 = 0.06.$$

This is a correct bound, but it clearly plays too safe and fails to convey how well the calculation has actually worked. □

Thus, while the error bounds reassure us that we can get accurate results, they are not always useful in gauging the real accuracy. If we used these theoretical bounds to decide how fine the subintervals should be, we would end up doing much more work than is really required. For this reason, practitioners simply increase the value of n and if the first 4 decimal places of the estimate do not change they decide they have achieved that much accuracy! However, once a certain accuracy is observed, the form of the theoretical bound helps estimate how to get further accuracy. For example, suppose $n = 6$ gives 4 places of accuracy and we need 6 places of accuracy. The $(\triangle x)^2$ factor in the error bound indicates that to reduce error a hundred times, we should increase n by a factor of 10. So, we should investigate $n = 60$ as a likely candidate for producing the desired accuracy.

Let us apply this to our example of $\int_0^3 e^{-x^2}\,dx$. Increasing n to 60 gives the estimate

$$\sum_{i=1}^{60} f(c_i) \times 0.05 = 0.88620742\ldots,$$

which is indeed accurate to 6 decimal places.

Simpson's Rule

We can view the midpoint rule as being based on locally linear approximations to the function being integrated. We can hope to get better results by using a quadratic approximation as that can follow the curve of f more closely. We shall first see how to fit a polynomial of degree at most 2 to any three data points.

To simplify calculations, suppose the values y_0, y_1, and y_2 of a function f are known at $x = -h, 0, h$ respectively. We wish to find a quadratic function $q(x)$ such that $q(-h) = y_0$, $q(0) = y_1$, and $q(h) = y_2$. It must have the form $q(x) = y_1 + xp(x)$ where $p(x)$ is linear. Then p has to satisfy $y_0 = y_1 - hp(-h)$ and $y_2 = y_1 + hp(h)$. This implies

$$p(-h) = \frac{y_1 - y_0}{h},\ p(h) = \frac{y_2 - y_1}{h} \text{ and } p(0) = \frac{y_2 - y_0}{2h}.$$

From these values we obtain the following expressions for p and q:

$$p(x) = \frac{y_2 - y_0}{2h} + \frac{y_0 - 2y_1 + y_2}{2h^2}x,$$
$$q(x) = y_1 + \frac{y_2 - y_0}{2h}x + \frac{y_0 - 2y_1 + y_2}{2h^2}x^2.$$

This gives the following approximation for the integral of f:

$$\int_{-h}^{h} f(x)\,dx \approx \int_{-h}^{h} q(x)\,dx = \left(y_0 + 4y_1 + y_2\right)\frac{h}{3}.$$

Now, if f has domain $[a,b]$, we take a partition $P = \{x_0, \ldots, x_{2n}\}$ where the subintervals have equal width $\triangle x = (b-a)/2n$. Let $y_i = f(x_i)$. On each interval $[x_{2i-2}, x_{2i}]$ we apply the above approximation. This gives **Simpson's rule**:

$$\int_a^b f(x)\,dx \approx S_a^b(f)$$

where

$$S_a^b(f) = \left(y_0 + 4y_1 + 2y_2 + 4y_3 + 2y_4 + \cdots + 2y_{2n-2} + 4y_{2n-1} + y_{2n}\right)\frac{\triangle x}{3}$$
$$= \left(y_0 + 2\sum_{i=1}^{n-1} y_{2i} + 4\sum_{i=1}^{n} y_{2i-1} + y_{2n}\right)\frac{\triangle x}{3}.$$

As these expressions are not Riemann sums, it is not yet clear if they will approach the integral when we increase n. However, the rule can be expressed in a different way:

$$S_a^b(f) = \frac{1}{6}\left(2\triangle x\sum_{i=0}^{n-1} y_{2i}\right) + \frac{1}{6}\left(2\triangle x\sum_{i=1}^{n} y_{2i}\right) + \frac{2}{3}\left(2\triangle x\sum_{i=1}^{n} y_{2i-1}\right).$$

Each bracketed term is a Riemann sum and converges to $\int_a^b f(x)\,dx$, hence so does their combination. This is reassuring, but what we really seek are explicit error bounds.

Simpson's rule has one surprising property: although it uses a quadratic approximation, it is exact even for cubics. This is quite easy to see over the $[-h,h]$ interval, since $\int_{-h}^{h} x^3\,dx = S_{-h}^{h}(x^3) = 0$. Our error analysis will exploit this, and the result will show why Simpson's rule is a favorite for numerical work.

Let $q(x)$ be a cubic that matches $f(x)$ at $x = -h, 0, h$ and also satisfies $q'(0) = f'(0)$. We will use the notation $y_0 = f(-h)$, $y_1 = f(0)$, $y_1' = f'(0)$, $y_2 = f(h)$. It must have the form $q(x) = y_1 + xy_1' + x^2p(x)$, where $p(x)$ is linear. We can solve for $p(x)$ as we did earlier, and the result is

$$q(x) = y_1 + y_1'x + \frac{y_0 - 2y_1 + y_2}{2h^2}x^2 + \frac{y_2 - y_0 - 2hy_1'}{2h^3}x^3.$$

Integrating q gives us the original Simpson's formula again. Now we can use the following error formula:

Theorem 6.5.2

Let $f\colon [a,b] \to \mathbb{R}$ be four times continuously differentiable. Let $a = x_0 < x_1 < x_2 = b$. Let $q(x)$ be a polynomial of degree three or less such that $q(x_0) = f(x_0)$, $q(x_1) = f(x_1)$, $q'(x_1) = f'(x_1)$, and $q(x_2) = f(x_2)$. Then for every $x \in [a,b]$ there is a $\xi_x \in (a,b)$ such that

$$f(x) - q(x) = \frac{f^{(4)}(\xi_x)}{4!}(x - x_0)(x - x_1)^2(x - x_2).$$

Proof. We can assume $x \neq x_i$ for $i = 0,1,2$. Define a function g by

$$g(t) = f(t) - q(t) - M(t - x_0)(t - x_1)^2(t - x_2),$$

where M is chosen such that $g(x) = 0$. Then we have four distinct zeroes of g:

$$g(x_0) = g(x_1) = g(x_2) = g(x) = 0.$$

Rolle's theorem gives us three distinct zeroes of g' in the open intervals created by x_0, x_1, x_2, and x. We also have $g'(x_1) = 0$. So, we again have four distinct zeroes of g'. Applying Rolle's theorem repeatedly, we get three distinct zeroes of g'', then two of g''', and finally one of $g^{(4)}$, which we shall call ξ_x. If we now differentiate g four times using its definition, we obtain $M = f^{(4)}(\xi_x)/4!$. Substituting this in $g(x) = 0$ gives the result. ■

Applying this error formula to our cubic approximation to f over $[-h,h]$ gives

$$f(x) - q(x) = \frac{f^{(4)}(\xi_x)}{4!} x^2(x^2 - h^2).$$

Therefore, the error for the integral is

$$\begin{aligned}\int_{-h}^{h} f(x)\,dx - \int_{-h}^{h} q(x)\,dx &= \int_{-h}^{h} \frac{f^{(4)}(\xi_x)}{4!} x^2(x^2 - h^2)\,dx \\ &= \frac{f^{(4)}(\xi)}{4!} \int_{-h}^{h} x^2(x^2 - h^2)\,dx \\ &= \frac{f^{(4)}(\xi)}{90} h^5.\end{aligned}$$

Applying this to the rule for a partition of $[a,b]$ into $2n$ equal subintervals, we get the error

$$\begin{aligned}\int_a^b f(x)\,dx - S_a^b(f) &= \sum_{i=1}^{n} \frac{f^{(4)}(\xi_i)}{90} (\triangle x)^5 \\ &= \frac{(\triangle x)^5 n}{90} f^{(4)}(\xi) = \frac{(\triangle x)^4}{180} (b-a) f^{(4)}(\xi).\end{aligned}$$

The $(\triangle x)^4$ factor implies that evaluating at more points will have a much greater impact on the error, in comparison with the midpoint rule.

Example 6.5.3

If we apply Simpson's rule to $\int_0^3 e^{-x^2}\,dx$ by cutting $[0,3]$ into 6 equal subintervals, we get the estimate $0.886172\ldots$, which has similar accuracy to the midpoint rule with the same number of intervals. The difference becomes visible when we increase the number of intervals by a factor of 10. Simpson's rule now gives $0.88620734\ldots$, which is accurate to 8 decimal places. With the same number of function evaluations the midpoint rule was accurate to 6 decimal places. □

Gaussian Quadrature

The midpoint and Simpson's rules have a common format: they are weighted averages of certain function values. We can create new rules by varying the points where we sample the function as well as the weight attached to each sampling location. In the Gaussian quadrature method we attempt to fix the sampling locations and the weights so as to get exact results for polynomials of as high a degree as possible. If we use n sampling points, we also have n weights to play with, and can hope to get exact results for polynomials up to degree $2n-1$. Let us see how this works for small n and over an interval $[-h,h]$.

First, consider $n=1$. We hope to get exact results for linear polynomials. If the sampling point is a and the weight is w_a, we get two equations by considering the integrals of 1 and x: $w_a = \int_{-h}^{h} 1\,dx = 2h$ and $w_a a = \int_{-h}^{h} x\,dx = 0$. This gives $a=0$ and $w_a = 2h$. In other words, we have recovered the midpoint rule.

Now let $n=2$. Let the sample points be a,b with corresponding weights w_a, w_b. We get four equations by matching with 1, x, x^2, and x^3:

$$\begin{aligned} w_a + w_b &= 2h \\ w_a a + w_b b &= 0 \\ w_a a^2 + w_b b^2 &= \frac{2}{3}h^2 \\ w_a a^3 + w_b b^3 &= 0 \end{aligned}$$

The second and fourth equations give $w_a a(b^2 - a^2) = 0$ and hence $a=0$ or $-b$. Now $a=0$ is ruled out by the second equation, so we have $a=-b \neq 0$. But then the second equation gives $w_a = w_b$ and the first one gives $w_a = w_b = h$. Substituting all this in the third equation gives $a = -h/\sqrt{3}$ and $b = h/\sqrt{3}$. Thus, the rule is

$$\int_{-h}^{h} f(x)\,dx \approx G_2(f) = h(f(-h/\sqrt{3}) + f(h/\sqrt{3})).$$

Example 6.5.4

Let us apply the basic integration rules to $\int_0^{\pi/2} \sqrt{x}\cos x\,dx = 0.704$. Write $g(x) = \sqrt{x}\cos x$.

The midpoint rule uses one function value and gives the value

$$g(\pi/4)\cdot \pi/2 = 0.984.$$

Simpson's rule uses three function values and gives

$$(g(0) + 4g(\pi/4) + g(\pi/2))\cdot \pi/12 = 0.656.$$

Gaussian quadrature uses two function values and gives

$$\Big(g((1-1/\sqrt{3})\pi/4) + g((1+1/\sqrt{3})\pi/4)\Big)\cdot \pi/4 = 0.712. \qquad \square$$

We can develop an error formula for Gaussian quadrature with $n=2$ using the same approach as for Simpson's rule. First, suppose $q(x)$ is a cubic polynomial such that $q(h/\sqrt{3}) = f(h/\sqrt{3})$, $q(-h/\sqrt{3}) = f(-h/\sqrt{3})$, $q'(h/\sqrt{3}) = f'(h/\sqrt{3})$, and $q'(-h/\sqrt{3}) = f'(-h/\sqrt{3})$.

Task 6.5.5

Suppose $a \neq b$ and y_a, y_b, y'_a, y'_b are given real numbers. Show that there is a unique polynomial of degree at most three such that $q(a) = y_a$, $q(b) = y_b$, $q'(a) = y_a$ and $q'(b) = y_b$.

Since q is cubic, and matches f on $\pm h/\sqrt{3}$, we have $\int_{-h}^{h} q(x)\,dx = G(q) = G(f)$. Therefore,

$$\int_{-h}^{h} f(x)\,dx - G_2(f) = \int_{-h}^{h} f(x)\,dx - \int_{-h}^{h} q(x)\,dx = \int_{-h}^{h} (f(x) - q(x))\,dx.$$

Theorem 6.5.6

Let $f\colon [a,b] \to \mathbb{R}$ be four times continuously differentiable. Let $a \le x_1 < x_2 \le b$. Let $q(x)$ be a polynomial of degree three or less such that $q(x_1) = f(x_1)$, $q'(x_1) = f'(x_1)$, $q(x_2) = f(x_2)$, and $q'(x_2) = f'(x_2)$. Then for every $x \in [a,b]$ there is a $\xi_x \in (a,b)$ such that

$$f(x) - q(x) = \frac{f^{(4)}(\xi_x)}{4!}(x - x_1)^2(x - x_2)^2.$$

Proof. Exercise. Similar to Theorem 6.5.2. ∎

Integrate this error formula to get

$$\begin{aligned}
\int_{-h}^{h} f(x)\,dx - G_2(f) &= \int_{-h}^{h} (f(x) - q(x))\,dx \\
&= \frac{f^{(4)}(\xi)}{4!} \int_{-h}^{h} (x + h/\sqrt{3})^2 (x - h/\sqrt{3})^2\,dx \\
&= \frac{f^{(4)}(\xi)}{4!} h^5 \int_{-1}^{1} (x^2 - 1/3)^2\,dx = \frac{f^{(4)}(\xi)}{135} h^5.
\end{aligned}$$

So, Gaussian quadrature achieves similar accuracy to Simpson's rule, while using two function values instead of three.

Exercises for § 6.5

1. Use the midpoint and Simpson's rules to estimate $\log 2 = \int_1^2 (1/x)\,dx$. (Use five function evaluations.)

2. Consider the following table of midpoint and Simpson's rules applied to estimate $\int_0^1 \sqrt{1 - x^2}\,dx$, with n being the number of intervals used:

n	2	20	200	2000
Midpoint rule	0.81...	0.786...	0.78542...	0.785399...
Simpson's rule	0.74...	0.784...	0.78535...	0.785397...

The actual value of $\int_0^1 \sqrt{1-x^2}\,dx$ is $0.785398\ldots$. Does the increase of accuracy with n follow the expected pattern? If not, what could be the reason?

3. (Trapezoidal Rule) Another common approach is to approximate the function over a small interval by a line that matches it at the endpoints. If the interval is $[0,h]$, this line is given by $\ell(x) = f(0) + \dfrac{f(h)-f(0)}{h}x$ and the resulting approximation is $\int_0^h f(x)\,dx \approx \int_0^h \ell(x)\,dx = \frac{h}{2}(f(0)+f(h))$. Now we take a general interval $[a,b]$, partition it into n equal subintervals $[x_{i-1},x_i]$, $i = 1,\ldots,n$, and apply this rule to each subinterval to get

$$\int_a^b f(x)\,dx \approx T_a^b(f) = h\left(\frac{f(a)}{2} + \sum_{i=1}^{n-1} f(x_i) + \frac{f(b)}{2}\right).$$

The following steps give an error analysis for this rule, assuming f is twice continuously differentiable.

(a) Show that for each $x \in [0,h]$ there is a $\xi_x \in [0,h]$ such that $f(x) - \ell(x) = \dfrac{x(x-h)}{2} f''(\xi_x)$.

(b) Show that $\int_0^h f(x)\,dx - \int_0^h \ell(x)\,dx = -\frac{h^3}{12} f''(\xi)$ for some $\xi \in [0,h]$.

(c) Show that $\int_a^b f(x)\,dx - T_a^b(f) = -\frac{h^2}{12}(b-a)f''(\xi)$ for some $\xi \in [a,b]$.

4. Note that the error of the trapeziodal rule is opposite and roughly twice of that of the midpoint rule. This suggests taking $\frac{2}{3}M_a^b(f) + \frac{1}{3}T_a^b(f)$ as a new rule for numerical integration. Show that this also leads to Simpson's rule.

5. Show that the Gaussian quadrature rule for $\int_{-h}^h f(x)\,dx$ with $n = 3$ is

$$G_3(f) = \left(\frac{5}{9}f(-\sqrt{3/5}h) + \frac{8}{9}f(0) + \frac{5}{9}f(\sqrt{3/5}h)\right)h.$$

Apply this rule to $\int_0^{\pi/2} \sqrt{x}\cos x\,dx$.

6. Show that the error formula for Gaussian quadrature with $n = 3$ is

$$\int_{-h}^h f(x)\,dx - G_3(f) = \frac{8}{7!\,25} f^{(6)}(\xi)h^7.$$

Thematic Exercises

Curve Fitting: Error Analysis

In the supplementary exercises on *Curve Fitting* following Chapter 1, we developed a formula for the unique interpolating polynomial of least degree for n data points.

Suppose the data is $(x_0,y_0),\ldots,(x_n,y_n)$ with $x_i \neq x_j$ when $i \neq j$. Then the **Lagrange interpolating polynomial** is given by

$$p(x) = \sum_{i=0}^{n} y_i \left(\prod_{\substack{k=0 \\ k\neq i}}^{n} (x - x_k) \Bigg/ \prod_{\substack{k=0 \\ k\neq i}}^{n} (x_i - x_k) \right).$$

The interpolating polynomial p is an approximation to the actual function f that produced the data. We would like to know how closely p approximates f. The first exercise, below, provides bounds on the error of this approximation, assuming f is smooth enough. Before attempting this exercise, you may find it helpful to review the steps of Exercise 12, §6.1.

A1. Let f be n times differentiable on an interval I. Let $x_1,\ldots,x_n$ be distinct points in I, and $p(x)$ a polynomial of degree $n-1$ or less such that $p(x_i) = f(x_i)$ for every i. Show that for any $x \in I$ there is a $\xi \in I$ such that

$$f(x) - p(x) = \frac{f^{(n)}(\xi)}{n!}(x - x_1)\cdots(x - x_n).$$

Next, suppose we have data for both a function and its derivative. Given (x_i, y_i, y_i'), with $i = 0,\ldots,n$, the corresponding **Hermite polynomial** is the polynomial H of least degree such that $H(x_i) = y_i$ and $H'(x_i) = y_i'$ for each i. (Of course, the x_i must be distinct from each other.)

Let L_j be the unique polynomial of degree n such that $L_j(x_k) = 0$ if $k \neq j$ and $L_j(x_j) = 1$.

A2. Verify that L_j^2 satisfies $L_j^2(x_k) = (L_j^2)'(x_k) = 0$ if $k \neq j$, $L_j^2(x_j) = 1$, and $(L_j^2)'(x_j) = 2L_j'(x_j)$.

We wish to obtain polynomials H_j, K_j of degree at most $2n+1$ such that

- $H_j(x_k) = 0$ if $k \neq j$, $H_j(x_j) = 1$, and $H_j'(x_k) = 0$ for every k.
- $K_j(x_k) = 0$ for every k, $K_j'(x_k) = 0$ if $k \neq j$, and $K_j'(x_j) = 1$.

We look for them in the form $(a + bx)L_j^2(x)$.

A3. Show that $H_j(x) = (1 - 2(x - x_j)L'(x_j))L_j^2(x)$ and $K_j(x) = (x - x_j)L_j^2(x)$.

A4. Show that $H(x) = \sum_{j=0}^{n} y_j H_j(x) + \sum_{j=0}^{n} y_j' K_j(x)$ is the unique polynomial of degree at most $2n+1$ that satisfies the requirements of the Hermite polynomial for the data (x_i, y_i, y_i'), $i = 0,\ldots,n$.

A5. Let $H(x)$ be the Hermite polynomial for data $(x_i, f(x_i), f'(x_i)), i = 0, \ldots, n$, where f is $2n+2$ times continuously differentiable and all $x_i \in [a,b]$. Show that for each $x \in [a,b]$ there is a $\xi_x \in [a,b]$ such that

$$f(x) - H(x) = \frac{(x-x_0)^2 \cdots (x-x_n)^2}{(2n+2)!} f^{(2n+2)}(\xi_x).$$

Riemann Integral

In Riemann's original approach, Riemann sums approach the integral not by taking finer partitions but by narrowing the widths of the subintervals created by partitions. The two processes are closely related but are not the same. Nevertheless, the definitions of Darboux and Riemann turn out to be equivalent.

First, given a partition $P = \{x_0 < \cdots < x_n\}$ of $[a,b]$, its **mesh** $\triangle_P$ is the width of its widest interval:

$$\triangle_P = \max\{ x_i - x_{i-1} \mid i = 1, \ldots, n \}.$$

We say $\lim\limits_{\triangle_P \to 0} R(f,P) = I$ if for every $\epsilon > 0$ there is a $\delta > 0$ such that every tagged partition P^* with $\triangle_P < \delta$ satisfies $|I - R(f,P^*)| < \epsilon$. In this case, we say that f is **Riemann integrable** over $[a,b]$ and call I the **Riemann integral** of f over $[a,b]$. Let us temporarily denote it by $\mathcal{R}_a^b f(x)\,dx$, reserving $\int_a^b f(x)\,dx$ for the Darboux integral.

In the following exercises, f is a bounded function on an interval $[a,b]$.

B1. Suppose f is Riemann integrable and $\epsilon > 0$.

(a) Show that there is $n \in \mathbb{N}$ such that every tag of the partition P of $[a,b]$ into n equal subintervals satisfies

$$|\mathcal{R}_a^b f(x)\,dx - R(f,P^*)| < \epsilon/2.$$

(b) Show that there are tags of P such that

$$R(f,P^*) - L(f,P) < \epsilon/2 \quad \text{and} \quad U(f,P) - R(f,P^{**}) < \epsilon/2.$$

(c) Conclude that f is Darboux integrable and $\int_a^b f(x)\,dx = \mathcal{R}_a^b f(x)\,dx$.

B2. Let P^* be a tagged partition of $[a,b]$. Show that

$$L(f,P) \le R(f,P^*) \le U(f,P).$$

where $L(f,P)$ and $U(f,P)$ are the lower and upper Darboux sums.

B3. Suppose f is Darboux integrable and $\epsilon > 0$.

(a) Show there is a partition $Q = \{x_0, \ldots, x_m\}$ of $[a,b]$ such that $U(f,Q) - L(f,Q) < \epsilon$.

(b) Let $|f|$ be bounded by M and define $\delta = \epsilon/(2Mm)$. If P is a partition of $[a,b]$ with $\triangle_P < \delta$, show that $L(f,Q) - \epsilon \leq L(f,P) \leq U(f,P) \leq U(f,Q) + \epsilon$.

(c) Conclude that f is Riemann integrable and $\int_a^b f(x)\,dx = \mathcal{R}_a^b f(x)\,dx$.

7 | Sequences and Series

In calculus, we mainly study continuous change. However, there are situations where discrete changes have to be considered. For example, when we try to describe a number such as π by its decimal representation, we actually create an iterative process of successively better approximations: 3, 3.1, 3.14, 3.141, 3.1415, 3.14159, and so on. A similar situation arises when we work with the Taylor polynomials of a function—we successively approximate a function by polynomials of increasing degree. What is common to the two examples is that there is a first stage, a second stage, a third stage, and we are interested in what happens as we keep going. Clearly, we need to develop a theory of limits for this context. We shall do so in this chapter. Further, we shall work out in detail the situation when discrete changes accumulate and we are interested in the total. This will have many similarities as well as a direct relation with integration.

As an example, let us consider a geometry problem that leads to an iterative method for approximating square roots by fractions. It is named Heron's method after a Greek mathematician, but the evidence is strong that this kind of reasoning was carried out earlier in ancient Iraq and India, three to four thousand years ago. The statement of the problem is: "Given a rectangle, construct a square with the same area." Now, if the rectangle has sides a and b, the square needs to have side $\sqrt{ab}$. To us, this may be a triviality, but what if the only numbers you know are the fractions? Then the problem will, in general, have only approximate solutions. How do we find good fractional approximations to $\sqrt{ab}$? Consider the following steps.

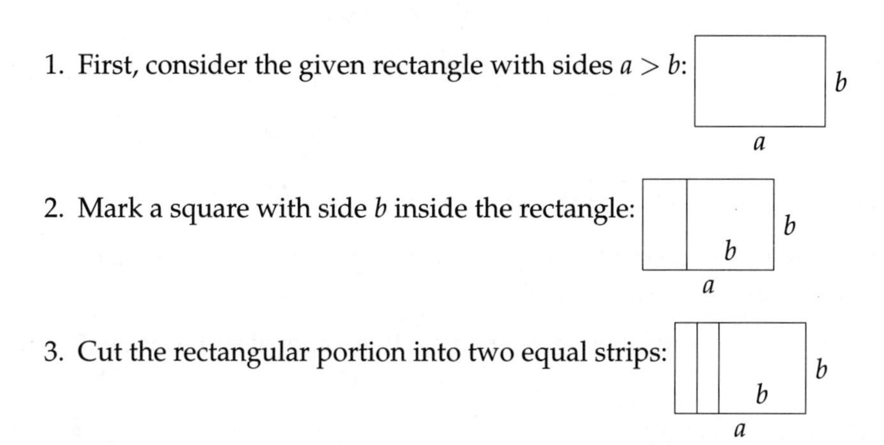

1. First, consider the given rectangle with sides $a > b$:
2. Mark a square with side b inside the rectangle:
3. Cut the rectangular portion into two equal strips:

4. Move one strip to an adjacent side of the inner square: 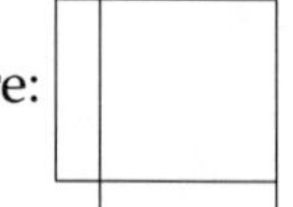

5. Fill in the missing part to complete a square: 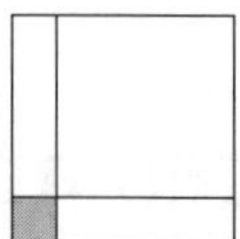

The final square is obviously a bit too big. Nevertheless, its side of $(a+b)/2$ is visibly better than the initial sides of a and b. If we have $ab = N$, we can repeat the process with a rectangle whose sides are $a' = (a+b)/2$ and $b' = N/a'$. This will lead to a new and further improved square with side $(a'+b')/2$.

We have obtained the following procedure for approximating $\sqrt{N}$. Take any overestimate a_1, and set $b_1 = N/a_1$. Then define a new estimate by $a_2 = \frac{1}{2}(a_1 + b_1) = \frac{1}{2}\left(a_1 + \frac{N}{a_1}\right)$. Repeat.

Suppose $N = 2$. Then we can take $a_1 = 2$. This leads to $a_2 = \frac{1}{2}\left(2 + \frac{2}{2}\right) = \frac{3}{2}$, $a_3 = \frac{1}{2}\left(\frac{3}{2} + \frac{2}{3/2}\right) = \frac{17}{12}$, and $a_4 = \frac{1}{2}\left(\frac{17}{12} + \frac{2}{17/12}\right) = \frac{577}{408}$. The last estimate is pretty good, as $(577/408)^2 = 2\frac{1}{166,464}$.

It appears that we have a successful technique. We need to subject it to more rigorous tests. Are we sure that the estimates can be made as accurate as desired? Can we say how many iterations would be needed to reach a desired accuracy? These are the questions which we now take up and place in a general setting.

7.1 Limit of a Sequence

A **sequence** is an unending list of real numbers. Let us consider some simple examples.

(a) $1, 2, 3, 4, \ldots$

(b) $1, 1, 1, 1, \ldots$

(c) $1, \frac{1}{2}, \frac{1}{3}, \frac{1}{4}, \ldots$

(d) $\sqrt{1}, -\sqrt{2}, \sqrt{3}, -\sqrt{4}, \ldots$

(e) $3, 1, 4, 1, 5, 9, \ldots$

(f) $0.1, -0.23, \pi, \sqrt{2}, e, \ldots$

These examples were chosen to illustrate certain features, which we list below.

1. A sequence *may* follow a simple pattern, and this pattern may be evident from the first few entries, as in examples (a) to (d).
2. The entries may be any mix of positive and negative, rational and irrational, as in (d) and (f).
3. Entries may repeat, as in (b).

4. All the entries should be known, in principle. For example, (e) consists of the digits in the decimal representation of π. These are known in principle, in the sense that if one wants to know the digit in the 10^{-15} place there is only one answer, even if it has not been worked out yet.
5. Example (f) is acceptable as a sequence only if it is part of some complete assignment of real numbers to positions in the sequence.

The general notation for a sequence is to label its members by their position, such as: $a_1, a_2, a_3, \ldots$. A more compact representation is $(a_n)_{n=1}^{\infty}$ or even just (a_n).

Example 7.1.1

Here are some examples of describing a sequence by giving the form of its nth term:

$$\begin{array}{ll} 1,2,3,4,\ldots & a_n = n \\ 1,1,1,1,\ldots & a_n = 1 \\ 1,-1,1,-1,\ldots & a_n = (-1)^{n+1} \\ 1,\frac{1}{2},\frac{1}{3},\frac{1}{4},\ldots & a_n = \frac{1}{n} \\ \sqrt{1},\sqrt{2},\sqrt{3},\sqrt{4},\ldots & a_n = \sqrt{n} \end{array}$$

This is the most satisfactory way of describing a sequence, although it is not always possible. □

Formally, a sequence is a function $f\colon \mathbb{N} \to \mathbb{R}$. Such a function generates numbers $a_1 = f(1)$, $a_2 = f(2)$, $a_3 = f(3)$, ...

Convergence and Divergence

Let (a_n) be a sequence of real numbers, and L a real number. We say that (a_n) **converges** to L if for every real number $\epsilon > 0$, there is $N \in \mathbb{N}$ such that $n \geq N$ implies $|a_n - L| < \epsilon$. The number L is called the **limit** of the sequence.

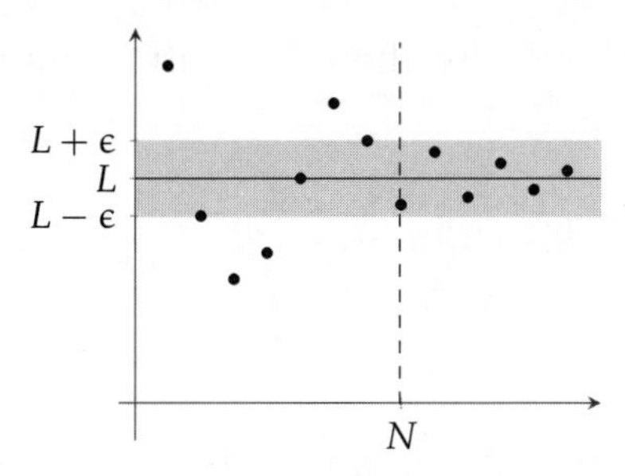

If (a_n) converges to L, we say $a_n \to L$ as $n \to \infty$, or $\lim_{n\to\infty} a_n = L$. More briefly, we may just say $a_n \to L$ or $\lim a_n = L$. If a sequence does not converge, we say it **diverges**.

Remark. Compare with the definition of $\lim_{x\to\infty} f(x) = L$.

Let us consider two fundamental examples. These will underlie most of our results.

Example 7.1.2

Let us show that $\lim_{n\to\infty} \frac{1}{n} = 0$. Consider any $\epsilon > 0$. Then $1/\epsilon > 0$. By the Archimedean Property, there is a natural number N such that $N > 1/\epsilon$. Hence, $\frac{1}{N} < \epsilon$. This N works for us: If $n > N$, then

$$|a_n - L| = \left|\frac{1}{n} - 0\right| = \frac{1}{n} < \frac{1}{N} < \epsilon.$$

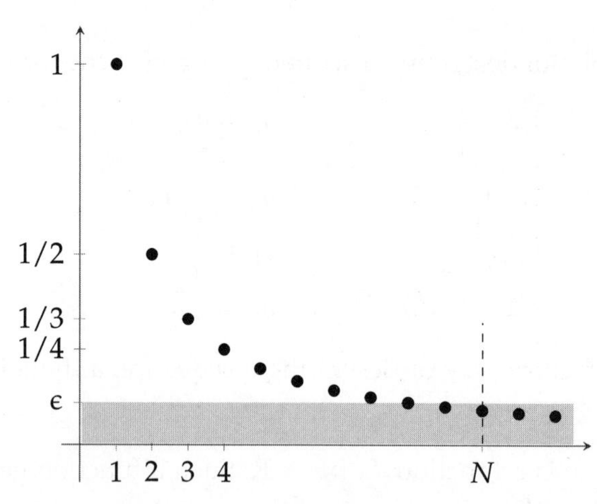

□

Example 7.1.3

Let us show that $\lim_{n\to\infty} r^n = 0$ if $|r| < 1$. Note that $|r| < 1$ implies $\frac{1}{|r|} > 1$. So we can write $\frac{1}{|r|} = 1 + h$ with $h > 0$. Hence, $\frac{1}{|r|^n} = (1+h)^n > nh$ and so $|r|^n < \frac{1}{nh}$. Consider any $\epsilon > 0$. By the Archimedean property, there is a natural number N such that $\frac{1}{N} < h\epsilon$. This N works: If $n > N$, then

$$|a_n - L| = |r^n - 0| = |r|^n < \frac{1}{nh} < \frac{1}{Nh} < \epsilon.$$

□

Task 7.1.4

Show that the limit of a sequence is unique, if it exists.

Task 7.1.5

Let $a_n = c$ be a constant sequence. Show that $a_n \to c$.

Now we consider an example of divergence. Proofs of divergence have a very different flavor.

Example 7.1.6

Consider the sequence given by $a_n = (-1)^n$. The entries $-1, 1, -1, 1, \ldots$ keep switching between ± 1 so the sequence is not settling down and does not have a limit. How do we establish this formally? We use the idea that if the sequence entries were to approach a certain number L, then they would also approach each other. For example, if some numbers are each within 1 unit of L, then they are also all within 2 units of each other.

So, suppose the given sequence has a limit L. Take $\epsilon = 1$. There will be an N such that $n \geq N$ implies $|a_n - L| < 1$. In particular, $|a_N - L| < 1$ and $|a_{N+1} - L| < 1$. Therefore, $|a_N - a_{N+1}| \leq |a_N - L| + |a_{N+1} - L| < 2$, which is false as consecutive entries actually have a gap of 2. This contradiction informs us that the sequence diverges. □

Task 7.1.7

Show that the sequence given by $a_n = n$ diverges.

Task 7.1.8

Let (a_n) be a given sequence and k a fixed natural number. Define a sequence (b_n) by $b_n = a_{n+k}$. (That is, we drop the first k terms of the given sequence to create a new sequence.) Show that $\lim b_n = L$ if and only if $\lim a_n = L$.

Limit Theorems

Task 7.1.9

Suppose (a_n) is a converging sequence and $m \leq a_n \leq M$ for every n. Then $m \leq \lim_{n\to\infty} a_n \leq M$.

Theorem 7.1.10 (Sandwich or Squeeze Theorem)

Let $(a_n), (b_n), (c_n)$ be sequences such that for every n, $a_n \leq b_n \leq c_n$. If $\lim_{n\to\infty} a_n = \lim_{n\to\infty} c_n = L$, then $\lim_{n\to\infty} b_n = L$.

Proof. Consider any $\epsilon > 0$. Then

$$a_n \to L \implies \text{there is } N_a \text{ such that if } n > N_a \text{ then } L - \epsilon < a_n < L + \epsilon,$$
$$c_n \to L \implies \text{there is } N_c \text{ such that if } n > N_c \text{ then } L - \epsilon < c_n < L + \epsilon.$$

Define $N = \max\{N_a, N_c\}$. This N works for (b_n). ■

Our main tool will be the sandwich theorem combined with the basic results $1/n \to 0$ and $r^n \to 0$ (for $|r| < 1$).

Example 7.1.11

Consider the sequence $\left(\frac{r^n}{n!}\right)$ where $r > 0$ is fixed. Fix $M \in \mathbb{N}$ such that $M > r$. For $n > M$,

$$0 < \frac{r^n}{n!} = \frac{r}{n} \cdots \frac{r}{M+1} \cdot \frac{r^M}{M!} < \frac{1}{n} \cdot \frac{r^{M+1}}{M!} \to 0.$$

Hence $\frac{r^n}{n!} \to 0$. □

Task 7.1.12

Let $|a_n| \to 0$. Show that $a_n \to 0$.

Theorem 7.1.13 (Algebra of Limits)

Let $a_n \to L$ and $b_n \to M$. Also, let $c \in \mathbb{R}$. Then we have the following limits:

1. $|a_n| \to |L|$.
2. $c\,a_n \to c\,L$.
3. $a_n + b_n \to L + M$.
4. $a_n - b_n \to L - M$.
5. $a_n b_n \to LM$.
6. $a_n / b_n \to L/M$ *if* $M \neq 0$.

Proof. The proofs are similar to the algebra of limits for functions. ■

Task 7.1.14

Find the following limits:

(a) $\lim_{n\to\infty} \frac{5n^2 - 1}{n^2 + 3n - 1000}$.

(b) $\lim_{n\to\infty} \frac{\sin n}{n}$.

Limit Equal to Infinity

We say that $\lim_{n\to\infty} a_n = \infty$ if for every real number M, there is an $N \in \mathbb{N}$ such that $n \geq N$ implies $a_n > M$. Similarly, we say $\lim_{n\to\infty} a_n = -\infty$ if for every real number M, there is an $N \in \mathbb{N}$ such that $n \geq N$ implies $a_n < M$.

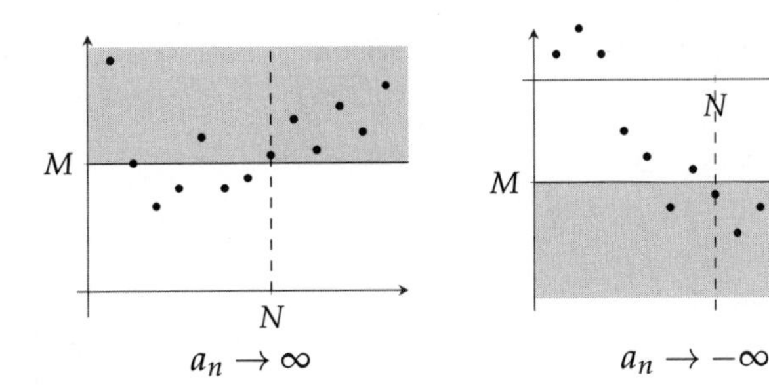

Example 7.1.15

We will show $2^n/n \to \infty$. We have

$$\frac{2^n}{n} = \frac{(1+1)^n}{n} > \frac{n(n-1)}{2n} = \frac{n-1}{2}.$$

So, for any given M, choose $N = 2M + 1$. □

Task 7.1.16

Prove the following:

(a) $\lim n = \infty$.

(b) *If* $a_n \geq b_n$ *for every* n*, and* $b_n \to \infty$*, then* $a_n \to \infty$.

(c) *Suppose* $a_n \neq 0$ *for every* n*. Then* $a_n \to 0$ *if and only if* $|1/a_n| \to \infty$.

Bounded Sequences

Consider a sequence (a_n). It is called

- **Bounded above** if there is a real number U such that $a_n \leq U$ for every n (U is called an **upper bound**),
- **Bounded below** if there is a real number L such that $a_n \geq L$ for every n (L is called a **lower bound**),
- **Bounded** if it is both bounded above and bounded below, and
- **Unbounded** if it is not bounded.

Task 7.1.17

For each given sequence, put a ✔ in each correct category and a ✖ in each incorrect category:

a_n	*Bounded Above*	*Bounded Below*	*Bounded*	*Unbounded*
n				
$-n$				
$(-1)^n$				
$(-1)^n n$				
$1/n$				

Theorem 7.1.18

Every convergent sequence is bounded.

If a sequence has a limit, the entries cluster around that value and cannot keep growing without bound.

Proof. Take $\epsilon = 1$. There will be an N such that $n \geq N$ implies $|a_n - L| < 1$ and so $L - 1 < a_n < L + 1$. In addition, the entries $a_1, \ldots, a_{N-1}$ are finitely many and have a maximum value M and a minimum value m. Then the entire sequence (a_n) lies between $\min\{m, L-1\}$ and $\max\{M, L+1\}$. ■

Monotone Sequences

Consider a sequence (a_n). It is called

- **Increasing** if $a_{n+1} \geq a_n$ for every n,
- **Decreasing** if $a_{n+1} \leq a_n$ for every n, and
- **Monotone** if it is either increasing or decreasing.

Task 7.1.19

For each given sequence, put a ✔ in each correct category and a ✖ in each incorrect category:

a_n	*Increasing*	*Decreasing*	*Monotone*
n			
$-n$			
$(-1)^n$			
1			
$1/n$			

Theorem 7.1.20 (Monotone Convergence Theorem)

Every bounded and monotone sequence is convergent.

Proof. Suppose (a_n) is increasing and bounded. We will show it converges to $L = \sup\{a_n : n \in \mathbb{N}\}$.

Consider any $\epsilon > 0$. Then $L - \epsilon$ is not an upper bound for $\{a_n : n \in \mathbb{N}\}$. Hence, there is $N \in \mathbb{N}$ such that $L - \epsilon < a_N \leq L$. This N works.

Similarly, if (a_n) is decreasing and bounded, it converges to $\inf\{a_n : n \in \mathbb{N}\}$. ■

Example 7.1.21

We offer another proof that $r^n \to 0$ if $|r| < 1$.

It is enough to show that $|r|^n \to 0$. Since $|r| < 1$, the sequence $|r|^n$ is a decreasing sequence, and it is bounded below by 0. So, it converges. Suppose it converges to L.

Now $|r|^{n+1}$ will have the same limit L. But $|r|^{n+1} = |r||r|^n \to |r|L$. This gives $L = |r|L$ and hence $L = 0$. □

Example 7.1.22

Let (a_n) be a decreasing sequence that converges to 0. We shall show that $2^{a_n} \to 1$. First, since (a_n) is decreasing, so is 2^{a_n}. Second, since $a_n \geq 0$, $2^{a_n} \geq 1$. Hence, $2^{a_n} \to L \geq 1$. To complete the proof, we need to show that 1 is the greatest lower bound of the set $\{2^{a_n}\}$. We already know it is a lower bound.

So, consider any number $1 + \epsilon$ with $\epsilon > 0$. Then $\log_2(1 + \epsilon) > 0$. Since $a_n \to 0$, we have an N such that $a_N < \log_2(1 + \epsilon)$. Hence, $2^{a_N} < 1 + \epsilon$. Therefore, $1 + \epsilon$ is not a lower bound for $\{2^{a_n}\}$. □

Example 7.1.23

Consider the sequence defined recursively by $a_1 = \sqrt{2}$ and $a_{n+1} = \sqrt{2a_n}$. We shall consider two approaches to investigate its limit. In the first approach, we try to obtain a direct formula for a_n. The first few terms are

$$\begin{aligned} a_1 &= 2^{1/2}, \\ a_2 &= \sqrt{2}\sqrt{a_1} = 2^{3/4}, \\ a_3 &= \sqrt{2}\sqrt{a_2} = 2^{7/8}. \end{aligned}$$

The pattern is $a_n = 2^{1-1/2^n}$. We leave it for you to verify this by mathematical induction. We can now calculate, using the previous example and the fact that $1/2^n \to 0$, that

$$\lim a_n = \lim 2^{1-1/2^n} = \frac{2}{\lim 2^{1/2^n}} = 2.$$

In the second approach, we try to establish whether the sequence is monotone and bounded. To see if the terms grow or shrink, consider the ratio of successive terms,

$$\frac{a_{n+1}}{a_n} = \sqrt{\frac{2}{a_n}}$$

We need to compare a_n with 2. Since the first few terms were less than 2, we conjecture that all are less than 2. This can be proven by mathematical induction:

$$\text{(a) } a_1 = \sqrt{2} < 2, \qquad \text{(b) } a_n < 2 \implies a_{n+1} = \sqrt{2a_n} < \sqrt{2 \times 2} = 2.$$

Hence, the sequence is increasing as well as bounded above (by the number 2). Therefore, it is convergent. Suppose it converges to L. From the defining relation $a_{n+1} = \sqrt{2a_n}$, we get $a_{n+1}^2 = 2a_n$ and hence $L^2 = 2L$. This implies $L = 0$ or 2. As the sequence has positive and increasing terms it cannot have 0 as a limit. Hence $L = 2$. ☐

This example shows that, sometimes, we can settle the issue of convergence of a sequence in the absence of a direct formula for its terms. An additional advantage to establishing monotonicity, and not merely computing the limit, is that it helps track how close we are getting to the limit. Once a monotone sequence gets within a distance ϵ from the limit, it never retreats.

Let us apply the same approach to estimating square roots.

Example 7.1.24

Consider Heron's method, the algorithm we created at the start of this chapter to estimate the square root of a number N. We start with an overestimate a_1 of $\sqrt{N}$. We then generate other terms by using the relation $a_{n+1} = \frac{1}{2}\left(a_n + \frac{N}{a_n}\right)$. Observe that for any x, y we have $(x+y)^2 \geq 4xy$, since $(x+y)^2 - 4xy = x^2 + y^2 - 2xy = (x-y)^2 \geq 0$. Hence,

$$a_{n+1}^2 = \frac{1}{4}\left(a_n + \frac{N}{a_n}\right)^2 \geq a_n\left(\frac{N}{a_n}\right) = N.$$

So, the sequence (a_n) is bounded below by $\sqrt{N}$. It follows that $\frac{N}{a_n} \leq a_n$ and hence $a_{n+1} \leq a_n$. Therefore, the sequence is monotone and converges to some L. From the relationship between a_{n+1} and a_n we obtain $L = \frac{1}{2}(L + N/L)$, and so $L^2 = N$. ☐

Subsequences

Given a sequence, a **subsequence** is created by dropping some of the terms of the sequence, as long as infinitely many terms still remain. For example, we may drop the first three terms, or all the terms indexed by even numbers. Consider a sequence whose first few terms are 1, 3, 2, 5, 4, 7, 6, 9, ... By dropping the first three terms we create the subsequence $5, 4, 7, 6, 9, \ldots$. Again, by dropping the terms in the even positions, we create the subsequence $1, 2, 4, 6, \ldots$.

Thus, consider a sequence (a_n). Let the first term that is retained be a_{n_1}. Let the second term that is retained be a_{n_2}, with $n_2 > n_1$. In this way we create a new sequence with terms $b_i = a_{n_i}$, and call it a subsequence of the original one.

Theorem 7.1.25

If a sequence converges to L, then each of its subsequences also converges to L.

Proof. Let $a_n \to L$. Consider a subsequence $b_k = a_{n_k}$ with $n_1 < n_2 < \cdots$. First, note that $n_k \geq k$. Now, for any $\epsilon > 0$, there is $N \in \mathbb{N}$ such that $n \geq N$ implies $|a_n - L| < \epsilon$. Then $k \geq N$ implies $n_k \geq k \geq N$ implies $|b_k - L| = |a_{n_k} - L| < \epsilon$. ■

It may happen that a sequence involves two or more different patterns. For example, the odd terms $a_1, a_3, \ldots$ may follow one rule while the even terms $a_2, a_4, \ldots$ follow another rule. The concept of subsequences helps in such situations.

Example 7.1.26

Consider $1, 1, 2, 1/2, 3, 1/3, 4, 1/4, \ldots$. The subsequence $1, 2, 3, 4, \ldots$ diverges and so the original sequence diverges.

Again, consider $1, -1, 1, -1, \ldots$. The subsequence $1, 1, \ldots$ converges to 1. The subsequence $-1, -1, \ldots$ converges to -1. Since the two subsequences have different limits, the original sequence diverges. □

Task 7.1.27

We have $\lim a_n = L$ *if and only if* $\lim a_{2n+1} = \lim a_{2n} = L$.

Task 7.1.28

Evaluate $\displaystyle\lim_{n\to\infty} \frac{(-1)^n n}{n+1}$.

Exercises for § 7.1

1. Prove that $\displaystyle\lim_{n\to\infty} \frac{1}{\sqrt{n}} = 0$.

2. Find the following limits:

(a) $\displaystyle\lim_{n\to\infty} \frac{(-1)^n}{n(n+1)}$.

(b) $\displaystyle\lim_{n\to\infty} \frac{n^2}{n+1}$.

(c) $\displaystyle\lim_{n\to\infty} (\sqrt{n+1} - \sqrt{n})$.

(d) $\displaystyle\lim_{n\to\infty} \frac{2^n}{n^2}$.

3. Find the following limits:

(a) $\lim_{n\to\infty} \dfrac{e^n - e^{-n}}{e^n + e^{-n}}$.

(b) $\lim_{n\to\infty} \dfrac{n!}{n^n}$.

(c) $\lim_{n\to\infty} \cos(n\pi/2)$.

(d) $\lim_{n\to\infty} n^{(-1)^n}$.

4. Prove that if $a_n \geq 0$ and $a_n \to 0$ then $a_n^{1/2} \to 0$.

5. Evaluate the following argument:

"Consider the sequence $a_n = (-1)^n$. Suppose $a_n \to L$. Then $a_{n+1} = -a_n \to -L$. However, we also have $a_{n+1} \to L$ since we only dropped the first term of (a_n). Therefore, $L = -L$, and so $L = 0$. Hence, the given sequence converges to 0."

6. If (a_n/b_n) converges and $b_n \to 0$, then $a_n \to 0$.

7. (Ratio Test) Let (a_n) be a sequence of non-zero numbers. Show that if $|a_{n+1}/a_n| \to L < 1$, then $a_n \to 0$. (Hint: Fix r such that $L < r < 1$. There will be an N so that $n \geq N$ implies $|a_{n+1}/a_n| < r$. Consider the sequence $|a_{N+n}|$.)

8. (Root Test) Show that if $|a_n|^{1/n} \to L < 1$, then $a_n \to 0$.

9. Prove that if $a_n \to L$, then $\bar{a}_n = \dfrac{a_1 + \cdots + a_n}{n} \to L$. Is the converse true? (Hint: $\bar{a}_{M+k} = \dfrac{M}{M+k}\bar{a}_M + \dfrac{k}{M+k}\dfrac{a_{M+1} + \cdots + a_{M+k}}{k}$.)

10. Prove that if $a_n \to 0$, then $a_n^n \to 0$. (Hint: For large enough n, $|a_n| < 1$.)

11. Show that the sequence defined by $a_1 = 2$, $a_{n+1} = \dfrac{1}{3 - a_n}$, satisfies $0 < a_n \leq 2$ and is decreasing. Deduce that the sequence is convergent and find its limit.

7.2 Sequences and Functions

Functions Applied to Sequences

Theorem 7.2.1

Let a real function f be continuous at $x = L$ and let $a_n \to L$. Then $f(a_n) \to f(L)$.

Proof. Take $\epsilon > 0$. First, by the continuity of f there is a $\delta > 0$ such that $|x - L| < \delta$ implies $|f(x) - f(L)| < \epsilon$. Next, by the convergence of (a_n) there is N such that $n \geq N$ implies $|a_n - L| < \delta$, and so $|f(a_n) - f(L)| < \epsilon$. ■

Example 7.2.2

Take a positive number c and consider the sequence $(c^{1/n})$. Now, the function $f(x) = c^x$ is continuous at every x. Hence,

$$\lim c^{1/n} = \lim f(1/n) = f(\lim 1/n) = f(0) = c^0 = 1.$$

□

Task 7.2.3

Show that $\log a_n \to L \implies a_n \to e^L$.

Another use of this theorem is to connect certain sequence limits with derivatives.

Theorem 7.2.4

Let $f(x)$ *be differentiable at* $x = L$. *Then*

$$\lim_{n\to\infty} n\Big(f(L+1/n) - f(L)\Big) = f'(L).$$

Proof. The function g defined below is continuous at $h = 0$.

$$g(h) = \begin{cases} \dfrac{f(L+h) - f(L)}{h} & \text{if } h \neq 0, \\ f'(L) & \text{if } h = 0. \end{cases}$$

Now, $n\Big(f(L+1/n) - f(L)\Big) = g(1/n)$. Hence,

$$\lim_{n\to\infty} n\Big(f(L+1/n) - f(L)\Big) = \lim_{n\to\infty} g(1/n) = g(0) = f'(L).$$ ■

Example 7.2.5

Consider the sequence $(1+1/n)^n$. First, we apply the log function to convert it into a product, which we can evaluate by the last theorem.

$$\lim \log(1+1/n)^n = \lim n\Big(\log(1+1/n) - \log 1\Big) = \log' 1 = 1.$$

And now, by the continuity of the exponential function,

$$\lim(1+1/n)^n = \lim e^{n\log(1+1/n)} = e^{\lim n\log(1+1/n)} = e^1 = e.$$ □

Task 7.2.6

Show that $\lim(1+2/n)^n = e^2$ *and* $\lim(1-1/n)^n = e^{-1}$.

Task 7.2.7

True or False: If $f(x) \to L$ *as* $x \to a$, *and* $a_n \to a$, *then* $f(a_n) \to L$.

Theorem 7.2.1 also gives a way to establish that a function is not continuous. If we find a sequence x_n such that $x_n \to a$ but $f(x_n) \not\to f(a)$ then f is not continuous at a. For example, consider $f(x) = \cos(\pi/x)$ if $x \neq 0$ and $f(0) = 0$. Now $1/n \to 0$ but $f(1/n) = (-1)^n$ diverges. So, f is not continuous at 0.

Task 7.2.8

Show that the function $\cos(1/x) + \sin(1/x)$ *cannot be assigned a value at* $x = 0$ *that makes it continuous.*

Sequences from Real Functions

Theorem 7.2.9

Let $f(x)$ be a real function with domain $[1,\infty)$ and let $\lim_{x\to\infty} f(x) = L$. Suppose $a_n = f(n)$ for $n \in \mathbb{N}$. Then $\lim_{n\to\infty} a_n = L$.

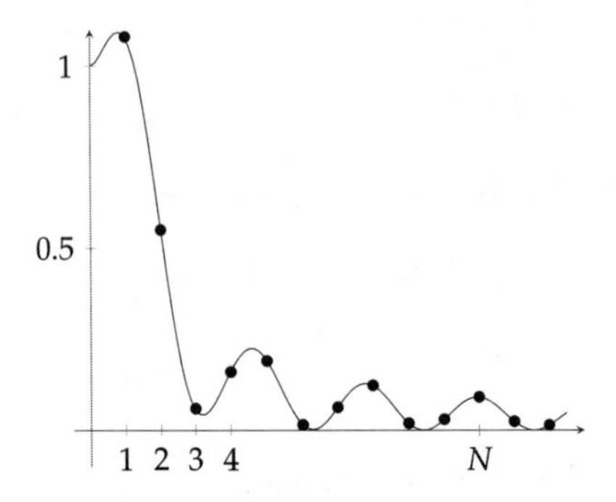

Proof. Consider any $\epsilon > 0$. There is a $c \in \mathbb{R}$ such that $x > c$ implies $|f(x) - L| < \epsilon$. Define $N = [c] + 1$. Then $n \geq N$ implies $|a_n - L| = |f(n) - L| < \epsilon$. ■

Example 7.2.10

We will calculate the limit of $a_n = n\sin(1/n)$. Consider $f(x) = x\sin(1/x)$. Then $f(n) = a_n$ and

$$\lim_{x\to\infty} f(x) = \lim_{x\to\infty} x\sin(1/x) = \lim_{y\to 0+} \frac{\sin y}{y} = 1.$$

Therefore, $\lim_{n\to\infty} a_n = 1$. □

Example 7.2.11

Consider the sequence $1/(\arctan n)^n$. First, we note that $\lim \arctan n = \lim_{x\to\infty} \arctan x = \pi/2$. Hence $1/(\arctan n) \to 2/\pi < 1$. Since we know that $r^n \to 0$ if $|r| < 1$, we suspect that $1/(\arctan n)^n \to 0$. However, the example of $(1 - 1/n)^n \to 1/e$ shows that we could have $0 < a_n < 1$ with $a_n^n \not\to 0$. More work is needed before this possibility can be confirmed.

Choose any real number r such that $2/\pi < r < 1$. There is an N such that $n \geq N$ implies $1/(\arctan n) < r$ and hence $0 < 1/(\arctan n)^n < r^n$. Now, $r^n \to 0$ and the sandwich theorem gives us $1/(\arctan n)^n \to 0$. □

A major gain from the last theorem is that the results for functions, such as L'Hôpital's rule, can be applied to sequence calculations.

Example 7.2.12

Consider the sequence $(n^{1/n})$. We start by applying log to convert to a ratio: $a_n = \log(n^{1/n}) = \dfrac{\log n}{n}$. Since $\lim_{x\to\infty} \dfrac{\log x}{x} = 0$, we have $\lim \dfrac{\log n}{n} = 0$. Hence, $\lim n^{1/n} = e^0 = 1$. ☐

Task 7.2.13

Find the limits of the following sequences:

(a) $\dfrac{e^n}{n^{100}}$.

(b) $\dfrac{\log n}{\sqrt{n}}$.

Stirling's Approximation

Our later study of "infinite series" will bring up the sequence $(n!)^{1/n}$. Let us plot it:

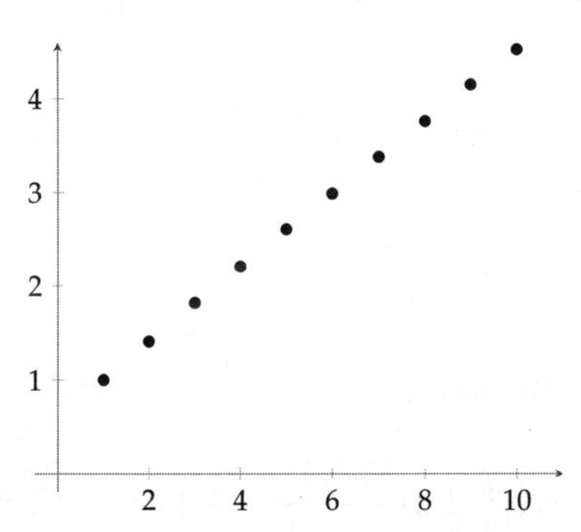

It looks very close to a straight line! Let us tabulate the slopes $(n!)^{1/n}/n$.

n	10	100	1000	10000
slope	0.453	0.3799	0.3695	0.3681
1/slope	2.12	2.63	2.706	2.717

The reciprocals could be approaching $e \approx 2.718\ldots$.

Theorem 7.2.14

$\lim_{n\to\infty} \dfrac{(n!)^{1/n}}{n/e} = 1$, *that is,* $(n!)^{1/n} \approx \dfrac{n}{e}$ *for large* n.

Proof. Consider $\log n! = \sum_{k=1}^{n} \log k$ as a lower/upper sum for integrals of $\log x$. The graphs drawn below show that $\sum_{k=1}^{n} \log k$ is an upper sum for $\int_1^n \log x\, dx$ and a lower sum for $\int_1^{n+1} \log x\, dx$.

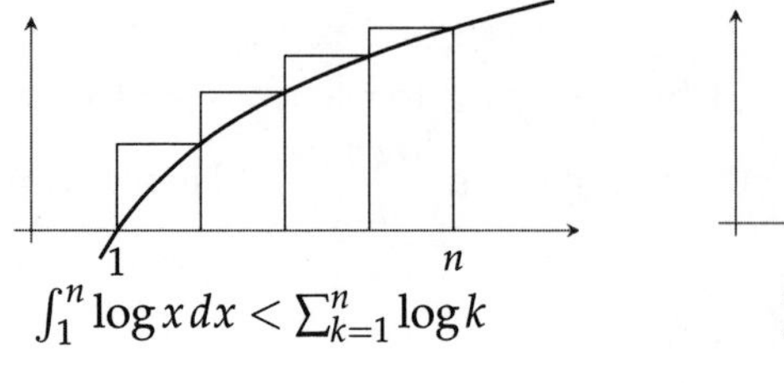

$\int_1^n \log x\,dx < \sum_{k=1}^n \log k$

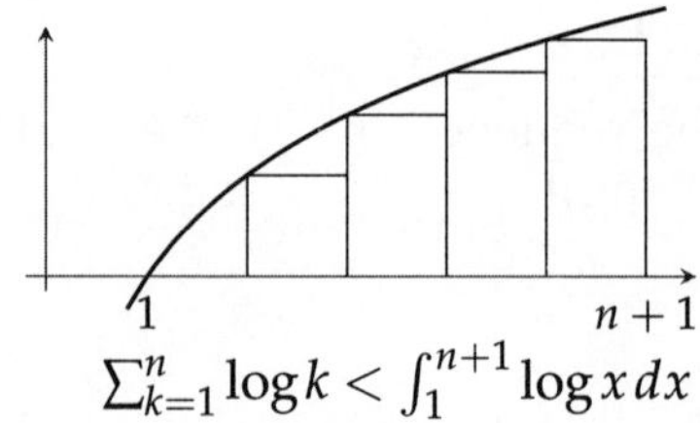

$\sum_{k=1}^n \log k < \int_1^{n+1} \log x\,dx$

$$\begin{aligned}
\int_1^n \log x\,dx < &\quad \log n! &< \int_1^{n+1} \log x\,dx \\
\implies n\log n - n + 1 < &\quad \log n! &< (n+1)\log(n+1) - n \\
\implies \log\left(\frac{n^n}{e^{n-1}}\right) < &\quad \log n! &< \log\left(\frac{(n+1)^{n+1}}{e^n}\right) \\
\implies \frac{n^n}{e^{n-1}} < &\quad n! &< \frac{(n+1)^{n+1}}{e^n} \\
\implies e^{1/n} < &\quad \frac{(n!)^{1/n}}{n/e} &< (1+1/n)(n+1)^{1/n}.
\end{aligned}$$

Clearly, $e^{1/n} \to 1$ and $1 + \frac{1}{n} \to 1$. Further,

$$\begin{aligned}
\lim_{x\to\infty} (1+x)^{1/x} = \lim_{x\to\infty} \exp\left(\frac{\log(1+x)}{x}\right) &= \exp\left(\lim_{x\to\infty} \frac{\log(1+x)}{x}\right) \\
= \exp\left(\lim_{x\to\infty} \frac{1/(1+x)}{1}\right) &= \exp 0 = 1.
\end{aligned}$$

■

Example 7.2.15

Here is an application of Stirling's approximation.

$$(n!)^{1/n^2} = \left(\frac{n!^{1/n}}{n/e}\right)^{1/n} \left(\frac{n}{e}\right)^{1/n} = \left(\frac{n!^{1/n}}{n/e}\right)^{1/n} \frac{n^{1/n}}{e^{1/n}}.$$

Now $a_n = \dfrac{n!^{1/n}}{n/e} \to 1$. Hence, $a_n^{1/n} \to 1$ (To prove this, apply log). We already know that $n^{1/n} \to 1$ and $e^{1/n} \to 1$. Hence, $(n!)^{1/n^2} \to 1$. □

The fact that $n!^{1/n}$ behaves like n/e does not imply that $n!$ behaves like $(n/e)^n$. In fact, another approximation due to Stirling addresses this:

$$n! \approx \left(\frac{n}{e}\right)^n \sqrt{2\pi n}.$$

This improved version is called **Stirling's formula**.

n	Factorial	Stirling's Formula
10	3.628×10^6	3.599×10^6
100	9.332×10^{157}	9.325×10^{157}
1000	4.0239×10^{2567}	4.0235×10^{2567}

We will not prove this formula here, as we do not have an immediate need for it. A proof is outlined in the supplementary exercises following this chapter. Stirling's formula is very useful in probability and its applications in physics and chemistry.

Newton–Raphson Method

The Newton–Raphson method uses differentiation to create a sequence of estimates for a solution of an equation. Suppose we have an equation such as $x^3 - 3 = x^2 + x$ that we have to solve for x. We move every term to the left side to put it in the form $f(x) = 0$. Now, suppose the function $f(x)$ is continuously differentiable. We visualize the situation as follows:

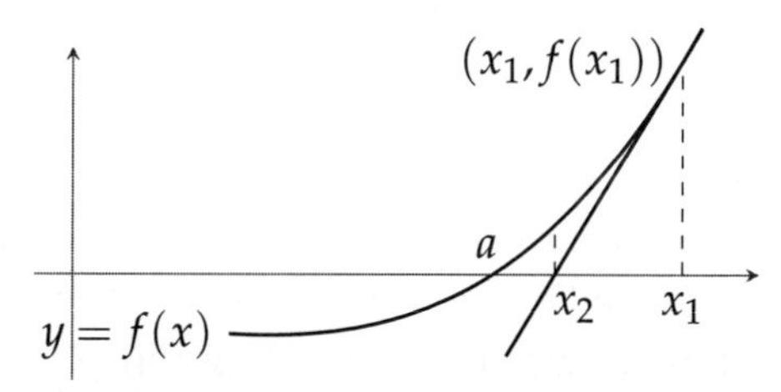

We wish to estimate the point a where $f(a) = 0$. Imagine you are at a point $(x_1, f(x_1))$ on the graph of the function f. In which direction should you move to move towards a? One idea is to generate the tangent line at $(x_1, f(x_1))$ and see where it cuts the x-axis. If it does so at x_2, we repeat the process from the point $(x_2, f(x_2))$.

The equation of the tangent line at $(x_1, f(x_1))$ is $y = f(x_1) + f'(x_1)(x - x_1)$. To see where it cuts the x-axis we put $y = 0$. This gives $x_2 = x_1 - f(x_1)/f'(x_1)$. Therefore, the general iterative step is

$$x_{n+1} = x_n - \frac{f(x_n)}{f'(x_n)}.$$

We have created a sequence $x_1, x_2, x_3, \ldots$. Does it converge? And does it converge to a? The good news is that if it converges at all, then it converges to a solution. If $x_n \to L$, then

$$x_{n+1} = x_n - \frac{f(x_n)}{f'(x_n)} \implies L = L - \frac{f(L)}{f'(L)} \implies f(L) = 0.$$

For the sequence to be generated, we must never have an x_n such that $f'(x_n) = 0$. And for the argument given above to work, we also need $f'(L) \neq 0$.

Example 7.2.16

Suppose the equation we wish to solve is $x^2 = N$, that is, we want to estimate a square root. We rearrange this as $f(x) = 0$, where $f(x) = x^2 - N$. Then $f'(x) = 2x$, and the Newton–Raphson iteration is

$$x_{n+1} = x_n - \frac{x_n^2 - N}{2x_n} = \frac{x_n^2 + N}{2x_n} = \frac{1}{2}\left(x_n + \frac{N}{x_n}\right).$$

We have again been led to Heron's method! □

On the other hand, the Newton–Raphson sequence may not converge at all. Here are two examples:

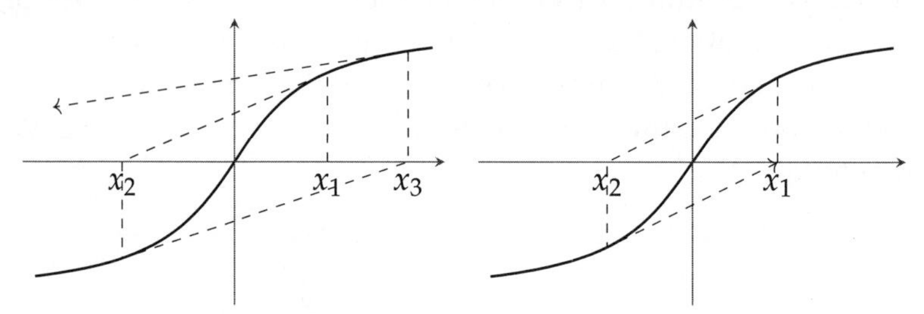

In the first example, the sequence moves away from the solution of $f(x) = 0$. In the second example, it cycles repeatedly through the same two values.

On comparing our initial diagram for the Newton–Raphson method with the examples of failure shown above, we can formulate a sufficient condition for convergence.

Theorem 7.2.17

Suppose f is continuously differentiable on (a,b), f', and f'' are never zero on (a,b) and there is a $c \in (a,b)$ such that $f(c) = 0$. Then there is a $\delta > 0$ such that the Newton–Raphson method applied to any initial point $x_1 \in (c-\delta, c+\delta)$ generates a sequence x_n that converges to c.

Proof. By Darboux's theorem, f' and f'' do not change sign in (a,b). Hence, f is strictly monotonic, c is the only zero of f, and the convexity does not change. So, we have the following cases:

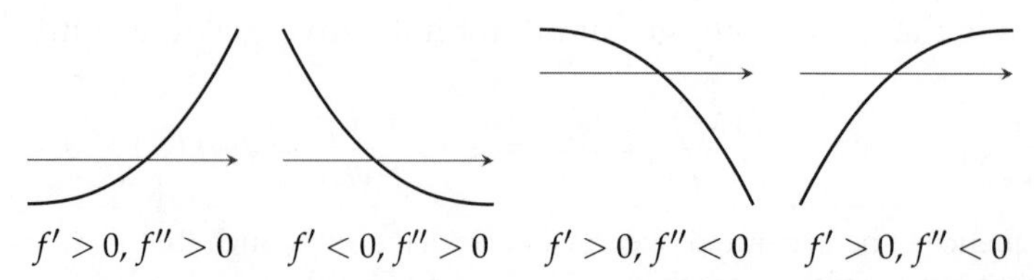

These situations can be converted into each other by horizontal and vertical reflections, which do not affect the convergence of the Newton–Raphson method. Therefore, we can assume that $f', f'' > 0$ throughout (a,b).

Suppose $x_1 \in [c,b)$. Then $f(x_1) \geq 0$ and $x_2 = x_1 - f(x_1)/f'(x_1) \leq x_1$. Further, the equation of the tangent line $y = \ell(x)$ at $(x_1, f(x_1))$ is $\ell(x) = f(x_1) + f'(x_1)(x - x_1)$. By the remainder theorem, we have a ξ between x and x_1 such that $f(x) - \ell(x) = f''(\xi)(x - x_1)^2$. Hence, $f(x) - \ell(x) \geq 0$ and the tangent line at $(x_1, f(x_1))$ lies below the graph of f. Therefore, it cuts the x-axis to the right of c. Therefore, $x_2 \in [c,b)$. It follows that the Newton–Raphson sequence (x_n) is decreasing and bounded below by c. Hence, it converges. By our earlier argument, it converges to c.

If $x_1 \in (a,c)$, then the above argument again gives $x_2 \geq c$ and then the previous case applies, provided that $x_2 < b$. The last is not guaranteed, but by the continuity of $x - f(x)/f'(x)$ it will happen for x_1 close to c. ■

Example 7.2.18

The problem of finding the cube root of 2 can be framed as solving $f(x) = 0$ with $f(x) = x^3 - 2$. We can see that f satisfies the hypothesis of the Theorem 7.2.17 on $(0,\infty)$. By comparing with the proof of the theorem, we see that the Newton–Raphson method will converge for any $x_1 \in (0,\infty)$. Here is an instance:

$$x_1 = 1, \quad x_2 = 4/3, \quad x_3 = 91/72 = 1.26, \quad x_4 = 1.2599.$$

The last iterate is accurate to 4 decimal places. And $x_5 = 1.25992105$ is accurate to 8 decimal places! This accelerating convergence is a special trait of Newton–Raphson that makes it popular. □

Exercises for § 7.2

1. Find the following limits:

(a) $\lim_{n\to\infty} \cos(\sqrt{n+1} - \sqrt{n})$.

(b) $\lim_{n\to\infty} \arctan(n^{1/n})$.

(c) $\lim_{n\to\infty} (2n)^{1/n}$.

(d) $\lim_{n\to\infty} \dfrac{\log n}{n^p} \quad (p > 0)$.

(e) $\lim_{n\to\infty} \dfrac{(\log n)^p}{n} \quad (p > 0)$.

(f) $\lim_{n\to\infty} \dfrac{e^n}{n^p} \quad (p > 0)$.

2. Prove that $e^r = \lim_{n\to\infty} (1 + r/n)^n$.

3. Let $a_n \to 1$. Give examples to show that we can still have any of the following: $a_n^n \to 1$, $a_n^n \to e$, $a_n^n \to 0$, $a_n^n \to \infty$.

4. Find the limits of the given sequences:

(a) $n\Big(\cos(\pi/4 + 1/n) - \cos(\pi/4)\Big)$.

(b) $n\Big(\arctan(1 + 1/n) - \pi/4\Big)$.

(c) $n\Big(a^{1/n} - 1\Big)$.

(d) $n\Big((x + 1/n)^r - x^r\Big)$.

5. Evaluate the limits of the given sequences by viewing them as upper/lower sums for an integral of a monotonic function.

(a) $\dfrac{1}{n+1}+\dfrac{1}{n+2}+\cdots+\dfrac{1}{2n}$.

(b) $\dfrac{n}{(n+1)^2}+\dfrac{n}{(n+2)^2}+\cdots+\dfrac{n}{(2n)^2}$.

6. Use Stirling's approximation to find the limits of the given sequences:

(a) $\left(\dfrac{n!}{n^{n+1}}\right)^{1/n}$. (b) $\left(\dfrac{(2n)!}{(n!)^2}\right)^{1/n}$.

7. Suppose $f\colon [0,1] \to [0,1]$ is continuous and increasing. Choose any $x_1 \in [0,1]$. Define a sequence by $x_{n+1} = f(x_n)$ for $n \geq 1$. Show that this sequence is monotonic and converges to a fixed point of f.

8. Use the intermediate value theorem to show that the equation $\cos x = x$ has a solution in the interval $[0,\pi/2]$. Use the Newton–Raphson method with initial point $x_1 = \pi/4$ to approximate the solution to 4 decimal places.

9. Consider the equation $e^x - x - 2 = 0$.

(a) Show that it has exactly two solutions, one positive and one negative.

(b) Let (x_n) be a sequence obtained by the Newton–Raphson method. Show that it converges to the positive solution if $x_1 > 0$ and to the negative solution if $x_1 < 0$.

10. Consider the equation $\arctan x = 0$.

(a) Use the intermediate value theorem to show there is a $c \in (1,2)$ such that the Newton–Raphson sequence for this equation starting at $x_1 = c$ is $c,-c,c,-c,\ldots$.

(b) Use the Newton–Raphson method to show that $c \approx 1.392$.

7.3 Sum of a Series

Consider a sequence $a_1,a_2,a_3,\ldots$. The corresponding expression

$$a_1 + a_2 + a_3 + \cdots \qquad \text{or} \qquad \sum_{n=1}^{\infty} a_n$$

is called a **series**. The very notation raises a question: How can we add infinitely many numbers?

Example 7.3.1

Consider the geometric progression $1, \frac{1}{2}, \frac{1}{4}, \frac{1}{8}, \frac{1}{16}, \ldots$. The corresponding series is $1+\frac{1}{2}+\frac{1}{4}+\frac{1}{8}+\frac{1}{16}+\cdots$ or $\sum\limits_{n=1}^{\infty}\frac{1}{2^{n-1}}$. This is called a **geometric series**.

We cannot add *all* these numbers in one go, but what happens if we add them one by one?

$$1+\frac{1}{2}=\frac{3}{2}, \qquad 1+\frac{1}{2}+\frac{1}{4}=\frac{7}{4}, \qquad 1+\frac{1}{2}+\frac{1}{4}+\frac{1}{8}=\frac{15}{8}.$$

In general, recalling that $1+x+\cdots+x^n=\dfrac{1-x^{n+1}}{1-x}$ for $x\neq 1$,

$$1+\frac{1}{2}+\frac{1}{4}+\cdots+\frac{1}{2^n}=\frac{2^{n+1}-1}{2^n}=2-\frac{1}{2^n}.$$

We see that the sums are converging to 2. We choose to take this 2 as the overall sum of all the terms of this geometric series. □

Consider a series $\sum_{n=1}^{\infty} a_n$. Its **partial sums** are

$$\begin{aligned} S_1 &= a_1 \\ S_2 &= a_1+a_2 \\ S_3 &= a_1+a_2+a_3 \\ &\vdots \\ S_n &= a_1+\cdots+a_n=\sum_{i=1}^{n} a_i \end{aligned}$$

We are interested in the limit of the partial sums. If $s=\lim_{n\to\infty} S_n$ exists, we call this limit the **sum of the series** and we write $\sum_{n=1}^{\infty} a_n = s$. The series is called

- **Convergent** if $\lim_{n\to\infty} S_n$ exists,
- **Divergent** if $\lim_{n\to\infty} S_n$ does not exist.

Example 7.3.2

Following on from the example of $\sum_{n=1}^{\infty} 1/2^{n-1}$, let us consider the general geometric series $\sum_{n=1}^{\infty} r^{n-1}$. If $r=1$, the partial sums are $S_n=n$ and the series diverges. If $r\neq 1$, the partial sums are

$$S_n=\sum_{i=1}^{n} r^{i-1}=\frac{1-r^n}{1-r}.$$

We know that r^n converges to 0 if $|r|<1$ and diverges if $r=-1$ or $|r|>1$. Hence, $\sum_{n=1}^{\infty} r^{n-1}=\dfrac{1}{1-r}$ if $|r|<1$ and diverges if $|r|\geq 1$. □

Here are some examples of divergence:

- $\sum_{n=1}^{\infty} 1$: $S_n=1+\cdots+1=n\to\infty$.

- $\sum_{n=1}^{\infty}(-1)^n$: $S_n = \begin{cases} 0 & \text{if } n \text{ even,} \\ -1 & \text{if } n \text{ odd,} \end{cases}$ diverges.
- $\sum_{n=1}^{\infty}\frac{1}{\sqrt{n}}$: $S_n = 1 + \frac{1}{\sqrt{2}} + \cdots + \frac{1}{\sqrt{n}} \geq \frac{n}{\sqrt{n}} = \sqrt{n} \to \infty$.

The last of these is the most important. It shows that terms that are decreasing to zero can still accumulate in an unbounded way.

Example 7.3.3

Consider $\sum_{n=1}^{\infty}\frac{1}{n(n+1)}$. We have the partial fraction decomposition $\frac{1}{n(n+1)} = \frac{1}{n} - \frac{1}{n+1}$. Hence,

$$S_n = \left(1 - \frac{1}{2}\right) + \left(\frac{1}{2} - \frac{1}{3}\right) + \cdots + \left(\frac{1}{n} - \frac{1}{n+1}\right) = 1 - \frac{1}{n+1} \to 1.$$

Therefore, $\sum_{n=1}^{\infty}\frac{1}{n(n+1)} = 1$. This is called a **telescoping series** due to the cancellations in the partial sums. □

Example 7.3.4

$\sum_{n=1}^{\infty} 1/n$ is called the **harmonic series**. Here are some of its partial sums:

$$S_1 = 1, \quad S_2 = 1.5, \quad S_4 = 2.1, \quad S_8 = 2.7, \quad S_{16} = 3.4.$$

We see that while the growth of the partial sums is slowing down, there seems to be an increase of at least $1/2$ every time the number of terms doubles. First, we verify this:

$$S_{2n} - S_n = \frac{1}{n+1} + \cdots + \frac{1}{2n} > \frac{n}{2n} = \frac{1}{2}.$$

Now suppose the partial sums S_n converge to S. Letting $n \to \infty$ gives $S - S \geq 1/2$, a contradiction! So the harmonic series diverges. □

Task 7.3.5

Show that $\sum_{k=1}^{n} 1/k > \int_1^{n+1} dx/x = \log(n+1)$. *Use this observation for another proof that the harmonic series diverges.*

Theorem 7.3.6 (Algebra of Series)

Consider convergent series $\sum_{n=1}^{\infty} a_n = L$ *and* $\sum_{n=1}^{\infty} b_n = M$. *Then we have the following:*

1. *For any* $c \in \mathbb{R}$, $\sum_{n=1}^{\infty}(c a_n) = cL$.

2. $\sum_{n=1}^{\infty}(a_n + b_n) = L + M$.

3. $\sum_{n=1}^{\infty}(a_n - b_n) = L - M$.

Proof. Apply the algebra of limits to the partial sums.

$$\sum_{n=1}^{\infty}(c a_n) = \lim_{N\to\infty}\sum_{n=1}^{N}(c a_n) = \lim_{N\to\infty}\left(c\sum_{n=1}^{N} a_n\right) = c\left(\lim_{N\to\infty}\sum_{n=1}^{N} a_n\right) = c\sum_{n=1}^{\infty} a_n.$$

The other two statements are left as an exercise. ■

Task 7.3.7

Do the following series converge?

(a) $\sum_{n=1}^{\infty}\frac{2^n}{3^{n+1}}$. (b) $\sum_{n=1}^{\infty}\frac{2^n + 3^n}{5^n}$. (c) $\sum_{n=1}^{\infty}\left(\frac{1}{n} - \frac{1}{2^n}\right)$.

The notation $\sum_{n=k}^{\infty} a_n$ refers to the expression $a_k + a_{k+1} + \cdots$. This is also a series, with first term $b_1 = a_k$, second term $b_2 = a_{k+1}$, and so on. Given an initial series $\sum_{n=1}^{\infty} a_n$, a series of the form $\sum_{n=k}^{\infty} a_n$ is called its **tail**.

Task 7.3.8

Show that $\sum_{n=1}^{\infty} a_n$ *converges if and only if the* ***tail*** $\sum_{n=k}^{\infty} a_n$ *converges. Further,*

$$\sum_{n=1}^{\infty} a_n = S_{k-1} + \sum_{n=k}^{\infty} a_n.$$

When we are only discussing convergence and not the actual sum, we can drop the range of the index and just write $\sum_n a_n$, or even $\sum a_n$.

Theorem 7.3.9 (Divergence Test)

If $\sum a_n$ *is a convergent series, then* $a_n \to 0$.

Proof. Let $\sum_{n=1}^{\infty} a_n = L$. Consider the partial sums $S_n = a_1 + \cdots + a_n$. We have $a_n = S_n - S_{n-1}$ for $n \geq 2$. Hence, $\lim_{n\to\infty} a_n = \lim_{n\to\infty} S_n - \lim_{n\to\infty} S_{n-1} = L - L = 0$. ■

Thus, if $a_n \not\to 0$, then $\sum a_n$ certainly diverges. However, if $a_n \to 0$ then we do not learn anything. The series may converge (for example, $\sum 1/2^n$ and $\sum 1/n(n+1)$) or diverge (for example, $\sum 1/n$ and $\sum 1/\sqrt{n}$).

Task 7.3.10

Show that the following series diverge:

(a) $\sum_{n=1}^{\infty} \sin(n\pi/2)$. (b) $\sum_{n=1}^{\infty} (-1)^n \frac{n-1}{n}$.

When we look at the partial sums of a series with an eye to their convergence, we realize that we will have a much easier time if all the terms of the series are non-negative. For then the partial sums form an increasing sequence, and we just have to check if they are bounded above. If they are bounded, they converge. If they are not bounded, they diverge. For this reason all our important convergence tests, bar one, will be about series with non-negative terms.

Theorem 7.3.11 (Comparison Test)

Let $\sum a_n$ and $\sum b_n$ satisfy $0 \le a_n \le b_n$ for every n. Then

1. *If $\sum b_n$ converges, so does $\sum a_n$.*
2. *If $\sum a_n$ diverges, so does $\sum b_n$.*

Proof. Consider the partial sums $S_n = \sum_{i=1}^{n} a_i$ and $T_n = \sum_{i=1}^{n} b_i$. The non-negativity of the terms implies that (S_n) and (T_n) are increasing sequences.

First, suppose that $\sum b_n$ is convergent. Then $T_n \to T = \sup\{T_n\}$. Now $S_n \le T_n \le T$ for each n, so (S_n) is increasing and bounded above. Hence, it is convergent.

The second claim is just a rewording of the first. ∎

Example 7.3.12

Consider the series $\sum_{n=1}^{\infty} 1/n^2$. To carry out a comparison, we need to think of a series whose convergence or divergence is known. For example, we might try $\sum 1/n$. We have $0 < 1/n^2 \le 1/n$. If $\sum 1/n$ were convergent, the comparison test would give us the convergence of $\sum 1/n^2$. Unfortunately, $\sum 1/n$ is divergent and so we learn nothing about $\sum 1/n^2$.

Now let us consider the telescoping series $\sum_{n=1}^{\infty} 1/n(n+1)$, which converges. We can be more optimistic about this choice because it has a closer resemblance to $\sum 1/n^2$. As n increases, the contribution of 1 becomes tiny compared to that of n and so we expect $\sum 1/n^2$ and $\sum 1/n(n+1)$ to have the same behavior. Now let us expand these two series:

$$\sum_{n=1}^{\infty} \frac{1}{n^2} = 1 + \frac{1}{4} + \frac{1}{9} + \cdots,$$

$$\sum_{n=1}^{\infty} \frac{1}{n(n+1)} = \frac{1}{2} + \frac{1}{6} + \frac{1}{12} + \cdots.$$

A direct comparison again fails to help. We have $0 \le \frac{1}{n(n+1)} \le \frac{1}{n^2}$, and the convergence of the series with smaller terms says nothing about the behavior of the series with larger terms. However, in this case, we have a way out. Suppose we drop the first term of the $\sum 1/n^2$ series, and also shift the index of the telescoping series:

$$\sum_{n=2}^{\infty} \frac{1}{n^2} = \frac{1}{4} + \frac{1}{9} + \frac{1}{16} \cdots,$$

$$\sum_{n=2}^{\infty} \frac{1}{n(n-1)} = \frac{1}{2} + \frac{1}{6} + \frac{1}{12} + \cdots.$$

Now the inequalities reverse: $\frac{1}{(n-1)n} \ge \frac{1}{n^2} > 0$. The comparison test applies and implies that $\sum_{n=2}^{\infty} 1/n^2$ converges. Hence, $\sum_{n=1}^{\infty} 1/n^2$ converges. □

Example 7.3.13

Consider $\sum_{n=1}^{\infty} 1/n^{3/4}$. Let us compare with all the series of form $\sum 1/n^p$ whose behavior is known.

$$0 < \underbrace{\frac{1}{n^2}}_{\sum \text{ converges}} \le \underbrace{\frac{1}{n}}_{\sum \text{ diverges}} \le \frac{1}{n^{3/4}} \le \underbrace{\frac{1}{\sqrt{n}}}_{\sum \text{ diverges}}.$$

The comparison with $\sum 1/n^2$ does not help because it features a smaller converging series. The one with $\sum 1/\sqrt{n}$ does not help because it features a larger diverging series. But the comparison with $\sum 1/n$ produces a useful combination: a smaller series that diverges. Hence, $\sum_{n=1}^{\infty} 1/n^{3/4}$ diverges. □

Example 7.3.14

Consider $\sum_{n=1}^{\infty} \frac{1}{2^n + 3^n}$. We know $\sum_{n=1}^{\infty} \frac{1}{2^n}$ converges. And $0 < \frac{1}{2^n + 3^n} \le \frac{1}{2^n}$. So $\sum_{n=1}^{\infty} \frac{1}{2^n + 3^n}$ converges. □

Theorem 7.3.15 (Limit Comparison Test)

Let $\sum a_n$ and $\sum b_n$ be series whose terms are all positive, and suppose

$$\lim_{n\to\infty} \frac{a_n}{b_n} = c \ne 0.$$

Then $\sum b_n$ converges if and only if $\sum a_n$ converges.

If $a_n/b_n \to c$, then in the long run a_n is essentially cb_n and so should have the same behavior.

Proof. Take two positive numbers m, M such that $m < c < M$. There is $N \in \mathbb{N}$ such that $m < \dfrac{a_n}{b_n} < M$ for every $n \geq N$. So, $m b_n < a_n$ and $M b_n > a_n$ for every $n \geq N$. Now apply the comparison test.

If $\sum b_n$ converges, then $\sum (M b_n)$ converges, so $\sum a_n$ converges.

If $\sum b_n$ diverges, then $\sum (m b_n)$ diverges, so $\sum a_n$ diverges. ■

Example 7.3.16

Consider the series $\sum_{n=1}^{\infty} \dfrac{\pi}{n^2+4n+3}$. The fastest growing term is the n^2 in the denominator. So, let us compare with $\sum_{n=1}^{\infty} \dfrac{1}{n^2}$:

$$\lim_{n\to\infty} \frac{\pi/(n^2+4n+3)}{1/n^2} = \lim_{n\to\infty} \frac{\pi n^2}{n^2+4n+3} = \pi.$$

Hence, both series have the same convergence behavior. Therefore, $\sum_{n=1}^{\infty} \dfrac{\pi}{n^2+4n+3}$ converges. □

Example 7.3.17

Consider $\sum_{n=1}^{\infty} \dfrac{2^n}{3^n - 2^n}$. Compare with $\sum_{n=1}^{\infty} \left(\dfrac{2}{3}\right)^n$:

$$\lim_{n\to\infty} \frac{2^n/(3^n-2^n)}{2^n/3^n} = \lim_{n\to\infty} \frac{3^n}{3^n-2^n} = \lim_{n\to\infty} \frac{1}{1-(2/3)^n} = 1.$$

So $\sum_{n=1}^{\infty} \dfrac{2^n}{3^n-2^n}$ converges. □

Theorem 7.3.18 (Integral Test)

Consider a series $\sum_{n=k}^{\infty} a_n$ *whose terms can be expressed as* $a_n = f(n)$, *where* $f\colon [k,\infty) \to \mathbb{R}$ *is positive and decreasing. Then* $\sum_{n=k}^{\infty} a_n$ *converges if and only if the improper integral* $\int_k^{\infty} f(x)\,dx$ *converges.*

Proof. We may assume $k = 1$. Since f is decreasing it is integrable on every closed interval $[1,b]$ with $b > 1$. The partial sums of $\sum_{n=k}^{\infty} a_n$ give upper and lower sums for the integral of f over $[1,n]$:

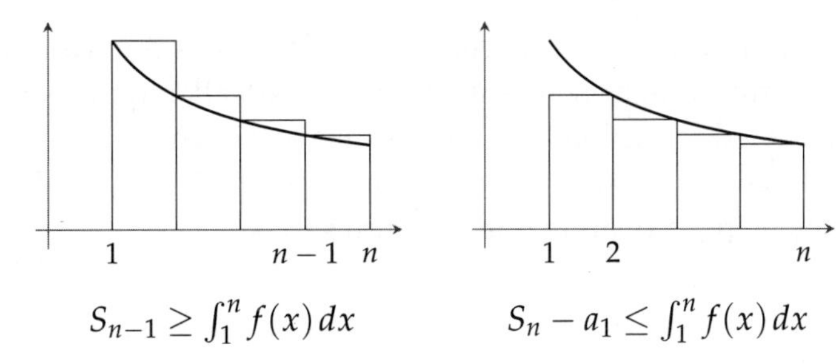

$S_{n-1} \geq \int_1^n f(x)\,dx$ $\qquad$ $S_n - a_1 \leq \int_1^n f(x)\,dx$

First, suppose $\int_1^\infty f(x)\,dx$ diverges. Then

$$S_{n-1} = \sum_{i=1}^{n-1} a_i \cdot 1 \geq \int_1^n f(x)\,dx \to \infty \implies S_n \to \infty.$$

Next, suppose $\int_1^\infty f(x)\,dx$ converges.

1. $a_n = f(n) > 0$ implies the partial sums (S_n) are increasing.
2. $S_n = a_1 + \sum_{i=2}^{n} a_i \cdot 1 \leq a_1 + \int_1^n f(x)\,dx \leq a_1 + \int_1^\infty f(x)\,dx.$

Since (S_n) is increasing and bounded above, it is convergent. ∎

Example 7.3.19

Consider $\sum_{n=2}^{\infty} \frac{\log n}{n^2}$. The corresponding function $f(x) = \frac{\log x}{x^2}$ is positive on $[2,\infty)$. Is it decreasing?

$$f'(x) = \frac{1 - 2\log x}{x^3} \implies f'(x) \leq 0 \text{ for } x > \sqrt{e} = 1.6\ldots.$$

So $f(x)$ is decreasing on $[2,\infty)$. Apply the integral test:

$$\int_2^\infty \frac{\log x}{x^2}\,dx = \frac{1}{2}(1 + \log 2) \implies \sum_{n=2}^{\infty} \frac{\log n}{n^2} \text{ converges.}$$ □

Theorem 7.3.20 (The p-Series Test)

*The **p-series** $\sum_{n=1}^{\infty} \frac{1}{n^p}$ converges if and only if $p > 1$.*

Proof. If $p \leq 0$ the terms are greater than or equal to 1, so the series diverges. For $p > 0$, the function $\frac{1}{x^p}$ is positive and decreasing. And

$$\int_1^\infty \frac{1}{x^p}\,dx = \begin{cases} \dfrac{1}{p-1} & \text{if } p > 1, \\ \infty & \text{if } 0 < p \leq 1. \end{cases}$$ ∎

We close this section with the observation that the significant results so far (comparison test, limit comparison test, integral test) all require series with non-negative terms. We need to explore techniques that work when the terms have mixed signs, and we also need to explore the extent to which knowledge about series with non-negative terms helps in dealing with general series.

Exercises for § 7.3

1. Show that the following series converge:

(a) $\sum_{n=1}^{\infty} \frac{1}{3^n+2}$.

(b) $\sum_{n=1}^{\infty} \frac{1}{3^n-2}$.

(c) $\sum_{n=1}^{\infty} \frac{n}{2^n}$.

(d) $\sum_{n=1}^{\infty} \frac{n^{10}}{n!}$.

2. Express the partial sums in telescoping form to show that $\sum_{n=1}^{\infty} \frac{1}{n(n+2)}$ converges and find its sum.

3. Show that the following series diverge:

(a) $\sum_{n=1}^{\infty} \frac{3}{2+n}$.

(b) $\sum_{n=1}^{\infty} \left(\frac{1}{n} - \frac{1}{n^2}\right)$.

(c) $\sum_{n=2}^{\infty} \frac{1}{\log n}$.

(d) $1 + \frac{1}{3} + \frac{1}{5} + \frac{1}{7} + \cdots$.

(e) $\sum_{n=1}^{\infty} (\sqrt{n+1} - \sqrt{n})$.

(f) $\sum_{n=1}^{\infty} \cos(1/n)$.

4. Determine the convergence or divergence of the following series:

(a) $\sum \frac{n^2}{n^3+n-5}$.

(b) $\sum \frac{1}{n \log n}$.

(c) $\sum \frac{1}{(\log n)^k} \quad (k > 1)$.

(d) $\sum \frac{n!}{n^n}$.

(e) $\sum \sin \frac{1}{n}$.

(f) $\sum \sqrt{n} \sin \frac{1}{n^2}$.

(g) $\sum \frac{\log n}{n^k} \quad (k > 1)$.

(h) $\sum \frac{n^2}{e^n}$.

(i) $\sum \frac{1}{n^{1+1/n}}$.

(j) $\sum \log\left(n \sin \frac{1}{n}\right)$.

5. For which values of p does $\sum_{n=1}^{\infty} \frac{\sqrt{n+1}-\sqrt{n}}{n^p}$ converge?

6. Prove that if $a_n \geq 0$ and $\sum_{n=1}^{\infty} a_n$ converges, then $\sum_{n=1}^{\infty} a_n^2$ converges.

7. Suppose $\sum a_n$ converges and $\sum b_n$ diverges. Show that $\sum(a_n + b_n)$ diverges.

8. State whether true or false. Justify your answer.

(a) If $\sum a_n$ converges, then $\sum(a_n + a_{n+1})$ converges.

(b) If $\sum(a_n + a_{n+1})$ converges, then $\sum a_n$ converges.

9. Suppose $\sum a_n$ and $\sum b_n$ converge and $a_n, b_n \geq 0$. Show that $\sum a_n b_n$ converges.

10. Let $\sum a_n$ and $\sum b_n$ be series whose terms are all positive and $a_n/b_n \to 0$. Show that if $\sum b_n$ converges, then $\sum a_n$ converges.

11. Let the terms of $\sum a_n$ be positive and decreasing. Show that if $\sum a_n$ converges, then $na_n \to 0$. (Hint: Show $b_n = a_{n+1} + a_{n+2} + \cdots + a_{2n} \to 0$.)

12. (Cauchy Condensation Test) Let (a_n) be a positive decreasing sequence. Prove that $\sum a_n$ converges if and only if $\sum 2^n a_{2^n}$ converges. (Hint: Group the terms as $a_1 + (a_2 + a_3) + (a_4 + \cdots + a_7) + \cdots$. Alternative Hint: Apply the integral test and do a suitable substitution.)

13. Show that $\int_1^{\infty} \frac{e^x}{x^x}\,dx$ converges by considering the series $\sum_{n=1}^{\infty} \left(\frac{e}{n}\right)^n$.

14. Show that the harmonic series requires approximately 10^{43} terms for the partial sums to exceed 100. The current estimate of the age of the universe is about 10^{18} *seconds*. (Hint: Use the comparison with the corresponding integral.)

15. (From G. E. Andrews [46]) "A problem attributed by R. Raimi to a Professor Sleator at the University of Michigan in 1941: Two trees are one mile apart. A drib (it is not necessary that you know what a drib is) flies from one tree to the other and back, making the first trip at 10 miles per hour, the return at 20 miles per hour, the next at 40, and so on, each trip at twice the speed of the preceding. When will the drib be in both trees at the same time? Do not spend time wondering or arguing about the drib, but solve the problem."

7.4 Absolute and Conditional Convergence

Consider the **alternating harmonic series**

$$\sum_{n=1}^{\infty} \frac{(-1)^{n+1}}{n} = 1 - \frac{1}{2} + \frac{1}{3} - \frac{1}{4} + \frac{1}{5} \cdots.$$

None of our previous tests help with this series. Since the terms converge to 0, the divergence test is inconclusive. And as the terms are not non-negative, the other tests do not apply either. We need to go back to the definition of convergence. Let us plot the partial sums.

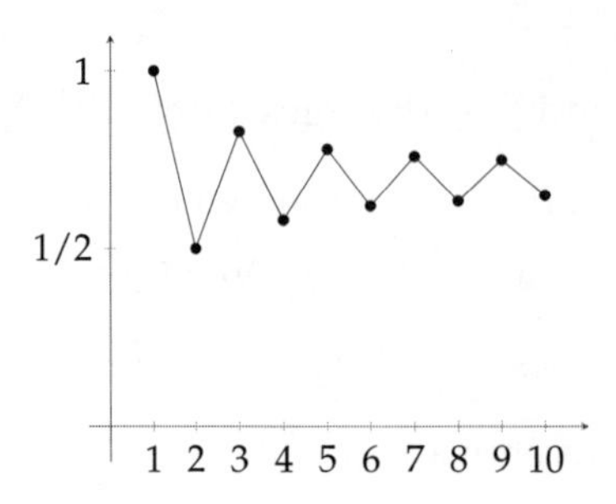

The odd partial sums $S_1, S_3, \ldots$ decrease and are bounded below by $1/2$, while the even partial sums $S_2, S_4, \ldots$ increase and are bounded above by 1. Hence, these two sequences converge. Let $S_{2k+1} \to L$ and $S_{2k} \to M$. Then

$$S_{2k+1} - S_{2k} = \frac{1}{2k+1} \to 0 \text{ as } k \to \infty \implies L = M.$$

Since the odd and even partial sums converge to the same limit L, we have $S_n \to L$. Therefore, the alternating harmonic series converges.

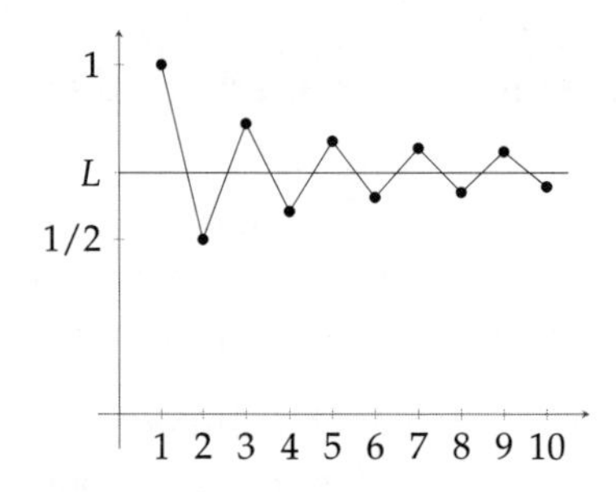

We formulate a general result along these lines.

Theorem 7.4.1 (Alternating Series Test)

Consider an ***alternating series*** $\sum_{n=1}^{\infty} (-1)^{n+1} b_n$ *with each* $b_n \geq 0$. *Suppose the sequence* (b_n) *is decreasing and has limit 0. Then the following hold:*

1. $\sum_{n=1}^{\infty} (-1)^{n+1} b_n$ *converges.*

2. *Let* $S = \sum_{n=1}^{\infty} (-1)^{n+1} b_n$ *and* $S_k = \sum_{n=1}^{k} (-1)^{n+1} b_n$. *Then* $|S - S_k| \leq b_{k+1}$.

Proof. Consider the consecutive odd partial sums:

- $S_{2k+1} - S_{2k-1} = (-1)^{2k+1} b_{2k} + (-1)^{2k+2} b_{2k+1} = b_{2k+1} - b_{2k} \leq 0.$
- $S_{2k+1} = (b_1 - b_2) + (b_3 - b_4) + \cdots + (b_{2k-1} - b_{2k}) + b_{2k+1} \geq 0.$

Thus, the odd partial sums are decreasing and bounded below by 0, hence convergent to some L. Similarly the even partial sums are increasing and bounded above (by b_1), hence convergent to some M. Finally,

$$S_{2k+1} - S_{2k} = b_{2k+1} \implies L - M = 0 \implies L = M = S.$$

For the error estimate we note that the limit S lies between S_n and S_{n+1}. Hence, $|S - S_n| \leq |S_{n+1} - S_n| = |b_{n+1}|$. ■

Example 7.4.2

Consider $\sum_{n=1}^{\infty} \frac{(-1)^{n+1}}{n!}$ It has the form $\sum_{n=1}^{\infty} (-1)^{n+1} b_n$ with $b_n = 1/n! > 0$. The sequence (b_n) decreases and converges to 0, hence by the alternating series test the given series converges.

Suppose we want to estimate the sum to 3 decimal places, that is, with an error less than 0.0005. Observe that $1/7! = 0.0002$. So, $S_6 = \sum_{n=1}^{6} \frac{(-1)^{n+1}}{n!} = 0.368$ is accurate to 3 decimal places. □

The other way to deal with a general series is to convert the terms to non-negative ones and then explore convergence. We say that a series $\sum a_n$ is **absolutely convergent** if $\sum |a_n|$ converges.

Example 7.4.3

Consider $\sum_{n=1}^{\infty} \frac{(-1)^{n+1}}{n^2}$. We know that $\sum_{n=1}^{\infty} \frac{1}{n^2}$ converges. Hence, $\sum_{n=1}^{\infty} \frac{(-1)^{n+1}}{n^2}$ is absolutely convergent. □

This conversion is useful because it can establish the convergence of the original series.

Theorem 7.4.4

If a series is absolutely convergent, then it is also convergent.

Proof. Suppose $\sum |a_n|$ converges. Then $0 \leq a_n + |a_n| \leq 2|a_n|$. Now,

$$\begin{aligned} \sum |a_n| \text{ converges} &\implies \sum 2|a_n| \text{ converges} \\ &\implies \sum (a_n + |a_n|) \text{ converges} \\ &\implies \sum a_n = \sum ((a_n + |a_n|) - |a_n|) \text{ converges.} \end{aligned}$$ ■

Example 7.4.5

Consider $\sum_{n=1}^{\infty} \frac{\cos n}{n^2}$. Here is a plot of the *terms* of the series:

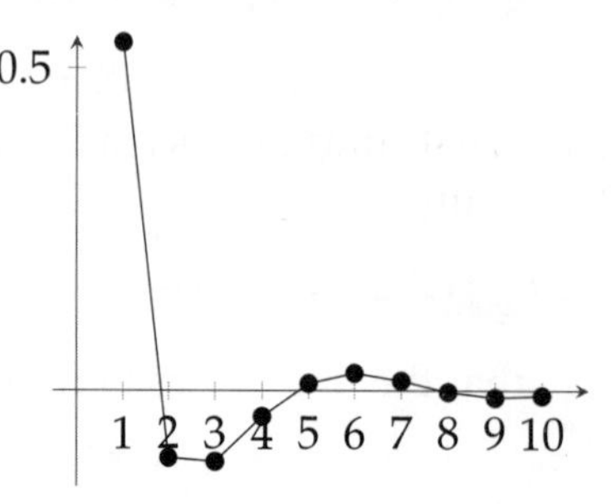

The terms are neither always positive, nor always negative, nor alternating. No convergence test applies directly.

But we can use $0 \leq \frac{|\cos n|}{n^2} \leq \frac{1}{n^2}$ to show absolute convergence! □

A series is called **conditionally convergent** if it is convergent but not absolutely convergent. For example, the alternating harmonic series is conditionally convergent. Any series has three possibilities, as depicted below.

$$\sum a_n \begin{cases} \text{Convergent} \begin{cases} \text{Absolutely Convergent} \\ \text{Conditionally Convergent} \end{cases} \\ \text{Divergent} \end{cases}$$

The behavior of conditionally and absolutely convergent series is dramatically different in one respect. We shall illustrate this with the alternating harmonic series $1 - \frac{1}{2} + \frac{1}{3} \cdots$.

By adding the first three terms we can see that its sum is between 0.5 and 0.83. (We will see later that it is exactly $\log 2 \approx 0.69$.) Let us rearrange it as follows, by moving the positive terms forward:

$$1 + \underbrace{\frac{1}{3} - \frac{1}{2} + \frac{1}{5}}_{>0} + \underbrace{\frac{1}{7} - \frac{1}{4} + \frac{1}{9}}_{>0} + \underbrace{\frac{1}{11} - \frac{1}{6} + \frac{1}{13}}_{>0} \cdots .$$

So, the rearrangement is either divergent or has a sum > 1.

To further investigate rearrangements, we look at the contributions of the positive and negative terms separately. For any $x \in \mathbb{R}$, define $x^+ = \max\{x, 0\}$ and $x^- = \max\{-x, 0\}$. Then $x^\pm \geq 0$, $x = x^+ - x^-$, and $|x| = x^+ + x^-$.

Task 7.4.6

Prove that $\sum a_n$ is absolutely convergent if and only if both $\sum a_n^+$ and $\sum a_n^-$ converge.

Task 7.4.7

Show that if $\sum a_n$ is conditionally convergent, then both $\sum a_n^+$ and $\sum a_n^-$ diverge to ∞.

Theorem 7.4.8 (Riemann Rearrangement Theorem)

1. *A conditionally convergent series can be rearranged so that its sum equals* any given real number *or even* diverges.
2. *Any rearrangement of an absolutely convergent series is absolutely convergent and converges to the same sum.*

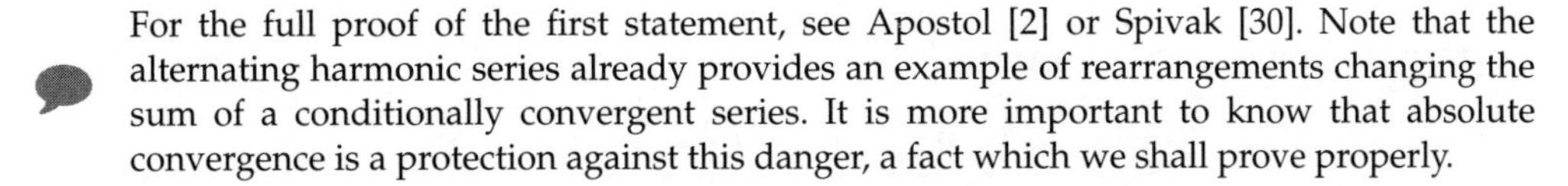

For the full proof of the first statement, see Apostol [2] or Spivak [30]. Note that the alternating harmonic series already provides an example of rearrangements changing the sum of a conditionally convergent series. It is more important to know that absolute convergence is a protection against this danger, a fact which we shall prove properly.

Proof. The idea for (1) is that if a series is conditionally convergent, then the positive terms must sum to infinity and the negative terms to minus infinity, while individual terms tend to zero. Given a number L, we first add positive terms until we just cross to above L. Then we throw in negative terms until we are just below L. Repeat to get sums that oscillate around L. Since the individual terms tend to zero, these oscillations settle to L!

We shall prove (2) in two stages.

(a) First, suppose that $\sum a_n$ is a convergent series with non-negative terms. Let $\sum a_n = L$. Then the sum of any finite selection from the a_n's is bounded by L. In fact, L is the supremum of the set A of these sums. If $\sum b_n$ is a rearrangement of $\sum a_n$, the sums of finite selections from the b_n's constitute the same set A, and hence $L = \sum b_n$ as well.

(b) Suppose $\sum a_n$ is absolutely convergent and $\sum b_n$ is its rearrangement. Then, by (a), $\sum |b_n| = \sum |a_n|$, $\sum b_n^+ = \sum a_n^+$ and $\sum b_n^- = \sum a_n^-$. Hence,

$$\sum b_n = \sum (b_n^+ - b_n^-) = \sum b_n^+ - \sum b_n^- = \sum a_n^+ - \sum a_n^- = \sum a_n.$$

∎

Sometimes the terms of a series do not come in any special order, and in such a situation only absolute convergence can lead to any definite conclusions. Thus, in applications of series to topics such as probability, you will find assumptions of absolute convergence rather than only convergence.

As an application of absolute convergence, let us see what happens if we multiply two series. We have to make sense of the following expression:

$$(a_1 + a_2 + a_3 + \cdots)(b_1 + b_2 + b_3 + \cdots).$$

If the expressions were finite, we would use the distributive law to expand this product, and the result would be the sum of all the products $a_i b_j$. If we try to do the same when the sums are infinite, the first problem we run into is that the a_ib_j terms have to be put into a particular order. This amounts to ordering $\mathbb{N} \times \mathbb{N}$ as the terms a_ib_j are identified by the pair (i,j) of natural numbers.

Here is one way in which $\mathbb{N} \times \mathbb{N}$, and hence the a_ib_j, can be ordered. We first group the (i,j) pairs by the sum $i+j$ of their entries. First we write the pairs whose indices sum to 2, then the ones whose indices sum to 3, and so on. Within each group we order them by i. Here are the first few terms:

$$\underbrace{(1,1)}_{i+j=2},\ \underbrace{(1,2),\ (2,1)}_{i+j=3},\ \underbrace{(1,3),\ (2,2),\ (3,1)}_{i+j=4},\ \ldots.$$

The corresponding ordering of the a_ib_j is

$$\underbrace{a_1b_1}_{i+j=2},\ \underbrace{a_1b_2,\ a_2b_1}_{i+j=3},\ \underbrace{a_1b_3,\ a_2b_2,\ a_3b_1}_{i+j=4},\ \ldots.$$

Once the terms have been arranged in a sequence, we can define their sum in the usual way via partial sums. However, we face another problem: the result may depend on the particular arrangement that we choose. Again, absolute convergence saves us.

Theorem 7.4.9

Let $\sum_{n=1}^{\infty} a_n$ and $\sum_{n=1}^{\infty} b_n$ be absolutely convergent. Let (c_n) be any arrangement of all the a_ib_j combinations. Then $\sum_{n=1}^{\infty} c_n$ is absolutely convergent and $\left(\sum_{n=1}^{\infty} a_n\right)\left(\sum_{n=1}^{\infty} b_n\right) = \sum_{n=1}^{\infty} c_n$.

Proof. We follow the pattern of our proof of the rearrangement theorem. First, we assume that all $a_n, b_n \geq 0$. Then all $c_n \geq 0$. Let $\sum_{n=1}^{\infty} a_n = L$ and $\sum_{n=1}^{\infty} b_n = M$. We have

$$\begin{aligned} LM &= \sup\{\text{finite sums of } a_n\text{'s}\} \cdot \sup\{\text{finite sums of } b_n\text{'s}\} \\ &= \sup\{(\text{finite sum of } a_n\text{'s})(\text{finite sum of } b_n\text{'s})\} \\ &= \sup\{\text{finite sums of } a_nb_m\text{'s}\} \\ &= \sum_{n=1}^{\infty} c_n. \end{aligned}$$

The second equality is a property of supremum (Exercise 11 of §1.2). For the third equality, note that the first set is contained in the second, giving $\leq$. Further, for every element of the second set, there is a larger element in the first set. This gives $\geq$.

Now, we allow $\sum_{n=1}^{\infty} a_n$ and $\sum_{n=1}^{\infty} b_n$ to be arbitrary absolutely convergent series. From the first part, we see that $\sum_{n=1}^{\infty} c_n$ is absolutely convergent. Now,

$$\begin{aligned} \left(\sum a_n\right)\left(\sum b_n\right) &= \left(\sum a_n^+ - \sum a_n^-\right)\left(\sum b_n^+ - \sum b_n^-\right) \\ &= \left(\sum a_n^+ \sum b_n^+ + \sum a_n^- \sum b_n^-\right) - \left(\sum a_n^+ \sum b_n^- + \sum a_n^- \sum b_n^+\right) \\ &= \sum c_n^+ - \sum c_n^- = \sum c_n. \end{aligned}$$

The third equality uses the observation that $c_n = a_ib_j > 0$ when a_i and b_j have the same sign. ■

Having shown the importance of absolute convergence, we take up two more methods for testing for it.

Theorem 7.4.10 (Ratio Test)

Consider a series $\sum a_n$ with non-zero terms. Let $\lim_{n\to\infty}\left|\frac{a_{n+1}}{a_n}\right| = L$.

1. *$L < 1$ implies absolute convergence.*
2. *$L > 1$ or $L = \infty$ implies divergence.*
3. *$L = 1$ is inconclusive.*

The intuition behind this theorem is that for large n we have $|a_{n+1}| \approx L|a_n|$ and so the tail of the series $\sum |a_n|$ resembles a geometric series with common ratio L. Hence, it converges for $L < 1$. If $L > 1$, then the terms a_n do not go to zero.

Proof. First, suppose $L < 1$. Fix r such that $L < r < 1$. There is $N \in \mathbb{N}$ such that $n \geq N$ implies $|a_{n+1}| < r|a_n|$. Hence, $|a_{N+k}| \leq |a_N| r^k$ for $k = 1, 2, \ldots$. By the comparison test, $\sum_{k=1}^{\infty} a_{N+k}$ converges absolutely. Hence, $\sum_{n=1}^{\infty} a_n$ converges absolutely.

Next, let $L > 1$. There is $N \in \mathbb{N}$ such that $n \geq N$ implies $|a_{n+1}| > |a_n|$. So $|a_{N+1}| > 0$ and $n > N \implies |a_n| \geq |a_{N+1}|$. Therefore, $|a_n| \not\to 0$. Hence, $a_n \not\to 0$ and $\sum a_n$ diverges.

For the inconclusiveness of $L = 1$, consider $\sum 1/n$ and $\sum 1/n^2$. ■

Example 7.4.11

Consider the series $\sum_{n=1}^{\infty} \frac{x^n}{n!}$, for any given $x \in \mathbb{R}$.

$$\left|\frac{a_{n+1}}{a_n}\right| = \left|\frac{x^{n+1}/(n+1)!}{x^n/n!}\right| = \left|\frac{x^{n+1} n!}{x^n (n+1)!}\right| = \left|\frac{x}{n+1}\right| \to 0.$$

$L = 0 < 1$ implies the series converges absolutely. □

Theorem 7.4.12 (Root Test)

Consider a series $\sum a_n$. Let $\lim_{n\to\infty} |a_n|^{1/n} = L$.

1. *$L < 1$ implies absolute convergence.*
2. *$L > 1$ or $L = \infty$ implies divergence.*
3. *$L = 1$ is inconclusive.*

For large n we have $|a_n| \approx L^n$ and so the tail of the series $\sum |a_n|$ resembles a geometric series with common ratio L. Hence, it should converge for $L < 1$. If $L > 1$, then the tail of the series $\sum a_n$ features terms that grow in magnitude.

Proof. Start by assuming $L < 1$. Fix r such that $L < r < 1$. There is $N \in \mathbb{N}$ such that $n \geq N$ implies $|a_n|^{1/n} < r$. Hence, $|a_n| < r^n$ for $n = N, N+1, \ldots$. By the comparison test, $\sum_{n=N}^{\infty} a_n$ converges absolutely, hence so does $\sum_{n=1}^{\infty} a_n$.

Now assume $L > 1$. There is $N \in \mathbb{N}$ such that $n \geq N$ implies $|a_n|^{1/n} > 1$, hence $|a_n| > 1$. Therefore, $a_n \not\to 0$ and $\sum a_n$ diverges.

For the inconclusiveness of $L = 1$, consider $\sum 1/n$ and $\sum 1/n^2$. ■

Example 7.4.13

Consider the series $\displaystyle\sum_{n=1}^{\infty} \frac{n^2}{2^n}$. Let us apply the root test:

$$\lim_{n\to\infty} |a_n|^{1/n} = \lim_{n\to\infty} \left| \frac{n^{2/n}}{2} \right| = \frac{1}{2}.$$

$L = 1/2 < 1$ implies the series converges absolutely. □

Example 7.4.14

Consider the series $\displaystyle\sum_{n=1}^{\infty} a_n$ with $a_n = \begin{cases} n/2^n & \text{if } n \text{ odd,} \\ 1/2^n & \text{if } n \text{ even.} \end{cases}$

$$\text{Ratio test}: \left| \frac{a_{n+1}}{a_n} \right| = \begin{cases} 1/(2n) & \text{if } n \text{ odd,} \\ (n+1)/2 & \text{if } n \text{ even,} \end{cases} \quad \text{diverges.}$$

So, the ratio test gives no result.

$$\text{Root test}: |a_n|^{1/n} = \begin{cases} n^{1/n}/2 & \text{if } n \text{ odd} \\ 1/2 & \text{if } n \text{ even} \end{cases} \to \frac{1}{2}.$$

$L = 1/2 < 1$ implies the series converges absolutely. □

Example 7.4.15

In a root test limit calculation, $(n!)^{1/n}$ can be replaced by n/e due to Stirling's approximation. Consider the series $\displaystyle\sum_{n=1}^{\infty} \frac{(n!)^2}{2^{n^2}}$. Apply the root test and Stirling's approximation,

$$\lim_{n\to\infty} \left[\frac{(n!)^2}{2^{n^2}} \right]^{1/n} = \lim_{n\to\infty} \frac{(n!)^{2/n}}{2^n} = \lim_{n\to\infty} \frac{n^2}{e^2 2^n} = 0.$$

So, the series converges. □

Exercises for § 7.4

1. Determine the convergence or divergence of these series:

(a) $\sum_{n=1}^{\infty}(-1)^n\frac{\sqrt{n}}{n+1000}$.

(b) $\sum_{n=1}^{\infty}(-1)^n\frac{n^3}{e^n}$.

(c) $\sum_{n=1}^{\infty}(-1)^n\left(\frac{2n+100}{3n+1}\right)^n$.

(d) $\sum_{n=1}^{\infty}(-1)^n\frac{n^2}{1+n^2}$.

2. Estimate $\sum_{n=0}^{\infty}\frac{(-0.5)^n}{(2n+1)!}$ to an accuracy of within 0.005.

3. Give an example of an alternating series $\sum(-1)^n b_n$ such that $b_n \geq 0$, $b_n \to 0$, but the series diverges. (This will show why the alternating series test had to include the assumption that b_n decreases.)

4. Do the following series converge absolutely, or conditionally, or diverge?

(a) $\sum_{n=1}^{\infty}(-1)^n\frac{n}{4n-1}$.

(b) $\sum_{n=1}^{\infty}(-1)^n\sin\left(\frac{1}{n^2}\right)$.

(c) $\sum_{n=2}^{\infty}(-1)^{n+1}\frac{n^2}{n^3+1}$.

(d) $\sum_{n=1}^{\infty}\frac{(-1)^n}{(\arctan n)^n}$.

5. Show the following:

(a) If $\sum a_n$ and $\sum b_n$ are absolutely convergent, so is $\sum(a_n+b_n)$.

(b) If $\sum a_n$ is absolutely convergent and $\sum b_n$ is conditionally convergent, then $\sum(a_n+b_n)$ is conditionally convergent.

(c) What can we say about $\sum(a_n+b_n)$ if both $\sum a_n$ and $\sum b_n$ are conditionally convergent?

6. Suppose $\sum a_n$ is absolutely convergent. Prove that $|\sum a_n| \leq \sum|a_n|$.

7. Apply the ratio or root test to study the convergence of these series:

(a) $\sum_{n=1}^{\infty}\frac{(-1)^n n^3}{3^n}$.

(b) $\sum_{n=1}^{\infty}\frac{n^n}{n!}$.

(c) $\sum_{n=1}^{\infty}\frac{n^{2n}}{(1+2n^2)^n}$.

(d) $\sum_{n=1}^{\infty}\frac{(-1)^n}{(\arctan n)^n}$.

(e) $\sum_{n=1}^{\infty}\frac{n^n}{(2n)!}$.

(f) $\sum_{n=1}^{\infty}\frac{(n!)^n}{n^{n^2}}$.

(g) $\sum_{n=2}^{\infty}\frac{1}{(\log n)^n}$.

(h) $\sum_{n=1}^{\infty}\left(\frac{n}{n+1}\right)^{n^2}$.

8. Let $\sum_{n=1}^{\infty} a_n$ be a series with non-negative terms. Prove that

$$\sum_{n=1}^{\infty} a_n = \sum_{n=1}^{\infty} a_{2n-1} + \sum_{n=1}^{\infty} a_{2n}.$$

9. Let $\sum a_n$ and $\sum b_n$ be two convergent series.

(a) Show that if $\sum a_n$ converges absolutely, then $\sum a_n b_n$ converges absolutely.

(b) Give an example where $\sum a_n b_n$ diverges.

10. (Dirichlet's Test) Let (a_n) and (b_n) be sequences such that b_n decreases with limit 0 and the partial sums $S_n = a_1 + \cdots + a_n$ are bounded. The following tasks will establish that $\sum a_n b_n$ converges:

(a) $\sum_{n=1}^{m} a_n b_n = \sum_{n=1}^{m-1} (b_n - b_{n+1}) S_n + b_m S_m$.

(b) $b_m S_m \to 0$.

(c) If $|S_k| \leq M$ for every k, then $\sum_{n=1}^{m-1} |(b_n - b_{n+1}) S_n| \leq M b_1$.

11. (Abel's Test) Let (a_n) and (b_n) be sequences such that b_n is monotonic and bounded and the series $\sum a_n$ converges. Then $\sum a_n b_n$ converges. (Hint: Assume b_n is decreasing and use Dirichlet's test.)

The next two exercises make a connection with linear algebra. We begin with the **Cauchy–Schwarz inequality** for finite sums.

12. For $x = (x_1, \ldots, x_n) \in \mathbb{R}^n$, define $||x|| = \left(\sum_{k=1}^{n} x_k^2 \right)^{1/2}$.

(a) If $||x|| = ||y|| = 1$, show that $|\sum_{k=1}^{n} x_k y_k| \leq 1$. (Hint: Consider $\sum_{k=1}^{n} (x_k - y_k)^2$.)

(b) For arbitrary $x, y \in \mathbb{R}^n$, show that $|\sum_{k=1}^{n} x_k y_k| \leq ||x|| \cdot ||y||$.

13. We extend these results to infinite series.

(a) Suppose $\sum a_n^2$ and $\sum b_n^2$ converge. Prove that $\sum a_n b_n$ is absolutely convergent and $|\sum a_n b_n|^2 \leq \left(\sum a_n^2 \right) \left(\sum b_n^2 \right)$.

(b) Let $\ell^2 = \{ (a_n) \mid \sum a_n^2 < \infty \}$. Prove ℓ^2 is a real vector space, with coordinate-wise addition and scalar multiplication.

(c) The map $\ell^2 \times \ell^2 \to \mathbb{R}$, defined by $\langle x, y \rangle = \sum x_n y_n$, has the same properties as the dot product on $\mathbb{R}^n$.

- $\langle x, x \rangle \geq 0$ and $\langle x, x \rangle = 0$ if and only if $x = 0$.
- $\langle x, y \rangle = \langle y, x \rangle$.
- $\langle \alpha x + \beta y, z \rangle = \alpha \langle x, z \rangle + \beta \langle y, z \rangle$.
- $|\langle x, y \rangle|^2 \leq \langle x, x \rangle \langle y, y \rangle$.

What is the importance of absolute convergence in this context?

Thematic Exercises

Stirling's Formula

We had proven that $(n!)^{1/n}$ behaves like n/e for large n, and remarked that this does not imply that $n!$ behaves like $(n/e)^n$. Let us plot their ratio.

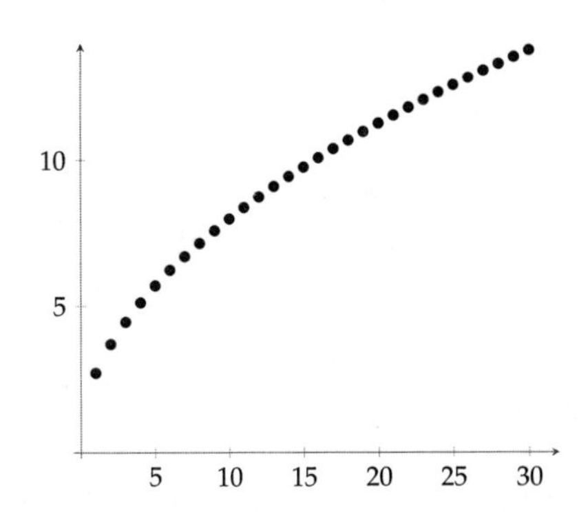

This resembles the graph of a constant multiple of $\sqrt{n}$. Therefore, we shall study $a_n = \dfrac{n!}{(n/e)^n\sqrt{n}}$.

A1. Show that $\log(1+\frac{1}{n}) > \frac{2}{2n+1}$ and hence $(1+\frac{1}{n})^{n+1/2} > e$.

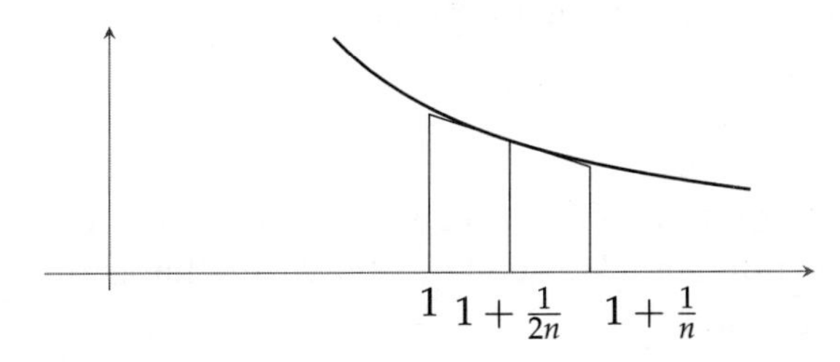

A2. Show that $\dfrac{a_{n+1}}{a_n} < 1$ for every n, and hence the sequence a_n converges.

A3. Show that $1 + \log 2 + \log 3 + \cdots + \log(n-1) + \frac{1}{2}\log n > \int_1^n \log x\, dx$.

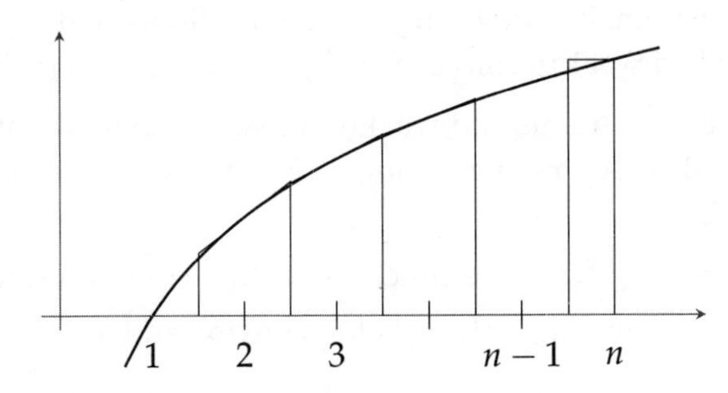

A4. Show that each $a_n > 1$ and hence $L = \lim a_n > 0$.

We will now make a connection with integrals of powers of sine functions. You should review Example 5.3.11 before proceeding.

A5. Observe that $\dfrac{a_n^2}{a_{2n}} = \dfrac{4^n (n!)^2\sqrt{2}}{(2n)!\sqrt{n}} = \sqrt{2\pi(1+\frac{1}{2n})}\left[\dfrac{\int_0^{\pi/2}\sin^{2n+1}x\,dx}{\int_0^{\pi/2}\sin^{2n}x\,dx}\right]^{1/2}$.

A6. Show that $\dfrac{\int_0^{\pi/2}\sin^{2n+1}x\,dx}{\int_0^{\pi/2}\sin^{2n}x\,dx} \to 1$. Use the inequalities

$$\int_0^{\pi/2}\sin^{2n+2}x\,dx \le \int_0^{\pi/2}\sin^{2n+1}x\,dx \le \int_0^{\pi/2}\sin^{2n}x\,dx$$

to create a sandwich.

A7. Conclude that $\dfrac{n!}{(n/e)^n\sqrt{n}} \to \sqrt{2\pi}$.

Our final calculation shows that $n! \approx \sqrt{2\pi n}\left(\dfrac{n}{e}\right)^n$ for large n. This is called **Stirling's formula**.

Gamma and Beta Functions

Recall that the **gamma function** was defined in §5.5 by $\Gamma(x) = \int_0^\infty e^{-t}t^{x-1}\,dt$ for $x > 0$. We checked that it has the following properties:

1. $\Gamma(x+1) = x\Gamma(x)$ for every $x > 0$,
2. $\Gamma(1) = 1$.

Hence, $\Gamma(n+1) = n!$ if $n \in \mathbb{W}$. Of course Γ is not the only function that interpolates the factorial at whole numbers. We shall see, however, that is unique once we impose a certain convexity criterion.

B1. Show that if f, g are convex functions with a common domain, then $f + g$ is convex.

B2. Suppose that $f, f_1, f_2, f_3, \ldots$ are functions with a common domain, $\lim_{n\to\infty} f_n(x) = f(x)$ for each x in the domain, and each f_n is convex. Show that f is convex.

A function f is called **weakly convex** if $f(\frac{x+y}{2}) \le \frac{1}{2}(f(x) + f(y))$ for every x, y.

B3. Show that if f is continuous and weakly convex, then it is convex.

A function f is called **log convex** if $\log \circ f$ is convex. This requires $f > 0$ so that $\log(f(x))$ is defined.

B4. Suppose that $f, f_1, f_2, f_3, \ldots$ are functions with a common domain, $\lim_{n\to\infty} f_n(x) = f(x)$ for each x in the domain, each f_n is log convex, and $f > 0$. Show that f is log convex.

B5. Let $Q(x,y) = ax^2 + 2cxy + by^2$. Show that $ab - c^2 \geq 0$ if and only if $Q(x,y) \geq 0$ for every x, y.

B6. Show that if f, g are log convex functions with a common domain, then $f + g$ is log convex. (Hint: Show $\log(f+g)$ is weakly convex.)

B7. Consider $f(x) = \int_0^b \phi(t)t^{x-1}\,dt$ where ϕ is continuous and positive, with $x > 0$.

(a) Show that $g_n(x) = \dfrac{1}{n}\left(\phi\left(\frac{b}{n}\right)\left(\frac{b}{n}\right)^{x-1} + \phi\left(\frac{2b}{n}\right)\left(\frac{2b}{n}\right)^{x-1} \cdots + \phi(b)b^{x-1}\right)$ is log convex.

(b) Show that $f(x)$ is log convex.

(c) Show that $f(x)$ is log convex even if $b = \infty$, assuming the improper integral converges for every x.

This shows that Γ is log convex.

B8. Let $f \colon (0,\infty) \to \mathbb{R}$ be a log convex function such that $f(x+1) = xf(x)$ for every $x > 0$ and $f(1) = 1$. The following tasks will establish that $f = \Gamma$.

(a) Show that it is enough to prove $f(x) = \Gamma(x)$ for $0 < x \leq 1$.

(b) For $0 < x \leq 1$ and $n \geq 2$, show that

$$\log(f(n)) - \log(f(n-1)) \leq \frac{\log(f(n+x)) - \log(f(n))}{x}$$
$$\leq \log(f(n+1)) - \log(f(n)).$$

(c) Show that the above rearranges to

$$\frac{(n-1)!(n-1)^x}{x(x+1)\cdots(x+n-1)} \leq f(x) \leq \frac{(n-1)!n^x}{x(x+1)\cdots(x+n-1)}.$$

(d) Replace n by $n+1$ in the leftmost expression and rearrange to get

$$\frac{n}{x+n}f(x) \leq \frac{n!n^x}{x(x+1)\cdots(x+n)} \leq f(x).$$

(e) Conclude that $f(x) = \lim\limits_{n\to\infty} \dfrac{n!n^x}{x(x+1)\cdots(x+n)}$ and hence $f(x)$ is uniquely determined by the given properties of f.

B9. The **beta function** is defined by

$$B(x,y) = \int_0^1 t^{x-1}(1-t)^{y-1}\,dt \qquad (x,y>0).$$

Prove the following:

(a) The improper integral that defines $B(x,y)$ converges for all $x,y>0$.

(b) $B(x,y) = B(y,x)$ for every $x,y>0$.

(c) $B(x,y) = 2\int_0^{\pi/2}(\sin\theta)^{2x-1}(\cos\theta)^{2y-1}\,d\theta$.

(d) $B(x+1,y) = \dfrac{x}{x+y}B(x,y)$. (Hint: Integration by parts.)

B10. For a fixed y, let $f(x) = B(x,y)\Gamma(x+y)$. Show that

(a) f is log convex,

(b) $f(x+1) = xf(x)$,

(c) $f(1) = \Gamma(y)$.

It follows from the above that $f(x) = \Gamma(y)\Gamma(x)$. Therefore, we have

$$B(x,y) = \frac{\Gamma(x)\Gamma(y)}{\Gamma(x+y)}.$$

B11. Compute $B(1/2,1/2) = \pi$. Hence, $\Gamma(1/2) = \sqrt{\pi}$. This gives the value of the Gaussian integral,

$$\int_{-\infty}^{\infty} e^{-x^2}\,dx = \sqrt{\pi}.$$

These exercises related to the factorial function are based on the presentations in Artin [3] and Widder [37].

8 | Taylor and Fourier Series

The decimal expansion of a real number is an instance of an infinite series. For example, 3.14159... can be viewed as the sum of the series created by its digits: $3 + \frac{1}{10} + \frac{4}{10^2} + \frac{1}{10^3} + \frac{5}{10^3} + \frac{9}{10^4} + \cdots$. The convergence of such a series is established by a comparison with the geometric series $\sum_{n=0}^{\infty} 1/10^n$. Isaac Newton realized that by replacing the powers of $1/10$ with powers of a variable x, one can create real functions that can be easily manipulated in analogy with the rules of decimal expansions. He developed rules for their differentiation and integration as well as a method for expanding inverse functions in this manner. Today's historians believe that for Newton, calculus consisted of working with these "power series." (See Stillwell [32, pp 167–70].) In this approach, the main task is to express a given function as a power series, after which it becomes trivial to perform the operations of calculus on it. In the first two sections of this chapter we shall study the general properties of power series, and then the problem of expressing a given function as a power series.

A century after Newton, Joseph Fourier replaced the powers x^n with the trigonometric functions $\sin nx$ and $\cos nx$ to create new ways of describing functions. The "Fourier series" could model much wilder behavior than power series, and forced mathematicians to revisit their notions of what is a function, and especially the definition of integration. We give a brief introduction to this topic in the third section.

In our final section, we introduce sequences and series of complex numbers. These bring further clarity to power and Fourier series, and even unify them through the famed identity of Euler: $e^{ix} = \cos x + i \sin x$.

8.1 Power Series

Let us recall our study of Taylor polynomials in §6.3. Given a function f that can be differentiated n times at $x = a$, we define the Taylor polynomial

$$T_n(x) = \sum_{k=0}^{n} \frac{f^{(k)}(a)}{k!}(x-a)^k.$$

T_n is intended to be an approximation for f near $x = a$, with the hope that the approximation improves when we increase n. These hopes are not *always* realized, but in many cases they are. (The remainder theorem gives us a way to assess them.) It is natural to make the jump from polynomials to series, and consider the expression

$$T(x) = \sum_{k=0}^{\infty} \frac{f^{(k)}(a)}{k!}(x-a)^k.$$

The questions that arise here are: (a) For which x does this series converge, and (b) When it converges, does it sum to $f(x)$? To tackle these questions, we initiate a general study of series of this form.

A **power series** is an expression of the form

$$c_0 + c_1(x-a) + c_2(x-a)^2 + \cdots \quad \text{or} \quad \sum_{n=0}^{\infty} c_n(x-a)^n$$

in which a and the c_n are constants and x is a variable.

- The c_n are called the **coefficients** of the power series.
- And a is called the **center** of the power series.

Main Question: For which values of x will a power series converge?

Example 8.1.1

The geometric series $\sum_{n=0}^{\infty} x^n$ is a power series centered at 0 and with all coefficients equal to 1. It converges absolutely for $|x| < 1$ and diverges for $|x| \geq 1$. □

Example 8.1.2

$\sum_{n=0}^{\infty} x^n/n!$ is centered at 0 and has coefficients $1/n!$. We saw earlier, using the ratio test, that it converges absolutely for every x. □

Example 8.1.3

$\sum_{n=0}^{\infty} n!\,x^n$ converges only at $x = 0$. □

Example 8.1.4

Consider the power series $\sum\limits_{n=1}^{\infty} \frac{(-1)^n}{n}(x-1)^n$. It is centered at 1 and has coefficients $(-1)^n/n$. Note that this series has $c_0 = 0$.

Apply the ratio test: $\left|\frac{a_{n+1}}{a_n}\right| = \left|\frac{(x-1)^{n+1}n}{(n+1)(x-1)^n}\right| \to |x-1|$.

So this series converges absolutely if $|x-1| < 1$ and diverges if $|x-1| > 1$.

At $x = 0$ it becomes the harmonic series and diverges, while at $x = 2$ it becomes the alternating harmonic series and converges. So, it converges for $0 < x \leq 2$, and diverges elsewhere. □

Before we start proving facts about power series, we note that is it enough to consider the $a = 0$ case, since the substitution $y = x - a$ converts a power series with center a to one with center 0.

Our examples of power series have a common pattern. The points at which a power series converges appear to always form an interval whose midpoint is the center a of the power series. We shall first establish this symmetry about the center.

Theorem 8.1.5

Suppose $\sum_{n=0}^{\infty} c_n x^n$ converges at $x = r \neq 0$. Then it converges absolutely at every x that satisfies $|x| < |r|$.

Proof. The convergence of $\sum_{n=0}^{\infty} c_n r^n$ implies $c_n r^n \to 0$, hence $|c_n r^n|$ is bounded by some real M. Now consider any x such that $|x| < |r|$. Then

$$|c_n x^n| = |c_n r^n| \left|\frac{x}{r}\right|^n \leq M \left|\frac{x}{r}\right|^n.$$

By the comparison theorem, $\sum_{n=0}^{\infty} |c_n x^n|$ converges. ■

Theorem 8.1.6

Consider a power series $\sum_{n=0}^{\infty} c_n (x-a)^n$. One of the following cases will occur:

1. *The series converges only if $x = a$.*
2. *The series converges absolutely for every $x \in \mathbb{R}$.*
3. *There is $R \in \mathbb{R}$ such that the series converges absolutely for $|x - a| < R$ and diverges for $|x - a| > R$.*

R is called the **radius of convergence**. In the first case we say $R = 0$. In the second case we say $R = \infty$.

Proof. We write the proof for $a = 0$. Let $S = \left\{ h \in \mathbb{R} : \sum_{n=0}^{\infty} c_n h^n \text{ converges} \right\}$. We know $0 \in S$, so $S \neq \varnothing$.

If S is not bounded above, then $S = \mathbb{R}$: For any $t \in \mathbb{R}$, there is $h \in S$ such that $h > |t|$. Therefore, $t \in S$. This is the $R = \infty$ case.

If S is bounded above, take $R = \sup S$. Suppose $|x| < R$. There is $h \in S$ such that $|x| < h < R$. Since $\sum c_n h^n$ converges, $\sum c_n x^n$ converges absolutely. Hence, the series converges for every $x \in (-R, R)$.

Finally, suppose the series converges for some x with $|x| > R$. Take any t such that $R < t < |x|$. Then the series converges at t, violating the definition of R. ■

The **interval of convergence** consists of all the x for which the power series converges. It can be $\{a\}$, $(-\infty, \infty)$, $(a - R, a + R)$, $(a - R, a + R]$, $[a - R, a + R)$, or $[a - R, a + R]$.

The ratio and root tests usually suffice to find R. Some other test would have to be applied to the end-points $a \pm R$.

Power Series as Functions

Let $\sum_{n=0}^{\infty} c_n(x-a)^n$ have a radius of convergence $R > 0$. Then it defines a function $f\colon (a-R, a+R) \to \mathbb{R}$ by $f(x) = \sum_{n=0}^{\infty} c_n(x-a)^n$. The endpoints $a \pm R$ can also be included in the domain, provided the series converges there.

Is $f(x)$ differentiable? The obvious candidate for its derivative is its **derived power series** $\sum_{n=1}^{\infty} nc_n(x-a)^{n-1}$ obtained by **term-by-term differentiation**. We will prove this is the right candidate. First, we establish its convergence.

Theorem 8.1.7

Let $\sum_{n=0}^{\infty} c_n(x-a)^n$ have radius of convergence $R > 0$. Then its derived series $\sum_{n=1}^{\infty} nc_n(x-a)^{n-1}$ also has radius of convergence R.

Proof. We write the proof for $a = 0$. Let $0 < x < R$. Pick h such that $0 < x < x+h < R$. Then

$$\frac{f(x+h)-f(x)}{h} = \sum_{n=1}^{\infty} c_n \frac{(x+h)^n - x^n}{h} = \sum_{n=1}^{\infty} n\, c_n\, y_n^{n-1}$$

for some $x < y_n < x+h$, by the mean value theorem. Since the series for $f(x+h)$ and $f(x)$ converge absolutely, so does $\sum n\, c_n\, y_n^{n-1}$. By comparison test, $\sum n\, c_n\, x^{n-1}$ converges absolutely.

We have proved that the derived series converges for $0 < x < R$. Hence, its radius of convergence is at least R. On the other hand, for $n > x$ we have $|n\, c_n\, x^{n-1}| \geq |c_n\, x^n|$ and so the derived series diverges for $x > R$. Hence, the radius of convergence is exactly R. ∎

Theorem 8.1.8

Let $\sum_{n=0}^{\infty} c_n(x-a)^n$ have radius of convergence $R > 0$, and let $f(x)$ be the function defined by this power series. Then $f'(x) = \sum_{n=1}^{\infty} nc_n(x-a)^{n-1}$ for $|x-a| < R$.

Proof. Again let $a = 0$. We investigate the gap between the difference quotient and the candidate derivative $g(x) = \sum_{n=1}^{\infty} nc_n(x-a)^{n-1}$.

$$\begin{aligned}
\frac{f(y)-f(x)}{y-x} - g(x) &= \sum_{n=1}^{\infty} c_n \frac{y^n - x^n}{y-x} - \sum_{n=1}^{\infty} nc_n x^{n-1} \\
&= \sum_{n=1}^{\infty} c_n \left[\frac{y^n - x^n}{y-x} - nx^{n-1}\right] \\
&= \sum_{n=1}^{\infty} c_n \left[\sum_{k=0}^{n-1} y^k x^{n-k-1} - nx^{n-1}\right]
\end{aligned}$$

$$= \sum_{n=1}^{\infty} c_n \left[\sum_{k=1}^{n-1} x^{n-k-1}(y^k - x^k) \right].$$

Fix a number ρ such that $|x|, |y| < \rho < R$. Then

$$|y^k - x^k| = |y - x| \left| \sum_{j=0}^{k-1} y^j x^{k-1-j} \right| \leq k|y - x|\rho^{k-1}.$$

Hence, $\left| \dfrac{f(y) - f(x)}{y - x} - g(x) \right| \leq |y - x| \sum_{n=1}^{\infty} |c_n| \left[\sum_{k=1}^{n-1} k\rho^{n-2} \right]$

$$= |y - x| \sum_{n=1}^{\infty} |c_n| \frac{n(n-1)}{2} \rho^{n-2}$$

$$= M|y - x|.$$

Therefore, $\lim_{y \to x} \left(\dfrac{f(y) - f(x)}{y - x} - g(x) \right) = 0$ and $f'(x) = g(x)$. ■

Example 8.1.9

We have seen that the power series $\sum_{n=0}^{\infty} x^n/n!$ converges for every $x \in \mathbb{R}$. Thus, it defines a differentiable function $f \colon \mathbb{R} \to \mathbb{R}$. Note that

$$f(0) = 1 \quad \text{and} \quad f'(x) = \sum_{n=1}^{\infty} n \frac{x^{n-1}}{n!} = \sum_{n=1}^{\infty} \frac{x^{n-1}}{(n-1)!} = f(x).$$

Therefore, $f(x)$ is the exponential function, and we get

$$e^x = \sum_{n=0}^{\infty} \frac{x^n}{n!} = 1 + x + \frac{x^2}{2!} + \frac{x^3}{3!} + \cdots.$$

In particular, $e = \sum_{n=0}^{\infty} \dfrac{1}{n!}$. □

Our result about term-by-term differentiation leads to one about **term-by-term integration.**

Theorem 8.1.10

Suppose $f(x) = \sum_{n=0}^{\infty} c_n (x - a)^n$ with radius of convergence $R > 0$. Then

$$\int f(x)\,dx = \sum_{n=0}^{\infty} \int c_n (x - a)^n\,dx = \sum_{n=0}^{\infty} \frac{c_n}{n+1}(x - a)^{n+1} + C.$$

Further, for any $b \in (c - R, c + R)$,

$$\int_a^b f(x)\,dx = \sum_{n=0}^{\infty} \int_a^b c_n (x - a)^n\,dx = \sum_{n=0}^{\infty} \frac{c_n}{n+1}(b - a)^{n+1}.$$

Proof. Consider the power series defined below:

$$h(x) = \sum_{n=0}^{\infty} \frac{c_n}{n+1}(x-a)^{n+1}.$$

As $f(x)$ is the derived series of $h(x)$, $h(x)$ also has radius of convergence R, and it is the anti-derivative of $f(x)$. This proves the first part. The second part follows from the second fundamental theorem. ∎

Consider the function defined by the geometric series

$$\frac{1}{1-x} = \sum_{n=0}^{\infty} x^n = 1 + x + x^2 + \cdots, \qquad \text{for } |x| < 1.$$

We can use various substitutions to get power series expansions of other functions.

$$\frac{1}{1+x} = \sum_{n=0}^{\infty} (-1)^n x^n = 1 - x + x^2 - + \cdots, \qquad \text{for } |x| < 1.$$

$$\frac{1}{1+x^2} = \sum_{n=0}^{\infty} (-1)^n x^{2n} = 1 - x^2 + x^4 - + \cdots, \qquad \text{for } |x| < 1.$$

$$\frac{1}{2-x} = \frac{1}{2}\sum_{n=0}^{\infty} \frac{x^n}{2^n} = \frac{1}{2} + \frac{x}{4} + \frac{x^2}{8} + \cdots, \qquad \text{for } |x| < 2.$$

Differentiating the geometric series repeatedly gives more power series expansions, each valid for $|x| < 1$.

$$\frac{1}{(1-x)^2} = \sum_{n=1}^{\infty} n x^{n-1} = 1 + 2x + 3x^2 + \cdots.$$

$$\frac{1}{(1-x)^3} = \sum_{n=2}^{\infty} \frac{n(n-1)}{2} x^{n-2} = 1 + 3x + 6x^2 + \cdots.$$

$$\frac{1}{(1-x)^k} = \sum_{n=k-1}^{\infty} \binom{n}{k-1} x^{n-k+1}.$$

Integration gives other interesting series.

$$\begin{aligned}\log(1+x) &= \int_0^x \frac{1}{1+t}\,dt = \int_0^x \left(\sum_{n=0}^{\infty} (-1)^n t^n\right) dt \\ &= \sum_{n=0}^{\infty} \frac{(-1)^n}{n+1} x^{n+1} = x - \frac{x^2}{2} + \frac{x^3}{3} - + \cdots \qquad \text{for } |x| < 1.\end{aligned}$$

Task 8.1.11

Prove that $\arctan x = x - \frac{x^3}{3} + \frac{x^5}{5} - + \cdots$ *for* $|x| < 1$.

Power series expansions enable us to go beyond convergence and find the actual sum of certain series. For example, substituting $x = 1/2$ in the power series expansion of $1/(1-x)^2$ gives us

$$\sum_{n=1}^{\infty} \frac{n}{2^n} = 2.$$

The series for $\log(1+x)$ converges at $x = 1$ by the alternating series test; in fact it is the alternating harmonic series. It is natural to expect that its sum should be $\log 2$, but our results so far do not establish this. The following theorem will help us:

Theorem 8.1.12 (Abel's Theorem)

Suppose a power series with center at $x = a$ has radius of convergence $R > 0$. If the power series converges at $a + R$, then the function defined by it is left-continuous at that point.

Proof. We may assume the power series has the form $\sum_{n=0}^{\infty} c_n x^n$, $R = 1$, and $\sum_{n=0}^{\infty} c_n$ converges. Now take any x such that $0 < x < 1$. Denote $S_n = c_0 + \cdots + c_n$ and $S = \lim S_n$. First we have

$$\sum_{n=0}^{m} c_n x^n = x^m S_m + \sum_{n=0}^{m-1} (x^n - x^{n+1}) S_n = x^m S_m + (1-x) \sum_{n=0}^{m-1} x^n S_n,$$

obtained by substituting $c_n = S_n - S_{n-1}$ and regrouping. By letting $m \to \infty$ we obtain

$$\sum_{n=0}^{\infty} c_n x^n = (1-x) \sum_{n=0}^{\infty} x^n S_n.$$

Hence,

$$\sum_{n=0}^{\infty} c_n x^n - \sum_{n=0}^{\infty} c_n = (1-x) \sum_{n=0}^{\infty} x^n (S_n - S).$$

Given $\epsilon > 0$, we first choose N such that $n \geq N$ implies $|S_n - S| < \epsilon/2$. Then

$$\begin{aligned} \left| \sum_{n=0}^{\infty} c_n x^n - \sum_{n=0}^{\infty} c_n \right| &\leq |1-x| \sum_{n=0}^{N-1} |x|^n |S_n - S| + |1-x| \sum_{n=N}^{\infty} |x|^n |S_n - S| \\ &\leq |1-x| \sum_{n=0}^{N-1} |S_n - S| + \epsilon/2. \end{aligned}$$

The choice of N was independent of x, hence $\sum_{n=0}^{N-1} |S_n - S|$ is also independent of x. Therefore, for x close enough to 1, we shall have $|1-x| \sum_{n=0}^{N-1} |S_n - S| < \epsilon/2$. This completes the proof. ■

It follows from Abel's theorem that $f(x) = \sum_{n=0}^{\infty} \frac{(-1)^n}{n+1} x^{n+1}$ is left-continuous at 1. Hence,

$$\sum_{n=0}^{\infty} \frac{(-1)^n}{n+1} = f(1) = \lim_{x \to 1-} f(x) = \lim_{x \to 1-} \log(1+x) = \log 2.$$

Task 8.1.13

Prove that

$$\pi = 4\sum_{n=0}^{\infty}\frac{(-1)^n}{2n+1} = 4\left(1-\frac{1}{3}+\frac{1}{5}-+\cdots\right).$$

This is called the **Gregory** or **Leibniz formula** for π. But its discovery appears to be due to Madhava in Kerala in the fourteenth century.

The first few partial sums for this series are 4, 2.67, 3.47, 2.89, 3.33, 2.98, 3.28, 3.01, 3.25, converging (slowly) towards 3.14.... We have to add about 300 terms to reach two decimal places of accuracy! Madhava found ways of estimating the "remainder" rather accurately and was able to calculate π to 11 decimal places. Later generations of his students and grand-students extended the accuracy to 20 places (we know how, but not why).

Exercises for § 8.1

1. Find the values of x for which the given power series converge:

(a) $\displaystyle\sum_{n=1}^{\infty} nx^n$.

(b) $\displaystyle\sum_{n=0}^{\infty}(-1)^n\frac{x^{2n}}{(2n)!}$.

(c) $\displaystyle\sum_{n=1}^{\infty}\frac{x^n}{n3^n}$.

(d) $\displaystyle\sum_{n=2}^{\infty}\frac{x^n}{n(\ln n)^2}$.

2. Exhibit a power series that converges exactly on the interval $[-1,1]$.

3. Find the radius of convergence of $\displaystyle\sum_{n=0}^{\infty}\frac{(n!)^k}{(kn)!}x^n$, $k\in\mathbb{N}$.

4. Prove the following:

(a) $\displaystyle\frac{1}{1-x}=\sum_{n=0}^{\infty}\frac{(x-3)^n}{(-2)^{n+1}}$, for $|x-3|<2$.

(b) $\displaystyle e^x=\sum_{n=0}^{\infty}e^3\frac{(x-3)^n}{n!}$ for every $x\in\mathbb{R}$.

5. Show that $\displaystyle\frac{1}{e}=\frac{1}{2!}-\frac{1}{3!}+\frac{1}{4!}-\frac{1}{5!}\cdots$. Use this to show that $2\frac{2}{3}<e<2\frac{8}{11}$.

6. We shall use the series expansion of $1/e$ to prove that e is irrational. Assuming $e=p/q$ with $p,q\in\mathbb{N}$, we have $1/e=s+r$, where $s=\sum_{n=2}^{q}(-1)^n/n!$ and $r=\sum_{n=q+1}^{\infty}(-1)^n/n!$. Carry out the following two calculations to obtain a contradiction and complete the proof

(a) The number $q!(1/e-s)$ is an integer.

(b) By alternating series test, $\displaystyle\frac{1}{q+2}\le|q!r|\le\frac{1}{q+1}$.

7. Find the sum of each of the following series:

(a) $\sum_{n=2}^{\infty} \frac{n(n-1)}{2^n}$. (b) $\sum_{n=1}^{\infty} \frac{n^2}{2^n}$. (c) $\sum_{n=1}^{\infty} \frac{n^3}{2^n}$.

8. Suppose $\sum_{n=0}^{\infty} a_n x^n$ has $R = 1$ and $\lim_{x\to 1-} \sum_{n=0}^{\infty} a_n x^n$ exists. Can we conclude that $\sum_{n=0}^{\infty} a_n$ converges?

9. Consider the differential equation $y'' + y = \sin x$ with initial conditions $y(0) = 0$ and $y'(0) = 1$. Substitute a power series $y = \sum_{n=0}^{\infty} c_n x^n$ and solve for c_n.

(a) Show that $c_0 = 1$ and $c_1 = 1$.

(b) Show that $c_{2n} = 0$ for every $n \in \mathbb{N}$.

(c) Show that $c_{2n+1} = \frac{n-1}{(2n+1)!}(-1)^{n+1}$ for every $n \in \mathbb{N}$.

(d) Show that $\sum_{n=0}^{\infty} c_n x^n$ converges for every x.

8.2 Taylor Series

We now take up the task of expressing a given function f as a power series with a given center. Recall that the derivative of a power series is also a power series, and has the same radius of convergence. Hence, a power series has derivatives of every order. A function is called C^∞ or **smooth** if it has derivatives of every order. Only smooth functions can be expressed as power series. However, this condition is not *sufficient*. There are smooth functions that cannot be expressed as power series about certain points. (An example is given in Exercise 7 of §6.3.)

Suppose a smooth function $f(x)$ is expressible as a power series with center a and radius of convergence $R > 0$,

$$f(x) = \sum_{n=0}^{\infty} c_n (x-a)^n, \qquad \text{for } |x-a| < R.$$

By differentiating repeatedly at $x = a$, we get the following:

$$f(x) = \sum_{n=0}^{\infty} c_n (x-a)^n \implies f(a) = c_0,$$

$$f'(x) = \sum_{n=1}^{\infty} n c_n (x-a)^{n-1} \implies f'(a) = c_1,$$

$$f''(x) = \sum_{n=2}^{\infty} n(n-1) c_n (x-a)^{n-2} \implies f''(a) = 2c_2.$$

Differentiating k times gives

$$f^{(k)}(x) = \sum_{n=k}^{\infty} n(n-1)\cdots(n-k+1)c_n(x-a)^{n-k} \quad \Longrightarrow f^{(k)}(a) = k!\,c_k$$

$$\Longrightarrow c_k = \frac{f^{(k)}(a)}{k!}.$$

The power series $T_f(x) = \sum_{n=0}^{\infty} \frac{f^{(n)}(a)}{n!}(x-a)^n$ is called the **Taylor series** of $f(x)$ with center at a. If $a = 0$, it is also called the **Maclaurin series**. Every smooth function $f(x)$ has a unique Taylor series $T_f(x)$. But they may not be equal to each other.

Our calculations show that any power series expansion is automatically an equality of a function and its Taylor series. Hence, we already know the following Maclaurin series expansions:

$$\begin{aligned}
\frac{1}{(1-x)^k} &= 1 + \binom{k}{k-1}x + \binom{k+1}{k-1}x^2 + \cdots && \text{for } |x| < 1,\ k \in \mathbb{N}, \\
\log(1+x) &= x - \frac{x^2}{2} + \frac{x^3}{3} - + \cdots && \text{for } |x| < 1, \\
\arctan x &= x - \frac{x^3}{3} + \frac{x^5}{5} - + \cdots && \text{for } |x| < 1, \\
e^x &= 1 + x + \frac{x^2}{2!} + \frac{x^3}{3!} + \cdots && \text{for } x \in \mathbb{R}.
\end{aligned}$$

Theorem 8.2.1 (Binomial Series)

Consider the function $f(x) = (1+x)^r$ for any $r \in \mathbb{R}$. The Maclaurin series of f is given by

$$T(x) = \sum_{n=0}^{\infty} \binom{r}{n} x^n$$

and equals $f(x)$ for $|x| < 1$. We have used the notation

$$\binom{r}{n} = \frac{r(r-1)\cdots(r-n+1)}{n!}, \text{ with the convention } \binom{r}{0} = 1.$$

Proof. We compute the Maclaurin series by repeated differentiation:

$$\begin{aligned}
f(x) &= (1+x)^r &&\Longrightarrow f(0) = 1 &&\Longrightarrow c_0 = 1, \\
f'(x) &= r(1+x)^{r-1} &&\Longrightarrow f'(0) = r &&\Longrightarrow c_1 = r, \\
f''(x) &= r(r-1)(1+x)^{r-2} &&\Longrightarrow f''(0) = r(r-1) &&\Longrightarrow c_2 = \frac{r(r-1)}{2!}.
\end{aligned}$$

Continuing in this fashion gives $c_n = \frac{r(r-1)\cdots(r-n+1)}{n!} = \binom{r}{n}$ and we obtain the desired form of the Maclaurin series. Now, we have to address its convergence. First apply the ratio test:

$$\begin{aligned}\lim_{n\to\infty}\left|\frac{c_{n+1}x^{n+1}}{c_n x^n}\right| &= |x|\lim_{n\to\infty}\left|\frac{r(r-1)\cdots(r-n)}{r(r-1)\cdots(r-n+1)}\cdot\frac{n!}{(n+1)!}\right| \\ &= |x|\lim_{n\to\infty}\left|\frac{r-n}{n+1}\right| = |x|.\end{aligned}$$

Thus, the Maclaurin series has radius of convergence 1 and defines a function $T(x)$ for $|x|<1$. To obtain the equality of f and T we consider their derivatives:

$$f'(x) = r(1+x)^{r-1} \implies (1+x)f'(x) = rf(x),$$

and

$$\begin{aligned}(1+x)T'(x) &= \sum_{n=1}^{\infty} n\binom{r}{n}x^{n-1} + \sum_{n=1}^{\infty} n\binom{r}{n}x^n \\ &= r + \sum_{n=1}^{\infty}\left[(n+1)\binom{r}{n+1} + n\binom{r}{n}\right]x^n \\ &= r + \sum_{n=1}^{\infty} r\binom{r}{n}x^n = rT(x).\end{aligned}$$

Hence, $\left(\dfrac{T(x)}{f(x)}\right)' = \dfrac{T'(x)f(x)-T(x)f'(x)}{f(x)^2} = r\dfrac{T(x)f(x)-T(x)f(x)}{(1+x)f(x)^2} = 0.$

It follows that $T(x) = cf(x)$. Checking the values at $x=0$ gives $T(x) = f(x)$, for $|x|<1$. ■

Task 8.2.2

Prove that the binomial series expansion of $(1+x)^r$ is a finite sum if and only if $r \in \mathbb{W}$.

Recall that the Taylor polynomials of $f(x)$ are defined by

$$T_N(x) = \sum_{n=0}^{N}\frac{f^{(n)}(a)}{n!}(x-a)^n.$$

The Taylor polynomials of a smooth function are the partial sums for its Taylor series. Hence,

$$T_f(x) = \lim_{n\to\infty} T_n(x).$$

So, the question whether $f(x)$ equals $T_f(x)$ can be addressed by checking whether $\lim\limits_{N\to\infty} R_N(x) = 0$, where $R_N(x)$ is the remainder after N terms.

Example 8.2.3

From the Taylor polynomial calculations in §6.3 we know that the Maclaurin series of the sine function is

$$T_{\sin}(x) = \sum_{k=0}^{\infty} (-1)^k \frac{x^{2k+1}}{(2k+1)!} = x - \frac{x^3}{3!} + \frac{x^5}{5!} - + \cdots.$$

The remainder terms are given by

$$R_{2n+1}(x) = \sin x - \sum_{k=0}^{n} (-1)^k \frac{x^{2k+1}}{(2k+1)!} = \sin^{(2k+2)}(c) \frac{x^{2k+2}}{(2k+2)!}.$$

Hence,

$$|R_{2n+1}(x)| \leq \frac{|x|^{2k+2}}{(2k+2)!} \to 0 \implies R_{2n+1}(x) \to 0.$$

Therefore,

$$\sin x = \sum_{k=0}^{\infty} (-1)^k \frac{x^{2k+1}}{(2k+1)!} = x - \frac{x^3}{3!} + \frac{x^5}{5!} - + \cdots.$$

□

Task 8.2.4

Prove that the cosine function equals its Maclaurin series:

$$\cos x = \sum_{k=0}^{\infty} (-1)^k \frac{x^{2k}}{(2k)!} = 1 - \frac{x^2}{2!} + \frac{x^4}{4!} - + \cdots.$$

Task 8.2.5

Let $f\colon (c-R, c+R) \to \mathbb{R}$ be a C^∞ function such that there is a constant M with $|f^{(n)}(x)| \leq M^n$ for all x and n. Prove that $f(x) = \sum_{n=0}^{\infty} \frac{f^{(n)}(c)}{n!}(x-c)^n$ for every x.

Taylor Series of Products

Let f and g be smooth functions, with a common domain. From the product rule, we obtain

$$(fg)^{(n)}(x) = \sum_{k=0}^{n} \binom{n}{k} f^{(k)}(x) g^{(n-k)}(x).$$

Hence,

$$\frac{(fg)^{(n)}(x)}{n!} = \sum_{k=0}^{n} \frac{f^{(k)}(x)}{k!} \frac{g^{(n-k)}(x)}{(n-k)!}.$$

Now, suppose $T_f(x) = \sum_{n=0}^{\infty} a_n(x-a)^n$ and $T_g(x) = \sum_{n=0}^{\infty} b_n(x-a)^n$ are the respective Taylor series of f and g, with common center a. Our calculation shows that the Taylor series of the product fg is given by

$$T_{fg}(x) = \sum_{n=0}^{\infty} \Big(\sum_{k=0}^{n} a_k b_{n-k} \Big) (x-a)^n.$$

This is called the **Cauchy product** of the series $T_f(x)$ and $T_g(x)$.

Theorem 8.2.6

Let T_f and T_g be the Taylor series of f and g respectively, with common center a. Then we have the following:

1. *If T_f has radius of convergence R_f and T_g has radius of convergence R_g, then the radius of convergence of their Cauchy product T_{fg} is at least* $\min\{R_f, R_g\}$.
2. *If $f = T_f$ and $g = T_g$ then $fg = T_{fg}$.*

Proof. Apply Theorem 7.4.9. ■

Example 8.2.7

Let $f(x) = 1/(1-x)$ and $g(x) = 1/(2-x)$. Then, for $|x| < 1$,

$$f(x)g(x) = \frac{1}{2} \sum_{n=0}^{\infty} x^n \sum_{n=0}^{\infty} \frac{x^n}{2^n} = \frac{1}{2} \sum_{n=0}^{\infty} \Big(\sum_{k=0}^{n} \frac{1}{2^k} \Big) x^n = \sum_{n=0}^{\infty} \Big(1 - \frac{1}{2^{n+1}} \Big) x^n. \qquad \square$$

Taylor Series of Reciprocals

Suppose f is smooth and $f(a) \neq 0$. Then $1/f$ is smooth in an open interval centered at a. If we know the Taylor series of f with center at a, can we obtain that of $1/f$ with the same center? Suppose the two Taylor series are $T_f(x) = \sum_{n=0}^{\infty} a_n(x-a)^n$ and $T_{1/f}(x) = \sum_{n=0}^{\infty} b_n(x-a)^n$. We use the Cauchy product to obtain

$$\sum_{n=0}^{\infty} \Big(\sum_{k=0}^{n} a_k b_{n-k} \Big) (x-a)^n = 1.$$

This gives a recursive way of finding the Taylor coefficients of $1/f$.

$$\begin{aligned}
a_0 b_0 &= 1 && \Longrightarrow b_0 = \frac{1}{a_0}, \\
a_0 b_1 + a_1 b_0 &= 0 && \Longrightarrow b_1 = -\frac{a_1 b_0}{a_0} = -\frac{a_1}{a_0^2}, \\
a_0 b_2 + a_1 b_1 + a_2 b_0 &= 0 && \Longrightarrow b_2 = -\frac{a_1 b_1 + a_2 b_0}{a_0} = -\frac{1}{a_0} \Big(-\frac{a_1^2}{a_0} + \frac{a_2}{a_0} \Big).
\end{aligned}$$

In general, once $b_0, \ldots, b_{n-1}$ have been found, we obtain b_n by

$$b_n = -\frac{1}{a_0} \sum_{k=1}^{n} a_k b_{n-k}.$$

Task 8.2.8

Use the Maclaurin series of $\cos x$ *to find the first few terms of the Maclaurin series of* $\sec x$:

$$1 + \frac{x^2}{2} + \frac{5x^4}{24} + \cdots.$$

Taylor Series and Limits

Substituting a Taylor series for the corresponding function can often be useful in computing limits. For example,

$$\lim_{x\to 0} \frac{\sin x - x}{x^3} = \lim_{x\to 0} \frac{(x - x^3/3! + x^5/5! \cdots) - x}{x^3} = \lim_{x\to 0} \frac{-x^3/3! + x^5/5! \cdots}{x^3}$$

$$= \lim_{x\to 0} \left(-\frac{1}{3!} + \frac{x^2}{5!} \cdots\right) = -\frac{1}{3!}.$$

To compute a limit at a, we use Taylor series expansions centered at a, provided the series equal the corresponding functions. Here is another example:

$$\lim_{x\to 1} \frac{\log x}{\sqrt{x} - 1} = \lim_{x\to 1} \frac{(x-1) - (x-1)^2/2 \cdots}{(x-1)/2 - (x-1)^2/8 \cdots}$$

$$= \lim_{x\to 1} \frac{1 - (x-1)/2 \cdots}{1/2 - (x-1)/8 \cdots} = 2.$$

Exercises for § 8.2

1. Prove the following for a smooth function f:

(a) If f is even, its Maclaurin series has only even powers of x.

(b) If f is odd, its Maclaurin series has only odd powers of x.

2. The following calculations constitute an alternate proof that the sine function equals its Maclaurin series:

(a) The power series $\sum_{k=0}^{\infty} (-1)^k \frac{x^{2k+1}}{(2k+1)!}$ converges for every $x \in \mathbb{R}$.

(b) $S(x) = \sum_{k=0}^{\infty} (-1)^k \frac{x^{2k+1}}{(2k+1)!}$ satisfies $S''(x) = -S(x)$, $S(0) = 0$, and $S'(0) = 1$.

3. Prove the following:

(a) $\sqrt{1+x} = \sum_{n=0}^{\infty} (-1)^{n-1} \frac{(2n)!}{4^n (n!)^2 (2n-1)} x^n$ for $|x| < 1$.

(b) $\sqrt{2} = \sum_{n=0}^{\infty} (-1)^{n-1} \frac{(2n)!}{4^n (n!)^2 (2n-1)}$ and $0 = \sum_{n=0}^{\infty} \frac{(2n)!}{4^n (n!)^2 (2n-1)}$.

4. Use the binomial series to find the Maclaurin series of

(a) $\sin^{-1} x$,

(b) $\sinh^{-1} x$.

5. Find the first three non-zero terms of the Maclaurin series of the given functions by manipulating known series:

(a) $\cosh x$.

(b) $\tanh^{-1} x = \frac{1}{2} \log\left(\frac{1+x}{1-x}\right)$.

(c) $e^{x^2} \sin x$.

(d) $\frac{\sqrt{1+x}}{1-x}$.

6. Use Taylor series to compute the given limits:

(a) $\lim_{x \to 0} \frac{2(\tan x - \sin x) - x^3}{x^5}$.

(b) $\lim_{x \to 0} \frac{\sinh(x^2) - x^2}{(\sin x - x)^2}$.

(c) $\lim_{x \to 0} \frac{\log(1 + x + x^2) + \log(1 - x + x^2)}{x(e^x - 1)}$.

(d) $\lim_{x \to \infty} \left(x - x^2 \log\left(1 + \frac{1}{x}\right)\right)$.

8.3 Fourier Series

As we approach the end of this transition from school calculus to university analysis, our concern with shoring up our foundations changes to curiosity about what lies ahead. This short section on Fourier series will therefore have a quite different purpose than what came before: to raise questions more than to provide answers. Fourier series are an apt topic for this as, ever since their inception, they have provided problems and techniques that have led to fresh developments in all areas of mathematics.

Taylor polynomials and series give us a way of approximating smooth functions by polynomials. More generally, we could decide on any convenient type of function and use members of this type to approximate more general functions. The concept that we require is of a sequence of functions f_n with a limit function f. What can it mean to take a limit of a sequence of functions? We have already been practicing this with Taylor polynomials: We fix an x and ask whether $T_n(x) \to f(x)$.

Suppose (f_n) is a sequence of real functions with common domain A, and f is another such function. We say (f_n) **converges pointwise** to f on A if $f_n(x) \to f(x)$ for every x in A. Note that once we fix x, $(f_n(x))$ is a sequence of numbers and we know the definition of its limit. To indicate pointwise convergence, we write any of the following:

$$\lim_{n\to\infty} f_n \overset{pw}{=} f, \quad \lim_{n\to\infty} f_n = f \text{ (pw)}, \quad f_n \overset{pw}{\longrightarrow} f, \quad f_n \to f \text{ (pw)}.$$

The reason for introducing a special notation for pointwise convergence is that there are other kinds of convergence as well. Rather than looking at each point individually, we may look at a measure of the overall gap between the functions, such as the maximum gap between them or the area between their graphs. Each such measure induces a different notion of convergence, with different properties. (See the exercises on uniform convergence at the end of this chapter.)

Example 8.3.1

Define $f_n\colon [0,1] \to \mathbb{R}$ by $f_n(x) = x^n$. Let us calculate the pointwise limit:

$$\lim_{n\to\infty} f_n(x) = \lim_{n\to\infty} x^n = \begin{cases} 0 & \text{if } 0 \le x < 1, \\ 1 & \text{if } x = 1. \end{cases}$$

Hence, $f_n \overset{pw}{\longrightarrow} f$ where $f(x) = 0$ if $x \in [0,1)$ and $f(1) = 1$.

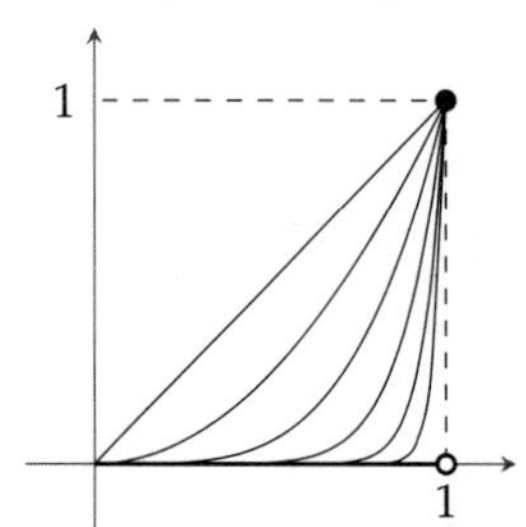

Although each f_n is individually smooth, as a collective these are moving towards a function that is discontinuous! □

Example 8.3.2

An interesting fact is that the rational numbers are in bijection with the natural numbers. A proof is given in the supplementary exercises on cardinality following Chapter 1. Let $T\colon \mathbb{N} \to \mathbb{Q}$ be such a bijection. Define $f_n\colon [0,1] \to \mathbb{R}$ by

$$f_n(x) = \begin{cases} 1 & \text{if } x \in T(\{1,\ldots,n\}), \\ 0 & \text{else.} \end{cases}$$

If $x \in [0,1]$ is irrational, then $f_n(x) = 0$ for every n. If it is rational, then there is N such that $T(N) = x$, hence $f_n(x) = 1$ for every $n \ge N$. This shows that $f_n \overset{pw}{\longrightarrow} D$, where D is the Dirichlet function. Now each f_n is a step function and hence integrable. But the pointwise limit D is not integrable. □

Example 8.3.3

Define $f_n\colon [0,2] \to \mathbb{R}$ by

$$f_n(x) = \begin{cases} n & \text{if } x \in [1/n, 2/n], \\ 0 & \text{else.} \end{cases}$$

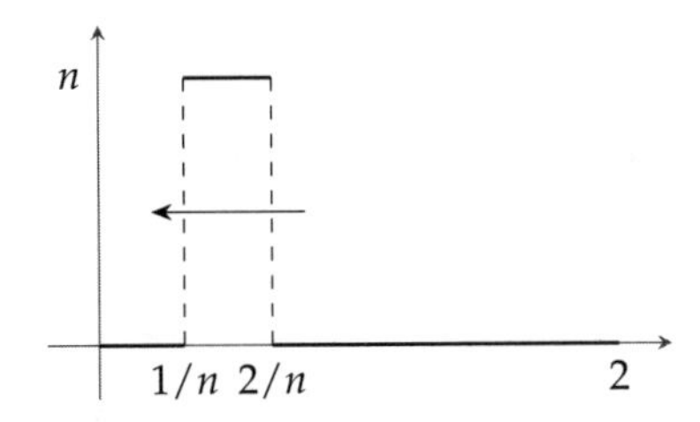

Then $f_n \xrightarrow{pw} 0$. Here there is no loss of "niceness." Yet there is a striking mismatch. We have $\int_0^1 f_n(x)\,dx = 1$ for each n but $\int_0^1 f(x)\,dx = 0$. □

These examples illustrate that the properties of the approximating functions may not carry over to the limit function. Our appreciation of power series should grow manifold after absorbing these examples, as not only were the limit functions smooth but their derivatives and integrals could be obtained as limits of those of the approximating functions. At the same time, this is a limitation. We do have important functions that are not differentiable or even continuous, and it would be useful to decompose them into (nice) parts. In this section we shall study one such decomposition.

We shall take up the goal of expressing periodic functions in terms of the sine and cosine functions. To begin, we consider functions whose period is 2π. It is enough to analyse such functions on the interval $[-\pi,\pi]$. The functions $\sin mx$ and $\cos mx$, with $m \in \mathbb{Z}$, have period 2π, so we can consider them as basic functions from which we would like to construct others. Note that it is enough to consider the following sub-collection of these functions: $\{1\} \cup \{\sin mx \mid m = 1,2,3,\ldots\} \cup \{\cos mx \mid m = 1,2,3,\ldots\}$. A **trigonometric polynomial** is a linear combination of such functions:

$$T(x) = \frac{a_0}{2} + \sum_{n=1}^{M} a_n \cos nx + \sum_{n=1}^{M} b_n \sin nx.$$

An immediate question is: if we are able to process T in terms of finding information like its integrals, can we recover the defining numbers a_n and b_n? This is part of a general approach of trying to recover the original data from knowledge of averages. We start with the following observations, whose verification is left as a task for you:

Task 8.3.4

Suppose $m, n \in \mathbb{N}$. Show the following:

$$\int_{-\pi}^{\pi} \cos mx \cos nx \, dx = \int_{-\pi}^{\pi} \sin mx \sin nx \, dx = \begin{cases} 0 & \text{if } m \neq n, \\ \pi & \text{if } m = n, \end{cases}$$

$$\int_{-\pi}^{\pi} \cos mx \sin nx \, dx = 0,$$

$$\int_{-\pi}^{\pi} 1 \cdot \cos mx \, dx = \int_{-\pi}^{\pi} 1 \cdot \sin mx \, dx = 0,$$

$$\int_{-\pi}^{\pi} 1 \cdot 1 \, dx = 2\pi.$$

From these relations, we can conclude that

$$\frac{1}{\pi} \int_{-\pi}^{\pi} T(x) \cos nx \, dx = a_n, \text{ (for } n \neq 0)$$

$$\frac{1}{\pi} \int_{-\pi}^{\pi} T(x) \sin nx \, dx = b_n,$$

$$\frac{1}{\pi} \int_{-\pi}^{\pi} T(x) \cdot 1 \, dx = a_0.$$

Setting the constant part of T to $a_0/2$ has enabled a uniform formula for all a_n.

Example 8.3.5

An example of a trigonometric polynomial is

$$T(x) = \sin x + \frac{\sin 3x}{3} + \frac{\sin 5x}{5} + \cdots + \frac{\sin 11x}{11}.$$

We plot its graph and observe that it appears to be approaching a square wave shape:

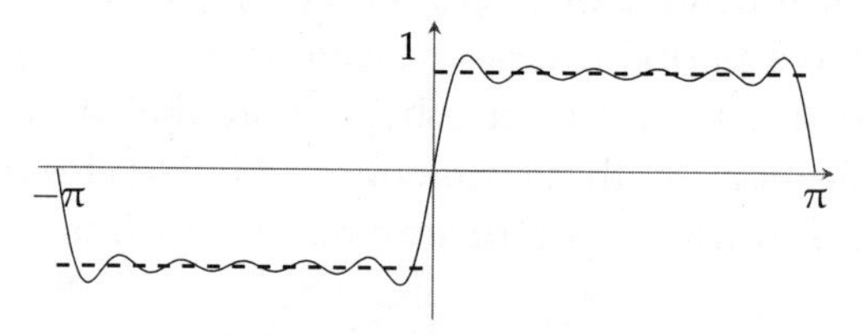

□

This example shows that trigonometric polynomials have the potential to approximate discontinuous periodic functions. Therefore, we define a **trigonometric series** to be an expression of the form

$$T(x) = \frac{a_0}{2} + \sum_{n=1}^{\infty} a_n \cos nx + \sum_{n=1}^{\infty} b_n \sin nx.$$

The goal is to generate a trigonometric series for a given periodic function f, and then establish that the series converges to the given function. Our work with trigonometric polynomials suggests we generate the coefficients a_n and b_n through integration.

Let f be a function with period 2π and which is integrable on $[-\pi,\pi]$. Then the functions $f(x)\cos nx$ and $f(x)\sin nx$ are also integrable. (See Exercise 10 of §2.2.) Hence, we can define

$$a_n = \frac{1}{\pi}\int_{-\pi}^{\pi} f(x)\cos nx\,dx, \quad b_n = \frac{1}{\pi}\int_{-\pi}^{\pi} f(x)\sin nx\,dx.$$

The numbers a_n and b_n are called the **Fourier coefficients** of f. The corresponding trigonometric series is called the **Fourier series** of f and we write:

$$f(x) \sim \frac{a_0}{2} + \sum_{n=1}^{\infty} a_n \cos nx + \sum_{n=1}^{\infty} b_n \sin nx.$$

The use of the symbol $\sim$ indicates that the right-hand side is the Fourier series of f but we have not established its convergence to f.

Example 8.3.6

Consider the **square wave function** defined by $S(x) = -1$ if $x \in (-\pi,0]$ and $S(x) = 1$ if $x \in \{-\pi\} \cup (0,\pi]$. Then the Fourier coefficients of S are:

$$\begin{aligned}
a_n &= \frac{1}{\pi}\int_{-\pi}^{0}(-1)\cos nx\,dx + \frac{1}{\pi}\int_{0}^{\pi}(1)\cos nx\,dx = 0+0 = 0,\\
b_n &= \frac{1}{\pi}\int_{-\pi}^{0}(-1)\sin nx\,dx + \frac{1}{\pi}\int_{0}^{\pi}(1)\sin nx\,dx\\
&= \frac{2}{\pi}\int_{0}^{\pi}(1)\sin nx\,dx = -\frac{2}{\pi}\cdot\left.\frac{\cos nx}{n}\right|_0^{\pi} = \frac{2}{\pi}\cdot\frac{1-(-1)^n}{n}\\
&= \begin{cases} 0 & \text{if } n \text{ even},\\ 4/\pi n & \text{if } n \text{ odd}.\end{cases}
\end{aligned}$$

Hence, the Fourier series is $S(x) \sim \frac{4}{\pi}\left(\sin x + \frac{\sin 3x}{3} + \frac{\sin 5x}{5} + \cdots\right)$. Let us see Fourier's own method for establishing the convergence of this series (See Fourier [8, p. 145–153]). Let $S_m(x)$ be the partial sum of the first m terms of the series,

$$\frac{\pi}{4}S_m(x) = \sin x + \frac{\sin 3x}{3} + \cdots + \frac{\sin(2m-1)x}{2m-1}.$$

Differentiate, and multiply both sides by $2\sin 2x$:

$$\begin{aligned}
\frac{\pi}{2}S_m'(x)\sin 2x &= 2\cos x\sin 2x + 2\cos 3x\sin 2x + \cdots + 2\cos(2m-1)x\sin 2x\\
&= (\sin 3x - \sin(-x)) + (\sin 5x - \sin x) + \cdots\\
&\quad + (\sin(2m+1)x - \sin(2m-3)x)\\
&= \sin(2m+1)x + \sin(2m-1)x = 2\sin(2mx)\cos x.
\end{aligned}$$

Hence, $S'_m(x) = \dfrac{2}{\pi}\dfrac{\sin 2mx}{\sin x}$. Note that $S_m(\pi/2) = (4/\pi)(1 - 1/3 + 1/5 + \cdots + (-1)^{m-1}/(2m-1)) \to 1$, by the Gregory–Leibniz formula that we observed earlier. Now, for any $a \in [\pi/2, \pi)$,

$$\begin{aligned}
S_m(a) &= S_m(\pi/2) + \int_{\pi/2}^{a} S'_m(x)\,dx = S_m(\pi/2) + \frac{2}{\pi}\int_{\pi/2}^{a} \sin 2mx \csc x\,dx \\
&= S_m(\pi/2) - \frac{2}{\pi}\frac{\cos 2mx \csc x}{2m}\bigg|_{\pi/2}^{a} + \frac{1}{m\pi}\int_{\pi/2}^{a} \cos 2mx \csc' x\,dx \\
&= S_m(\pi/2) - \frac{\cos 2ma \csc a - (-1)^m}{m\pi} + \frac{\cos 2m\xi_m}{m\pi}\int_{\pi/2}^{a} \csc' x\,dx \\
&= S_m(\pi/2) - \frac{1}{m\pi}\Big(\cos 2ma \csc a - \cos 2m\xi_m(\csc a - 1) - (-1)^m\Big) \\
&\to 1 \text{ as } m \to \infty.
\end{aligned}$$

The argument also works for $a \in (0, \pi/2]$. We have to separate these cases because in the fourth equality we have used the mean value theorem for weighted integration (Theorem 3.6.7), which requires the weight function to not change sign. Further, for $x \in (-\pi, 0)$, we have $S_m(x) \to -1$ because these are odd functions.

Thus, the Fourier series converges to $S(x)$ at the points where it is continuous. At the points of discontinuity ($-\pi$, 0, π) the series gives zero, which is the mean of the left- and right-hand limits. □

Task 8.3.7

Show that $\dfrac{\pi}{4} = \dfrac{1}{\sqrt{2}}\left(1 + \dfrac{1}{3} - \dfrac{1}{5} - \dfrac{1}{7} + \cdots\right)$.

Task 8.3.8

Let S be the square wave function of the last example. Show that the Fourier series of $g(x) = S(x-a)$ converges to 0 at $x = a$.

The example of the square wave gives some idea of the results and proofs of Fourier series. In the remaining part of this section, we shall do a few more examples and partial calculations to bring greater clarity to the questions at hand. The first thing we do is to take Fourier's idea of collapsing partial sums into a compact expression and apply it in a slightly different way. Consider a Fourier series expansion

$$f(x) \sim \frac{a_0}{2} + \sum_{n=1}^{\infty} a_n \cos nx + \sum_{n=1}^{\infty} b_n \sin nx.$$

Let f_m represent the following partial sum,

$$f_m(x) = \frac{a_0}{2} + \sum_{n=1}^{m} a_n \cos nx + \sum_{n=1}^{m} b_n \sin nx.$$

Substitute the formulas for the Fourier coefficients:

$$\begin{aligned}
f_m(x) &= \frac{1}{2\pi}\int_{-\pi}^{\pi} f(s)\,ds + \sum_{n=1}^{m}\left(\frac{1}{\pi}\int_{-\pi}^{\pi} f(s)\cos ns\,ds\right)\cos nx \\
&\quad + \sum_{n=1}^{m}\left(\frac{1}{\pi}\int_{-\pi}^{\pi} f(s)\sin ns\,ds\right)\sin nx \\
&= \frac{1}{\pi}\int_{-\pi}^{\pi} f(s)\left(\frac{1}{2} + \sum_{n=1}^{m}(\cos ns\cos nx + \sin ns\sin nx)\right)ds \\
&= \frac{1}{\pi}\int_{-\pi}^{\pi} f(s)\left(\frac{1}{2} + \sum_{n=1}^{m}\cos n(s-x)\right)ds.
\end{aligned}$$

Task 8.3.9

Show that $\frac{1}{2} + \sum_{n=1}^{m}\cos n\theta = \frac{\sin(m+1/2)\theta}{2\sin\theta/2}$. *Give a justification for claiming equality even when* $\theta = 2n\pi$.

The collection of functions $D_m(x) = \frac{\sin(m+1/2)x}{2\pi\sin x/2}$ is called the **Dirichlet kernel**. Assuming that f extends beyond $[-\pi,\pi]$ with a period of 2π, the partial sums of the Fourier series can be expressed as

$$f_m(x) = \int_{-\pi}^{\pi} f(s)D_m(s-x)\,ds = \int_{-\pi}^{\pi} f(s+x)D_m(s)\,ds.$$

If we take $f(x) = 1$ then all its partial sums are also 1, and we get

$$1 = \int_{-\pi}^{\pi} D_m(s)\,ds \qquad \text{for every } n.$$

The following graphs depict D_5 and D_{10}:

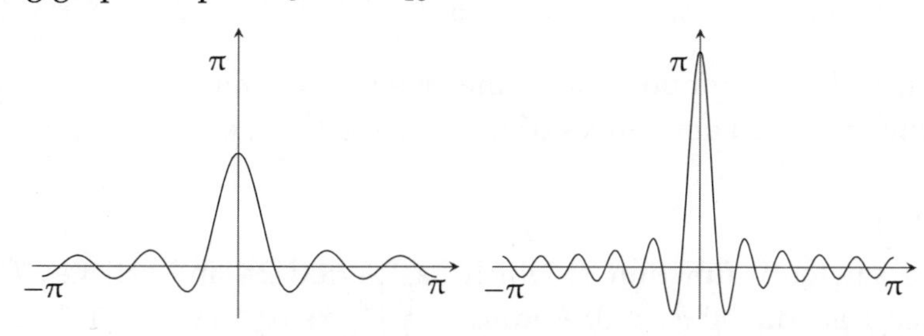

As n increases, the Dirichlet kernel gets more concentrated near origin. Consequently, the values of $\int_{-\pi}^{\pi} f(s+x)D_n(s)\,ds$ depend more and more on the values of f near x (corresponding to $s = 0$). If f is continuous at x, this should make the integral approach $f(x)$. If f has a jump discontinuity, then the left and right limits contribute equally (due to the symmetry of D_n) and so we expect the integral to converge to their mean.

These arguments are not proofs but they tell us what is feasible, and we can look for extra conditions under which they can be strengthened into proofs. Fourier himself, working in a time when functions were not formally defined, talked of "pieces of arcs" and applied integration by parts, thus implicitly assuming functions defined piecewise, with each piece being continuously differentiable. We state below the simplest modern version of his result.

Theorem 8.3.10

Let $f\colon [-\pi,\pi] \to \mathbb{R}$ be "piecewise differentiable" in the following sense:

1. *The interval $[-\pi,\pi]$ has a partition $-\pi = x_0 < \cdots < x_n = \pi$ such that f is differentiable on each subinterval (x_i, x_{i+1}).*
2. *The left and right limits of f and f' exist at every point.*

Then we have the following convergence results:

1. *If f is continuous at x, the Fourier series at x converges to $f(x)$.*
2. *If f has a jump discontinuity at x, the Fourier series at x converges to the mean of the left- and right-hand limits at x.*

The proof requires some facts about the behavior of the Fourier coefficients, and we defer it to later in this section (p. 330). We make the following observation in advance:

Suppose f satisfies the hypotheses of the theorem and is differentiable on (a,b). Let us denote the right-hand limit of f at a by $f(a+)$, and the left-hand limit at b by $f(b-)$. Define $g\colon [a,b] \to \mathbb{R}$ by $g(x) = f(x)$ if $a < x < b$, $g(a) = f(a+)$, $g(b) = f(b-)$. Then the mean value theorem applies to g. Hence,

$$\lim_{h\to 0+} \frac{f(a+h) - f(a+)}{h} = \lim_{h\to 0+} f'(\xi_h) = \lim_{x\to a+} f'(x) = f'(a+),$$
$$\lim_{h\to 0-} \frac{f(b+h) - f(b-)}{h} = \lim_{h\to 0-} f'(\eta_h) = \lim_{x\to b-} f'(x) = f'(b-).$$

In particular, if f is continuous at x, the one-sided limits of f' are equal to the corresponding one-sided derivatives of f: $f'(x+) = f'_+(x)$ and $f'(x-) = f'_-(x)$.

Example 8.3.11

Consider the **sawtooth function** f, which has period 2π and satisfies $f(x) = x$ on $(-\pi,\pi)$. Since f is odd, all $a_n = 0$. And $b_n = \frac{1}{\pi}\int_{-\pi}^{\pi} x \sin nx\, dx = (-1)^{n+1} 2/n$. Hence, the Fourier series is

$$f(x) \sim 2\left(\sin x - \frac{\sin 2x}{2} + \frac{\sin 3x}{3} - \cdots\right)$$

We plot two of the partial sums below, with 3 and 10 terms respectively.

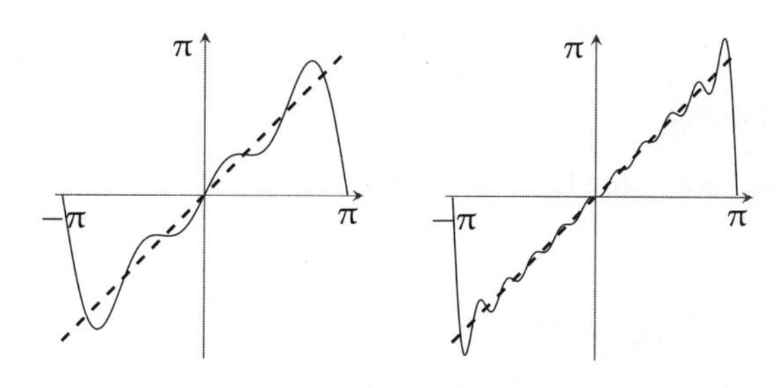

□

Example 8.3.12

Let us try out a function that is continuous but fails to be differentiable at one point: $C(x) = \pi - |x|$ for $x \in [-\pi,\pi]$. We ask you to calculate, using integration by parts, that

$$a_n = \begin{cases} \pi & \text{if } n = 0, \\ 4/\pi n^2 & \text{if } n \text{ odd}, \\ 0 & \text{else}, \end{cases}$$
$$b_n = 0.$$

Therefore, $C(x) \sim \frac{\pi}{2} + \frac{4}{\pi}\left(\cos x + \frac{\cos 3x}{3^2} + \frac{\cos 5x}{5^2} + \cdots\right)$. Convergence of the Fourier series to $C(x)$ is guaranteed by the last theorem. The partial sum $\frac{\pi}{2} + \frac{4}{\pi}\left(\cos x + \frac{\cos 3x}{3^2} + \frac{\cos 5x}{5^2}\right)$ is plotted below.

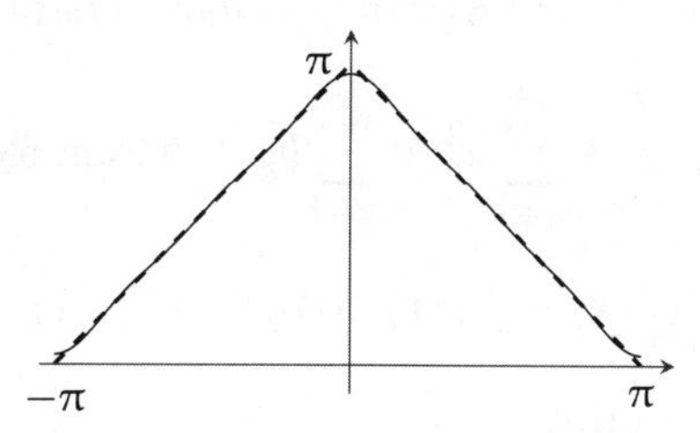

Note how quickly the series converges when the function has no discontinuity. □

Task 8.3.13

Show that $\frac{\pi^2}{8} = 1 + \frac{1}{3^2} + \frac{1}{5^2} + \cdots$.

Let us use the name $\mathcal{I}$ for the class of integrable functions on $[-\pi,\pi]$. For $f,g \in \mathcal{I}$ we define $\langle f,g\rangle = \frac{1}{\pi}\int_{-\pi}^{\pi} f(x)g(x)\,dx$. Note that the Fourier coefficients of f are given by $a_n = \langle f(x), \cos nx\rangle$ and $b_n = \langle f(x), \sin nx\rangle$.

Task 8.3.14

Let $f,g,h \in \mathcal{I}$. Show the following:

(a) $\langle f,f\rangle \geq 0$. *(Hence we can define* $||f|| = \langle f,f\rangle^{1/2}$.*)*

(b) $\langle f,g\rangle = \langle g,f\rangle$.

(c) *If* $c \in \mathbb{R}$ *then* $\langle cf,g\rangle = \langle f,cg\rangle = c\langle f,g\rangle$.

(d) $\langle f+g,h\rangle = \langle f,h\rangle + \langle g,h\rangle$ *and* $\langle f,g+h\rangle = \langle f,g\rangle + \langle f,h\rangle$.

(e) $\langle f,g\rangle = 0 \implies ||f+g||^2 = ||f||^2 + ||g||^2$.

Theorem 8.3.15 (Bessel's Inequality)

Let $f \in \mathcal{I}$ *and* $f(x) \sim \frac{a_0}{2} + \sum_{n=1}^{\infty} a_n \cos nx + \sum_{n=1}^{\infty} b_n \sin nx$. *Then*

$$\frac{a_0^2}{2} + \sum_{n=1}^{\infty} a_n^2 + \sum_{n=1}^{\infty} b_n^2 \leq \frac{1}{\pi}\int_{-\pi}^{\pi} f(x)^2\,dx.$$

Proof. Let f_m be a partial sum of the Fourier series:

$$f_m(x) = \frac{a_0}{2} + \sum_{k=1}^{m} a_k \cos kx + \sum_{k=1}^{m} b_k \sin kx.$$

Due to the orthogonality relations, we have for $1 \leq n \leq m$,

$$\langle f_m(x),1\rangle = a_0, \quad \langle f_m(x),\cos nx\rangle = a_n, \quad \langle f_m(x),\sin nx\rangle = b_n.$$

Hence, $||f_m||^2 = \langle f_m,f_m\rangle = \dfrac{a_0^2}{2} + \sum\limits_{n=1}^{\infty} a_n^2 + \sum\limits_{n=1}^{\infty} b_n^2$. Further, for $1 \leq n \leq m$,

$$\langle f(x) - f_m(x),1\rangle = \langle f(x) - f_m(x),\cos nx\rangle = \langle f(x) - f_m(x),\sin nx\rangle = 0.$$

Hence, $\langle f - f_m,f_m\rangle = 0$. Therefore,

$$||f||^2 = ||(f - f_m) + f_m||^2 = ||f - f_m||^2 + ||f_m||^2 \geq ||f_m||^2.$$

Since this is true for all m, the result follows. ■

Task 8.3.16

(Riemann Lemma) Let $f \in \mathcal{I}$ *with Fourier coefficients* a_n,b_n. *Prove that* $a_n,b_n \to 0$.

We are now in a position to prove Theorem 8.3.10 on the convergence of Fourier series.

Proof. Let f satisfy the hypothesis of Theorem 8.3.10. Let $f_m(x) = \frac{a_0}{2} + \sum_{n=1}^{m} a_n \cos nx + \sum_{n=1}^{m} b_n \sin nx$ be a partial sum of its Fourier series. Then

$$\begin{aligned} f_m(x) - f(x) &= \int_{-\pi}^{\pi} f(x+s)D_m(s)\,ds - \int_{-\pi}^{\pi} f(x)D_m(s)\,ds \\ &= \int_{-\pi}^{\pi} \Big(f(x+s) - f(x)\Big) D_m(s)\,ds \\ &= \frac{1}{2\pi}\int_{-\pi}^{\pi} \frac{f(x+s) - f(x)}{\sin s/2} \sin(m+1/2)s\,ds. \end{aligned}$$

Suppose f is continuous at x. The function $\phi(s) = \dfrac{f(x+s) - f(x)}{\sin s/2} = \dfrac{f(x+s) - f(x)}{s}\dfrac{s}{\sin s/2}$ has left and right limits at $s = 0$ because f has left and right derivatives at x. Consequently ϕ is bounded and piecewise continuous, hence integrable. Now we continue our calculations:

$$\begin{aligned} f_m(x) - f(x) &= \frac{1}{2\pi}\int_{-\pi}^{\pi} \phi(s)\sin(m+\frac{1}{2})s\,ds \\ &= \frac{1}{2\pi}\int_{-\pi}^{\pi} (\phi(s)\cos\frac{s}{2})\sin ms\,ds + \frac{1}{2\pi}\int_{-\pi}^{\pi} (\phi(s)\sin\frac{s}{2})\cos ms\,ds \\ &\to 0 + 0 = 0. \qquad \text{(Riemann lemma)} \end{aligned}$$

Thus, the Fourier series converges to f at points of continuity.

Next, suppose f has a jump discontinuity at x. Let $f(x+) = \lim_{t\to x+} f(t)$ and $f(x-) = \lim_{t\to x-} f(t)$. Define $c = f(x+) - f(x-)$ and $g(t) = f(t) - \frac{c}{2}S(t-x)$ for $t \neq x$, where S is the square wave function of Example 8.3.6. Then $g(x+) = g(x-)$ and we define $g(x) = \lim_{t\to x} g(t)$.

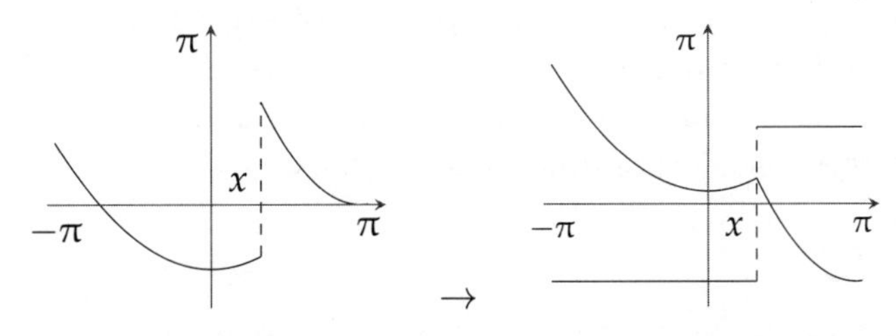

It follows that g satisfies the hypotheses of the theorem and is continuous at x. Hence, the Fourier series of g converges to $g(x)$ at x. Further, the Fourier series of $S(t-x)$ converges to 0 at x. Hence, the Fourier series of $f(t) = g(t) + \frac{c}{2}S(t-x)$ converges to $g(x) = g(x+) = f(x+) - \frac{c}{2}S(0+) = \dfrac{f(x+) + f(x-)}{2}$ at x. This concludes the proof. ∎

For a concise description of Fourier series, see Hamming [12]. Tolstov [36] provides a full development, accessible after completing this book. Finally, Körner [15] will take you beyond the Fourier series to the Fourier transform and to the interaction of these topics with a wide variety of mathematics and its applications.

Let us conclude this section with a sample of questions that arise from studying Fourier series, and the resulting topics you would likely come across in subsequent courses on analysis:

1. Given that pointwise convergence is not well coordinated with differentiation and integration, can we come up with notions of convergence that do better? This leads to different measures of the "size" of a function (most commonly the L^p-norms) and using them to define distance between two functions.
2. Given that integration is a tool for obtaining the coefficients and studying convergence, can we create a better version of integration that would expand the class of "integrable" functions as well as have a better relationship with pointwise convergence? This leads to the Lebesgue integral.
3. We arrived at a convergence result when a function has finitely many points of discontinuity. How many discontinuities can we tolerate, and with what distribution, and still have positive results? This leads to the Lebesgue measure, which generalizes the notion of length of an interval to arbitrary sets, and is closely tied to the Lebesgue integral. It further leads to abstract measure theory, which is the appropriate language for studying probability.
4. The Fourier series is one means of generating a trigonometric series that may match a given function. Could a function be expanded in trigonometric series in multiple ways? This question led Cantor to study sets with the property that if two trigonometric series agree on them, then they have the same coefficients, and thus seeded the modern theory of sets.
5. Integrals of the form $\int_{-\pi}^{\pi} f(x)g(x)\,dx$ can be viewed as generalizations of the dot product $\sum x_i y_i$ on $\mathbb{R}^n$. Then the process of extracting Fourier coefficients is analogous to finding the orthogonal components of a vector. This connects the theory to linear algebra, and pushes us to define and study infinite dimensional vector spaces, the subject of functional analysis.

Exercises for § 8.3

1. Find the Fourier series of the following functions defined on $(-\pi, \pi)$:

(a) $f(x) = x$.

(b) $g(x) = x^2$.

(c) $h(x) = x^3$.

(d) $k(x) = \cos(x/2)$.

2. Suppose $f(x) \sim a_0/2 + \sum_{n=1}^{\infty} a_n \cos nx + \sum_{n=1}^{\infty} b_n \sin nx$. Show that

$$f(x-a) \sim \frac{a_0'}{2} + \sum_{n=1}^{\infty} a_n' \cos nx + \sum_{n=1}^{\infty} b_n' \sin nx,$$

where $a_n' = a_n \cos na - b_n \sin na$ and $b_n' = a_n \sin na + b_n \cos na$.

3. Suppose $f(x) \sim a_0/2 + \sum_{n=1}^{\infty} a_n \cos nx + \sum_{n=1}^{\infty} b_n \sin nx$ and f' is integrable. Show that $f'(x) \sim a_0'/2 + \sum_{n=1}^{\infty} a_n' \cos nx + \sum_{n=1}^{\infty} b_n' \sin nx$, where $a_n' = n\, b_n$ and $b_n' = -n\, a_n$. (Hint: integration by parts.)

4. Show that the Fourier series $a_0/2 + \sum_{n=1}^{\infty} a_n \cos nx + \sum_{n=1}^{\infty} b_n \sin nx$ can be expressed as

$$f(x) \sim \frac{A_0}{2} + \sum_{n=1}^{\infty} A_n \cos(nx + \theta_n),$$

where $A_n = \sqrt{a_n^2 + b_n^2}$.

5. Use the Fourier series from the examples and previous exercises to obtain the following formulas:

(a) $\dfrac{\pi^2}{6} = 1 + \dfrac{1}{2^2} + \dfrac{1}{3^2} + \cdots.$

(b) $\dfrac{\pi^2}{12} = 1 - \dfrac{1}{2^2} + \dfrac{1}{3^2} - \cdots.$

(c) $\dfrac{\pi^3}{32} = 1 - \dfrac{1}{3^3} + \dfrac{1}{5^3} - \cdots.$

(d) $\dfrac{\pi^4}{90} = 1 + \dfrac{1}{2^4} + \dfrac{1}{3^4} + \cdots.$

6. Show that $\dfrac{\pi^4}{96} = 1 + \dfrac{1}{3^4} + \dfrac{1}{5^4} + \cdots$. (Hint: Split (d) of Exercise 5 into powers of even and odd numbers.)

7. Find a function $f \colon [-\pi, \pi] \to \mathbb{R}$ whose Fourier expansion is $\sum_{n=1}^{\infty} \dfrac{\cos nx}{n^2}$. (Hint: Start with the Fourier series of x^2.)

8. Show that the trigonometric series $\sum_{n=1}^{\infty} \dfrac{\cos nx}{\sqrt{n}}$ converges for every x but is not the Fourier series of an integrable periodic function. (Hint: See Exercise 10 of §7.4.)

8.4 Complex Series

In the previous section, we used Fourier series as a lens to bring into focus various questions about analysis with real numbers. In this section, we look at some instances of the benefits from roping in complex numbers. We begin with a brief review of complex numbers, and then take up limits of complex sequences and sums of complex series. We show how the study of both power series and Fourier series gains clarity when we carry it out with complex numbers. These are the first steps of an exciting journey with a very different flavor, into the realms of "complex analysis."

Consider $\mathbb{R}^2 = \mathbb{R} \times \mathbb{R}$ with the binary operations given below.

$$\text{Addition: } (x,y) + (u,v) = (x+u, y+v).$$
$$\text{Multiplication: } (x,y) * (u,v) = (xu - yv, xv + yu).$$

We denote the cartesian plane with these operations by $\mathbb{C}$ and call it the **complex plane**. First we observe that $\mathbb{C}$ is a field.

1. The operations $+$ and $*$ are commutative and associative.
2. $(0,0)$ is the additive identity, and $(-x,-y)$ is the additive inverse of (x,y).
3. $(1,0)$ is the multiplicative identity, and a nonzero (x,y) has multiplicative inverse $\left(\frac{x}{x^2+y^2}, \frac{-y}{x^2+y^2}\right)$.
4. Multiplication distributes over addition.

Now we identify $(1,0)$ with the real number 1 and denote $(0,1)$ by i. Then $(x,y) = x(1,0)+y(0,1) = x+iy$. In this notation, the rules for $+$ and $*$ become

$$(x+iy)+(u+iv) = (x+u)+i(y+v).$$
$$(x+iy)(u+iv) = (xu-yv)+i(xv+yu).$$

If $z = x+iy$ with $x,y \in \mathbb{R}$, we say x is the **real part** of z and y is the **imaginary part** of z. We shall use "Let $z = x+iy$" as an abbreviation for "Let $z = x+iy$ with $x,y \in \mathbb{R}$."

The **absolute value** or **modulus** of z is $|z| = \sqrt{x^2+y^2}$ and the **conjugate** of z is $\overline{z} = x-iy = x+i(-y)$. We have the following properties:

1. $z = 0 \iff |z| = 0$,
2. $z\overline{z} = |z|^2$,
3. $z \neq 0 \implies z^{-1} = \overline{z}/|z|^2$,
4. $\overline{\overline{z}} = z$,
5. $\overline{z+w} = \overline{z}+\overline{w}$, $\overline{zw} = \overline{z}\overline{w}$,
6. $|z+w| \leq |z|+|w|$,
7. $|z-w| \geq ||z|-|w||$,
8. $|zw| = |z||w|$.

The operations involving complex numbers can be visualized in the Cartesian plane, which is then called the **Argand plane**.

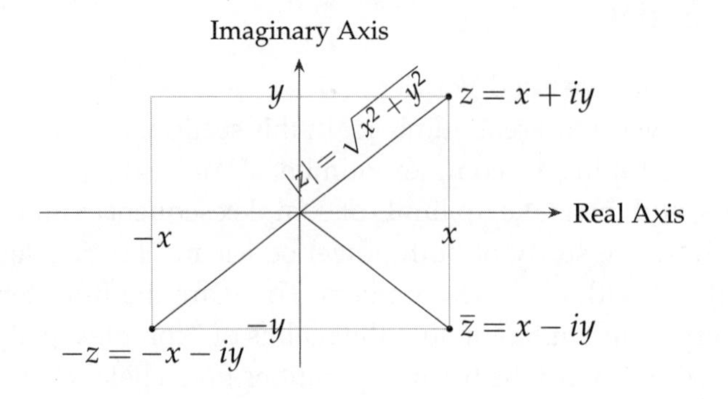

The process of addition is depicted by the "Parallelogram Law" in the Argand plane:

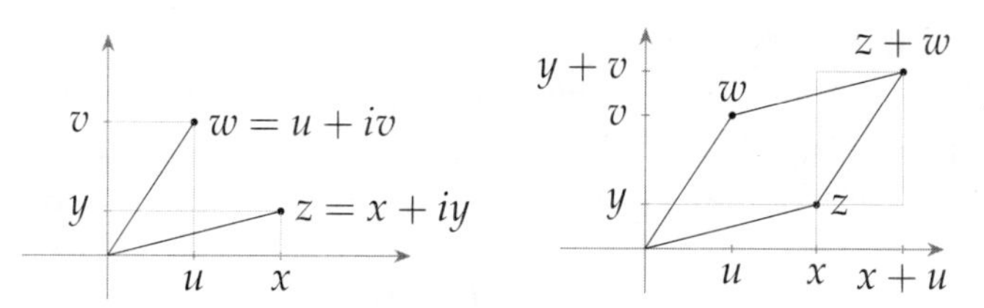

To visualize multiplication, let us first view what happens when a complex number is multiplied by i. We have $i(x+iy) = -y+ix$.

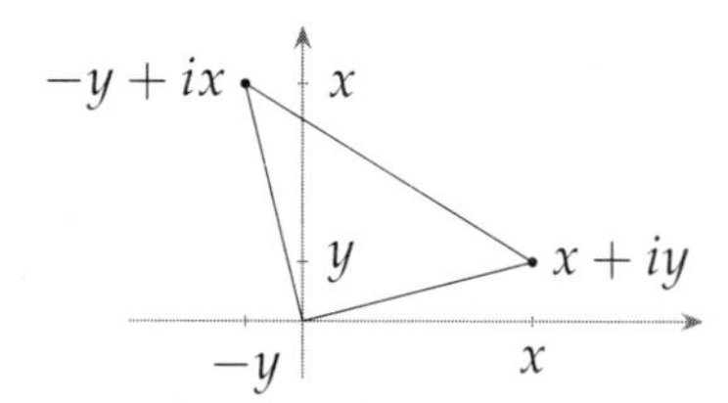

We see that the points $x+iy$, $-y+ix$, and 0 form a right angled triangle. Thus, multiplication by i causes a counter-clockwise rotation by a right angle around the origin. Multiplying by i *twice* gives a total rotation of 180 degrees and hence sends z to $-z$. This is the geometric description of i being the square root of -1.

Any complex number $z = x+iy$ can be expressed as $|z|(\cos\theta + i\sin\theta)$ where θ is the angle between the rays starting at origin and passing through the points $(1,0)$ and (x,y). Let $w = |w|(\cos\phi + i\sin\phi)$ be another complex number. Let us multiply them:

$$\begin{aligned} zw &= |z||w|(\cos\theta + i\sin\theta)(\cos\phi + i\sin\phi) \\ &= |z||w|((\cos\theta\cos\phi - \sin\theta\sin\phi) + i(\cos\theta\sin\phi + \sin\theta\cos\phi)) \\ &= |z||w|(\cos(\theta+\phi) + i\sin(\theta+\phi)) \end{aligned}$$

Thus, the effect of multiplying by z is to stretch the other number by a factor of $|z|$ and also rotate it by θ.

Task 8.4.1

(De Moivre's formula) If $z = |z|(\cos\theta + i\sin\theta) \neq 0$ *and* $n \in \mathbb{Z}$ *then*

$$z^n = |z|^n(\cos n\theta + i\sin n\theta).$$

Task 8.4.2

Compute $(1+i)^{1000}$.

Roots of Complex Numbers

Suppose $z^n = 1$ for $n \in \mathbb{N}$. Write $z = |z|(\cos\theta + i\sin\theta)$. Then $1 = |z|^n(\cos n\theta + i\sin n\theta)$. Hence, $|z| = 1$ and $n\theta = 2\pi k$ with $k \in \mathbb{Z}$. Taking $k = 1$, we get the root $w = \cos(2\pi/n) + i\sin(2\pi/n)$. If we let k vary over $0,\ldots,n-1$ we get n distinct **roots of unity**: $1, w, w^2, \ldots, w^{n-1}$. They are equally spaced out on the unit circle:

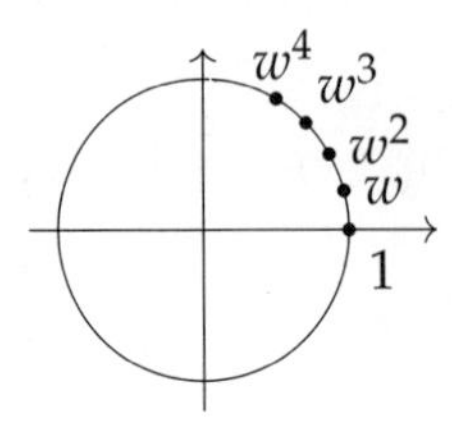

Task 8.4.3

Prove that $1 + w + w^2 + \cdots + w^{n-1} = 0$.

If $z = |z|(\cos\theta + i\sin\theta)$ is a non-zero complex number, then

$$\alpha = |z|^{1/n}(\cos(\theta/n) + i\sin(\theta/n))$$

is an nth root: $\alpha^n = z$. Let $1, w, \ldots, w^{n-1}$ be the roots of unity. Then $\alpha, \alpha w, \ldots, \alpha w^{n-1}$ are the nth roots of z.

Thus, every non-zero complex number has n distinct n^{th} roots, and they are equally spaced on a circle centered at origin. Notice how uniform and pleasant this situation is compared to taking roots of real numbers.

Example 8.4.4

Let us compute the 4th roots of i. We will need the 4th roots of unity: 1, $w = i$, $w^2 = -1$, $w^3 = -i$. We need the root $\alpha = \cos\pi/8 + i\sin\pi/8$ of i, which is obtained by taking $\theta = \pi/2$ and $n = 4$. Then all the roots of i are obtained by multiplying each root of unity by α, that is, by rotating each of them through an angle of $\pi/8$.

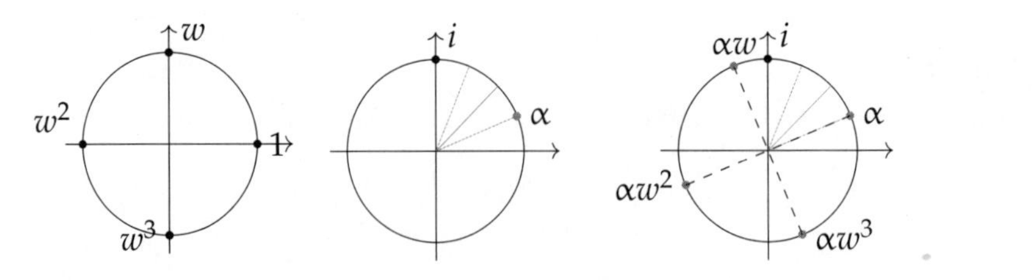

□

Convergence of Complex Sequences

A **complex sequence** is a function from $\mathbb{N}$ to $\mathbb{C}$, that is, an unending list of complex numbers. We can set up the notion of the limit of a complex sequence just as we did for real numbers, and have corresponding properties.

Let (z_n) be a complex sequence, and L a complex number. We say that (z_n) **converges** to L if for every real number $\epsilon > 0$ there is $N \in \mathbb{N}$ such that $n \geq N$ implies $|z_n - L| < \epsilon$. The number L is called the **limit** of (z_n), and we write $\lim_{n\to\infty} z_n = L$ or $\lim z_n = L$ or $z_n \to L$.

Task 8.4.5

Prove the following:

(a) $|z_n| \to 0$ *if and only if* $z_n \to 0$.

(b) $z_n \to L$ *if and only if* $z_n - L \to 0$.

(c) $z_n \to L$ *implies* $|z_n| \to |L|$.

Theorem 8.4.6

Consider a sequence (z_n) *and let* $z_n = x_n + iy_n$*. Further, let* $L = M + iN$*. Then* $z_n \to L$ *if and only if* $x_n \to M$ *and* $y_n \to N$*.*

Proof. We begin by noting that $|z_n - L|^2 = |x_n - M|^2 + |y_n - N|^2$. Hence,

$$\begin{aligned} x_n \to M \text{ and } y_n \to N &\implies |x_n - M| \to 0 \text{ and } |y_n - N| \to 0 \\ &\implies |z_n - L|^2 = |x_n - M|^2 + |y_n - N|^2 \to 0 \\ &\implies z_n \to L. \end{aligned}$$

In the other direction, noting that $|x_n - M|, |y_n - N| \leq |z_n - L|$, we apply the sandwich theorem for real sequences.

$$\begin{aligned} z_n \to L \implies |z_n - L| \to 0 &\implies |x_n - M|, |y_n - N| \to 0 \\ &\implies x_n \to M \text{ and } y_n \to N. \end{aligned}$$

■

Example 8.4.7

Consider a fixed complex number z and consider the sequence (z^n) of powers of z. If $|z| > 1$, we already know $|z|^n$ diverges, and hence z^n diverges. On the other hand, if $|z_n| < 1$, then $|z|^n \to 0$ and hence $z_n \to 0$. What happens if $|z| = 1$?

If $|z| = 1$ the sequence (z^n) rotates around the unit circle and so has no limit. The only exception, of course, is when $z = 1$.

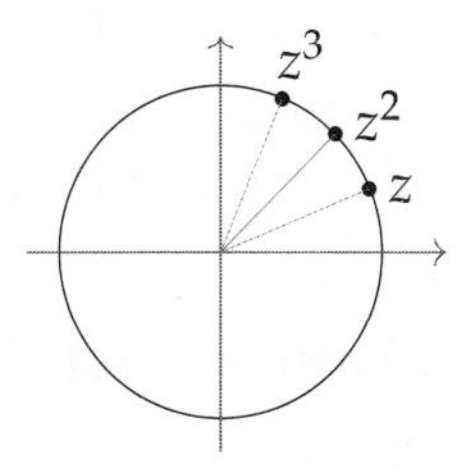

□

Complex sequences follow the same algebra of limits as real sequences. They lack the sandwich theorem as complex numbers are not an ordered field.

Convergence of Complex Series

Given a complex sequence (z_n), we create the corresponding series $\sum_{n=1}^{\infty} z_n$. It has the partial sums $S_n = \sum_{k=1}^{n} z_k$, and we define the **sum** of the series as $\lim S_n$. The series **converges** if this limit exists and **diverges** if it does not.

Example 8.4.8

Consider a geometric series $\sum_{n=1}^{\infty} z^{n-1} = 1 + z + z^2 + \cdots$ with $z \neq 1$. The partial sums are

$$S_n = 1 + z + \cdots + z^{n-1} = \frac{1 - z^n}{1 - z}.$$

By our earlier calculations we see that $\sum_{n=1}^{\infty} z^{n-1} = \frac{1}{1-z}$ if $|z| < 1$, and diverges otherwise. □

Task 8.4.9

(Divergence Test) If $\sum z_n$ *converges, then* $z_n \to 0$.

Task 8.4.10

(Algebra of Series) Consider convergent complex series $\sum a_n = L$ *and* $\sum b_n = M$. *Show that*

(a) *For any* $c \in \mathbb{R}$, $\sum(c a_n) = cL$,

(b) $\sum(a_n \pm b_n) = L \pm M$.

Task 8.4.11

Let $z_n = x_n + iy_n$. *Then* $\sum z_n$ *converges if and only if both* $\sum x_n$ *and* $\sum y_n$ *converge. If these series do converge, we have* $\sum z_n = \sum x_n + i \sum y_n$.

This gives us one way to study convergence of a complex series, by separating into real and imaginary parts. However, this is not always fruitful, as a simple complex expression may look complicated when viewed in terms of real and imaginary parts. The other way to connect with our existing results is to take the absolute value of each entry.

A complex series $\sum z_n$ is called **absolutely convergent** if $\sum |z_n|$ converges. For example, if $|z| < 1$, the geometric series $\sum z^n$ is absolutely convergent.

Theorem 8.4.12

Let $z_n = x_n + iy_n$. *Then* $\sum z_n$ *is absolutely convergent if and only if both* $\sum x_n$ *and* $\sum y_n$ *are absolutely convergent.*

Proof. First, suppose $\sum |z_n|$ converges. Then $|x_n|, |y_n| \leq |z_n|$ implies $\sum |x_n|$ and $\sum |y_n|$ converge.

Next, suppose $\sum |x_n|$ and $\sum |y_n|$ converge. Then $|z_n| \leq |x_n| + |y_n|$ implies $\sum |z_n|$ converges. ■

Task 8.4.13

If a series is absolutely convergent, then it is convergent.

All the tests that require non-negative terms can be applied to $\sum|z_n|$ and thus become tests for absolute convergence of a complex series. In this brief introduction, we shall primarily apply the comparison test in proofs and the ratio and root tests in calculations.

Task 8.4.14

Show that any rearrangement of an absolutely convergent complex series has the same sum.

Theorem 8.4.15

Let $\sum_{n=1}^{\infty} a_n$ and $\sum_{n=1}^{\infty} b_n$ be absolutely convergent complex series. Let (c_n) be any arrangement of all the $a_i b_j$ combinations. Then $\sum_{n=1}^{\infty} c_n$ is absolutely convergent and $\left(\sum_{n=1}^{\infty} a_n\right)\left(\sum_{n=1}^{\infty} b_n\right) = \sum_{n=1}^{\infty} c_n$.

Proof. We have already proved this for real series (Theorem 7.4.9). Let us fix any particular ordering of $\mathbb{N} \times \mathbb{N}$ and write $\sum\limits_{m,n}$ for a sum that proceeds according to this ordering. We assume (c_n) follows this ordering of the $a_m b_n$ terms. Let $a_m = x_m + iy_m$ and $b_n = u_n + iv_n$. Then

$$\begin{aligned}
\sum_m a_m \sum_n b_n &= (\sum_m x_m + i\sum_m y_m)(\sum_n u_n + i\sum_n v_n) \\
&= (\sum_m x_m \sum_n u_n - \sum_m y_m \sum_n v_n) + i(\sum_m x_m \sum_n v_n + \sum_m y_m \sum_n u_n) \\
&= (\sum_{m,n} x_m u_n - \sum_{m,n} y_m v_n) + i(\sum_{m,n} x_m v_n + \sum_{m,n} y_m u_n) \\
&= \sum_{m,n} \Big((x_m u_n - y_m v_n) + i(x_m v_n + y_m u_n)\Big) \\
&= \sum_{m,n} (x_m + iy_m)(u_n + iv_n) = \sum_{m,n} a_m b_n = \sum_n c_n.
\end{aligned}$$

■

Complex Power Series

A **complex power series** is an expression of the form

$$\sum_{n=0}^{\infty} c_n (z-a)^n = c_0 + c_1(z-a) + c_2(z-a)^2 + \cdots$$

in which a and the c_n are complex numbers and z is a complex variable. Our main question is: For which values of z will a power series converge?

Example 8.4.16

The geometric series $\sum_{n=0}^{\infty} z^n$ is a power series centered at $a = 0$. It converges absolutely on the open disc $|z| < 1$ and diverges for $|z| \geq 1$. □

We can replicate the basic results for real power series, with the same proofs.

Theorem 8.4.17

Suppose $\sum_{n=0}^{\infty} c_n z^n$ converges at $z_0 \neq 0$. Then it converges absolutely at every z that satisfies $|z| < |z_0|$.

Theorem 8.4.18

Consider a complex power series $\sum c_n(z-a)^n$. One of the following cases will occur:

1. *The series converges only if $z = a$.*
2. *The series converges absolutely for every $z \in \mathbb{C}$.*
3. *There is $R \in \mathbb{R}$ such that the series converges absolutely for $|z-a| < R$ and diverges for $|z-a| > R$.*

The **radius of convergence** R determines a **disc of convergence** inside which the series converges absolutely at every point. The bounding circle of this disc becomes an object of special interest, as the series may converge on some points of this circle and diverge on others.

Complex power series give a convenient procedure for extending real functions to the complex plane. We simply take the power series of a real function and convert the real variable to a complex one. For example, we define

$$\begin{aligned}\exp z &= 1 + z + \frac{z^2}{2!} + \frac{z^3}{3!} + \cdots,\\ \sin z &= z - \frac{z^3}{3!} + \frac{z^5}{5!} - \frac{z^7}{7!} + \cdots,\\ \cos z &= 1 - \frac{z^2}{2!} + \frac{z^4}{4!} - \frac{z^6}{6!} + \cdots.\end{aligned}$$

Task 8.4.19

Use the ratio or root test to prove that the power series that define the complex exponential, sine and cosine functions converge absolutely for every complex z.

Following the real notation, we shall write $e^z = \exp z$. The complex numbers reveal a close relationship between the exponential and trigonometric functions.

Theorem 8.4.20

$$\cos z = \frac{e^{iz} + e^{-iz}}{2}, \quad \sin z = \frac{e^{iz} - e^{-iz}}{2i}.$$

Proof. We have $$e^{iz} = 1 + iz - \frac{z^2}{2!} - i\frac{z^3}{3!} + \frac{z^4}{4!} + i\frac{z^5}{5!} \cdots,$$
$$e^{-iz} = 1 - iz - \frac{z^2}{2!} + i\frac{z^3}{3!} + \frac{z^4}{4!} - i\frac{z^5}{5!} \cdots.$$
Adding and subtracting these expressions gives
$$e^{iz} + e^{-iz} = 2\left(1 - \frac{z^2}{2!} + \frac{z^4}{4!} \cdots\right) = 2\cos z,$$
$$e^{iz} + e^{-iz} = 2i\left(z - \frac{z^3}{3!} + \frac{z^5}{5!} \cdots\right) = 2i\sin z.$$ ■

Task 8.4.21

Show that $e^{iz} = \cos z + i\sin z$. *In particular, we have* ***Euler's identity***: $e^{i\pi} = -1$.

Task 8.4.22

Let $\theta, \phi \in \mathbb{R}$. *Show that* $e^{i(\theta+\phi)} = e^{i\theta}e^{i\phi}$.

Theorem 8.4.23

For any $z, w \in \mathbb{C}$, $e^{z+w} = e^z e^w$.

Proof. The concerned series converge absolutely and so we can apply Theorem 8.4.15 after ordering the product terms by their degree.
$$e^z e^w = \sum_{n=0}^{\infty} \frac{z^n}{n!} \sum_{n=0}^{\infty} \frac{w^n}{n!} = \sum_{n=0}^{\infty} \left(\sum_{k=0}^{n} \frac{z^k}{k!} \frac{w^{n-k}}{(n-k)!} \right) = \sum_{n=0}^{\infty} \left(\sum_{k=0}^{n} \binom{n}{k} \frac{z^k w^{n-k}}{n!} \right)$$
$$= \sum_{n=0}^{\infty} \frac{(z+w)^n}{n!} = e^{z+w}.$$ ■

We can now express the exponential function in terms of the real and imaginary parts of z:
$$z = x + iy \implies e^z = e^{x+iy} = e^x e^{iy} = e^x(\cos y + i\sin y).$$

Task 8.4.24

Prove the following properties of the exponential function:

(a) *For every* $z \in \mathbb{C}$, $e^z \neq 0$ *and* $(e^z)^{-1} = e^{-z}$.

(b) *Every non-zero* $z \in \mathbb{C}$ *can be expressed as* $z = e^w$ *for some* $w \in \mathbb{C}$.

(c) *We have* $e^z = 1$ *if and only if* $z = 2n\pi i$, *with* $n \in \mathbb{Z}$. *Hence,* $e^z = e^w$ *if and only if* $w - z \in 2\pi i\mathbb{Z}$.

Complex numbers bring clarity to another phenomenon that seems mysterious when we work with real numbers alone. As an example, consider the real power series expansion
$$\frac{1}{1+x^2} = 1 - x^2 + x^4 - \cdots.$$

The power series has $R = 1$ and represents the function only on $(-1,1)$. Yet the function itself is smooth on all of $\mathbb{R}$, so it seems odd that the power series representation breaks down at ± 1. Now let us consider the complex version,

$$\frac{1}{1+z^2} = 1 - z^2 + z^4 - \cdots.$$

The function $1/(1+z^2)$ is not defined at $\pm i$. In fact, as z varies along the imaginary axis, the function takes the following appearance:

$$\frac{1}{1+(it)^2} = \frac{1}{1-t^2}.$$

It takes values that are arbitrarily large in magnitude as $t \to \pm 1$. Hence, its power series expansion *must* break down beyond a radius of 1.

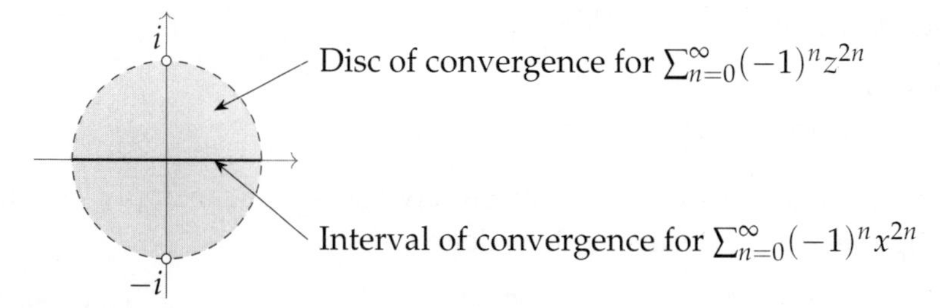

Complex Fourier Series

Complex numbers also bring better organization to Fourier series. Consider a function $f\colon \mathbb{R} \to \mathbb{C}$. We can express it in the form $f(x) = u(x) + iv(x)$ with $u,v\colon \mathbb{R} \to \mathbb{C}$. In future, we will just say "Let $f = u + iv$" to depict this situation. We say f has period T if $f(x+T) = f(x)$ for every x. This is equivalent to both u,v having period T. Further, we call f **integrable** on $[a,b]$ if u,v are integrable there, and we define

$$\int_a^b f(x)\,dx = \int_a^b u(x)\,dx + i\int_a^b v(x)\,dx.$$

We shall work with functions that have period 2π. We only need to understand these over the interval $[-\pi,\pi]$. The role of the integral $\int_a^b f(x)g(x)\,dx$ in the real case will be assumed by $\int_a^b f(x)\overline{g(x)}\,dx$ in the complex case.

Example 8.4.25

For $n \in \mathbb{Z}$, define $e_n(x) = e^{inx}$. Then

$$\int_{-\pi}^{\pi} e_n(x)\,dx = \int_{-\pi}^{\pi} \cos nx\,dx + i\int_{-\pi}^{\pi} \sin nx\,dx = \begin{cases} 0 & \text{if } n \neq 0, \\ 2\pi & \text{if } n = 0. \end{cases}$$

This leads to the following calculation:

$$\int_{-\pi}^{\pi} e_n(x)\overline{e_m(x)}\,dx = \int_{-\pi}^{\pi} e_{n-m}(x)\,dx = \begin{cases} 0 & \text{if } n \neq m, \\ 2\pi & \text{if } n = m. \end{cases} \qquad \square$$

Now consider a trigonometric polynomial $T(x)$. We can express it in terms of exponentials as follows:

$$\begin{aligned} T(x) &= \frac{a_0}{2} + \sum_{n=1}^{M} a_n \cos nx + \sum_{n=1}^{M} b_n \sin nx \\ &= \frac{a_0}{2} + \sum_{n=1}^{M} a_n \frac{e^{inx} + e^{-inx}}{2} - i \sum_{n=1}^{M} b_n \frac{e^{inx} - e^{-inx}}{2} \\ &= \frac{a_0}{2} + \sum_{n=1}^{M} \frac{a_n - ib_n}{2} e^{inx} + \sum_{n=1}^{M} \frac{a_n + ib_n}{2} e^{-inx} \\ &= \sum_{n=-M}^{M} c_n e_n(x). \end{aligned}$$

Similarly, a trigonometric series can be expressed as

$$\sum_{n=-\infty}^{\infty} c_n e_n(x) = \lim_{M\to\infty} \sum_{n=-M}^{M} c_n e_n(x).$$

We can use integration to extract the coefficients of a trigonometric polynomial:

$$\begin{aligned} \int_{-\pi}^{\pi} T(x)\overline{e_m(x)}\,dx &= \int_{-\pi}^{\pi} \sum_{n=-M}^{M} c_n e_n(x)\overline{e_m(x)}\,dx \\ &= \sum_{n=-M}^{M} c_n \int_{-\pi}^{\pi} e_n(x)\overline{e_m(x)}\,dx \\ &= 2\pi c_m. \end{aligned}$$

Now, given an integrable function f with period 2π we try to express it as a trigonometric series by first defining its **Fourier coefficients**

$$\hat{f}(n) = \frac{1}{2\pi} \int_{-\pi}^{\pi} f(x)\overline{e_n(x)}\,dx, \qquad (n \in \mathbb{Z})$$

and then its **Fourier series**

$$f(x) \sim \sum_{n=-\infty}^{\infty} \hat{f}(n)\, e_n(x).$$

We see that the use of complex numbers gives a much cleaner description of Fourier series, without a split into the sine and cosine coefficients.

Let us use the name $\mathcal{I}$ for the class of integrable functions on $[-\pi,\pi]$. For $f,g \in \mathcal{I}$ we define $\langle f,g\rangle = \frac{1}{2\pi}\int_{-\pi}^{\pi} f(x)\overline{g(x)}\,dx$ and $||f||^2 = \langle f,f\rangle$. Note that $\hat{f}(n) = \langle f,e_n\rangle$.

Task 8.4.26

Show that $\langle f,g\rangle = 0 \implies ||f+g||^2 = ||f||^2 + ||g||^2$.

Task 8.4.27

Prove Bessel's inequality: $\sum\limits_{n=-\infty}^{\infty} |\hat{f}(n)|^2 \le ||f||^2$.

Finally, let us observe that complex numbers also reveal a connection between trigonometric series and power series. Consider a complex power series $f(z) = \sum_{n=0}^{\infty} c_n z^n$ and let us describe z as $re^{i\theta}$. If we fix θ and vary r, we get a power series in the real variable r,

$$f^{\theta}(r) = \sum_{n=0}^{\infty} (c_n e^{in\theta}) r^n.$$

On the other hand, if we fix r and vary θ we get a trigonometric series,

$$f_r(\theta) = \sum_{n=0}^{\infty} (c_n r^n) e^{in\theta}.$$

This relationship is especially interesting because power series are closely tied to differentiation while Fourier series are related more to integration (think about how the coefficients are calculated). We conclude with an example of how this relationship can help us further understand the behavior of a power series.

Consider the function $f(z) = \sum_{n=1}^{\infty} z^n/n^2$. The series has $R = 1$ and also converges absolutely at each point of the unit circle. Since the sum of the series is defined at each point of the unit circle, one wonders what restricts the series from extending beyond it. So we look at the values on the unit circle:

$$f_1(\theta) = \sum_{n=1}^{\infty} \frac{e^{in\theta}}{n^2} = \sum_{n=1}^{\infty} \frac{\cos n\theta}{n^2} + i\sum_{n=1}^{\infty} \frac{\sin n\theta}{n^2}.$$

The real part of this series is the Fourier series of the even function given by $g(\theta) = \frac{1}{4}(\theta-\pi)^2 - \pi^2/12$ for $0 \le \theta \le \pi$. This function is not differentiable at $\theta = 0$. Below, we see the graph of g as well as that of the real part of $f(z)$, showing the kink corresponding to $r = 1$ and $\theta = 0$.

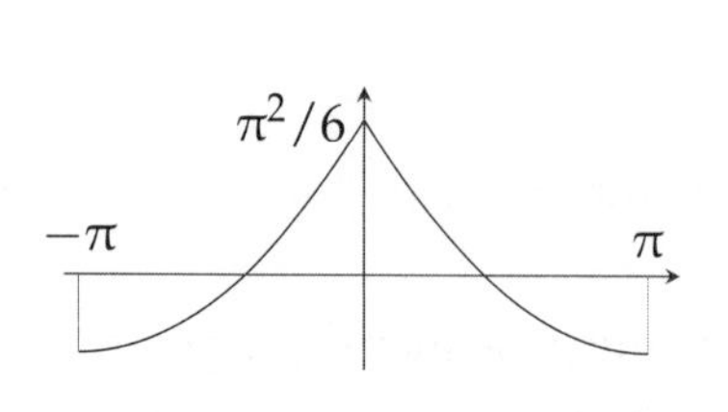

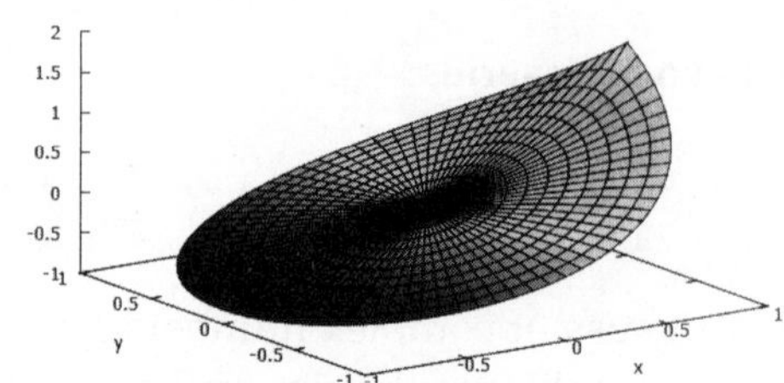

This lack of differentiability is what prevents the series from converging beyond the unit circle. To appreciate this fully, we need to develop calculus for complex functions. What we can sense from this brief discussion is that this calculus, which is called "complex analysis," will feature an even richer connection between differentiation and integration.

Exercises for § 8.4

1. Let $p(z)$ be a polynomial with real coefficients. Prove that $p(\alpha) = 0 \implies p(\overline{\alpha}) = 0$.

2. Prove the "parallelogram identity" for complex numbers z, w:

$$|z+w|^2 - |z-w|^2 = 2(|z|^2 + |w|^2).$$

3. Use De Moivre's formula to prove $\cos 3\theta = \cos^3\theta - 3\cos\theta\sin^2\theta$ and $\sin 3\theta = 3\cos^2\theta\sin\theta - \sin^3\theta$.

4. Use complex numbers to prove the following:

(a) $\sum_{k=1}^{n} \cos(kx) = \dfrac{\sin(n+1/2)x}{2\sin(x/2)} - \dfrac{1}{2}$.

(b) $\sum_{k=1}^{n} \sin(kx) = \dfrac{\cos(x/2) - \cos(n+1/2)x}{2\sin(x/2)}$.

5. Find all the points z at which $\sum_{n=0}^{\infty} \dfrac{n+1}{3^n}(z+2)^n$ converges.

6. Express the function $f(z) = 1/(1-z)^2$ as a power series with center at $a = i$. Sketch the set of points at which it converges.

7. For which $z \in \mathbb{C}$ does the series $\sum_{n=0}^{\infty} \left(\dfrac{z^n}{n!} + \dfrac{n^2}{z^n}\right)$ converge?

8. Sketch the images of the following subsets of $\mathbb{C}$ under the exponential map:

(a) $z = x + iy$ with $y \in \mathbb{Z}$.

(b) $z = x + iy$ with $x \in \mathbb{Z}$.

(c) $z = x + ix$.

(d) $|z| = \pi/2$.

9. Prove the following properties of the complex trigonometric functions:

(a) $\cos z = \cos w$ if and only if $z - w \in 2\pi\mathbb{Z}$ or $z + w \in 2\pi\mathbb{Z}$.

(b) $\sin z = \sin w$ if and only if $z - w \in 2\pi\mathbb{Z}$ or $z + w \in \pi(2\mathbb{Z}+1)$.

(c) $\cos z = 0$ if and only if $z \in \pi(\mathbb{Z} + 1/2)$.

(d) $\sin z = 0$ if and only if $z \in \pi\mathbb{Z}$.

(e) $\cos(z+w) = \cos z \cos w - \sin z \sin w$,
$\sin(z+w) = \sin z \cos w + \cos z \sin w$.

10. The hyperbolic functions are extended to the complex plane by $\cosh z = \dfrac{e^z + e^{-z}}{2}$ and $\sinh z = \dfrac{e^z - e^{-z}}{2}$. Show that $\cosh z = \cos iz$ and $\sinh z = -i \sin iz$. (This explains the similarities between the properties of the hyperbolic and trigonometric functions.)

11. Find and sketch the region consisting of all the points z such that the series $\sum_{n=1}^{\infty} \frac{1}{n^2} \exp\left(\frac{nz}{z-2}\right)$ converges.

Thematic Exercises

Uniform Convergence

Let us use $\mathcal{B}(A)$ to denote the collection of bounded real functions with domain A. If $A = [a,b]$, we will denote this collection by $\mathcal{B}[a,b]$. For any $f \in \mathcal{B}(A)$, we define its **sup norm** by $||f|| = \sup\{\,|f(x)| \mid x \in A\,\}$. This quantity works for bounded functions in much the same way that absolute value works for real numbers.

A1. Let $f, g \in \mathcal{B}(A)$ and $c \in \mathbb{R}$. Then

(a) $||f|| \geq 0$,

(b) $||f|| = 0$ if and only if $f = 0$,

(c) $||cf|| = |c|\,||f||$,

(d) (Triangle Inequality) $||f + g|| \leq ||f|| + ||g||$.

We use the sup norm as a measure of the gap between two bounded functions.

A2. Let $f, g, h \in \mathcal{B}(A)$. Then

(a) $||f - g|| \geq 0$,

(b) $||f - g|| = 0$ if and only if $f = g$,

(c) (Triangle Inequality) $||f - h|| \leq ||f - g|| + ||g - h||$.

Given a sequence $f_n \in \mathcal{B}(A)$ and a function $f \in \mathcal{B}(A)$, we say that (f_n) **converges uniformly** to f on A if $||f_n - f|| \to 0$, and we denote this by $f_n \xrightarrow{u} f$.

A3. Consider a sequence $f_n \in \mathcal{B}(A)$ and a function $f \in \mathcal{B}(A)$.

(a) Show that $f_n \xrightarrow{u} f \implies f_n \xrightarrow{pw} f$.

(b) Show that $f_n \xrightarrow{pw} f \not\Rightarrow f_n \xrightarrow{u} f$. (Hint: Consider $f_n(x) = x^n$ on $[0,1]$.)

A4. Let $f_n \xrightarrow{u} f$ on A. Show that the collection $\{\,f_n \mid n \in \mathbb{N}\,\}$ is bounded: There is an $M \in \mathbb{R}$ such that $||f_n|| \leq M$ for every n.

A5. Let $f_n \xrightarrow{u} f$ on $[a,b]$ and each f_n be integrable on $[a,b]$. The following tasks will show that f is integrable and $\int_a^b f(x)\,dx = \lim_{n\to\infty} \int_a^b f_n(x)\,dx$. Fix $\epsilon > 0$.

(a) Show there is an $N \in \mathbb{N}$ such that $n \geq N$ implies $|f(x) - f_n(x)| < \epsilon/(b-a)$ for every $x \in [a,b]$.

(b) Show that for any partition P, the upper and lower Darboux sums satisfy

$$L(f_n,P) - \epsilon \leq L(f,P) \leq U(f,P) \leq U(f_n,P) + \epsilon \qquad (n \geq N)$$

(c) Draw the desired conclusions.

A6. Let $f_n \xrightarrow{u} f$ on $[a,b]$ and each f_n be continuous on $[a,b]$. The following tasks will show that f is continuous. Fix an $x \in [a,b]$ and $\epsilon > 0$.

(a) Show there is an $N \in \mathbb{N}$ such that $|f(t) - f_N(t)| < \epsilon/3$ for every $t \in [a,b]$.

(b) Show there is a $\delta > 0$ such that $|x-y| < \delta$ implies $|f_N(x) - f_N(y)| < \epsilon/3$.

(c) Show that $|x-y| < \delta$ implies $|f(x) - f(y)| < \epsilon$.

Suppose $f_n \in \mathcal{B}(A)$. We say the series $\sum_{n=1}^{\infty} f_n$ **converges uniformly** to f if the sequence of partial sums $g_N = \sum_{n=1}^{N} f_n$ converges uniformly to f. We will denote this by $\sum_{n=1}^{\infty} f_n \overset{u}{=} f$.

A7. Suppose $\sum_{n=1}^{\infty} f_n \overset{u}{=} f$ on $[a,b]$. Prove the following:

(a) If each f_n is integrable, then f is integrable and $\int_a^b f(x)\,dx = \sum_{n=1}^{\infty} \int_a^b f_n(x)\,dx$.

(b) If each f_n is continuous, then f is continuous.

A8. (Weierstrass M-Test) Suppose $f_n \in \mathcal{B}(A)$, $||f_n|| \leq M_n \in \mathbb{R}$, and $\sum_{n=1}^{\infty} M_n$ converges. Show that $\sum_{n=1}^{\infty} f_n$ converges uniformly and absolutely, via the following steps:

(a) Show that for each $x \in A$, the series $\sum_{n=1}^{\infty} f_n(x)$ converges absolutely.

(b) Define $f(x) = \sum_{n=1}^{\infty} f_n(x)$. Show that

$$||f - \sum_{n=1}^{m} f_n|| \leq \sum_{n=m+1}^{\infty} M_n.$$

A9. Suppose $\sum a_n x^n$ converges absolutely at $x = c$. Show that $\sum a_n x^n$ converges uniformly on $[-c,c]$.

A10. Suppose a power series $\sum_n a_n(x-a)^n$ has radius of convergence $R > 0$. Show the following:

(a) The series converges uniformly on any closed and bounded interval $[a-\delta, a+\delta]$ with $0 \leq \delta < R$. (Hint: Pick r such that $\delta < r < R$.)

(b) If the series converges absolutely at $x = a + R$, then it converges uniformly on $[a-R, a+R]$.

A11. Let f be a continuously differentiable function with period 2π. We know that the Fourier series of f converges to it pointwise. The following tasks will show that the convergence is uniform. First, let the Fourier series of f be

$$f(x) \sim a_0/2 + \sum_{n=1}^{\infty} a_n \cos nx + \sum_{n=1}^{\infty} b_n \sin nx.$$

(a) Show that $\sum n^2 a_n^2$ and $\sum n^2 b_n^2$ converge. (Hint: Bessel's inequality.)

(b) Show that $\sum |a_n|$ and $\sum |b_n|$ converge. (Hint: $|a_n| \leq \max\{\frac{1}{n^2}, n^2 a_n^2\}$.)

(c) Apply the Weierstrass M-test.

Irrationality of Some Numbers

The question of whether a number is rational or not appears to be an algebraic one and one may expect to resolve it by algebra. The standard proof that $\sqrt{2}$ is not a rational number is of this type. It uses the properties related to integers being odd or even. However, $\sqrt{2}$ is a special number in that its very definition is algebraic – it is a number whose square is 2. Numbers like e and π are different. Their definitions involve limits and this indicates that calculus should have a role in establishing whether they are rational or not.

This set of exercises will show that e^r is irrational whenever r is a non-zero rational number. We begin with a basic result that helps demarcate irrationals from rationals.

B1. Let r be a real number r. If there are sequences p_n, q_n of integers such that $q_n r - p_n \neq 0$ and $q_n r - p_n \to 0$ then r is irrational.

B2. Show that if f is continuously differentiable, then

$$\int_0^a e^{-x} f(x)\,dx = -e^{-a} f(a) + f(0) + \int_0^a e^{-x} f'(x)\,dx.$$

B3. Show that if f is a polynomial of degree n, then

$$\int_0^a e^{-x} f(x)\,dx = \sum_{k=0}^{n} f^{(k)}(0) - e^{-a} \sum_{k=0}^{n} f^{(k)}(a).$$

This gives a way of creating rational approximations of e^{-a}. We need polynomials that converge uniformly to zero and whose derivatives at 0 and a are integers. We define

$$f_n(x) = \frac{x^n (a-x)^n}{n!}.$$

B4. Show the following:

(a) For $x \in [0,a]$, $0 \le f_n(x) \le \dfrac{a^{2n}}{n!}$.

(b) $0 < \int_0^a e^{-x} f_n(x)\, dx \to 0$ as $n \to \infty$.

B5. Show that if $a \in \mathbb{N}$, then all derivatives of $f_n(x)$ at $x = 0$ and $x = a$ are integers. This establishes that e^{-m} is irrational if $m \in \mathbb{N}$.

B6. Show that e^r is irrational if r is a non-zero rational.

B7. Show that $\log r$ is irrational if r is a positive rational.

You can find these and related results in Niven [25]. For example, similar arguments prove the irrationality of π as well as of $\cos r$ and $\sin r$ when r is a non-zero rational. For some highlights, consult the notes by Conrad [39].

Appendix | Solutions to Odd-Numbered Exercises

§ 1.1

1. By the cancellation law, we cannot have $1+2=1$ or $1+2=2$. Hence, $1+2=0=2+1$.

+	0	1	2
0	0	1	2
1	1		0
2	2	0	

By cancellation law, again, each member must show in each row exactly once. This completes the table.

+	0	1	2
0	0	1	2
1	1	2	0
2	2	0	1

Recall that multiplication by 0 must give 0. This leaves just one blank entry in the multiplication table. By the cancellation law, that entry must be 1.

·	0	1	2
0	0	0	0
1	0	1	2
2	0	2	1

3. (a) $x^2 - x > 0 \implies x(x-1) > 0 \implies x > 1$ or $x < 0$.

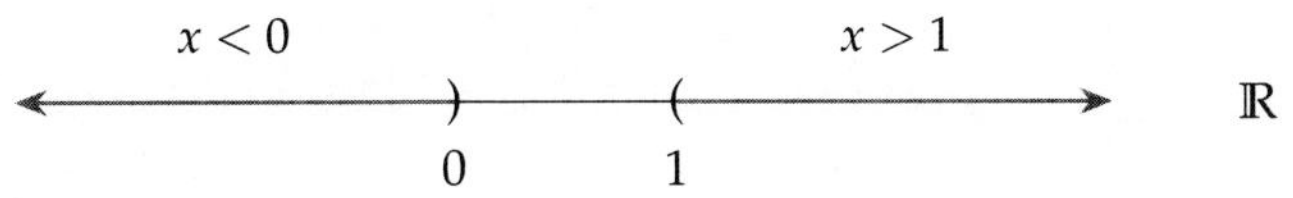

$x < 0$ $\quad$ $x > 1$

0 $\quad$ 1 $\quad$ $\mathbb{R}$

(b) $3x^2 + 2x - 1 \geq 0 \implies (x+1)(3x-1) \geq 0 \implies x \geq 1/3$ or $x \leq -1$.

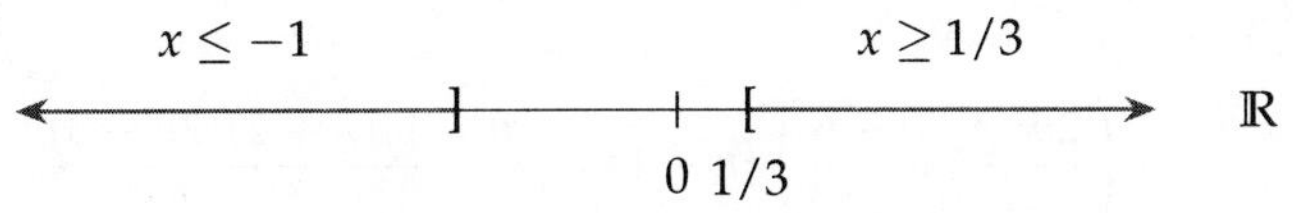

$x \leq -1$ $\quad$ $x \geq 1/3$

0 1/3 $\quad$ $\mathbb{R}$

(c) $x^2 - 5x + 6 < 0 \implies (x-2)(x-3) < 0 \implies 2 < x < 3.$

$2 < x < 3$

0 2 3 $\mathbb{R}$

5. We shall prove (a) $x^m x^n = x^{m+n}$. We apply mathematical induction to the following statements: $P(n): x^m x^n = x^{m+n}$ for every $m \in \mathbb{Z}$. For the truth of $P(1)$, we first note that $x^{m+1} = x^m x$ holds for $m \geq 0$ by definition. Now, consider $m < 0$. Then $-m-1 \geq 0$. Hence,

$$x^{m+1}x^{-1} = (x^{-1})^{-m-1}x^{-1} = (x^{-1})^{-m} = x^m.$$

Suppose $P(k)$ is true and consider $P(k+1)$. Now,

$$x^m x^{k+1} = x^m x x^k = x^{m+1}x^k = x^{m+1+k} = x^{m+(k+1)}.$$

So, the required identity has been proved for $m \in \mathbb{Z}$ and $n > 0$. It is trivial for $n = 0$. If $n < 0$, then $x^{m+n}x^{-n} = x^{m+n-n} = x^m$.

7. (a) $A \cap S = \varnothing$.

(b) If $n \in S$, then $1 \leq n$ implies $1 \notin A$. Since 1 is the least member of $\mathbb{N}$, it follows that $1 \in S$.

(c) If $1 \notin A$, then $1 \in S$. Hence $S = \varnothing$ implies $1 \in A$, and then of course 1 is the least element of A.

(d) Suppose $S \neq \varnothing$. Then $1 \in S$. Since $S \neq \mathbb{N}$, mathematical induction implies that there is $N \in \mathbb{N}$ such that $N \in S$ and $N + 1 \notin S$. The number $N + 1$ is the least element of A.

9. For $n = 1$ we have

$$RHS = (x-y)\sum_{i=0}^{1} x^i y^{1-i} = (x-y)(y+x) = x^2 - y^2 = LHS.$$

Assume it is also true for $n = k$. Then for $n = k+1$,

$$RHS = (x-y)\sum_{i=0}^{k+1} x^i y^{k+1-i} = (x-y)\left(y\sum_{i=0}^{k} x^i y^{k-i} + x^{k+1}\right)$$
$$= y(x^{k+1} - y^{k+1}) + (x-y)x^{k+1} = x^{k+2} - y^{k+2} = LHS.$$

11. We apply mathematical induction.

(a) For $n = 1$, $LHS = 1 = RHS$. Assume true for n. Then for $n+1$,

$$\sum_{k=1}^{n+1} k = \frac{n(n+1)}{2} + n + 1 = \left(\frac{n}{2} + 1\right)(n+1) = \frac{(n+1)(n+2)}{2}.$$

(b) For $n = 1$, $LHS = 1 = RHS$. Assume true for n. Then for $n+1$,

$$\sum_{k=1}^{n+1} k^2 = \frac{n(n+1)(2n+1)}{6} + (n+1)^2 = \left(\frac{n(2n+1)}{6} + n + 1\right)(n+1)$$
$$= \left(\frac{2n^2 + 7n + 6}{6}\right)(n+1) = \frac{(n+1)(n+2)(2n+3)}{6}.$$

13. We apply mathematical induction. For $n = 1$, we have $LHS = x + y = RHS$. Assume true for n. Then for $n + 1$,

$$\begin{aligned}(x+y)^{n+1} &= (x+y)\sum_{k=0}^{n}\binom{n}{k}x^k y^{n-k} = \sum_{k=0}^{n}\binom{n}{k}x^{k+1}y^{n-k} + \sum_{k=0}^{n}\binom{n}{k}x^k y^{n+1-k} \\ &= \sum_{k=1}^{n+1}\binom{n}{k-1}x^k y^{n+1-k} + \sum_{k=0}^{n}\binom{n}{k}x^k y^{n+1-k} \\ &= y^{n+1} + \sum_{k=1}^{n}\left(\binom{n}{k-1} + \binom{n}{k}\right)x^k y^{n+1-k} + x^{n+1} \\ &= y^{n+1} + \sum_{k=1}^{n}\binom{n+1}{k}x^k y^{n+1-k} + x^{n+1} = \sum_{k=0}^{n+1}\binom{n+1}{k}x^k y^{n+1-k}.\end{aligned}$$

§ 1.2

1. Manipulate the inequality $(\sqrt{x} - \sqrt{y})^2 \geq 0$.

3. We formalize Dedekind's statement as follows: "Suppose $\mathbb{R} = A \cup B$, A and B are non-empty, and if $x \in A$, $y \in B$ then $x < y$. Then there is a unique $\alpha \in \mathbb{R}$ such that $x \leq \alpha \leq y$ for every $x \in A$, $y \in B$." Let us call this 'Dedekind's axiom.'

If A, B satisfy the hypotheses of Dedekind's axiom, they also satisfy the hypotheses of the completeness axiom and there is $\alpha \in \mathbb{R}$ such that $x \leq \alpha \leq y$ for every $x \in A$, $y \in B$. Suppose α' also satisfies $x \leq \alpha' \leq y$ for every $x \in A$, $y \in B$. Let $\alpha < \alpha'$. Then $(\alpha + \alpha')/2 \in (\alpha, \alpha')$ cannot belong to either A or B, contradicting $\mathbb{R} = A \cup B$. So, α is unique. Thus, the completeness axiom implies Dedekind's axiom.

For the converse, let A, B be non-empty subsets of $\mathbb{R}$ such that $a \in A$, $b \in B$ implies $a \leq b$. Define $X, Y \subset \mathbb{R}$ by $X = \{x \in \mathbb{R} \mid x \text{ is a lower bound of } B\}$ and $Y = \mathbb{R} \setminus X$. Then X, Y satisfy the hypotheses of Dedekind's axiom and so we have $\alpha \in \mathbb{R}$ such that $x \leq \alpha \leq y$ for every $x \in X$, $y \in Y$. Now, $A \subseteq X$ implies $a \leq \alpha$ for every $a \in A$. Finally, let $b \in B$. If $b \leq \alpha$ then $x \leq b \leq \alpha \leq y$ for every $x \in X$, $y \in Y$. By uniqueness, $b = \alpha$. Therefore, $\alpha \leq b$.

5. $[a, b]$, $[a, b)$, $(a, b]$, (a, b), with $a, b \in \mathbb{R}$.

7. (a) The general member of the given set satisfies

$$1 \leq 1 + \frac{1}{2} + \frac{1}{2^2} + \cdots + \frac{1}{2^n} = \frac{1 - (1/2)^{n+1}}{1 - (1/2)} < 2.$$

So, 2 is an upper bound. We claim that it is also the least upper bound. Consider any $2 - \epsilon$ with $\epsilon > 0$. By the Archimedean property, there is an n such that $1/2^n > \epsilon$. Then

$$\left(1 + \frac{1}{2} + \frac{1}{2^2} + \cdots + \frac{1}{2^n}\right) > 2 - \epsilon,$$

hence $2 - \epsilon$ is not an upper bound of the set.

(b) Observe that the general member of the set has the form

$$\frac{1}{10} + \frac{1}{10^2} + \cdots \frac{1}{10^n}.$$

Proceed as in (a) to show that the set is bounded below by 0.1 and has supremum $1/9$.

9. It is clear that 1 is an upper bound of the given sets, so we concentrate on showing that lower numbers cannot be upper bounds.

(a) Suppose $\alpha < 1$ is an upper bound of $A = [0,1)$. Then $\alpha \geq 0$. And $(\alpha+1)/2 \in A$ is greater than α, a contradiction.

(b) Suppose $\alpha < 1$ is an upper bound of $B = \left\{1 - \frac{1}{n} : n \in \mathbb{N}\right\}$. By the Archimedean property, there is an $n \in \mathbb{N}$ such that $1/n < 1 - \alpha$. Then $1 - 1/n > \alpha$.

11. Let $A, B \subseteq \mathbb{R}$ be non-empty and bounded above.

(a) If $a \in A$ and $b \in B$, then $a \leq \sup(A), b \leq \sup(B)$ implies $a + b \leq \sup(A) + \sup(B)$. Hence, $\alpha = \sup(A) + \sup(B)$ is an upper bound of $A + B$. We shall show that for any $\epsilon > 0$, $\alpha - \epsilon$ is not an upper bound of $A + B$. We have $a \in A$ and $b \in B$ such that $a > \sup(A) - \epsilon/2$ and $b > \sup(B) - \epsilon/2$. Then $a + b > \alpha - \epsilon$.

(b) If $a \in A$ and $b \in B$, then $a \leq \sup(A)$, $b \leq \sup(B)$ implies $ab \leq \sup(A)\sup(B)$. Hence $\alpha = \sup(A)\sup(B)$ is an upper bound of AB. We shall show that for any $0 < \delta < 1$, $\delta\alpha$ is not an upper bound of AB. We have $a \in A$ and $b \in B$ such that $a > \sqrt{\delta}\sup(A)$ and $b > \sqrt{\delta}\sup(B)$. Then $ab > \delta\alpha$.

13. Of course, there are infinitely many ways to find a solution to this problem! We present one that has historical significance. We first note that $(17/12)^2 = 289/144 > 2$ implies $17/12 > \sqrt{2}$. Next,

$$\frac{17}{12} \cdot \frac{24}{17} = 2 \implies \frac{24}{17} < \sqrt{2}.$$

Let $\alpha = \frac{1}{2}\left(\frac{17}{12} + \frac{24}{17}\right)$. Then $\alpha < 17/12$. Further, by Exercise 1,

$$\alpha > \sqrt{17/12}\sqrt{24/17} = \sqrt{2}.$$

So, $\alpha = 577/408$ is a solution to our task. These two approximations to $\sqrt{2}$ appear in the *Sulvasutra* texts from India, dated to about 2,500 years ago, with the following easy to memorise expressions:

$$\frac{17}{12} = 1 + \frac{1}{3} + \frac{1}{3 \cdot 4}, \quad \frac{577}{408} = 1 + \frac{1}{3} + \frac{1}{3 \cdot 4} - \frac{1}{3 \cdot 4 \cdot 34}.$$

For the possible geometric origins of these approximations, see the introduction to Chapter 7.

§ 1.3

1. (a) $f\colon \mathbb{R} \setminus \{0\} \to \mathbb{R}$, $f(x) = 1/x$ is one-one but not onto. The image is $\mathbb{R} \setminus \{0\}$.

(b) $g\colon \mathbb{R} \setminus \{1\} \to \mathbb{R}$, $g(x) = \dfrac{x}{1-x}$. The function is one-one:

$$\frac{x}{1-x} = \frac{y}{1-y} \implies x(1-y) = y(1-x) \implies x = y.$$

Next, if y is in the image, we have $y = x/(1-x)$ for some $x \in \mathbb{R} \setminus \{1\}$. Now,

$$y = \frac{x}{1-x} \implies y(1-x) = x \implies (1+y)x = y \implies x = \frac{y}{1+y}.$$

Therefore, the pre-image x exists if $y \neq -1$, and the image of g is $\mathbb{R} \setminus \{-1\}$.

(c) $h \colon \mathbb{R} \setminus \{0,1\} \to \mathbb{R}$, $h(x) = \dfrac{1}{x(1-x)}$.

$$\frac{1}{x(1-x)} = \frac{1}{y(1-y)} \implies x(1-x) = y(1-y) \implies x - y = x^2 - y^2$$
$$\implies y = x \text{ or } x + y = 1.$$

For example, $h(1/4) = h(3/4) = 16/3$, and h is not one-one. To identify the image, we compute as follows:

$$y = \frac{1}{x(1-x)} \implies yx(1-x) = 1 \implies yx^2 - yx + 1 = 0.$$

We see that $y \neq 0$. When $y \neq 0$, the discriminant for this quadratic is $y^2 - 4y$, so y will be in the image if and only if $y^2 - 4y \geq 0$. Hence, the image is $(-\infty,0) \cup [4,\infty)$.

3. If you randomly pick two functions, chances are they will not commute! For example, take $f(x) = x+1$ and $g(x) = x^2$. Then $f \circ g(2) = 5$ and $g \circ f(2) = 9$.

5. The following f,g serve as a counterexample to all three converses!

$$f \colon [0,\infty) \to \mathbb{R},\ f(x) = \sqrt{x},$$
$$g \colon \mathbb{R} \to [0,\infty),\ g(x) = x^2.$$

7. (a) Define $f \colon \mathbb{N} \to \mathbb{W}$ by $f(n) = n - 1$.

(b) Define $f \colon \mathbb{N} \to \mathbb{Z}$ by $f(n) = \begin{cases} -n/2 & \text{if } n \text{ even}, \\ (n-1)/2 & \text{if } n \text{ odd}. \end{cases}$

(c) Define $f \colon [0,1) \to [0,\infty)$ by $f(x) = \dfrac{x}{1-x}$.

§ 1.4

1. (a) $(-\infty,\sqrt{2}] \cup [\sqrt{2},\infty)$.

(b) $(-\infty,0) \cup [1,\infty)$.

(c) $[-1,0] \cup [1,\infty)$.

(d) $\mathbb{R} \setminus \mathbb{Z}$.

3.

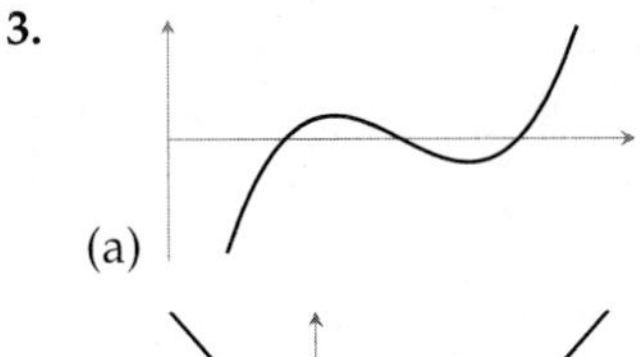

(a)

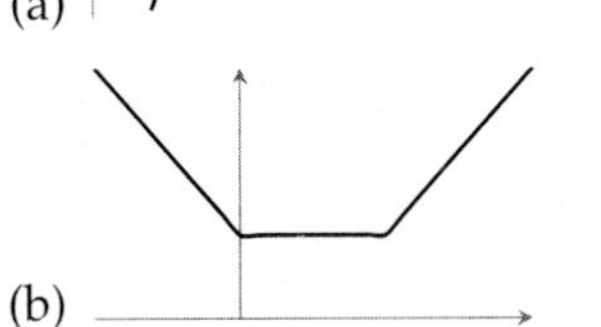

(b)

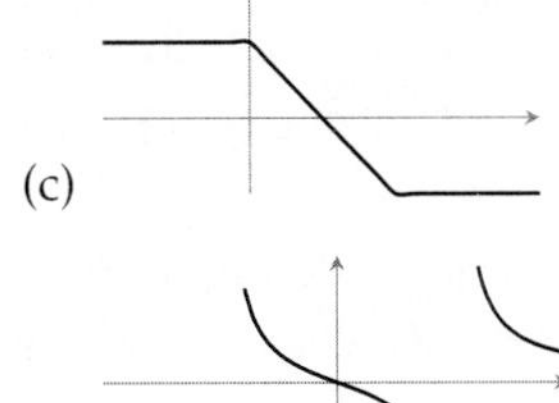

(c)

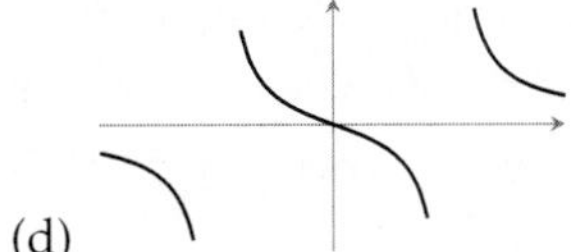

(d)

5. All the graphs are drawn to the same scale with x and y both varying over $[-2,2]$. The vertical lines have been drawn to emphasize the jumps, especially at the endpoints.

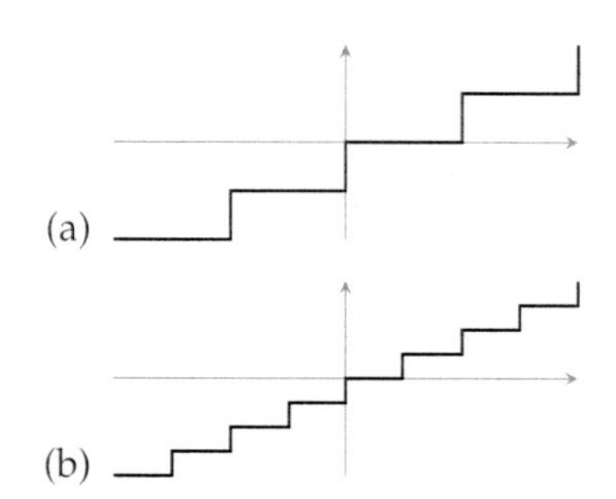

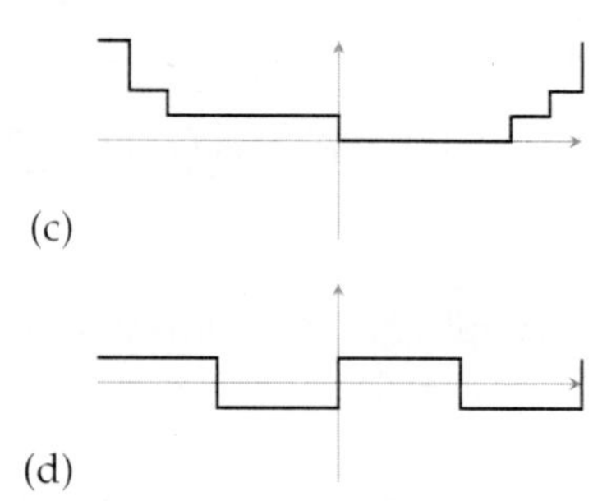

7. Extend the given graph to a suitable domain so that it represents an odd function:

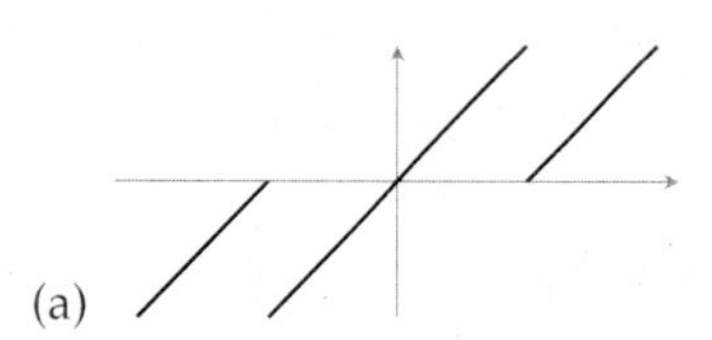

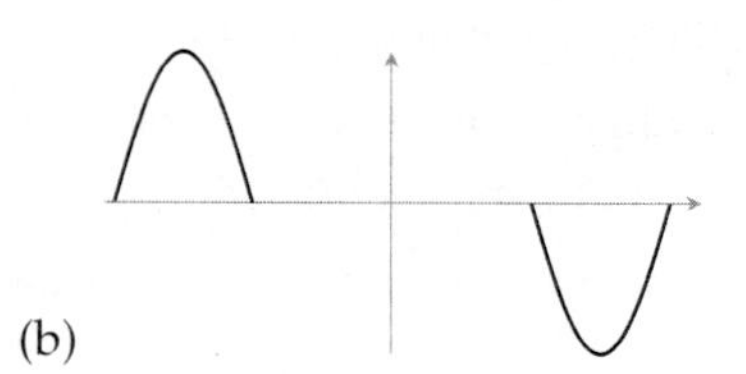

9. (a) Proof: $x < y \implies (f+g)(x) = f(x) + g(x) \leq f(y) + g(y) = (f+g)(y)$.

(b) Counter-example: $f(x) = g(x) = x$.

11.

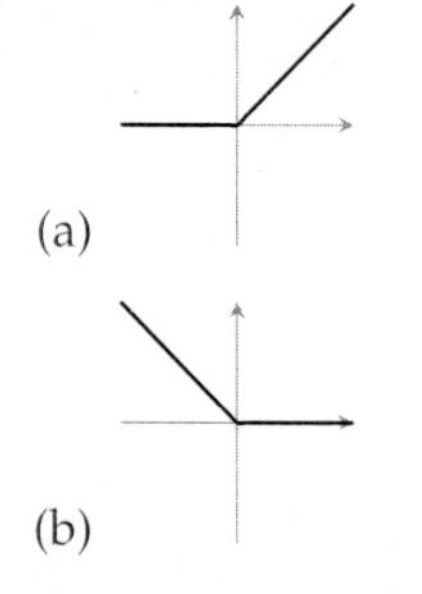

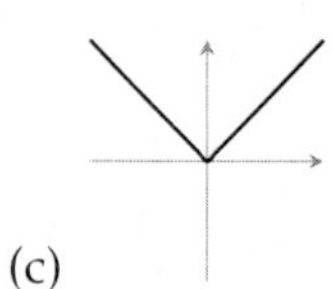

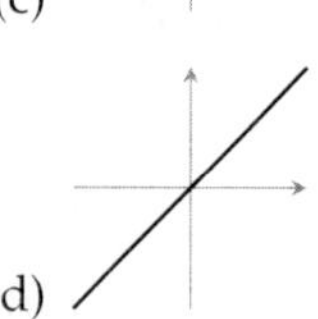

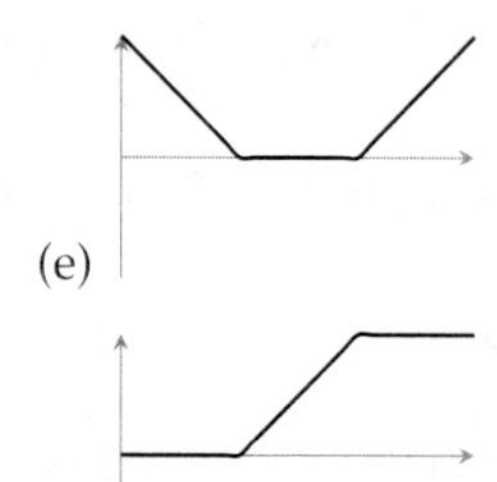

13. The proof of periodicity of $f+g$ is given below. The others are similar:

$$(f+g)(x+T) = f(x+T) + g(x+T) = f(x) + g(x) = (f+g)(x).$$

15. Suppose f has period T. It is enough to show that f is constant on $[0,T)$. Let $x,y \in [0,T)$ with $x < y$. We give the proofs when f is increasing.

(a) We have $-y < x$. Then $f(x) \leq f(y) = f(-y) \leq f(x)$.

(b) We have $x + T > y$. Then $f(x) \leq f(y) \leq f(x+T) = f(x)$.

§ 2.1

1. (a) 4. (b) 0.

3. (a) Let $P = \{x_0 = 0, \ldots, x_n = a\}$ be a partition of $[0,a]$ that is adapted to s. Let $Q = \{-x \mid x \in P\}$. Then Q is adapted to s over $[-a,0]$, and

$$\int_{-a}^{0} s(t)\,dt = \sum_{i=1}^{n} s_i(-x_{i-1} - (-x_i)) = 2\sum_{i=1}^{n} s_i(x_i - x_{i-1}) = \int_0^a s(t)\,dt.$$

(b) Similar.

5. (a) Consult Example 2.1.12.

(b) Similar to (a), using $1^2 + 2^2 + \cdots + n^2 = n(n+1)(2n+1)/6$.

7. (a) Use mathematical induction.

(b) Follow Example 2.1.12 and use (a).

9. (a) Let $I = \int_0^a f(x)\,dx$. There are step functions s,t on $[0,a]$ such that $s \le f \le t$ on $[0,a]$ and $\int_0^a t(x)\,dx - \int_0^a s(x)\,dx < \epsilon$. Define step functions $s', t' \colon [-a,0] \to \mathbb{R}$ by $s'(x) = s(-x)$ and $t'(x) = t(-x)$. Then $s' \le f \le t'$ on $[-a,0]$, $\int_{-a}^0 t'(x)\,dx - \int_{-a}^0 s'(x)\,dx < \epsilon$ and $\int_{-a}^0 s'(x)\,dx \le I \le \int_{-a}^0 t'(x)\,dx$.

(b) Similar.

11. Hint: For each $n \in \mathbb{N}$, consider the partition $P_n = \{0, \frac{1}{n}, \frac{1}{n-1}, \ldots \frac{1}{2}, 1\}$.

§ 2.2

1. (a) $\int_0^2 f(x)\,dx - \int_0^1 f(x)\,dx = 7$.

(b) $\int_0^2 f(x)\,dx - \int_0^{-2} f(x)\,dx = 16$.

(c) 2.

(d) 14.

3. (a) $c = 0, 3/2$.

(b) $c = 0$.

5. $-1/6$.

7. (a) Consider the cases $f(x) \ge 0$ and $f(x) \le 0$.

(b) If s,t are step functions such that $s \le f \le t$, then s^+, t^+ are step functions satisfying $s^+ \le f^+ \le t^+$, and $\int_a^b t^+(x)\,dx - \int_a^b s^+(x)\,dx < \int_a^b t(x)\,dx - \int_a^b s(x)\,dx$.

(c) $f^{\pm}$ are integrable, hence so is $|f| = f^+ + f^-$. Then $-|f| \le f \le |f|$ gives the desired inequality.

9. Hint: Modify the Dirichlet function.

11. Polynomials are linear combinations of monomials and monomials are piecewise monotonic.

13. Take a partition $P = \{x_0, \ldots, x_n\}$ of $[1,2]$ into n equal parts. Then $x_i = 1 + i/n$, with $i = 0, 1, \ldots, n$. Define corresponding step functions s,t by fixing the values on the subintervals by the left and right end-points, respectively. Then $s(x) \le 1/x \le t(x)$. Further,

$$\int_1^2 t(x)\,dx - \int_1^2 s(x)\,dx = \frac{1}{n}\left(\frac{1}{1} - \frac{1}{2}\right) = \frac{1}{2n}.$$

We need $1/2n \le 0.1$, so we take $n = 5$. Then we get the following upper and lower estimates for $\int_1^2 \frac{1}{x}\,dx$: $\int_1^2 t(x)\,dx = 0.745$, $\int_1^2 s(x)\,dx = 0.645$. Each is within 0.1 of the integral and is a solution to the question. Of course, we can do still better by taking their mean, which is 0.695. This is very close to the actual value of 0.693!

15. We give the proof that $f(x) = x^3$ is a surjection. From Exercise 5(b) of the previous section we know that $\int_0^x t^2\,dt = x^3/3$. Hence $\int_0^x 3t^2\,dt = x^3$. By Theorem 2.2.14 we see that $f(x) = x^3$ has the intermediate value property. Then $-(1+|x|)^3 < x < (1+|x|)^3$ completes the proof.

17. (a) By using shifts, we see f is integrable on any interval of the form $[nT, (n+1)T]$, $n \in \mathbb{Z}$. By additivity over intervals, we see it is integrable on any $[nT, mT]$. By the Archimedean property, it is integrable on any $[a, b]$.

(b) $F(x+T) = \int_0^{x+T} f(t)\,dt = \int_T^{x+T} f(t)\,dt = \int_0^x f(t+T)\,dt = \int_0^x f(t)\,dt = F(x)$.

§ 2.3

1. See Exercise 13 of § 2.2.

3. (a) $-\log 2$, (b) $3 + (e^2)^4$.

5. The shaded area is $\int_1^e \frac{1}{x}\,dx = \log e = 1$. The trapezium in the first figure has area $\frac{1}{2}(e-1)(\frac{1}{e} + 1) = \frac{e^2-1}{2e}$, giving $\frac{e^2-1}{2e} > 1$, or $(e-1)^2 > 2$. Hence, $e > \sqrt{2} + 1 > 2.4$. Similarly, add the areas of the rectangles in the second figure to get $e < 3.5$.

7.

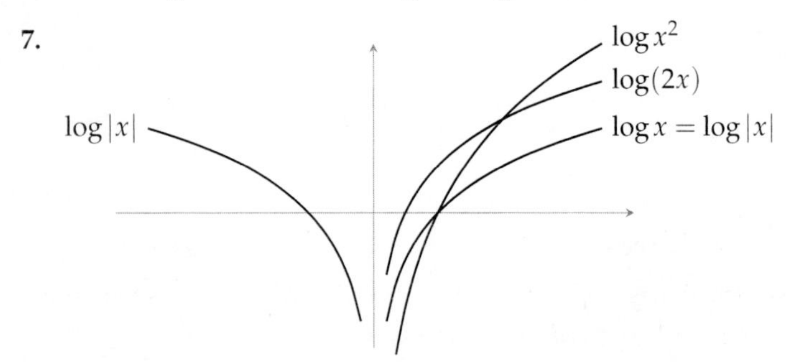

9. $a^x b^x = \exp(x \log a)\exp(x \log b) = \exp(x \log a + x \log b) = \exp(x \log(ab)) = (ab)^x$.

11.

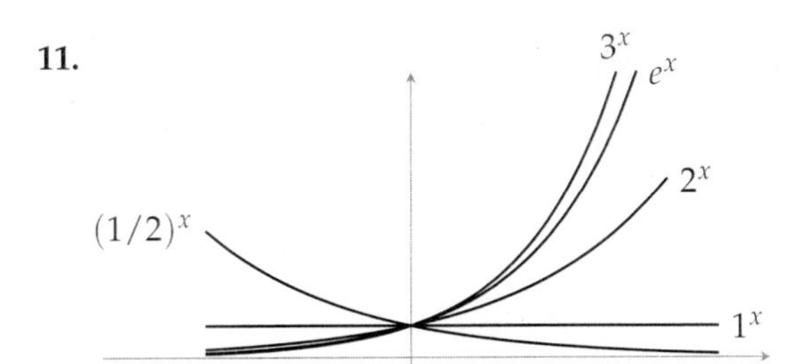

§ 2.4

1. Let $A = (x_1, y_1)$, $B = (x_2, y_2)$, $C = (x_3, y_3)$, $D = (x_4, y_4)$. The required area is the sum of the areas of $\triangle ABC$ and $\triangle ACD$. Hence, its value is

$$\frac{1}{2}\Big((x_1y_2 - x_2y_1) + (x_2y_3 - x_3y_2) + (x_3y_1 - x_1y_3)\Big)$$
$$+\frac{1}{2}\Big((x_1y_3 - x_3y_1) + (x_3y_4 - x_4y_3) + (x_4y_1 - x_1y_4)\Big).$$

The $x_3y_1 - x_1y_3$ terms cancel to give the desired result.

3.

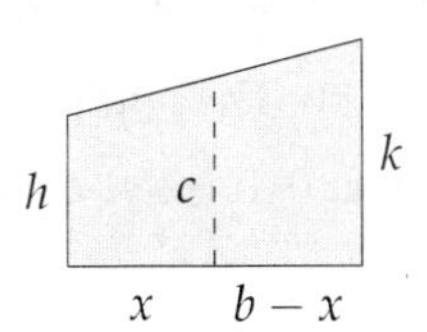

Let the location of the vertical line be given by x and its height by c. Obtain one equation by equating the areas of the two trapeziiums created by the line. Obtain another equation by equating the sum of their areas to the area of the whole trapezium. Now eliminate x from these equations to get $c^2 = (h^2 + k^2)/2$.

5. $\int_0^1 (x^2 - x^3)\,dx = 1/12$.

7. The point c such that $f(c) = (m + M)/2$.

§ 3.1

1. Consider f which takes value 1 on rationals and -1 on irrationals.

3. Let $S(n)$ be the statement that every polynomial $p(x)$ of degree n satisfies $\lim_{x\to a} p(x) = p(a)$. $S(0)$ is true because polynomials of degree 0 are constant. Suppose $S(n)$ is true and let $p(x)$ be a polynomial of degree $n+1$. Then $p(x) = xq(x) + c$ where $q(x)$ is a polynomial of degree n and $c \in \mathbb{R}$. Then

$$\lim_{x\to a} p(x) = \lim_{x\to a} x \lim_{x\to a} q(x) + \lim_{x\to a} c = aq(a) + c = p(a).$$

5. For both (a), (b), consider $f(x) = \text{sgn}(x)$ and $g(x) = -\text{sgn}(x)$.

7. (a) We can assume $|x| < 1$. Then $1 \le \exp(x^2) \le e$, and hence $-|x| \le x\exp(x^2) \le e|x|$. Apply the sandwich theorem.

(b) Consider the inequalities from Exercise 6 of § 2.3: $1-\frac{1}{x}\le \log x\le x-1$ for $x>0$. Apply the sandwich theorem.

(c) $\lim_{x\to a}\log x=\lim_{y\to 1}\log ay$.

(d) Again apply the inequalities from Exercise 6 of § 2.3.

9. (a) $$1-\frac{1}{x}\le \log x\le x-1 \implies \frac{h}{1+h}\le \log(1+h)\le h$$
$$\implies \frac{1}{1+h}\le \frac{\log(1+h)}{h}\le 1.$$

(b) $$\lim_{h\to 0}\frac{\log(x+h)-\log(x)}{h}=\frac{1}{x}=\lim_{h\to 0}\frac{\log(1+h/x)}{x\cdot h/x}=\frac{1}{x}\lim_{t\to 0}\frac{\log(1+t)}{t}.$$

11. $0\le |g(x)|\le M \implies 0\le |f(x)g(x)|\le M|f(x)|$. Apply the Sandwich theorem.

§ 3.2

1. (a) f has a jump discontinuity at every integer. At each integer it is continuous from the right but not from the left.

(b) g has a jump discontinuity at 0. It is not left- or right-continuous.

(c) S has an essential discontinuity at 0. The left- and right-limits do not exist.

(d) H has an essential discontinuity at 0. It is left-continuous, while the right limit does not exist.

3. (a) $\delta=\epsilon$ works. (b) Define $g(x)=x(1-x)$ if $x\in\mathbb{Q}$ and $g(x)=0$ if $x\notin\mathbb{Q}$.

5. Consider $f=g$ defined $f(x)=0$ if $x\ne 0$ and $f(0)=1$, with $p=0$.

7. Apply the completeness axiom to $A=\{x\in[a,b]\mid f(x)<L\}$ and $B=\{x\in[a,b]\mid f(x)>L\}$.

§ 3.3

1. Evaluate $f(x)=4x^3-6x^2-6x+2$ at various points until you find 3 non-overlapping intervals such that f takes on different signs at the endpoints of each interval.

3. We need to solve $x^3=7$. So let $f(x)=x^3-7$. We need to find a number which is within 0.005 of $\sqrt[3]{7}$. We compute $f(1.9)=-0.141$ and $f(2)=1$. Applying the bisection process, we compute in succession: $f(1.95)=0.41$, $f(1.925)=0.13$, $f(1.9125)=-0.005$, $f(1.91875)=0.064$. The solution is between 1.91875 and 1.9125. This gap is 0.0062, so the next midpoint will have the desired accuracy. Hence, we can take 1.915 as our answer.

5. Consider $a,b\in[-1,1]$ with $a<b$. If $0\notin[a,b]$, then S is continuous on $[a,b]$ and hence takes on the values between $f(a)$ and $f(b)$. If $0\in[a,b]$, then S takes on *all* values in $[-1,1]$ and in particular the ones between $f(a)$ and $f(b)$.

7. Let m,M be the minimum and maximum values of f over $[a,b]$. Note that $m\le \frac{f(c_1)+\cdots+f(c_n)}{n}\le M$.

9. Consider $f(x) - g(x)$.

11. (a) If $f(a)$ is not the minimum value, then there is $c \in (a,b)$ with $f(c) < f(a)$. By the intermediate value theorem, there will be $d \in (c,b)$ with $f(d) = f(a)$.

(b) Let $x,y \in [a,b]$ with $x < y$. By (a), $f(a) < f(y)$ and hence $f(y)$ is the maximum value of f over $[a,y]$.

(c) By (a), $f\colon [a,b] \to [f(a),f(b)]$ is a bijection and hence has an inverse $g\colon [f(a),f(b)] \to [a,b]$. Since f is strictly increasing so is g. By Theorem 3.2.9, g is continuous.

13. The following tasks will establish that a cubic polynomial $p(x) = x^3 + ax^2 + bx + c$ has at least one real root:

(a) $x^3\left(1 + \frac{a}{x} + \frac{b}{x^2} + \frac{c}{x^3}\right) \geq x^3\left(1 - \frac{|a|}{x} - \frac{|b|}{x^2} - \frac{|c|}{x^3}\right) \geq x^3\left(1 - \frac{|a|+|b|+|c|}{x}\right)$.

(b) Observe that $p(-x) = -q(x)$.

15. Proceed as in Exercise 13 to show that $p(x)$ must take a positive value.

§ 3.4

1. All graphs are over the domain $[-\pi,\pi]$:

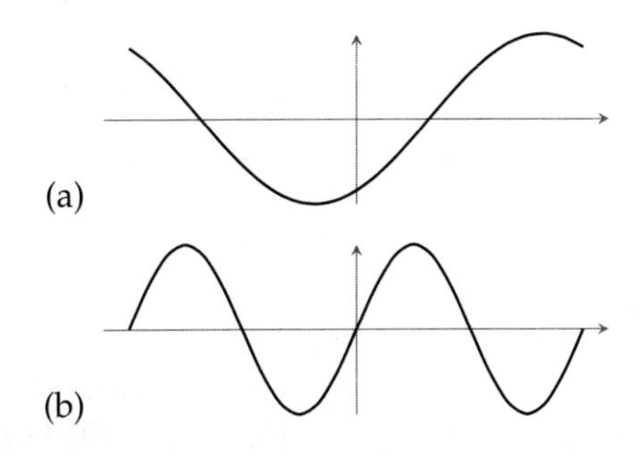

(a)

(b)

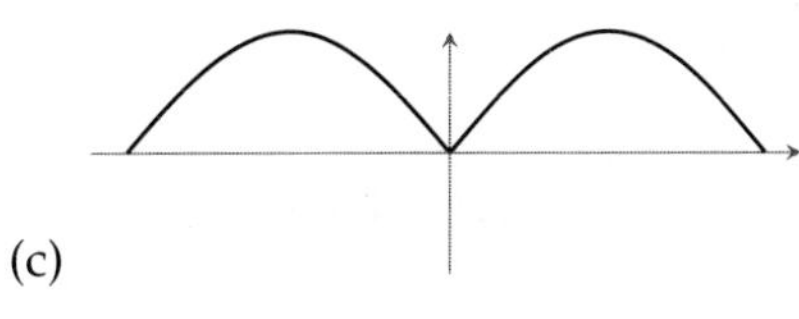

(c)

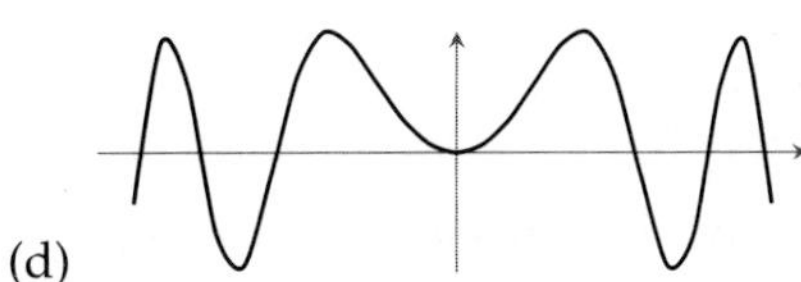

(d)

3. (a) 2, (b) $\cos a$, (c) 2, (d) 0.

5.

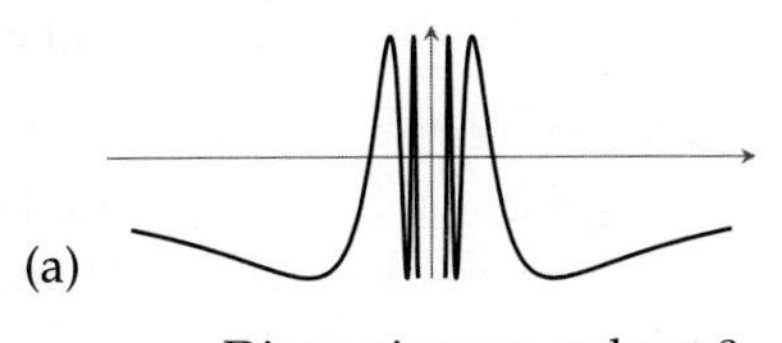

(a)

Discontinuous only at 0.

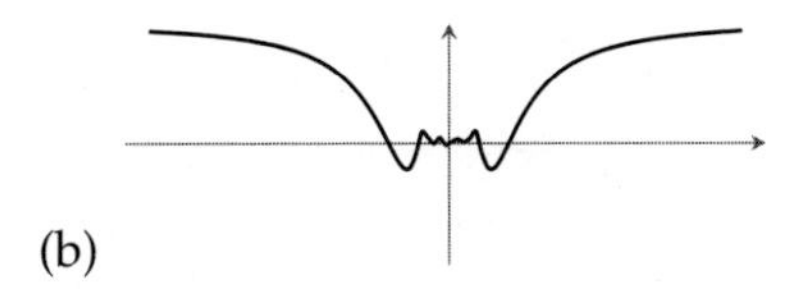

(b)

Continuous everywhere.

7. We shall prove surjectivity. For $m \in \mathbb{R}$, the intersection of the line $y = mx$ and the unit circle $x^2 + y^2 = 1$ is given by $x^2 + m^2x^2 = 1$, or $x^2 = 1/(1+m^2)$. Take $x = 1/\sqrt{1+m^2}$ and

$y = m/\sqrt{1+m^2}$. Let θ be the angle corresponding to the point (x,y) on the unit circle. Then $\tan\theta = m$.

9. $\mathbb{R} \setminus \{n\pi \mid n \in \mathbb{Z}\}, \mathbb{R} \setminus \{(n+\frac{1}{2})\pi \mid n \in \mathbb{Z}\}, \mathbb{R} \setminus \{n\pi \mid n \in \mathbb{Z}\}$ respectively.

11. (a) Begin with $\cos(\alpha+\beta) = \cos\alpha\cos\beta - \sin\alpha\sin\beta$,.

$$\cos(\alpha-\beta) = \cos\alpha\cos\beta + \sin\alpha\sin\beta.$$

Hence, $\cos(\alpha+\beta)+\cos(\alpha-\beta) = 2\cos\alpha\cos\beta$. Substitute $\alpha = (x+y)/2$ and $\beta = (x-y)/2$.

13.

(a) $\dfrac{4x(\pi-x)}{\pi^2}$. (b) $\dfrac{5}{4} - \dfrac{x(\pi-x)}{\pi^2}$. (c)

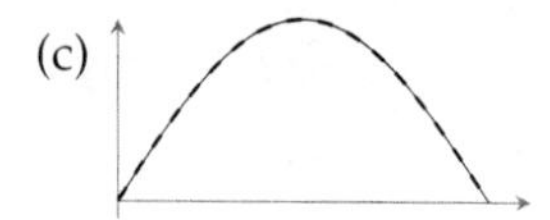

§ 3.5

1. (a) Unbounded on $[0,1]$.

(b) Does not attain its maximum on $[0,1]$.

(c) Does not attain the value 1.5, which is between $[e^0] = 1$ and $[e^1] = 2$.

3. (a) Write $p(x) = x^n\left(1 + \dfrac{a_{n-1}}{x} + \cdots + \dfrac{a_0}{x^n}\right)$. Now choose $R(y)$ large enough so that $|x| > R(y)$ implies

1. $\left|\dfrac{a_{n-1}}{x} + \cdots + \dfrac{a_0}{x^n}\right| < \dfrac{1}{2}$.

2. $x^n > 2y$.

5. By the extreme value theorem, f has a minimum value δ. Since δ is a value of f, we have $\delta > 0$.

§ 3.6

1. First, observe that $f^{-1}\colon [f(a), f(b)] \to [a,b]$ is also strictly increasing and continuous. The given equality is represented by the shaded part of the first diagram below. For the formal proof, create a lower sum for the integral of f as shown next. It automatically creates an upper sum for the integral of f^{-1}.

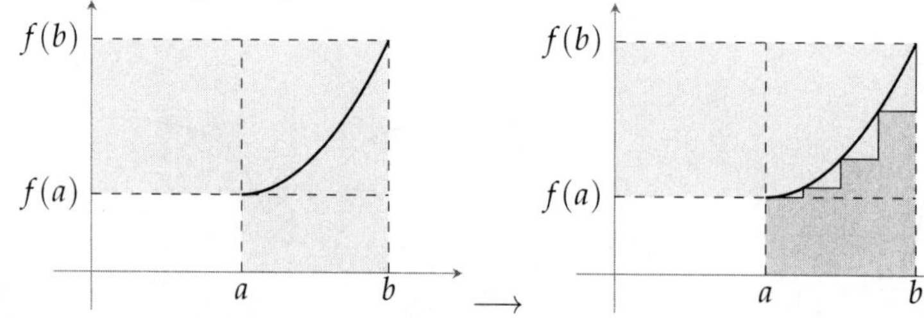

3. Follows from Exercise 12 of § 2.1.

5. Follows from the previous Exercise.

7. $\displaystyle\int_0^1 \frac{x^2}{\sqrt{1+x}}\,dx = \frac{1}{\sqrt{1+c}}\int_0^1 x^2\,dx = \frac{1}{3\sqrt{1+c}}$ for some $c \in [0,1]$.

§ 3.7

1. (a) For any $\epsilon > 0$ there is a $c \in \mathbb{R}$ such that $\arctan c = \pi/2 - \epsilon$. Then $x > c$ implies $\pi/2 > \arctan x > \arctan c > \pi/2 - \epsilon$.

(b) $\lim_{x\to\infty} \frac{\sinh x}{\cosh x} = \lim_{x\to\infty} \frac{e^x - e^{-x}}{e^x + e^{-x}} = \lim_{x\to\infty} \frac{1-e^{-2x}}{1+e^{-2x}} = 1.$

(c) $\lim_{x\to 0+} \frac{\cos x}{\sin x} = \lim_{t\to 0+} \frac{\sqrt{1-t^2}}{t} = \lim_{t\to 0+} \sqrt{\frac{1}{t^2} - 1} = \infty.$

(d) Apply Task 3.7.10, noting that $0 < a < 1 \implies 0 < a^n < a$.

3. (a) False, consider $\lim_{x\to 0} \frac{1}{x}$. (b) False, consider $\lim_{x\to 0+} \frac{-1}{x}$. (c) True.

5. The upward opening parabola straightens into a straight line with positive slope, so one root should move out to minus infinity. The calculations are:

$$\frac{-b+\sqrt{b^2-4ac}}{2a} = -\frac{2c}{b+\sqrt{b^2-4ac}} \to -\frac{c}{b},$$
$$\frac{-b-\sqrt{b^2-4ac}}{2a} = \frac{2c}{-b+\sqrt{b^2-4ac}} \to -\infty.$$

7. Let $-M \le g(x) \le M$ on $[a, \infty)$.

(a) For $L \in \mathbb{R}$, choose $c \in \mathbb{R}$ such that $x > c$ implies $f(x) > L + M$.

(b) Apply the sandwich theorem.

9. (a) Vertical asymptote $x = 1$, horizontal asymptote $y = 0$.

(b) Vertical asymptote $x = 0$, horizontal asymptote $y = 0$.

(c) Vertical asymptote $x = 0$, slant asymptote $y = x - 1$.

(d) Slant asymptote $y = -x + 2/3$.

§ 4.1

1. (a) 2, (b) 0, (c) 1, (d) 0.

3. Use the sandwich theorem to prove differentiability at $x = 0$. At other points, f is not even continuous.

5. Use $y - x = (y^{1/n} - x^{1/n}) \sum_{k=0}^{n-1} x^{k/n} y^{(n-1-k)/n}$.

7.
$$\lim_{x\to a} \frac{g(x)-g(a)}{x-a} = \lim_{x\to a} \frac{xf(x) - xf(a) + xf(a) - af(a)}{x-a}$$
$$= \lim_{x\to a} \left(x\frac{f(x)-f(a)}{x-a} + f(a) \right).$$

9. Use Exercise 7 and mathematical induction.

11. $f'(a+T) = \lim_{h\to 0} \frac{f(a+T+h) - f(a+T)}{h} = \lim_{h\to 0} \frac{f(a+h)-f(a)}{h} = f'(a).$

§ 4.2

1. (a) $\dfrac{4x}{(x^2+1)^2}$, (b) $2\cos 2x$, (c) $\log x$, (d) $\dfrac{1}{x}$.

3. Apply mathematical induction and use the product rule.

5. Consider the effect of repeated differentiation on sine and on polynomials.

7. Apply mathematical induction and use the product rule.

9. First use mathematical induction and the Leibniz rule to prove that $p(x) = (x-a)^k q(x)$ if and only if $p(a) = p'(a) = \cdots = p^{(k-1)}(a) = 0$. Then match the non-vanishing conditions.

11. Rearrange the condition $a^b = b^a$ to $\dfrac{b}{a} = \dfrac{\log b}{\log a}$. Put $b = a(1+t)$, to get

$$e^{t\log a} = 1 + t \leq e^t, \text{ hence } a = 2.$$

Now $b^2 = 2^b$ shows that b is a power of 2, and in fact must be $2^2 = 4$.

§ 4.3

1. (a) $\dfrac{x}{\sqrt{x^2+1}}$.

(b) $\sin 2x - 2x\cos x^2$.

(c) $\cos(\sin x)\cos x$.

(d) $\dfrac{2}{x^3}e^{-1/x^2}$.

(e) $\dfrac{2x}{1+x^2}$.

(f) $(\log\pi)\pi^x - \pi x^{\pi-1}$.

3. Use implicit differentiation to compute the slopes of the two curves through a point (a,b) and check that their product is -1. The x-axis has to considered separately.

5. $\operatorname{arccot}'(x) = \dfrac{1}{\cot'(\operatorname{arccot} x)} = \dfrac{-1}{\csc^2(\operatorname{arccot} x)} = \dfrac{-1}{1+\cot^2(\operatorname{arccot} x)} = \dfrac{-1}{1+x^2}$.

7. $\operatorname{arccsc}' x = \dfrac{1}{\csc'(\operatorname{arccsc} x)} = \dfrac{-1}{\csc(\operatorname{arccsc} x)\cot(\operatorname{arccsc} x)} = \dfrac{-1}{|x|\sqrt{x^2-1}}$.

9.
$$\begin{aligned}(f(x)^{g(x)})' &= \left(e^{g(x)\log f(x)}\right)' = e^{g(x)\log f(x)}(g(x)\log f(x))' \\ &= f(x)^{g(x)}\left(g'(x)\log f(x) + g(x)\frac{f'(x)}{f(x)}\right).\end{aligned}$$

§ 4.4

1. All the graphs are drawn over the domain $[-2,2]$:

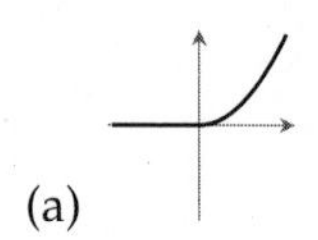

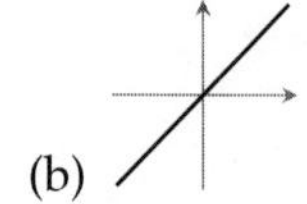

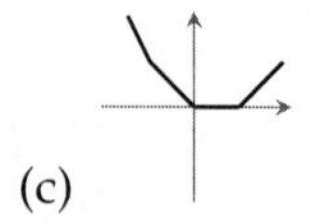

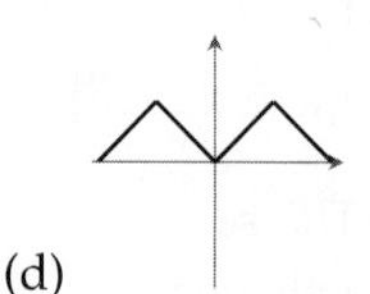

3. F' is C^k, hence F is C^{k+1}.

5. Fix $a \in I$ and let $F(x) = \int_a^x f(t)\,dt$. Then $\int_{g(x)}^{h(x)} f(t)\,dt = F(h(x)) - F(g(x))$. Hence,

$$\left(\int_{g(x)}^{h(x)} f(t)\,dt\right)' = F'(h(x))h'(x) - F'(g(x))g'(x) = f(h(x))h'(x) - f(g(x))g'(x).$$

§ 4.5

1. (a) $1/\sqrt{2}$, (b) absolute maximum is $f(1/\sqrt{2}) = 1/\sqrt{2e} \approx 0.43$, absolute minimum is $f(-1/2) = -e^{-1/4}/2 \approx -0.39$.

3. Let x be the base of the rectangle. Then the height is $1-x$, and the area is $A(x) = x(1-x)$ with $x \in [0,1]$. The absolute maximum is $A(1/2) = 1/4$.

5. Let $f(x) = x^2 - x\sin x - \cos x$. We have $f'(x) = 2x - x\cos x = x(2-\cos x)$. Then $f'(x) > 0$ for $x > 0$ and $f'(x) < 0$ for $x < 0$. So, f is strictly decreasing on $(-\infty,0)$ and strictly increasing on $(0,\infty)$. Since $f(0) = -1$, this means there are at most two solutions. We observe that $f(\pm 2) = 4 - 2\sin 2 - \cos 2 > 4 - 2 - 1 = 1$ and hence there are solutions in $(-2,0)$ and $(0,2)$.

7. We have $y(x) + y'(x)((x-1)-x) = 0$, hence $y'(x) = y(x)$. Therefore $y(x) = Ae^x$.

9. Symmetry gives $|f(x) - f(y)| \leq (x-y)^2$, hence $\left|\dfrac{f(x)-f(y)}{x-y}\right| \leq |x-y|$, and so $f'(x) = 0$ for every x. Therefore, f is constant.

11. By Darboux's theorem.

13. Monotonic functions with intermediate value property are continuous.

§ 4.6

1. Consider f defined by $f(x) = x$ when $x \leq 0$ and $f(x) = 2x$ when $x > 0$.

3. (a) Inflection at $\pm\pi/4$, $\pm 3\pi/4$.

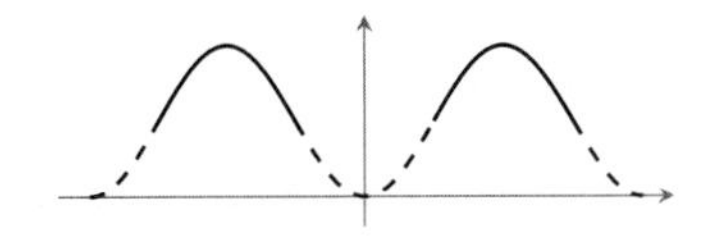

(b) Inflection at ± 1.

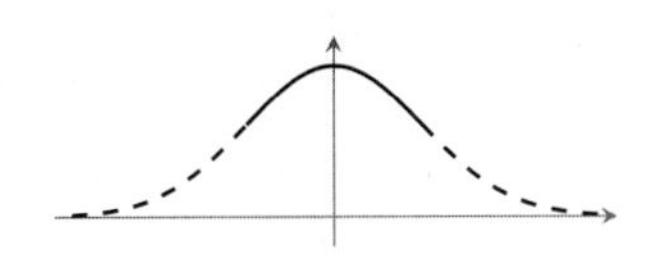

(c) Inflection at $0, \pm\sqrt{3}$.

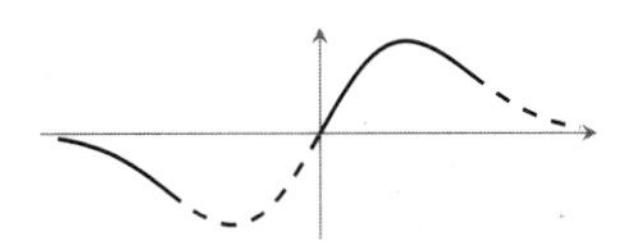

(d) Inflection at 1.

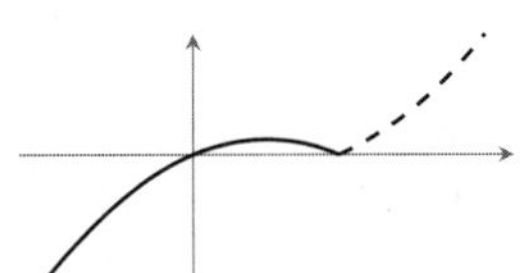

5. The second derivative of a cubic polynomial is linear with non-zero slope and hence has exactly one zero. The second derivative has opposite sign on either side of this zero.

7. Consider $g(x) = x^2 f(x)$. We have $g''(x) = x^2 f''(x) + 4xf'(x) + 2f(x) \geq 0$, hence g is convex. Then $g(a) = g(b) = 0$ implies $g \leq 0$. Therefore, $f \leq 0$.

9. The following observations give (a) and (b): Any point $a \in [x,y]$ can be uniquely expressed as $a = (1-t)x + ty$ with $t \in [0,1]$. If $y = \ell(x)$ gives the secant line joining $(x, f(x))$ and $(y, f(y))$, then $\ell(a) = (1-t)f(x) + tf(y)$.

§ 5.1

1. (a) $\int (3x^2 - 1)\,dx = 3\int x^2\,dx - \int 1\,dx = x^3 - x + C.$

(b) $\int \frac{x^2+3x+2}{x^2+x+1}\,dx = \int \left(1 + \frac{2x+1}{x^2+x+1}\right)dx = x + \log(x^2+x+1) + C.$

(c) $\int \cos^2 t\,dt = \int \frac{1}{2}(1+\cos 2t)\,dt = \frac{1}{2}(t + \frac{1}{2}\sin 2t) + C.$

3. (a) For $x \neq 0$, we have $F'(x) = \frac{3}{2}x^{1/2}\sin(1/x) - x^{-1/2}\cos(1/x)$. For $x = 0$, we have $F'(0) = \lim_{x\to 0} \frac{x^{3/2}\sin(1/x) - 0}{x} = 0$, using the Sandwich Theorem.

(b) We have $f(1/(2n\pi)) = -\sqrt{2n\pi}$ for $n \in \mathbb{N}$.

5. (a) $\int_0^{\pi/4} \tan t\,dt = \log(\sec t)\Big|_0^{\pi/4} = \log(\sec\pi/4) - \log(\sec 0) = \frac{1}{2}\log 2.$

(b) $\int_0^1 \frac{1}{x^2-4}\,dx = \frac{1}{4}\int_0^1 \left(\frac{1}{x-2} - \frac{1}{x+2}\right)dx = \frac{1}{4}\log\left|\frac{x-2}{x+2}\right|\Big|_0^1 = -\frac{\log 3}{4}.$

(c) $\int_{-1}^1 \arctan u\,du = 0$, since $\arctan u$ is an odd function.

7. (a) The curves meet at $x = \pm\sqrt{2}$. So, the area is $\int_{-\sqrt{2}}^{\sqrt{2}} (x^2 + 2 - x^4)\,dx = \frac{7\cdot 2^{5/2}}{15}.$

(b) We have $\sin(5\pi/6) = 1/2$, so the given line is $y = x/2$. Therefore, the area is $2\int_0^{5\pi/6} (\sin x - \frac{x}{2})\,dx = 2 - \sqrt{3} - \frac{25\pi^2}{72}.$

9. (a) $\int_0^{2\pi} \cos 3x \cos 2x\,dx = \frac{1}{2}\int_0^{2\pi} (\cos 5x + \cos x)\,dx = 0.$

(b) $\int_0^{\pi} \cos 3x \sin 2x\,dx = \frac{1}{2}\int_0^{\pi} (\sin 5x - \sin x)\,dx = -\frac{4}{5}.$

11. We have $p(x) = Ax + B$ and $q(x) = (x-\alpha)^2$.

(a) $\frac{Ax+B}{(x-\alpha)^2} = \frac{A(x-\alpha) + B + A\alpha}{(x-\alpha)^2} = \frac{A}{(x-\alpha)} + \frac{B + A\alpha}{(x-\alpha)^2}.$

13. (a) $\int \frac{3x-2}{(x+1)^2}\,dx = \int \frac{3}{x+1}\,dx - \int \frac{5}{(x+1)^2}\,dx = 3\log(x+1) + \frac{5}{x+1} + C.$

(b) $\int \frac{7x}{(x-1)(x-2)}\,dx = \int \frac{14}{x-2}\,dx - \int \frac{7}{x-1}\,dx = 7\log\frac{(x-2)^2}{x-1} + C.$

(c) $\int \frac{7x+3}{(x-1)^2+3^2}\,dx = \frac{7}{2}\log(x^2 - 2x + 10) + \frac{10}{3}\arctan\frac{x-1}{3} + C.$

(d) First divide x^3 by $x^2 - 2x + 1$.

§ 5.2

1. (a) Put $u = x+1$ to get $\frac{2}{7}(x+1)^{7/2} - \frac{4}{5}(x+1)^{5/2} + \frac{2}{3}(x+1)^{3/2} + C$.

(b) Multiply and divide the integrand by $\sqrt{x+1} - \sqrt{x-1}$, then use the substitutions $u = x-1$ and $v = x+1$, to get $\frac{1}{3}\left((x+1)^{3/2} - (x-1)^{3/2}\right) + C$.

(c) Substitute $u = \log x$ to get $(\log x)^3/3 + C$.

(d) Substitute $u = \sqrt{x}$ to get $2\sqrt{x}(\log\sqrt{x} - 1) + C$.

3. (a) Substitute $u = x^2$ to get $\dfrac{1}{2} - \dfrac{1}{2e}$.

(b) Substitute $u = x - 4$ to get $1976/105$.

(c) The integrand is odd, so the integral is zero. For an explicit integration, substitute $t = u^2 + 2$.

(d) Express $\tan^3\theta = \tan\theta\sec^2\theta - \tan\theta$ and use $u = \tan\theta$ to get $\dfrac{1}{6} + \log\dfrac{\sqrt{3}}{2}$.

5. Apply the half-angle formulas:

$$\cos^{2m}(ax) = \left(\frac{1+\cos 2ax}{2}\right)^m = \frac{1}{2^m}\sum_{k=0}^{m}\binom{m}{k}\cos^k(2ax).$$

7. (a) Substitute $x = \sin t$ to get $\dfrac{\pi}{12} + \dfrac{\sqrt{3}}{8}$.

(b) Substitute $x = \sec\theta$ to get $\log(3+\sqrt{8}) - \log(2+\sqrt{3})$.

(c) Substitute $x = \sec\theta$ to get $\arccos\dfrac{1}{3} - \dfrac{\pi}{3}$.

(d) Substitute $x = \sec^2\theta$ to get $\pi/6$.

(e) Substitute $x = 2\tan\theta$ to get $\frac{1}{2}\log(2+\sqrt{5}) - \frac{1}{2}\log(1+\sqrt{2})$.

(f) Write $\dfrac{1}{x^2+x+1} = \dfrac{1}{(x+1/2)^2 + 3/4}$ and substitute $x + \frac{1}{2} = \frac{\sqrt{3}}{2}\tan\theta$ to get $\pi/3^{3/2}$.

§ 5.3

1. (a) First substitute $u = \sqrt{x}$. Then $2u\,du = dx$ and

$$\begin{aligned}\int \sin\sqrt{x}\,dx &= 2\int u\sin u\,du = -2u\cos u + 2\int \cos u\,du \\ &= 2\sin u - 2u\cos u + C = 2\sin\sqrt{x} - 2\sqrt{x}\cos\sqrt{x} + C.\end{aligned}$$

(b) First substitute $u = \log x$. Then $e^u du = dx$ and

$$\begin{aligned}\int \sin(\log x)\,dx &= \int e^u \sin u\,du = \frac{1}{2}e^u(\sin u - \cos u) + C \\ &= \frac{1}{2}x(\sin(\log x) - \cos(\log x)) + C.\end{aligned}$$

(c) $\displaystyle\int \arcsin x\,dx = \int 1\cdot\arcsin x\,dx = x\arcsin x - \int \frac{x}{\sqrt{1-x^2}}\,dx$
$\displaystyle = x\arcsin x + \sqrt{1-x^2} + C.$

(d) Substitute $\tan\theta = x$ and then integrate by parts twice to get

$$\frac{1}{2}(x^2-1)\arctan^2 x - x\arctan x + \frac{1}{2}\log(x^2+1) + C.$$

3. (a) $\displaystyle\int_{-1}^{1}\sqrt{x^2+1}\,dx = 2\int_0^{\pi/4}\sec^3\theta\,d\theta = (\sec\theta\tan\theta + \log(\sec\theta+\tan\theta))\Big|_0^{\pi/4}.$
$\displaystyle = \sqrt{2} + \log(1+\sqrt{2})$

(b) $\displaystyle\int_0^{\pi} e^x\sin x\,dx = \frac{1}{2}e^x(\sin x - \cos x)\Big|_0^{\pi} = (e^{\pi}+1)/2.$

5. (a) $\displaystyle\int (\log x)^n\,dx = \int 1\cdot(\log x)^n\,dx = x(\log x)^n - n\int(\log x)^{n-1}\,dx.$

(b) Use (a) and mathematical induction.

(c) Substitute $u = e^x$ and use (b).

7. (a) Substitute $u = x+1$ and use Exercise 6(a).

(b) Use Exercise 6(c).

§ 5.4

1. (a) $\displaystyle\int \frac{1}{(x-1)(x+1)}\,dx = \frac{1}{2}\int\left(\frac{1}{x-1} - \frac{1}{x+1}\right)dx = \frac{1}{2}\log\left|\frac{x-1}{x+1}\right| + C.$

(b) $\displaystyle\int \frac{1}{x^3-1}\,dx = \frac{1}{3}\int\left(\frac{1}{x-1} - \frac{x+2}{x^2+x+1}\right)dx$
$\displaystyle = \frac{1}{3}\int\left(\frac{1}{x-1} - \frac{1}{2}\frac{2x+1}{x^2+x+1} - \frac{3}{2}\frac{1}{(x+1/2)^2+3/4}\right)dx$
$\displaystyle = \frac{1}{3}\log(x-1) - \frac{1}{6}\log(x^2+x+1) - \frac{1}{\sqrt{3}}\arctan\left(\frac{2x+1}{\sqrt{3}}\right) + C.$

(c) Write $x^2+x+1 = (x+1/2)^2 + 3/4$ and substitute $x + 1/2 = (\sqrt{3}/2)\tan t$. The final solution is $\displaystyle\frac{4}{3\sqrt{3}}\arctan\left(\frac{2x+1}{\sqrt{3}}\right) + \frac{1}{3}\frac{2x+1}{x^2+x+1}.$

(d) $\displaystyle\int \frac{x}{(x-1)^2}\,dx = \int\left(\frac{1}{x-1} + \frac{1}{(x-1)^2}\right)dx = \log(x-1) - \frac{1}{x-1} + C.$

3. $\displaystyle\frac{p(x)}{(x-a_1)\cdots(x-a_n)} = \sum_{k=1}^{n}\frac{A_k}{x-a_k} \implies p(x) = \sum_{k=1}^{n}\left(A_k\prod_{i:i\neq k}(x-x_i)\right).$ Put $x = x_k$. On the RHS, only one term survives.

5. (a) Substitute $u = \sqrt{x}$. (b) Substitute $u = e^{2x}$.

§ 5.5

1. We have seen $\int_1^\infty x^\alpha\,dx$ diverges if $\alpha \geq -1$, while $\int_0^1 x^\alpha\,dx$ diverges for $\alpha \leq -1$. Hence $\int_0^\infty x^\alpha\,dx$ always diverges.

3. (a) π, (b) -1, (c) -4.

5. (a) For large enough x, $e^{-x}x^{3/2} \leq e^{-x/2}$, whose integral converges.

(b) $\dfrac{1}{\sqrt{1+x^4}} \leq \dfrac{1}{x^2}$, whose integral converges.

(c) $\dfrac{2+\sin x}{1+x} \geq \dfrac{1}{1+x}$, whose integral diverges.

(d) $e^{-x}\log x \leq e^{-x/2}$, whose integral converges.

7. The denominator has no zeroes in $[2,\infty)$, so this is an improper integral of the first kind. A limit comparison with $1/x^2$ establishes its convergence.

9. (a) $\int_{-R}^R f(x)\,dx = \int_{-R}^0 f(x)\,dx + \int_0^R f(x)\,dx$ for every R.

(b) $\int_{-R}^R f(x)\,dx = 0$ for every R.

(c) $\int_{-R}^R f(x)\,dx = 2\int_0^R f(x)\,dx$ for every R.

§ 5.6

1. (a) $y' + 2xy^2 = 0 \implies \displaystyle\int \frac{dy}{y^2} = -2\int x\,dx \implies -\frac{1}{y} = -x^2 + c \implies y = \frac{1}{x^2+c}$.

(b) $y'\sin x = y\cos x \implies \displaystyle\int \frac{dy}{y} = \int \frac{\cos x}{\sin x}\,dx \implies \log|y| = \log A|\sin x| \implies \log y = A\sin x$.

(c) $\frac{2}{3}(1+y)^{3/2} + \frac{1}{2}\sin^{-1}x + \frac{1}{2}x\sqrt{1-x^2} + C = 0$.

(d) $y = \dfrac{1}{C - x - \log|x-1|}$.

3. For $0 \leq y_0 \leq M$, the domain is all of $\mathbb{R}$. For $y_0 < 0$, it is $x < \dfrac{1}{kM}\log(1 - M/y_0)$. For $y_0 > M$, it is $x > \dfrac{1}{kM}\log(1-M/y_0)$.

5. Solve the given initial value problem.

(a) Apply separation of variables to get the general equation $y = \dfrac{A}{x-1}e^{-x} - 1$. The initial condition gives $A = -1$.

(b) Separation of variables gives $\cos 2y\cos^2 x = A$. The initial condition gives $A = -1$.

(c) First substitute $u = y/x$ to convert to a separable ODE. The final solution is $\csc(y/x) = 1 + \log x$.

(d) This is a linear first-order ODE with general solution $r = A\sin\theta + (\sin\theta)\log|\sec\theta + \tan\theta|$. The initial condition gives $A = \log(\sqrt{2} - 1)$.

7. (a) First divide through by y^n to get $y^{-n}y' + p(x)y^{1-n} = q(x)$. The substitution $u = y^{1-n}$ gives $u' = (1-n)y^{-n}y'$, and so the ODE becomes $\dfrac{u'}{1-n} + p(x)u = q(x)$ or $u' + (1-n)p(x)u = (1-n)q(x)$.

(b) The substitution in (a) converts this ODE to $u' + u = -x$, with general solution $u = 1 - x + Ae^{-x}$. Hence, $y = \dfrac{1}{1 - x + Ae^{-x}}$.

(c) $y = \dfrac{5x^2}{x^5 + C}$.

§ 6.1

1. Assume there is a $c \in [a,b)$ such that $f(c) - g(c) < 0$. Observe $f'(x) - g'(x) \leq 0$ and apply the monotonicity theorem to get a contradiction.

3. Apply mathematical induction and the racetrack principle.

5. (a) Given any $a < b$ in the interval, by the mean value theorem there is a $c \in (a,b)$ such that $f(b) - f(a) = f'(c)(b-a) = 0$.

(b) Apply (a) to $f - g$.

(c) Given any $a < b$ in the interval, by the mean value theorem there is a $c \in (a,b)$ such that $f(b) - f(a) = f'(c)(b-a) \geq 0$.

7. By Rolle's theorem, there is $x_1 \in (0,1)$ such that $f'(x_1) = 0$. Again by Rolle's theorem, there is $x_2 \in (0,x_1)$ such that $f''(x_2) = 0$. Continue...

9. Define $g(x) = f(x)e^{-\lambda x}$. By Rolle's theorem, there is a $c \in (a,b)$ such that $g'(c) = 0$.

11. Let $a < b$ be points in I. Define $g(x) = f(x) - f(a) - \dfrac{f(b) - f(a)}{b - a}(x - a)$. We have to show $g(x) \leq 0$ for every $x \in [a,b]$. By Rolle's theorem, there is $c \in (a,b)$ such that $g'(c) = 0$. Then $g' \leq 0$ on $[a,c]$ and $g' \geq 0$ on $[c,b]$. Apply the monotonicity theorem or the racetrack inequality.

13. It is enough to verify this for $x \in [0,2]$. By the Cauchy mean value theorem, for any such x, there is a $c \in (0,x)$ such that

$$\frac{1 - \cos x}{x^2/2} = \frac{\sin c}{c} < 1.$$

§ 6.2

1. (a) $\displaystyle\lim_{x\to 1-} \frac{\log x}{\sqrt{1-x}} = \lim_{x\to 1-} \frac{1/x}{-\frac{1}{2}(1-x)^{-1/2}} = \lim_{x\to 1-} -\frac{2\sqrt{1-x}}{x} = 0.$

(b) $\lim_{x\to a} \frac{x^n - a^n}{\log x - \log a} = \lim_{x\to a} \frac{nx^{n-1}}{1/x} = \lim_{x\to a} nx^n = na^n.$

(c) $\lim_{t\to 0} \frac{b - \sqrt{b^2 - t^2}}{t^2} = \lim_{t\to 0} \frac{(b^2 - t^2)^{-1/2}t}{2t} = \frac{1}{2b}.$

(d)
$$\lim_{x\to a} \frac{a - x - a\log a + a\log x}{a - \sqrt{2ax - x^2}} = \lim_{x\to a} \frac{-1 + a/x}{(2ax - x^2)^{-1/2}(x-a)}$$
$$= \lim_{x\to a} -\frac{\sqrt{2ax - x^2}}{x} = -1.$$

(e) 2, (f) 1, (g) 1, (h) 1/2, (i) $\pi^2/6$, (j) -2.

3. (a) We already know that $\lim_{x\to 0} \frac{\cos x - 1}{x} = 0$. So, we compare $\cos x - 1$ to x^2:

$$\lim_{x\to 0} \frac{\cos x - 1}{x^2} = \lim_{x\to 0} \frac{-\sin x}{2x} = -\frac{1}{2} \implies \cos x - 1 \approx -\frac{x^2}{2} \text{ for } x \approx 0.$$

Next, we compare $\cos x - 1 + x^2/2$ with x^4:

$$\lim_{x\to 0} \frac{\cos x - 1 + x^2/2}{x^4} = \lim_{x\to 0} \frac{-\sin x + x}{4x^3} = \lim_{x\to 0} \frac{-\cos x + 1}{12x^2} = \frac{1}{4!}$$
$$\implies \cos x - 1 + x^2/2! \approx x^4/4! \text{ for } x \approx 0.$$

5. Equality is only expected when the limit of f'/g' exists. L'Hôpital's rule allows for f/g to have a limit when f'/g' fails to have one.

7. Since f is differentiable in an open interval containing a, we have

$$\lim_{h\to 0} \frac{f(a+2h) - 2f(a+h) + f(a)}{h^2} = \lim_{h\to 0} \frac{f'(a+2h) - f'(a+h)}{h}$$
$$= \lim_{h\to 0} \left(2\frac{f'(a+2h) - f'(a)}{2h} - \frac{f'(a+h) - f'(a)}{h} \right)$$
$$= 2f''(a) - f''(a) = f''(a).$$

Note that L'Hôpital's rule could not be applied a second time since f' is not given to be differentiable near a.

§ 6.3

1.
$$T_{f,n+1}(x) = f(a) + f'(a)(x-a) + \frac{f''(a)}{2!}(x-a)^2 + \cdots + \frac{f^{(n+1)}(a)}{(n+1)!}(x-a)^{n+1}$$
$$\implies T'_{f,n+1}(x) = f'(a) + f''(a)(x-a) + \cdots + \frac{f^{(n+1)}(a)}{n!}(x-a)^n = T_{Df,n}(a).$$

3. (a) $\sum_{i=0}^{n} x^i$. (b) $\sum_{i=0}^{n} x^{2i}$. (c) $\sum_{i=0}^{n} (i+1)x^i$.

(d) $-\sum_{i=1}^{n}\frac{x^i}{i}$. (e) $\sum_{i=0}^{n}(-1)^i x^{2i}$. (f) $\sum_{i=0}^{n}\frac{(-1)^i}{2i+1}x^{2i+i}$.

5. For an approximate solution, we solve $x = 1 - x^2/2$ to get $x = \sqrt{3} = 0.732\ldots$. (The precise solution is $0.739\ldots$)

7. (a) $f_+^{(n)}(0) = 0$,

(b) For $x > 0$, $f^{(n)}(x) = \frac{p(x)}{x^{2n}}e^{-1/x}$, where p is a polynomial with degree at most $n-1$.

Note that f is even and therefore it is enough to calculate the right-hand derivative at $x = 0$ and the two-sided derivatives for $x > 0$. We apply mathematical induction simultaneously to (a) and (b). For $n = 1$, we have

$$f'_+(0) = \lim_{x\to 0+}\frac{e^{-1/|x|}}{x} = \lim_{t\to\infty}\frac{t}{e^t} = 0 \quad\text{and}\quad x > 0 \Longrightarrow f'(x) = \frac{e^{-1/x}}{x^2}.$$

Now suppose (a) and (b) hold for n. Then for $n+1$,

$$f_+^{(n+1)}(0) = \lim_{x\to 0+}\frac{p(x)}{x^{2n+1}}e^{-1/x} = \lim_{x\to 0+}p(x)\lim_{x\to 0+}\frac{e^{-1/x}}{x^{2n+1}} = p(0)\lim_{t\to\infty}\frac{t^{2n+1}}{e^t} = 0.$$

And for $x > 0$,

$$f^{(n+1)}(x) = \left(\frac{p(x)}{x^{2n}}e^{-1/x}\right)' = \frac{(p'(x)x^2 + p(x))x^{2n-2} - 2np(x)x^{2n-1}}{x^4 n}e^{-1/x}$$
$$= \frac{p'(x)x^2 + p(x) - 2np(x)x}{x^{2(n+1)}}e^{-1/x}.$$

9. If n is even and $f^{(n)}(a) < 0$, then we have $f^{(n)}(\xi) < 0$ for every $\xi \in I\setminus\{a\}$. Therefore $f(x) < f(a)$ for every $x \in I\setminus\{a\}$ and there is a local maximum at a. If n is odd then $(x-a)^n$ changes sign at $x = a$, therefore $f(x) - f(a)$ also changes sign at $x = a$.

11. We have $p(a) = p'(a) = \cdots = p^{(n-1)}(a) = 0$ and $p^{(n)}(a) = n!q(a) \neq 0$. Apply Theorem 6.3.11.

13. The remainder theorem gives (a) and (b). For (c), we first note that

$$\lim_{x\to a}\frac{T_n(x) - P(x)}{(x-a)^n} = \lim_{x\to a}\frac{T_n(x) - f(x) + f(x) - P(x)}{(x-a)^n} = 0.$$

Hence, $\lim_{x\to a}\frac{T_n(x) - P(x)}{(x-a)^k} = 0$ for every $k = 0,1,\ldots,n$. Apply this to $T_n(x) - P(x) = \sum_{k=0}^{n} a_k(x-a)^k$.

§ 6.4

1. (a) Let $P = \{x_0,\ldots,x_n\}$. By the mean value theorem, for each $i = 1,\ldots,n$, there is $x_i^* \in (x_{i-1},x_i)$ such that $F(x_i) - F(x_{i-1}) = (x_i - x_{i-1})F'(x_i^*) = (x_i - x_{i-1})f(x_i^*)$. Using these x_i^* as tags, we get $R(f,P^*) = \sum_{i=1}^{n}(x_i - x_{i-1})f(x_i^*) = \sum_{i=1}^{n}\left(F(x_i) - F(x_{i-1})\right) = F(b) - F(a)$.

(b) Apply Theorem 6.4.2.

3. Note that f is not given to be integrable, or even bounded! Let us first show that f must be bounded. Take $\epsilon = 1$ and let $P_\epsilon = \{x_0, \ldots, x_n\}$. If f is not bounded, we can fix an interval $[x_{i-1}, x_i]$ on which f is unbounded. By choosing function values of high magnitude in this interval, we see that the Riemann sums $R(f, P_\epsilon^*)$ are unbounded. But this contradicts $|I - R(f, P_\epsilon^*)| < 1$.

Next, for any $\epsilon > 0$, let $P_\epsilon = \{x_0, \ldots, x_n\}$. Since f is bounded, we can define $m_i = \inf\{f(x) \mid x \in [x_{i-1}, x_i]\}$ and $M_i = \sup\{f(x) \mid x \in [x_{i-1}, x_i]\}$. By choosing points where f takes values close to m_i and M_i we create tagged partitions P_ϵ^* and P_ϵ^{**} that are within ϵ of $L = \sum_i m_i(x_i - x_{i-1})$ and $M = \sum_i M_i(x_i - x_{i-1})$ respectively. Then the lower and upper sums, L and M, are within 2ϵ of I. So, f is integrable and I is its integral.

5. The astroid is parametrised by $x(t) = a\cos^3 t$ and $y(t) = a\sin^3 t$, with $t \in [0, 2\pi]$.

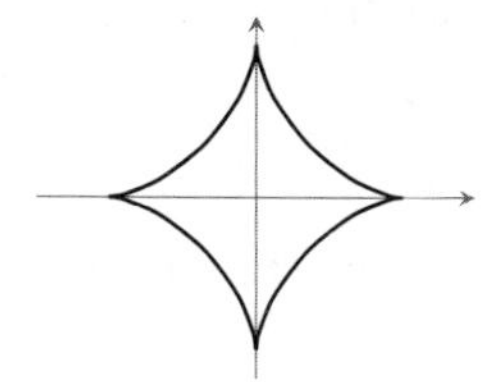

Its length is

$$L = 3a \int_0^{2\pi} \left(\cos^4 t \sin^2 t + \sin^4 t \cos^2 t\right)^{1/2} dt = 3a \int_0^{2\pi} \left(\cos^2 t \sin^2 t\right)^{1/2} dt$$
$$= \frac{3a}{2} \int_0^{2\pi} |\sin 2t|\, dt = 6a \int_0^{\pi/2} \sin 2t\, dt = 6a.$$

7. Use the description in Example 6.4.13. The resulting surface area is $4\pi^2 Rr$.

9. The volume is $\frac{4}{3}\pi ab^2$, an easy generalization of the formula for a sphere. The surface area S takes more work. First, we note that $S = 2\int_0^a 2\pi y\sqrt{1 + (y')^2}\, dx$, where $y = \frac{b}{a}\sqrt{a^2 - x^2}$. This gives

$$S = 4\pi b \int_0^a \sqrt{1 - (1 - b^2/a^2)(x/a)^2}\, dx = 4\pi b \int_0^a \sqrt{1 + ((b^2/a^2) - 1)(x/a)^2}\, dx.$$

The first expression is useful when $a > b$, the second when $a < b$. The final formulas are:

$$S = \begin{cases} 2\pi b \left(b + \dfrac{a^2}{\sqrt{a^2 - b^2}} \sin^{-1}\left(\dfrac{\sqrt{a^2 - b^2}}{a} \right) \right) & \text{if } a > b, \\[2ex] 2\pi b \left(b + \dfrac{a^2}{\sqrt{b^2 - a^2}} \sinh^{-1}\left(\dfrac{\sqrt{b^2 - a^2}}{a} \right) \right) & \text{if } a < b. \end{cases}$$

Ellipsoids like these, with circular cross-sections in one direction, are called **spheroids**. The $a > b$ ones are called **prolate spheroids** and the $a < b$ ones are called **oblate spheroids**. The bulging of the Earth at its equator makes it an oblate spheroid.

11. (a) $\displaystyle\int_0^{\pi} 2\pi \sin x \sqrt{1+\cos^2 x}\, dx = 2\pi(\sqrt{2} + \log(1+\sqrt{2}))$.

(b) $\displaystyle\int_0^{\pi} \pi \sin^2 x\, dx = \pi^2/2$.

(c) $\displaystyle\int_0^{\pi} 2\pi x \sin x\, dx = 2\pi^2$.

13. $\displaystyle\int_0^{\infty} \frac{\pi}{(1+x^2)^2}\, dx = \frac{\pi^2}{4}$.

§ 6.5

1. The actual value is $\log 2 = 0.69315$. The approximations are:

Midpoint Rule: $\displaystyle\sum_{k=0}^{4} \frac{1}{1.1+0.2k} \cdot \frac{1}{5} = 0.69191$.

Simpson's Rule: $\displaystyle\left(\frac{1}{1} + 4\cdot\frac{1}{1.25} + 2\cdot\frac{1}{1.5} + 4\cdot\frac{1}{1.75} + \frac{1}{2}\right)\frac{0.25}{3} = 0.69325$.

3. (a) Fix $x \in (0,h)$. Define a number M by $f(x) - \ell(x) = x(x-h)M$. Then define $g\colon [0,h] \to \mathbb{R}$ by $g(t) = f(t) - \ell(t) - Mt(t-h)$.

Now g is twice continuously differentiable and $g(0) = g(x) = g(h) = 0$. Rolle's theorem gives $\xi_1 \in (0,x)$ and $\xi_2 \in (x,h)$ such that $g'(\xi_1) = g'(\xi_2) = 0$. Apply Rolle's theorem again to get $\xi_x \in (\xi_1, \xi_2)$ such that $g''(\xi_x) = 0$. Now $g''(t) = f''(t) - 2M$ implies $M = f''(\xi_x)/2$.

(b) Note that $f''(\xi_x)$ is a continuous function of x. By the mean value theorem for weighted integration (Theorem 3.6.7), we have $\xi \in [0,h]$ such that

$$\begin{aligned}\int_0^h (f(x) - \ell(x))\, dx &= \int_0^h \frac{x(x-h)}{2} f''(\xi_x)\, dx = f''(\xi)\int_0^h \frac{x(x-h)}{2}\, dx \\ &= -\frac{h^3}{12} f''(\xi).\end{aligned}$$

(c) We divide $[a,b]$ in n equal parts and apply (b) over each of them. We have $h = (b-a)/n$, and we get

$$\begin{aligned}\int_a^b f(x)\, dx - T_a^b(f) &= -\sum_{i=1}^{n} \frac{h^3}{12} f''(\xi_i) = -\frac{h^2}{12}(b-a)\sum_{i=1}^{n} \frac{f''(\xi_i)}{n} \\ &= -\frac{h^2}{12}(b-a) f''(\xi).\end{aligned}$$

5. Let the distinct sampling points be a, b, c with corresponding weights w_a, w_b, w_c. Matching with the integrals of $1, x, \ldots, x^5$ over $[-h,h]$ gives the equations:

$$w_a + w_b + w_c = 2h \tag{1}$$

$$w_a a + w_b b + w_c c = 0 \tag{2}$$

$$w_a a^2 + w_b b^2 + w_c c^2 = 2h^3/3 \tag{3}$$
$$w_a a^3 + w_b b^3 + w_c c^3 = 0 \tag{4}$$
$$w_a a^4 + w_b b^4 + w_c c^4 = 2h^5/5 \tag{5}$$
$$w_a a^5 + w_b b^5 + w_c c^5 = 0 \tag{6}$$

Eliminate w_a between Equations 2 and 4, and also between 4 and 6:

$$w_b b(b^2 - a^2) + w_c c(c^2 - a^2) = 0 \tag{7}$$
$$w_b b^3(b^2 - a^2) + w_c c^3(c^2 - a^2) = 0 \tag{8}$$

Eliminate w_b between Equations 7 and 8:

$$w_c c(c^2 - a^2)(c^2 - b^2) = 0 \tag{9}$$

Eliminate w_c between Equations 7 and 8:

$$w_b b(b^2 - a^2)(b^2 - c^2) = 0 \tag{10}$$

Suppose $c \neq 0$. Then $c = -a$ or $c = -b$. We may assume $c = -a$. Then we cannot have $b = -c$ or $b = -a$, so we must have $b = 0$. We can further take the situation to be $a < b = 0 < c = -a$. Substituting this in 2 gives $w_a = w_c$. We leave the further (enjoyable) calculations to the reader. The final result is $a = -\sqrt{3/5}h$, $b = 0$, $c = \sqrt{3/5}h$, $w_a = w_c = 5h/9$, $w_b = 8h/9$.

To estimate the given integral, we first shift the interval of integration to $[-\pi/4, \pi/4]$. Then we put $h = \pi/4$ in the Gaussian quadrature formula:

$$\int_0^{\pi/2} \sqrt{x}\cos x\, dx = \int_{-\pi/4}^{\pi/4} \sqrt{x + \pi/4}\cos(x + \pi/4)\, dx \approx 0.709.$$

The precise value is $0.704\ldots$.

§ 7.1

1. Given $\epsilon > 0$, choose $N = [1/\epsilon^2] + 1$.

3. (a) $\displaystyle\lim_{n\to\infty} \frac{e^n - e^{-n}}{e^n + e^{-n}} = \lim_{n\to\infty} \frac{1 - e^{-2n}}{1 + e^{-2n}} = 1.$

(b) $\displaystyle 0 < \frac{n!}{n^n} \leq \frac{1}{n} \implies \lim_{n\to\infty} \frac{n!}{n^n} = 0.$

(c) The $n = 2k$ terms form the divergent sequence $(-1)^k$, therefore $\lim_{n\to\infty} \cos(n\pi/2)$ diverges.

(d) The $n = 2k$ terms form the divergent sequence $2k$, therefore $\lim_{n\to\infty} n^{(-1)^n}$ diverges.

5. The argument is correct insofar as it says that "If a_n converges then it converges to 0". However, we cannot conclude that a_n *does* converge, and in fact it does not.

7. Following the hint, we reach $|a_{N+n}| \leq r^n |a_N|$, hence $a_{N+n} \to 0$.

9. The intuition is that as we increase n, terms close to L will dominate in number and will move the average towards L. Given $\epsilon > 0$, fix M such that $n \geq M$ implies $|a_n - L| < \epsilon/2$. Then, using the decomposition given in the hint, we get

$$|\bar{a}_{M+k} - L| \leq \frac{M}{M+k}|\bar{a}_M - L| + \frac{k}{M+k}\frac{\epsilon}{2} \to \frac{\epsilon}{2} \text{ as } k \to \infty.$$

Therefore, there is k_0 such that $k \geq k_0$ implies $|\bar{a}_{M+k} - L| < \epsilon$.

11. We can prove by mathematical induction that for every n, $a_n \geq a_{n+1}$. For $n = 1$, we have $a_1 = 2$ and $a_2 = 1$. Assume the truth for n. Then for $n + 1$,

$$0 \leq \frac{1}{3-0} \leq a_{n+1} = \frac{1}{3-a_n} \leq \frac{1}{3-2} = 1 \leq 2.$$

We similarly prove by induction that a_n is decreasing. First, we have $a_2 = 1/2 \leq a_1$. Then, assuming $a_n \leq a_{n-1}$, we have

$$a_{n+1} = \frac{1}{3-a_n} \leq \frac{1}{3-a_{n-1}} = a_n.$$

Being decreasing and bounded below by 0, a_n must converge to some L. This L must satisfy $L = \dfrac{1}{3-L}$. Therefore, $L = \dfrac{3-\sqrt{5}}{2}$.

§ 7.2.

1. (a) First, $\lim_{n\to\infty}(\sqrt{n+1} - \sqrt{n}) = \lim_{n\to\infty}\dfrac{1}{\sqrt{n+1}+\sqrt{n}} = 0$. Since cosine is continuous, $\lim_{n\to\infty}\cos(\sqrt{n+1} - \sqrt{n}) = \cos 0 = 1$.

(b) By continuity of arctan, $\lim_{n\to\infty}\arctan(n^{1/n}) = \arctan(\lim_{n\to\infty} n^{1/n}) = \arctan 1 = \pi/4$.

(c) $\lim_{n\to\infty}(2n)^{1/n} = \lim_{n\to\infty} 2^{1/n} \lim_{n\to\infty} n^{1/n} = 1$.

(d) $\lim_{x\to\infty}\dfrac{\log x}{x^p} = \lim_{x\to\infty}\dfrac{1}{px^p} = 0$.

(e) $\lim_{n\to\infty}\dfrac{(\log n)^p}{n} = \lim_{n\to\infty}\left(\dfrac{\log n}{n^{1/p}}\right)^p = 0^p = 0$, since x^p is continuous.

(f) $\lim_{n\to\infty}\dfrac{e^n}{n^p} = \infty$. Apply L'Hôpital's Rule until the power of n becomes non-positive.

3. $1^n \to 1$, $(e^{1/n})^n \to e$, $(e^{-1/\sqrt{n}})^n \to 0$, $(e^{1/\sqrt{n}})^n \to \infty$.

5. Evaluate the limits of the given sequences by viewing them as upper/lower sums for an integral of a monotonic function:

(a) $\dfrac{1}{n}\left(\dfrac{1}{1+1/n} + \cdots + \dfrac{1}{1+n/n}\right) \to \displaystyle\int_1^2 \frac{dx}{x} = \log 2$.

(b) $\dfrac{1}{n}\left(\dfrac{1}{(1+1/n)^2} + \cdots + \dfrac{1}{(1+n/n)^2}\right) \to \displaystyle\int_1^2 \frac{dx}{x^2} = \frac{1}{2}$.

7. Use mathematical induction to show that the inequality between x_1 and x_2 determines the behavior of the sequence.

9. Consider the equation $e^x - x - 2 = 0$.

(a) Let $f(x) = e^x - x - 2$. We have $f(-2) = e^{-2} > 0$, $f(0) = -1 < 0$, $f(2) = e^2 - 4 > 0$. So, there is a solution in each of $(-2,0)$ and $(0,2)$. If $f(x) = 0$ had a third solution, then by Rolle's theorem $f'(x) = 0$ would have at least two solutions. But it has only one.

(b) Consult the proof of Theorem 7.2.17.

§ 7.3

1. (a) Compare with the geometric series $\sum_{n=1}^{\infty} 1/3^n$.

(b) Compare with the geometric series $\sum_{n=1}^{\infty} 1/2^{n-1}$.

(c) Hint: $1.5^n = (1+0.5)^n \geq n/2$.

(d) For $n \geq 12$, we have $\dfrac{n^{10}}{n!} \leq \dfrac{n^{10}}{n(n-1)\cdots(n-9)} \dfrac{1}{(n-10)(n-11)}$. We have $\lim_{n\to\infty} \dfrac{n^{10}}{n(n-1)\cdots(n-9)} = 1$, so, for large n, $\dfrac{n^{10}}{n!} \leq \dfrac{2}{(n-10)(n-11)}$.

3. (a) Limit comparison with $\sum_{n=1}^{\infty} 1/n$.

(b) By contradiction, based on $\sum_{n=1}^{\infty} \dfrac{1}{n} = \sum_{n=1}^{\infty} \left(\dfrac{1}{n} - \dfrac{1}{n^2}\right) + \sum_{n=1}^{\infty} \dfrac{1}{n^2}$.

(c) Comparison with $\sum_{n=2}^{\infty} 1/n$.

(d) Comparison with $\dfrac{1}{2} \sum_{n=2}^{\infty} \dfrac{1}{n}$.

(e) Limit comparison with $\sum_{n=1}^{\infty} 1/\sqrt{n}$.

(f) Divergence test.

5. Carry out a limit comparison with $\sum_{n=1}^{\infty} \dfrac{1}{n^{p+1/2}}$.

7. Hint: $b_n = (a_n + b_n) - a_n$.

9. Hint: Convergence of $\sum a_n$ implies $a_n \to 0$, so $0 \leq a_n \leq 1$ for large n.

11. Let $S_n = a_1 + \cdots + a_n$ and $b_n = a_{n+1} + a_{n+2} + \cdots + a_{2n}$. $S_{2n} = S_n + b_n$ implies $b_n \to 0$. By sandwich theorem, $na_{2n} \to 0$ and hence $(2n)a_{2n} \to 0$. Further, $(2n+1)a_{2n+1} \leq (2n)a_{2n} + a_{2n+1} \to 0$.

13. Differentiate e^x/x^x to show that it is decreasing and the integral test can be applied. Compare $\sum(e/n)^n$ with $\sum(e/3)^n$.

15. Share with your friends and see what you can come up with!

§ 7.4

1. (a) $\left(\dfrac{\sqrt{x}}{x+1000}\right)' = \dfrac{1000-x}{2\sqrt{x}(x+1000)^2} \leq 0$ for $x \geq 1000$. Apply the alternating series test to $\sum\limits_{n=1000}^{\infty} (-1)^n \dfrac{\sqrt{n}}{n+1000}$.

(b) $\left(\dfrac{x^3}{e^x}\right)' = \dfrac{3x^2-x^3}{e^x} \leq 0$ for $x \geq 3$. Apply the alternating series test to $\sum\limits_{n=3}^{\infty} (-1)^n \dfrac{n^3}{e^n}$.

(c) $\dfrac{2n+100}{3n+1} \to \dfrac{2}{3}$. Hence, for large n, $\left(\dfrac{2n+100}{3n+1}\right)^n < \left(\dfrac{3}{4}\right)^n$.

(d) Apply the divergence test.

3. Define $b_n = 0$ when n is odd, and $b_n = 1/n$ when n is even.

5. (a) Compare $\sum |a_n + b_n|$ with $\sum(|a_n| + |b_n|)$.

(b) Since $\sum a_n$ and $\sum b_n$ converge, so does $\sum(a_n + b_n)$. However, if $\sum |a_n + b_n|$ converges, then so will $\sum |b_n|$ by (a), since $b_n = (a_n + b_n) - a_n$.

(c) $\sum(a_n + b_n)$ will converge, but the convergence could be either absolute or conditional. Provide an example of each.

7. We have given a choice of root or ratio test and the corresponding value of L.

(a) Ratio test, $L = 1/3$.

(b) Ratio test, $L = e$.

(c) Root test, $L = 1/2$.

(d) Root test, $L = 2/\pi$.

(e) Ratio test, $L = 0$.

(f) Root test, $L = 0$.

(g) Root test, $L = 0$.

(h) Root test, $L = 1/e$.

9. (a) Since $\sum b_n$ converges, we have $|b_n \to 0$, and so $|b_n| < 1$ for large n. Now compare $\sum |a_n b_n|$ with $\sum |a_n|$.

(b) $a_n = b_n = (-1)^n / \sqrt{n}$.

11. First, since $\sum a_n$ converges, its partial sums are bounded. Second, since b_n is monotone and bounded, it converges to some L. Suppose b_n decreases. Then $b_n - L$ decreases to 0. By Dirichlet's test, $\sum a_n(b_n - L)$ converges. Hence, $\sum a_n b_n$ converges.

13. (a) Apply the previous Exercise to the partial sums.

(b) The only vector space property that needs a little work is closure under addition:

$$\sum(a_n + b_n)^2 = \sum a_n^2 + \sum b_n^2 + 2\sum a_n b_n,$$

the convergence of the first two sums on the right is given, and the third converges by (a).

§ 8.1

1. (a) $x \in (-1,1)$, (b) $x \in \mathbb{R}$, (c) $x \in [-3,3)$, (d) $x \in [-1,1]$.

3. We can do this either by ratio test or by root test. In applying the root test, we use Stirling's approximation.

$$\lim_{n\to\infty}\left|\frac{a_{n+1}}{a_n}\right| = \lim_{n\to\infty}\left|\frac{(n+1)^k x}{(kn+k)\cdots(kn+1)}\right| = \frac{|x|}{k^k}.$$

$$\lim_{n\to\infty}|a_n|^{1/n} = \lim_{n\to\infty}\left|\frac{(n!)^{k/n}x}{((kn)!)^{1/n}}\right| = \lim_{n\to\infty}\left|\frac{(n/e)^k x}{(kn/e)^k}\right| = \frac{|x|}{k^k}.$$

Both calculations lead to $R = k^k$.

5. In Example 8.1.9 we saw that $e^x = \sum_{n=0}^{\infty}\frac{x^n}{n!}$. Put $x=-1$ to get $\frac{1}{e} = \frac{1}{2!} - \frac{1}{3!} + \frac{1}{4!}\cdots$. The RHS is an alternating series, so its sum lies between any two consecutive partial sums. The partial sums ending in $1/4!$ and $1/5!$ give the desired bounds.

7. (a) Differentiate $\frac{1}{1-x} = \sum_{n=0}^{\infty} x^n$ twice to get $\frac{2}{(1-x)^3} = \sum_{n=2}^{\infty} n(n-1)x^{n-2}$. Hence, $\frac{2x^2}{(1-x)^3} = \sum_{n=2}^{\infty} n(n-1)x^n$, and this is valid for $|x|<1$. Put $x=1/2$ to get $\sum_{n=2}^{\infty}\frac{n(n-1)}{2^n} = 4$.

(b) Differentiate $\frac{1}{1-x} = \sum_{n=0}^{\infty} x^n$ and multiply by x to get $\frac{x}{(1-x)^2} = \sum_{n=1}^{\infty} nx^n$. Again differentiate and multiply by x to get $\frac{x(1+x)}{(1-x)^3} = \sum_{n=1}^{\infty} n^2x^n$. This gives $\sum_{n=1}^{\infty}\frac{n^2}{2^n} = 6$.

(c) 26.

9. (a) Put $x=y=0$ in $y = \sum_{n=0} c_n x^n$. Put $x=0$ and $y'=1$ in $y' = \sum_{n=1} nc_n x^{n-1}$.

(b) The ODE becomes $\sum_{n=0}^{\infty}(n+2)(n+1)c_{n+2}x^n + \sum_{n=0} c_n x^n = x - x^3/3! + x^5/5! \cdots$. This gives $(2n+2)(2n+1)c_{2n+2} + c_{2n} = 0$. Apply mathematical induction.

(c) Replace n by $2n+1$ in the series expansion of the ODE to get

$$(2n+3)(2n+2)c_{2n+3} + c_{2n+1} = \frac{(-1)^n}{(2n+1)!}.$$

Now apply mathematical induction.

(d) Apply the ratio test.

§ 8.2

1. Recall that if f is even then f' is odd, while if f is odd then f' is even. Also, if g is odd then $g(0) = 0$.

3. (a) Apply the binomial series expansion.

$$\binom{1/2}{n} = \frac{(1/2)(1/2-1)(1/2-2)\cdots(1/2-n+1)}{n!} = (-1)^{n-1}\frac{1\cdot 3\cdot (2n-3)}{2^n n!}$$
$$= (-1)^{n-1}\frac{(2n)!}{2^n n!(2n-1)(2n)(2n-2)\cdots 2} = (-1)^{n-1}\frac{(2n)!}{4^n (n!)^2(2n-1)}.$$

(b) The series in (a) converges at $x = \pm 1$, and the function $\sqrt{1+x}$ is continuous at these points. By Abel's theorem, the series converges to the function values at ± 1.

5. Exercises (a) and (b) involve sums of power series, while (c) and (d) utilize the Cauchy product.

(a) $\frac{1}{2}\left(\sum_{n=0}^{\infty}\frac{x^n}{n!} + \sum_{n=0}^{\infty}(-1)^n\frac{x^n}{n!}\right) \approx 1 + \frac{x^2}{2!} + \frac{x^4}{4!}.$

(b) $\frac{1}{2}\left(\sum_{n=1}^{\infty}(-1)^{n-1}\frac{x^n}{n} + \sum_{n=1}^{\infty}\frac{x^n}{n}\right) \approx x + \frac{x^3}{3} + \frac{x^5}{5}.$

(c) $\left(1 + x^2 + \frac{x^4}{2}\cdots\right)\left(x - \frac{x^3}{3!} + \frac{x^5}{5!}\cdots\right) \approx x + \frac{5}{6}x^3 + \frac{41}{120}x^5.$

(d) $\left(1 + \frac{x}{2} - \frac{x^2}{8}\cdots\right)\left(1 + x + x^2\cdots\right) \approx 1 + \frac{3}{2}x + \frac{11}{8}x^2.$

§ 8.3.

1. (a) $a_n = 0$, $b_n = (-1)^{n-1}\frac{2}{n}$.

(b) $a_0 = \frac{2}{3}\pi^2$, $a_n = (-1)^n\frac{4}{n^2}$ for $n \geq 1$, $b_n = 0$.

(c) $a_n = 0$, $b_n = (-1)^n 2\left(\frac{6}{n^3} - \frac{\pi^2}{n}\right).$

(d) $a_0 = \frac{4}{\pi}$, $a_n = (-1)^{n-1}\frac{4}{\pi}\frac{1}{4n^2-1}$ for $n \geq 1$, $b_n = 0$.

3. We show the computation of b_n':

$$b_n' = \frac{1}{\pi}\int_{-\pi}^{\pi} f'(x)\sin nx\,dx = \frac{1}{\pi}f(x)\sin nx\Big|_{-\pi}^{\pi} - \frac{n}{\pi}\int_{-\pi}^{\pi} f(x)\cos nx\,dx = -na_n.$$

5. (a) Use $x = \pi$ in the Fourier series of x^2 over $[-\pi, \pi]$.

(b) Use $x = 0$ in the Fourier series of x^2 over $[-\pi, \pi]$.

(c) Use $x = \pi/2$ in the Fourier series of $x^3 - \pi^2 x$ over $[-\pi, \pi]$.

(d) Use Exercise 3 to obtain the Fourier series of x^4 from that of x^3, put $x = \pi$ and also use (a).

7. For $x \in [-\pi,\pi]$, we have

$$x^2 = \frac{\pi^2}{3} + 4\sum_{n=1}^{\infty}(-1)^n\frac{\cos nx}{n^2} = \frac{\pi^2}{3} + 4\sum_{n=1}^{\infty}\frac{\cos nx\cos n\pi}{n^2} = \frac{\pi^2}{3} + 4\sum_{n=1}^{\infty}\frac{\cos n(x+\pi)}{n^2}.$$

Therefore, for $x \in [0,2\pi]$, $(x-\pi)^2 = \frac{\pi^2}{3} + 4\sum_{n=1}^{\infty}\frac{\cos nx}{n^2}$. Since the RHS is even, the following holds for $x \in [-2\pi,2\pi]$: $\frac{(|x|-\pi)^2}{4} - \frac{\pi^2}{12} = \sum_{n=1}^{\infty}\frac{\cos nx}{n^2}$.

§ 8.4

1. First show that $p(\bar{z}) = \overline{p(z)}$.

3.
$$\begin{aligned}\cos 3\theta + i\sin 3\theta &= e^{i3\theta} = (e^{i\theta})^3 = (\cos\theta + i\sin\theta)^3\\ &= (\cos^3\theta - 3\cos\theta\sin^2\theta) + i(3\cos^2\theta\sin\theta - \sin^3\theta).\end{aligned}$$

5. First apply the Ratio Test: $\frac{n+2}{3^{n+1}}|z+2|^{n+1}\cdot\frac{3^n}{(n+1)|z+2|^n} \to \frac{|z+2|}{3}$. So, the series converges for $|z+2| < 3$ and diverges for $|z+2| > 3$. If $|z+2| = 3$, then the series diverges by the divergence test.

7. We know $\sum_{n=0}^{\infty} z^n/n!$ converges for every z. So this series converges exactly when $\sum_{n=0}^{\infty} n^2/z^n$ converges, i.e., for $|z| > 1$.

9. (a)
$$\begin{aligned}\cos z = \cos w &\iff e^{iz} + e^{-iz} = e^{iw} + e^{-iw} \iff e^{iz} - e^{iw} = e^{-iw} - e^{-iz}\\ &\iff e^{iz} - e^{iw} = \frac{e^{iz} - e^{iw}}{e^{i(w+z)}} \iff e^{iz} = e^{iw} \text{ or } e^{i(w+z)} = 1\\ &\iff z - w \in 2\pi\mathbb{Z} \text{ or } z + w \in 2\pi\mathbb{Z}.\end{aligned}$$

(b) Similar to (a).

(c) Apply (a) with $w = \pi/2$.

(d) Apply (b) with $w = 0$.

(e) Use $e^{iz} = \cos z + i\sin z$.

11. As the final regular exercise of the book, this is deliberately quite challenging. It was asked in the Prelim Exam for the Berkeley PhD program [5]. The following tasks will take you to the solution. We will denote the real part of a complex number z by $\Re z$.

(a) $\sum_{n=1}^{\infty} w^n/n^2$ converges when $|w| \le 1$, so the given series converges when $|\exp(z/(z-2))| \le 1$.

(b) The given series converges when $\Re(z/(z-2)) \le 0$.

(c) Let $f(z) = z/(z-2)$. Its inverse function is given by $g(w) = 2w/(w-1)$.

(d) Show that g is a bijection from the imaginary axis onto the circle $|z-1| = 1$ except for $z = 2$.

(e) Show that $g(-1) = 1$ implies that g maps the half-plane given by $\Re w \leq 0$ onto the closed disk $|z-1| \leq 1$ except for $z = 2$.

Therefore the set of convergence is:

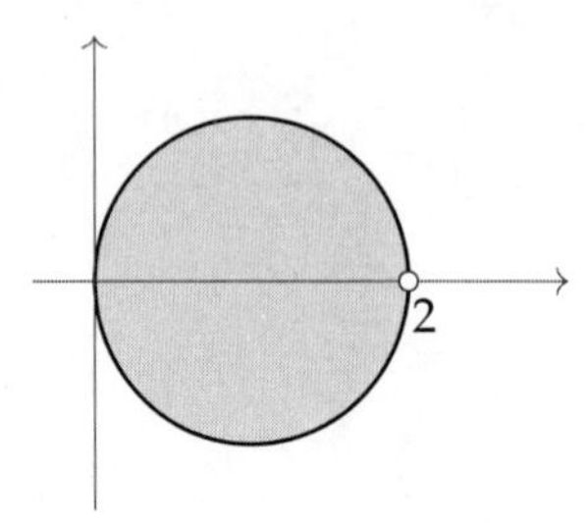

A solution like this leaves one dissatisfied. How did the thought of the circle $|z-1| = 1$ arise? Perhaps by computing f at several imaginary points? It could come about like that, but for genuinely satisfying answers you must read more about complex numbers and their geometry, specifically the topics of the Riemann sphere and Möbius transformations. Any book on complex functions would discuss this. My favorite is Sarason [29].

References

Books

[1] Apostol, T. M. (2002). *Mathematical Analysis*, 2nd ed., Narosa. ISBN: 978-8185015668.

[2] Apostol, T. M. (1967). *Calculus, Vol. I.* 2nd ed., Wiley. ISBN: 978-8126515196.

[3] Artin E. (1964; 2015 Dover Reprint). *The Gamma Function*. Holt, Rinehart & Winston. ISBN: 978-0486789781.

[4] Courant, R. and John, F. *Introduction to Calculus and Analysis*, Vol. I. Springer (India). ISBN: 978-3540650584.

[5] de Souza, P. N. and Silva, J-N. (2007). *Berkeley Problems in Mathematics*. 3rd ed. Springer (India). ISBN: 978-8181286345.

[6] Dedekind, R. (Translated by Wooster Woodruff Beman). (1901). *Essays on the Theory of Numbers*. Open Court Publishing Company. http://www.gutenberg.org/2/1/0/1/21016/.

[7] Diefenderfer, C. L. and Nelsen, R. B. eds. (2010) *The Calculus Collection: A Resource for AP* and Beyond*. The Mathematical Association of America. ISBN: 978-0883857618.

[8] Fourier, J. (Translated by Alexander Freeman). (1878). *The Analytical Theory of Heat*. Cambridge University Press. http://self.gutenberg.org/Get956uFile.aspx?&bookid=1623098.

[9] Ghorpade, S. R. and Limaye, B. V. (2006). *A Course in Calculus and Real Analysis—Undergraduate Texts in Mathematics*. Springer. ISBN: 978- 1493970810.

[10] Giaquinta, M. and Modica G. (2003). *Mathematical Analysis: Functions of One Variable*. Springer.

[11] Hamilton, N. and Landin, J. (1961). *Set Theory and the Structure of Arithmetic*. Allyn and Bacon. ISBN: 978-0486824727.

[12] Hamming, R.W. (2004). *Methods of Mathematics*. Dover. ISBN: 978- 0486439457.

[13] Hardy, G. H. (Foreword: Körner, T.W.). (2008). *A Course of Pure Mathematics; Centenary Edition*. Cambridge University Press. ISBN: 978-0521720557

[14] Hardy, G. H., Littlewood, J. E. and Pólya, G. (1952). *Inequalities*. 2nd ed. Cambridge University Press, 1952.

[15] Körner, T. W. (1989). *Fourier Analysis*. Cambridge University Press. ISBN: 978-0521389914.

[16] Kumar, A., Kumaresan, S. and Sarma, B. K. (2018). *A Foundation Course in Mathematics*. Narosa. ISBN: 978-8184876109.

[17] Landau, E. (2001). *Foundations of Analysis*. 3rd ed. Vol. 79. AMS Chelsea Publishing Series; American Mathematical Society. ISBN: 978-0821826935.

[18] Lang, S. (2002). *Short Calculus*. Springer, 2002. ISBN: 978-8181289742.

[19] Serge Lang. (2009). *A First Course in Calculus*. 5th ed. Springer. ISBN: 979-8181282407.

[20] Marsden, J. E. and Weinstein, A. (1985). *Calculus I*. 2nd ed. Springer-Verlag. ISBN: 0387909745. https://authors.library.caltech.edu/25030/.

[21] Marsden, J. E. and Weinstein, A. (1985). *Calculus II*. Springer-Verlag. ISBN: 0387909753. https://authors.library.caltech.edu/25036/.

[22] Marsden, J. E. and Weinstein, A. (1985). *Calculus III*. Springer-Verlag. ISBN: 9780387909851. https://authors.library.caltech.edu/25043/.

[23] Marsden, J. E. and Weinstein, A. (1985). *Calculus Unlimited*. Benjamin-Cummings Publishing Company, 1981. https: //authors.library.caltech. edu/25054/.

[24] Mendelson, E. (2008). *Number Systems and the Foundations of Analysis*. Dover; (originally published by.Academic Press 1973). ISBN: 978-0486457925.

[25] Niven, I. M. (1956). *Irrational Numbers*. The Carus Mathematical Monographs. The Mathematical Association of America. ISBN: 9780471641193.

[26] Piskunov, N. (1969). *Differential and Integral Calculus, Vol. I*. Moscow: Mir Publishers.

[27] Rajwade, A. R. and Bhandari, A. K. (2007). *Surprises and Counterexamples in Real Function Theory* (Texts and Readings in Mathematics). Hindustan Book Agency. ISBN: 978-9380250168.

[28] Rudin, W. (2017). *Principles of Mathematical Analysis*. 3rd ed. McGraw Hill (reprint). ISBN: 978-1259064784.

[29] Sarason, D. (2008). *Complex Function Theory*. Texts and Readings in Mathematics. 2nd ed. Hindustan Book Agency, 2008. ISBN: 978-8185931845.

[30] Spivak, M. (1994). *Calculus*. 3rd ed. Publish or Perish. ISBN: 978-0914098898.

[31] Stewart, J. (2014). *Essential Calculus–Early Transcendentals*. 2nd ed. Cengage. ISBN: 978-8131525494.

[32] Stillwell, J. (2010). *Mathematics and its History*. 3rd ed. Springer. ISBN: 978-1441960528.

[33] Tenenbaum, M. and Pollard, H. (1985). *Ordinary Differential Equations*. Dover. ISBN: 978-0486649405.

[34] Thomas, G. B. and Finney, R. L. (2010). *Calculus and Analytic Geometry*. 9th ed. Pearson Education India. ISBN: 978-8177583250.

[35] Toeplitz, O. (with new foreword by David M. Bressoud). (2007). *The Calculus—A genetic approach*. The University of Chicago Press. ISBN: 978-0226806686.

[36] Tolstov, G. P. (1976). *Fourier Series*. Dover. ISBN: 978-0486633176.

[37] Widder, D. V. (1989). *Advanced Calculus*. Dover. ISBN: 978-0486661032.

Websites and E-books

Note: Websites are inherently unstable as references as they are liable to either migrate or die out. However, the ones given below were in operation as of May 2022.

[38] Callahan, J. et al. (2019). *Calculus in Context—The Five College Calculus Project*. http://math.smith.edu/~callahan/intromine.html.

[39] Conrad, K. *Expository Papers*. https://kconrad.math.uconn.edu/blurbs/.

[40] The Euler Archive. *Foundations of Differential Calculus*. https://scholarlycommons.pacific.edu/euler-works/212/.

[41] Habib, A. *Undergraduate Mathematics: Foundations*. https://www.ams.org/open-math-notes/omn-view-listing? listingId = 110812.

[42] Richard Hammack. (2018). *Book of Proof*. https://www.people.vcu.edu/~rhammack/BookOfProof/.

[43] Jerison, D. et al. (2010). *18.01SC Single Variable Calculus*. Massachusetts Institute of Technology: MIT OpenCourseWare. https://ocw.mit.edu/courses/mathematics/.

[44] *Mathematical Training and Talent Search Programme. Expository Articles*. https://4dspace.mtts.org.in/ea.

[45] Thomson, B. S., Bruckner, J. B. and Bruckner A. M. *Elementary Real Analysis*. http://www.classicalrealanalysis.info/Elementary-Real-Analysis.php.

Articles

[46] Andrews, G. E. (1998). The Geometric Series in Calculus. *The American Mathematical Monthly*, 105(1), 36–40.

[47] Apostol, T. M. (1952). Term-Wise Differentiation of Power Series. *The American Mathematical Monthly*, 59(5), 323–6.

[48] Bers, L. (1967). Avoiding the Mean Value Theorem. *The American Mathematical Monthly*, 74(5), 583.

[49] Boas, R. L. (1969). L'Hôpital's Rule Without Mean-Value Theorems. *The American Mathematical Monthly*, 76(9), 1051–3.

[50] Boas, R. L. (1986). Counter-examples to L'Hôpital's Rule. *The American Mathematical Monthly* 94(7), 644–5.

[51] Daileda, R. C. (2011). Solutions of Autonomous First Order ODEs. http://ramanujan.math.trinity.edu/rdaileda/teach/m3366s11/ autonomous.pdf

[52] Dunham, W. (2009). When Euler Met l'Hôpital. *Mathematics Magazine*, 82(1), 16–25.

[53] Holden, L. (1973). The March of the Discoverer. *Educational Studies in Mathematics*, 5(2), 193–205.

[54] Khovanova, T. and Radul, A. (2012). Killer Problems. *The American Mathematical Monthly* 119(10), 815–23.

[55] Man, Yiu-Kwong. (2011). Partial Fraction Decomposition of Rational Functions with Irreducible Quadratic Factors in the Denominators. *Proceedings of the World Congress on Engineering 2011. Vol. I.* London. Newswood Limited.

[56] McCrimmon, K. (1960). Enumeration of Positive Rationals. *The American Mathematical Monthly* 67(9), 868.

[57] Raghavan, K. N. (2013). Partial Fractions: A Critical Look. 2013. https://www.imsc.res.in/ knr/past/part_frac.pdf

[58] Swann, H. (1997). Commentary on Rethinking Rigor in Calculus: The Role of the Mean Value Theorem. *The American Mathematical Monthly* 104(3), 241–5.

[59] Taylor, A. E. (1952). L'Hopital's Rule. *The American Mathematical Monthly* 59(1), 20–4.

[60] Teismann, H. (2013). Toward a More Complete List of Completeness Axioms. *The American Mathematical Monthly* 120(2), 99–114.

[61] Tucker, T. W. (1997). Rethinking Rigor in Calculus: The Role of the Mean Value Theorem. *The American Mathematical Monthly* 104(3), 231–40.

Index